Genetically Modified Organisms

Genetically Modified Organisms

Edited by Nigel Hogan

SYRAWOOD
PUBLISHING HOUSE

New York

Published by Syrawood Publishing House,
750 Third Avenue, 9th Floor,
New York, NY 10017, USA
www.syrawoodpublishinghouse.com

Genetically Modified Organisms
Edited by Nigel Hogan

International Standard Book Number: 978-1-68286-437-1 (Hardback)

Cataloging-in-publication Data

Genetically modified organisms / edited by Nigel Hogan.
   p. cm.
Includes bibliographical references and index.
ISBN 978-1-68286-437-1
1. Transgenic organism. 2. Genetic engineering. 3. Organisms. I. Hogan, Nigel.
QH442.6 .G46 2017
636.082 1--dc23

Printed in the United States of America.

TABLE OF CONTENTS

PREFACE

Genetically modified organisms are organisms whose genes have been altered using genetic engineering techniques. This book on genetically modified organisms discusses the process of modification along with developments in the fields of medicine and biology. Much of the information on genetics is still theoretical and long-term changes in terms of evolutionary patterns, disease recurrence and immunity are still areas of study and development. There has been rapid progress in this field and its applications are finding their way across multiple industries. This book covers in detail some existence theories and innovative concepts revolving around genetically modified organisms. It is a vital tool for all researching or studying genetic engineering as it gives incredible insights into emerging trends and concepts. Coherent flow of topics, student-friendly language and extensive use of examples make this book an invaluable source of knowledge for students and experts alike.

This book has been the outcome of endless efforts put in by authors and researchers on various issues and topics within the field. The book is a comprehensive collection of significant researches that are addressed in a variety of chapters. It will surely enhance the knowledge of the field among readers across the globe.

It gives us an immense pleasure to thank our researchers and authors for their efforts to submit their piece of writing before the deadlines. Finally in the end, I would like to thank my family and colleagues who have been a great source of inspiration and support.

Editor

Identification of Novel SHOX Target Genes in the Developing Limb Using a Transgenic Mouse Model

Katja U. Beiser[1], Anne Glaser[1], Kerstin Kleinschmidt[2], Isabell Scholl[1], Ralph Röth[1], Li Li[3], Norbert Gretz[3], Gunhild Mechtersheimer[4], Marcel Karperien[5], Antonio Marchini[1,6], Wiltrud Richter[2], Gudrun A. Rappold[1]*

1 Department of Human Molecular Genetics, Heidelberg University Hospital, Heidelberg, Germany, 2 Division of Experimental Orthopaedics, Orthopaedic University Hospital, Heidelberg, Germany, 3 Medical Research Center (ZMF), Medical Faculty Mannheim at Heidelberg University, Mannheim, Germany, 4 Institute of Pathology, Heidelberg University Hospital, Heidelberg, Germany, 5 Department of Developmental Bioengineering, University of Twente, Enschede, The Netherlands, 6 German Cancer Research Center (DKFZ), Heidelberg, Germany

Abstract

Deficiency of the human short stature homeobox-containing gene (*SHOX*) has been identified in several disorders characterized by reduced height and skeletal anomalies such as Turner syndrome, Léri-Weill dyschondrosteosis and Langer mesomelic dysplasia as well as isolated short stature. SHOX acts as a transcription factor during limb development and is expressed in chondrocytes of the growth plates. Although highly conserved in vertebrates, rodents lack a *SHOX* orthologue. This offers the unique opportunity to analyze the effects of human *SHOX* expression in transgenic mice. We have generated a mouse expressing the human *SHOXa* cDNA under the control of a murine *Col2a1* promoter and enhancer (*Tg(Col2a1-SHOX)*). *SHOX* and marker gene expression as well as skeletal phenotypes were characterized in two transgenic lines. No significant skeletal anomalies were found in transgenic compared to wildtype mice. Quantitative and *in situ* hybridization analyses revealed that *Tg(Col2a1-SHOX)*, however, affected extracellular matrix gene expression during early limb development, suggesting a role for *SHOX* in growth plate assembly and extracellular matrix composition during long bone development. For instance, we could show that the connective tissue growth factor gene *Ctgf*, a gene involved in chondrogenic and angiogenic differentiation, is transcriptionally regulated by SHOX in transgenic mice. This finding was confirmed in human NHDF and U2OS cells and chicken micromass culture, demonstrating the value of the *SHOX*-transgenic mouse for the characterization of SHOX-dependent genes and pathways in early limb development.

Editor: Andre van Wijnen, University of Massachusetts Medical, United States of America

Funding: This work was supported by the Deutsche Forschungsgemeinschaft (grant number RA 380/13-2; URL: www.dfg.de) and the Baden-Wurttemberg Foundation (grant number 1.1601.07; URL: www.bwstiftung.de). The funders had no role in study design, data collection and analysis, decision to publish, or preparation of the manuscript.

Competing Interests: The authors have declared that no competing interests exist.

* E-mail: gudrun.rappold@med.uni-heidelberg.de

Introduction

Height is a complex trait defined by multiple biological and environmental factors that are involved in bone formation and growth. The development of the long bones is characterized by coordinated gene expression from early embryonic stages until adulthood. Disturbances in bone development can affect growth and lead to clinical consequences. The homeodomain transcription factor SHOX is involved in different human short stature syndromes (Turner syndrome, Léri-Weill dyschondrosteosis LWD [MIM 127300] and Langer mesomelic dysplasia [MIM 249700]) and isolated (idiopathic) short stature [MIM 300582] [1,2,3,4,5,6,7]. Mutations and deletions of the *SHOX* gene and its enhancers have been identified as etiologic for the short stature and skeletal anomalies in these disorders [8,9,10,11,12,13]. Comprehensive case studies have shown that *SHOX* defects have also been identified in the more common nonsyndromic (isolated) forms of short stature with a prevalence of 5–17% in geographically different populations [6,12,14]. An overdosage of *SHOX* as in patients with Triple-X or Klinefelter syndrome results in tall stature [15].

Phenotypic characteristics are variable in *SHOX*-deficient patients and include disproportional (mesomelic) short stature, shortening of the forearms as well as Madelung deformity, a skeletal abnormality of the wrist characteristic for LWD [4,16]. Histopathological evaluation of LWD growth plates revealed a variable disruption of the architecture and an irregular chondrocyte stacking [17], and the SHOX protein was mainly detected in prehypertrophic and hypertrophic chondrocytes of fetal and childhood growth plates by immunohistochemistry [18,19,20]. Since clinical studies have demonstrated that growth hormone (somatropin) therapy before the onset of puberty effectively ameliorates the short stature in *SHOX*-deficient patients [21], a somatropin-based therapy is proposed in affected individuals.

Despite the high clinical relevance of *SHOX* mutations, surprisingly little is known about the molecular mechanisms that are governed by *SHOX* deficiency. This is mainly due to the limited availability of patient tissue samples (growth plate material) and the lack of cellular systems that reliably express *SHOX* endogenously at sufficiently high levels [22]. Mice do not have a *SHOX* orthologue, thus a knock-out model cannot be generated. Since the vast majority of genes that govern early developmental processes are highly conserved between human and mouse [23], characterization of genes that are divergent between the two species has not attracted much attention. SHOX has been shown to act as both a transcriptional activator and repressor of target genes [8,20,24,25,26]. Functional studies have also shown that overexpression of the SHOX protein can induce growth arrest and apoptosis, suggesting that SHOX may regulate chondrocyte hypertrophy by inducing apoptosis [19].

The clinical relevance of *SHOX* in short stature prompted us to generate a transgenic mouse to study the effect of the human *SHOX* gene during early chondrogenesis. While the phenotypic features are sparse in these animals, we demonstrate that *Ctgf*, among other genes, is regulated by SHOX in transgenic mice as well as in human and chicken cell cultures. In addition, microarray and molecular analyses revealed that the *SHOX*-transgene can effectively regulate genes important in early processes during limb formation.

Materials and Methods

Animals and genotyping

All animal experiments were conducted according to German animal protection laws and approved by the regional board of Baden Württemberg (permission No. 35–9185.81/G–64/05 and A-30/09). To express *SHOX* (genomic coordinates according to GRCh37: X:585,078-620,145) in mouse limbs, the *SHOXa* cDNA (CCDS14107.1) was cloned into the murine expression vector p1757 including the rat *Col2a1* promoter (1 kb), a Globin splicing sequence (640 bp) and the *Col2a1* enhancer (1.4 kb) [27,28,29] and a SV40 polyadenylation signal from pGL3 Basic (Promega). The construct (p1757 SHOX) was linearized with *AgeI* and microinjected into pronuclei of fertilized C57BL/6 x DBA/2 hybrid eggs to generated transgenic mice. Founders were identified by extraction of genomic DNA from tails followed by PCR using primers SHOX1 and XHO_REV (1-409 of the *SHOXa* cDNA) and SHOX_ECORI_FOR and LUMI-OSHOXCTER_REV (242-TGA of the *SHOXa* cDNA). Southern Blot was carried out according to standard procedures with a probe spanning nucleotides 1-409 to confirm the integration of the transgene at a single locus. Primer sequences are included in the Table S2 in File S1.

Limb preparation and RNA samples

Limbs of wildtype and transgenic littermates at E10.5-E14.5 were dissected and frozen in liquid nitrogen. RNA was isolated using the RNeasy Kit (Qiagen), following homogenization using a PT1300 D polytron (Kinematica). DNA was hydrolyzed using the RNAse-free DNAse Kit (Qiagen). RNA yield was measured using a NanoDrop 2000 spectrophotometer (Nanodrop technologies) and quality-checked on agarose gels. For microarray analysis, RNA from 2-4 E12.5 wildtype and transgenic littermates was pooled and the quality-checked on a 2100 Bioanalyzer (Agilent).

In vitro transcription and quantitative RT-PCR

In vitro transcription of 1 µg RNA was performed using the Superscript II First Strand Synthesis System for RT-PCR (Invitrogen). qRT-PCR was carried out using the Applied Biosystems 7500 Real-Time PCR System and Absolute SYBR Green ROX Mix (Abgene). Each sample and the housekeeping genes were run in duplicates. Relative mRNA levels were calculated according to the delta-delta Ct method [30] by normalization to mRNA expression of the housekeeping genes *Sdha* and *Adam9*. Primer sequences are included in Table S2 in File S1.

µCT imaging and analysis

Transgenic and wildtype littermates were anesthetized by i.p. injection of Ketamin (75 mg/kg) and Domitor (1 mg/kg) at the age of 4 (P28–30), 12 (P84–86) and 24 weeks (P168–170). Microcomputed tomography analyses on tibiae and femora of narcotized mice was performed using a Skyscan 1076 *in vivo* scanner (Skyscan, Antwerp, Belgium) at a resolution of 17.7 µm/pixel with an 0.5 mm aluminium filter. A source voltage of 48 kV, current of 200 µA, exposure time of 320 ms and a rotation step of 0.6 degree were used. Reconstructions (NRecon, Skyscan, Antwerp, Belgium) were made using an under-sampling factor of 1, a threshold for defect pixel mask of 30%, a beam hardening correction factor of 100%, minimum of 0.0061 and maximum of 0.0674 for CS to image conversion. Length of long bones and cortical thickness were measured manually using ruler tool function (CTAn, Skyscan, Antwerp, Belgium). Equal anatomical bone markers were used for reproducibility. For quantitative analysis of bone volume (BV) and bone mineral density (BMD) a region of interest was chosen that included the total bone and thresholds of 68-255 were used for binarisation. For BMD measurement mice were euthanized at the age of 24 weeks, legs were prepared and scanned again in water. Phantoms with known densities of 0.25 and 0.75 g/cm^3 and water were scanned for houndsfield unit calibration. Statistical analyses were carried out using Student's t-test and GraphPad Prism 5 software.

Microarray analysis

Gene expression profiling was performed using GeneChip Mouse Genome 430.2 from Affymetrix (Santa Clara, CA, USA). Duplicate Arrays were done for each genotype (transgene or wildtype). cDNA, cRNA synthesis and hybridization to arrays were performed according to the recommendations of the manufacturer. Microarray data were submitted to NCBI GEO, sample number GSE47902. Microarray data was analyzed based on ANOVA using the software package JMP Genomics, version 4.0 (SAS Institute, Cary, NC, USA). Values of perfect-matches were log transformed, quantile normalized and fitted with log-linear mixed models, with probe_ID and genotype considered to be constant and the sample ID random. Custom CDF version 13 with Entrez gene based gene/transcript definitions (http://brainarray.mbni.med.umich.edu/Brainarray/Database/CustomCDF/genomic_curated_CDF.asp) different from the original Affymetrix probe set definitions were used to annotate the arrays. Gene Set Enrichment analysis (GSEA 2.0) was applied to reveal biological pathways modulated between sample groups. Genes were ranked according to the expression change between genotypes. All Gene Ontology terms were examined using 1000 rounds of permutation of gene sets. Pathways with absolute NES (normalized enrichment score) more than 1.7 and NP (normalized p-value) <0.02 were considered to be differentially modulated.

The nCounter system assay

Assays were performed using 100 ng of total RNA plus reporter and capture probes for 10 genes (nanostring codeset). After over-night hybridization, sample purification and nCounter digital reading, counts for each RNA species were extracted and analyzed using a home-made Excel macro. Codesets include positive controls (spiked RNA at various concentrations) as well as negative controls (alien probes for background calculation). Background correction consisted of the subtraction of negative control average plus two SD from the raw counts. To avoid negative values, signals lower than one after correction were thresholded to one. The positive controls were used as a quality assessment. For each sample, the ratio between sample-related positive control average and the smallest positive control average was accepted when lower

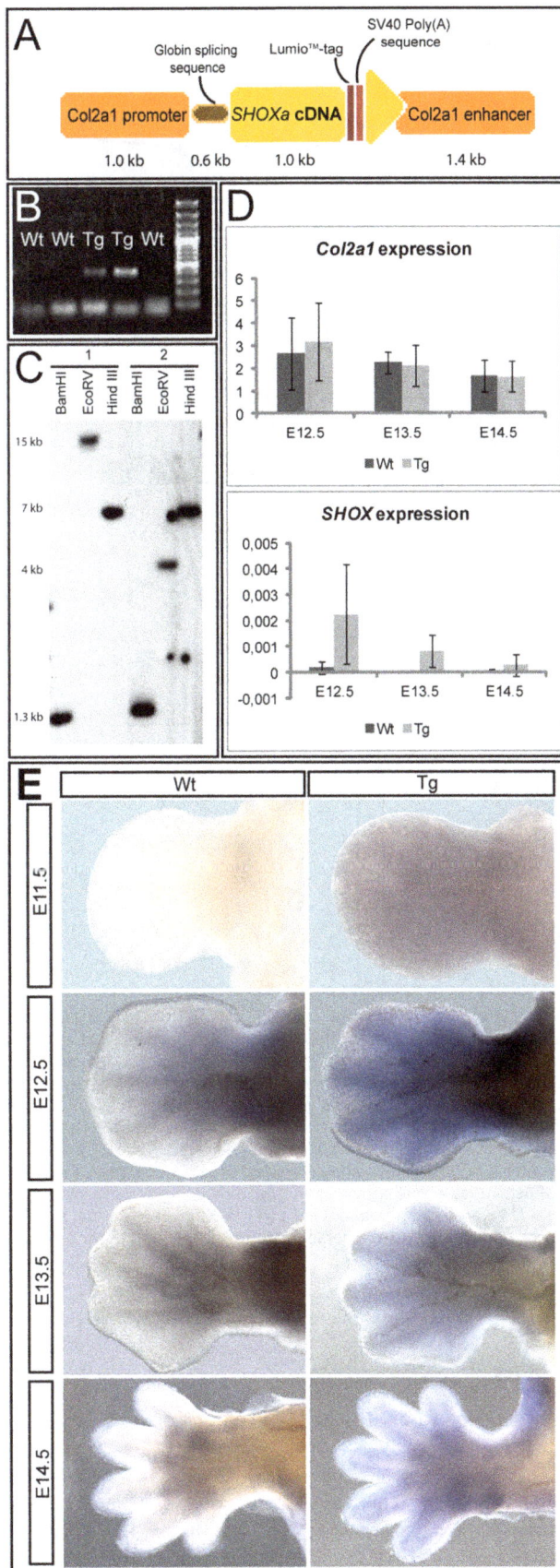

Figure 1. Generation and expression analysis of *SHOX*-transgenic mice. (A): The *SHOXa* cDNA was tagged with a Lumio and SV40 Poly(A) sequence and cloned under the control of a murine *Col2a1* promotor/enhancer expression cassette. (B): Genotyping was performed using specific primers spanning the first 409 nucleotides of the *SHOXa* cDNA. No PCR product was detected in wildtype animals. (C):-Southern Blot analysis of the two transgenic lines (1 and 2) used for our investigations. Genomic DNA was digested with *BamHI*, *EcoRV* and *Hind III*. *BamHI* digestion results in a 1.3 kb fragment that corresponds to the Lumio/SV40-tagged *SHOX* cDNA, which was flanked by *BamHI* sites. The presence of only one signal per lane indicates a single integration site of the transgene. (D): Relative quantitative expression of *Col2a1* and *SHOXa* transcripts in limbs of wildtype and transgenic littermates (N = 5–8 per litter) at E12.5, E13.5 and E14.5. The expression of the transgene corresponds to the expression dynamics of *Col2a1*. *SHOX* levels are generally low with highest expression at E12.5. Values are variable among individual animals as indicated by the standard deviation (SD). (E): WISH of wildtype (Wt) and transgenic (Tg) embryonic limbs from E11.5-E14.5 (N = 20 for each stage). The transgene is weakly expressed in the developing limb at E11.5 and becomes defined around the cartilaginous anlagen at E12.5. From E13.5 onwards, the expression is mainly seen in the mesenchyme around the developing cartilage and in the perichondrium and decreases during later stages.

than 3. To select adequate normalization genes from series of candidates included in the CodeSet, the geNorm method (5) was implemented. Therefore, the geometric mean of the selected normalization genes according to geNorm was calculated and used as normalization factor. Normalized values were then compared between samples. Probe sequences are included in Table S2 in File S1.

In situ hybridization

Whole-mount *in situ* hybridization using embryos fixed in 4% paraformaldehyde was performed according to standard procedures. Section *in situ* hybridisation was performed on 12 μm paraffin sections using standard protocols. Antisense riboprobe for *Ctgf* was cloned using the pSTBlue-1 AccepTor vector Kit (Novagen) with the primers Ctgf_ISH_FOR: AAA TGC TGC GAG GAG TGG GTG and Ctgf_ISH_REV: GTG CGT TCT GGC ACT GTG CGC. Antisense riboprobe for SHOX was generated from a *Bam/XhoI* fragment of pBSK SHOX, *Shox2* riboprobe was used as reported [31]. Templates for antisense *in vitro* transcription were digested and digoxigenin-labelled antisense RNA was synthesized using MEGAscript ® Kit (Ambion) as follows: SHOX: *KpnI*/Sp6; Ctgf: *BamHI*/Sp6; Ihh: *XbaI*/T7; Col10a1: *XhoI*/T3; Col2a1: *EcoRI*/T7; Fgfr3: *NdeI,*/T7; Shh: *HindIII*/T3; Runx2: *SpeI*/T7; Shox2: *SacI*/T7; Ogn: *XhoI*/T7.

Cell culture, transfections and luciferase assays

Cells were cultivated and transfections as well as reporter gene assays were carried out as reported before [26]. Primers used for the cloning of the reporter construct are included in Table S2 in File S1.

Electrophoretic Mobility Shift Assays (EMSA)

EMSA were carried out as described [10] using the probes sequences included in the Table S2 in File S1.

Immunohistochemistry

Immunohistochemistry was performed on growth plate sections from a pubertal 12 years old boy (tibial growth plate) as described [19] using anti SHOX- and anti-CTGF (clone L20, Santa Cruz) antibodies at the dilution of 1:25 and 1:100, respectively.

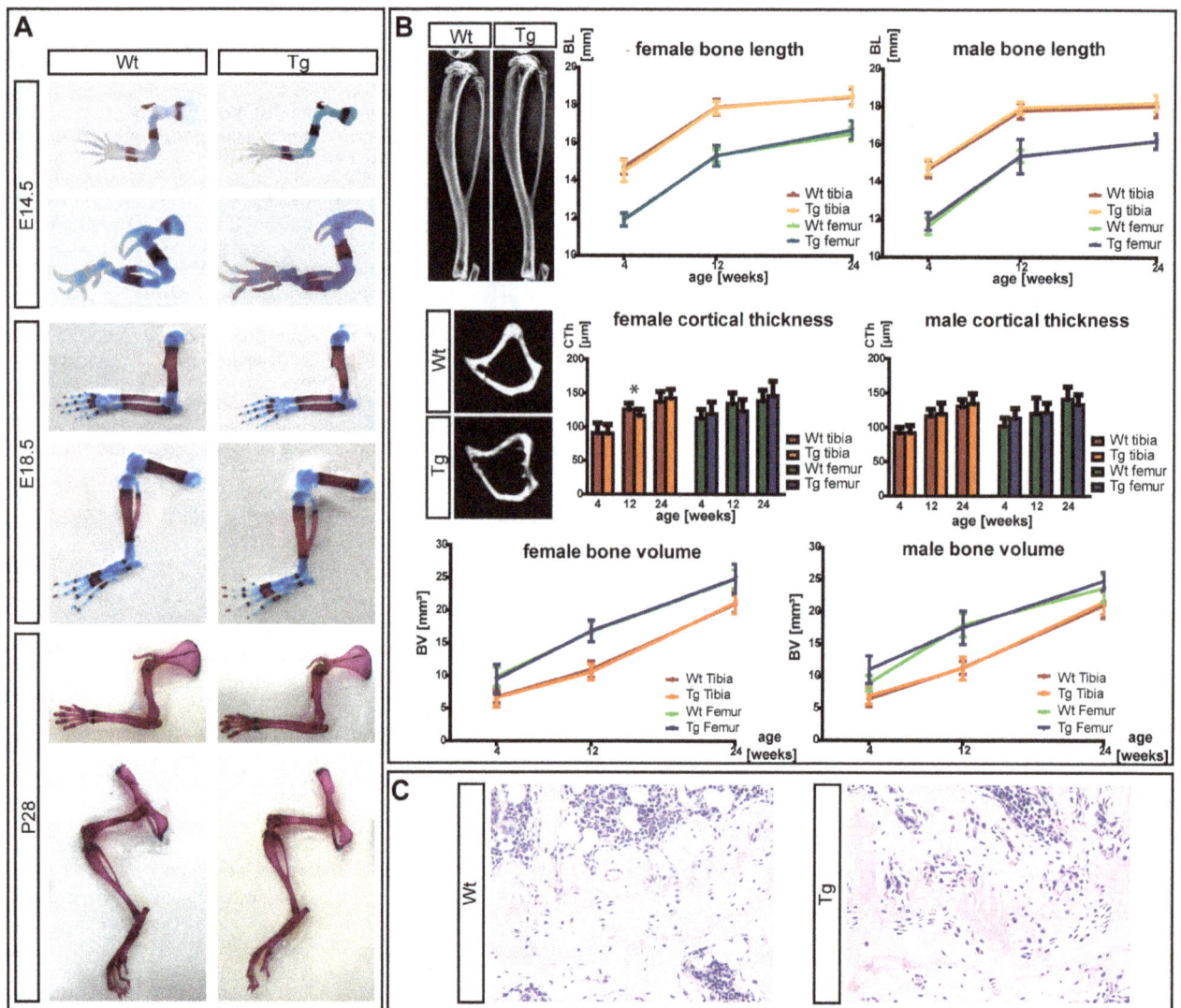

Figure 2. Analysis of postnatal bone parameters of *Col2a1-SHOX*-transgenic mice. (A): Alcian Blue/Alizarin Red S staining at different developmental (E14.5, E18.5) and postnatal (P28) stages does not reveal apparent differences between transgenic and wildtype skeletal elements. (B): Postnatal *in vivo* time-course analysis of bone growth in 65 animals of two transgenic lines by μ-CT analysis. Tibiae and femora of wildtype and *Tg(Col2a1-SHOX)* littermates at the age of 4, 12 and 24 weeks were scanned, female and male individuals were evaluated separately. Total bone length, cortical bone thickness and bone volume do not show significant differences between wildtype and transgenic females or males. Some transgenic animals presented longer bones and weaker structures of the cortical bone in the subcartilaginous region (indicated in the μ-CT images). Other micromorphological parameters (bone mineral density (BMD), trabecular volume and thickness) showed no significant differences. Statistical analyses were performed using student's t-test. (C): hematoxilin and eosin (H&E) stainings of the growth plate in wildtype and transgenic tibiae. Consistent differences between wildtype and *Tg(Col2a1-SHOX)* adult growth plates (24 weeks of age) did not exist (N = 8), but some transgenic tibiae showed a buckling, and the columns of chondrocytes became shorter and were not strictly oriented in a parallel assembly compared to the wildtype (right image).

Histology

For histological examination of growth plates, femora and tibiae of wildtype and transgenic mice (24 weeks of age) were fixed in 4% formalin and decalcified in 10% EDTA. The femora and tibiae were then bisected in the middle, and paraffin embedded. Subsequently, paraffin sections were cut at 4 μm intervals in the plane of the physis. The sections were stained with hematoxylin and eosin (H&E), periodic acid-Schiff (PAS) and Masson's trichrome (MT) by standard protocols.

Results

Generation and expression studies of *Col2a1-SHOX*-transgenic mice

To generate transgenic mice expressing the human *SHOX* gene, the *SHOXa* coding sequence was cloned into a murine transgene expression vector harbouring the rat *Collagen type II (Col2a1)* promoter and enhancer sequence (Fig. 1A). This system was previously used to drive the expression of transgenic constructs in proliferating chondrocytes [27,28,29]. Transgenic founders were identified by the presence of the construct *Tg(Col2a1-SHOX)* using

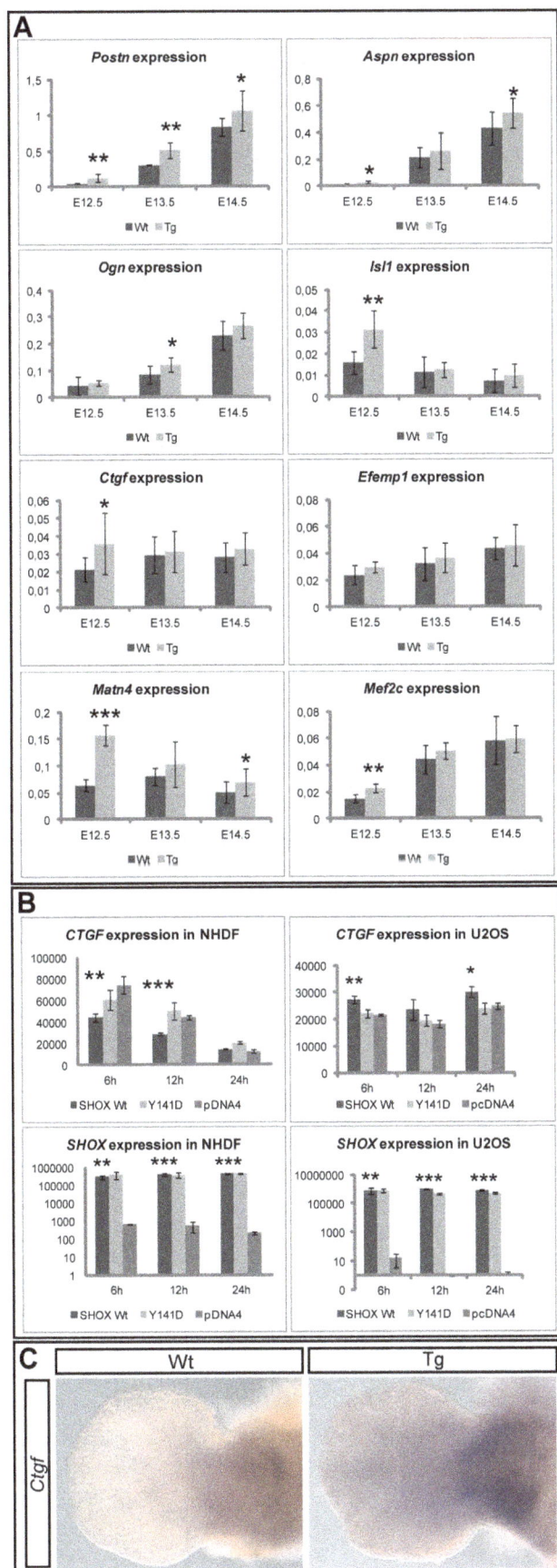

Figure 3. Regulated genes in transgenic mice and validation of Ctgf as a target. (A): qRT-PCR using limb RNA (E12.5-E14.5) from wildtype (Wt) and transgenic littermates (Tg) (N = 8–10 for each stage). Measurements were carried out individually, in duplicates, and normalized to *Adam9* and *Sdha*. Relative normalized values are presented on the y-axis. Significances are indicated in each diagram by asterisks (*: $p \leq 0.05$, **: $p \leq 0.01$, ***: $p \leq 0.001$). Variations are indicated by the standard deviation (SD). In 7/8 candidates an upregulation was confirmed as significant in at least one embryonic stage. (B): nCounter analysis of *CTGF* and *SHOX* expression in NHDF and U2OS cells after transient transfections of *SHOX* and *p.Y141D*. *CTGF* is significantly downregulated in NHDF cells, whereas it is significantly upregulated in U2OS cells. Values on y-axis represent absolute counts of mRNA, normalized to *ADAM9*, *HPRT1* and *SDHA*. Significancies are indicated by asterisks. (C): *In situ* hybridization using a *Ctgf* antisense riboprobe on embryonic limbs from wildtype and *SHOX*-transgenic littermates (N = 8) at stage E12.5. In transgenic embryos, enhanced and distalized expression of *Ctgf* was detected in the middle part of the developing limbs.

PCR and were mated with C57Bl/6 mice (Fig. 1B). Two independent heterozygous transgenic lines were investigated in more detail. Southern blot analysis using genomic DNA from animals of the two transgenic lines showed a single integration locus of the transgenic DNA (Fig. 1C). All transgenic animals were viable and fertile, and the *Tg(Col2a1-SHOX)* allele was transmitted according to Mendelian ratios.

Transgenic expression was analyzed by quantitative RT-PCR and whole mount *in situ* hybridization (WISH), demonstrating that *Tg(Col2a1-SHOX)* was expressed in the developing limbs (Fig. 1D–E). The expression started from E11.5 onwards (Fig. 1E) with a variable expression level among different transgenic mice. Following the expression dynamics of the endogenous *Col2a1*, *Tg(Col2a1-SHOX)* quantities were highest at around E12.5 and gradually decreased during later stages of embryonic development (Fig. 1D). The expression pattern of *Tg(Col2a1-SHOX)* at E12.5 resembled *Col2a1* expression which is transcribed at high levels in chondrogenic tissues [32] (Fig. 1E). During later embryonic stages (e.g. E14.5), transgenic expression was confined to the region around the developing cartilage including the perichondrium (Fig. 1E). Thus, the detected expression pattern of the *SHOX*-transgene was comparable to the endogenous *SHOX* expression domains reported in the developing limbs of human and chick embryos [33,34].

Analysis of skeletal parameters in *Col2a1-SHOX*-transgenic mice

Transgenic animals showed no obvious difference compared to their wildtype littermates. To investigate whether the *Col2a1-SHOX*-transgene has an effect on embryonic cartilage and bone development, E14.5 and E18.5 embryos were stained with Alcian Blue/Alizarin Red S (Fig. 2A). The transgenic embryos were indistinguishable from wildtype littermates at these stages, indicating that bone formation was grossly normal. As some phenotypic features in patients with SHOX deficiency (e.g. Madelung deformity) are sometimes not detectable before the onset of puberty [4], we also investigated the skeletal elements at postnatal stage P28. Again, no striking phenotype was detected in the transgenic animals (Fig. 2A).

To determine if bone length is increased in transgenic animals, we measured the postnatal bone length in 65 animals of two transgenic lines *in vivo* using micro-computed tomography (μ-CT), which enabled the analysis of different bone-specific parameters simultaneously (Fig. 2B). Tibiae and femora of anaesthetized wildtype and *SHOX*-transgenic mice were scanned *in vivo* in a time-

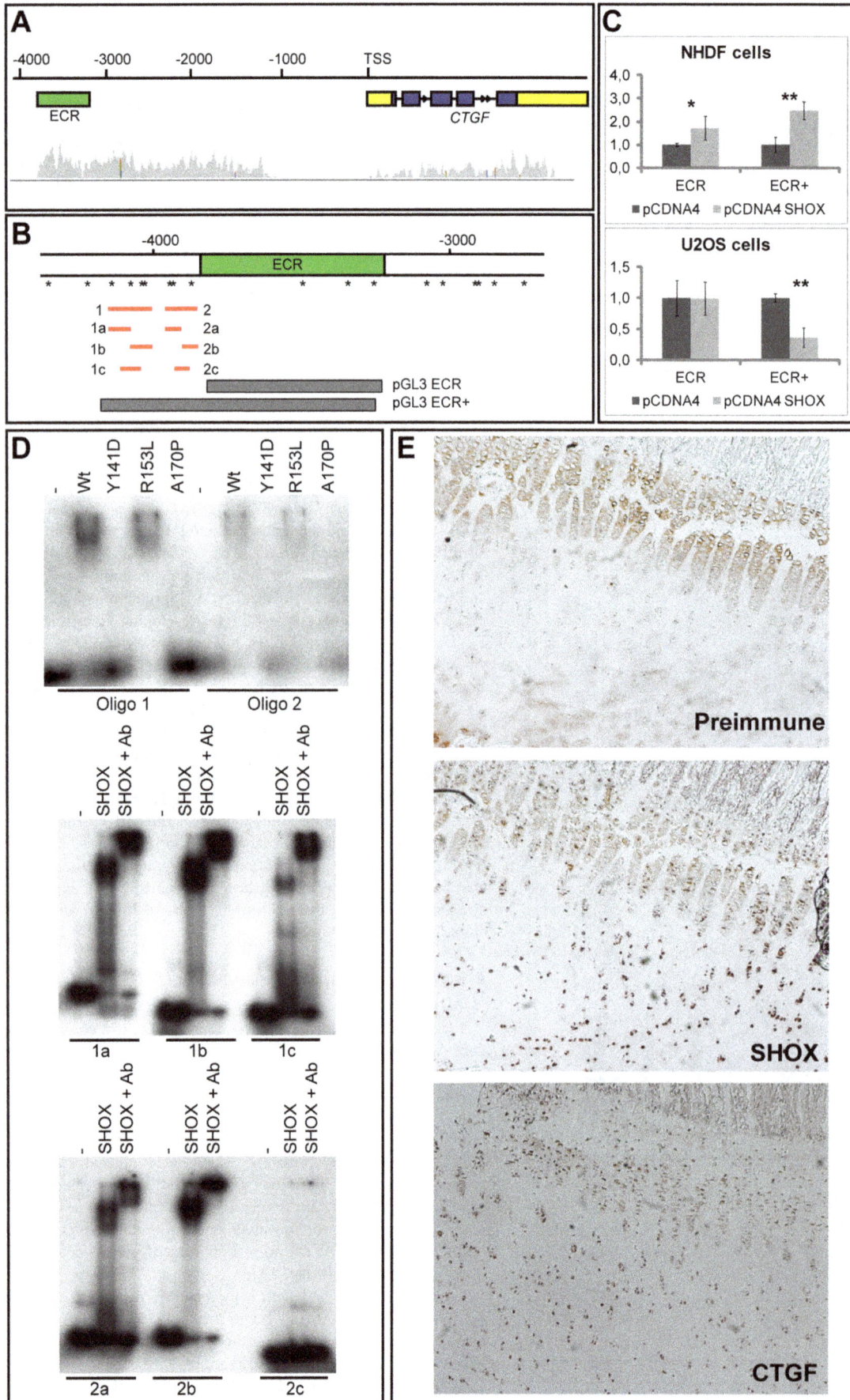

Figure 4. Analysis of *CTGF* as a direct transcriptional target of SHOX. (A): Genomic structure of the human *CTGF* region. ChIP-Seq analysis in ChMM cultures revealed an accumulation of Shox binding in the *Ctgf* promoter region (grey peaks), especially in a region 3–4 kb from the transcriptional start site (TSS) where an evolutionary conserved sequence (ECR) of 597 bp (human chr6:132317086-132318077) was identified (green bar). (B): Location of the pGL3 ECR and pGL3 ECR+ reporter constructs (grey bars) within the *CTGF* upstream region. The ECR+ construct encompasses the ECR and an upstream region including ATTA/TAAT motifs and palindromes. SHOX binding motifs (ATTA/TAAT sites and palindromes) in the *CTGF* 5′ region around the ECR are indicated by asterisks. Red bars represent the location of the generated oligonucleotides for EMSA. (C): Luciferase reporter gene assays in NHDF and U2OS cells. pcDNA4/TO *SHOX* was cotransfected with a luciferase reporter vector harbouring either the ECR or the ECR+ sequence. Transfections and measurements were carried out in triplicates. A significant activation in the luciferase activity was observed 24 h after *SHOX* transfection in NHDF cells using both reporter constructs (1.7-fold/2.5-fold with $p = 0.02/0.007$ for ECR/ECR+). In U2OS cells, an alteration was not observed for the ECR reporter, but a significant reduction was demonstrated for the ECR+ reporter construct (1.0-fold/2.8-fold with $p = 0.1/0.003$ for ECR/ECR+). (D): EMSA. The SHOX wildtype (Wt) and the mutant p.R153L proteins bind to oligonucleotides 1 and 2, whereas the defective proteins p.Y141D and p.A170P cannot. All fragments of oligonucleotides 1 and 2 containing an ATTA/TAAT site are sensitive to SHOX binding (1a–c, 2a–b). The fragment lacking this motif does not bind (oligonucleotide 2c). Using the SHOX-3 antibody (Ab), we demonstrate that the binding is SHOX-specific. (E): Immunohistochemistry performed on pubertal tibial growth plates. Staining was performed using preimmune serum as a negative control, SHOX antibody [19] and a CTGF-specific antibody. Both the SHOX and CTGF proteins were detected in growth plate chondrocytes.

course until 24 weeks of age. Data from female and male mice were analyzed separately to eliminate gender-specific effects. Even though we observed increases in bone length in some transgenic animals, these were not significant (Student's t-test). Significant differences in bone volume and bone mineral density were not found either, indicating that long bone development was largely normal upon *Tg(Col2a1-SHOX)* expression. A statistically significant decrease of the cortical bone thickness (CTh) was identified in 12 weeks old female transgenic mice, but not in males or at any other time points. Since the assessment of the growth plate in patients with LWD previously demonstrated a normal to disorganized morphology including abnormal chondrocyte stacking [17], we analyzed the femoral and tibial growth plate morphology of transgenic and wildtype mice (24 weeks of age) using hematoxylin and eosin (H&E), periodic acid-Schiff (PAS) and Masson's trichrome (MT) stainings. In some cases, a buckling of the growth plate was observed, and the columns of chondrocytes became shorter and were not strictly oriented in a parallel assembly (Fig 2C). However, these alterations were not consistently found in all transgenic samples.

Target gene expression and microarray analyses in *Col2a1-SHOX*-transgenic mice

We performed expression analysis of cartilage- and bone-specific markers from E11.5 to E14.5 using whole mount *in situ* hybridization (WISH) to identify whether limb specific markers show aberrant expression in the *Tg(Col2-SHOX)* embryos (Fig. S1A). We found that early genes such as *Shh* were not altered in the transgenic embryos, indicating that limb initiation and limb bud outgrowth were grossly normal. The expression of *Col2a1, Shox2, Runx2, Ihh* as well as *Col10a1* was similar in transgenic and wildtype embryos, suggesting that chondrocyte proliferation and maturation were largely unaffected. The expression levels of these marker genes were also quantified by qRT-PCR, but no significant differences in the amount of the respective transcripts could be detected.

A regulatory effect of SHOX on *FGFR3, AGC1* and *NPPB* (*BNP*) was recently reported using human cell lines [20,24,26]. We therefore analyzed whether the *SHOX*-transgene was able to alter the expression of the mouse *Fgfr3, Agc1* and *Nppb* genes. By using reversely transcribed RNA from E12.5-E14.5 wildtype and transgenic limbs, we detected no effect on *Fgfr3*, but an increasing effect on *Agc1* (in all three tested stages) and *Nppb* (at E13.5) (Fig. S1B). The finding that *Fgfr3* did not respond to *SHOX*-transgenic expression in mouse is consistent with the fact that the relevant SHOX-regulatory elements in the human *FGFR3* promoter do not exist in mouse, while they are present in *Agc1* and *Nppb*.

The altered expression of two known SHOX target genes in transgenic mice prompted us to perform microarray analyses of wildtype and transgenic limb RNA. Prior to hybridization, *Tg(Col2a1-SHOX)* expression was confirmed by qRT-PCR and pooled whole limb RNA of either E12.5 wildtype or transgenic littermates were hybridized to microarrays. Selection of differentially regulated genes was carried out using a significant change of expression in both experiments ($\log 2f > 0.2$ or < -0.2 and $p < 0.05$). According to these criteria, 189 genes (83%) were upregulated and 40 genes (17%) were downregulated, suggesting that the *Col2a1*-driven *SHOX*-transgene mainly exerted activating effects. A categorization of differentially expressed genes was performed by gene ontology-based pathway analysis and the most significantly regulated genes were identified in biological pathways associated with either the extracellular matrix or skeletal muscle. The eight most significantly upregulated candidate genes (*Postn, Aspn, Ogn, Isl1, Ctgf, Efemp1, Matn4, Mef2c*) that were either known to be involved in limb development, extracellular matrix or skeletal muscle pathways are summarized on Table S1 in File S1. qRT-PCR of the candidate genes was carried out using RNA from wildtype and transgenic limbs of stages E12.5-E14.5. An increase in expression of all candidate target genes including *CTGF* was detected in the transgenic embryos (Fig. 3A).

To further confirm the regulatory effects of SHOX on these genes, we carried out transient transfections of wildtype *SHOX* and a *SHOX* mutant (Y141D) in human U2OS and NHDF cell lines which have been previously used for the characterization of target genes [20,24,26]. The p.Y141D variant was identified in two short stature patients and functionally characterized as a defective SHOX protein [10]. For subsequent expression analysis, we applied the nCounter technology that allows direct RNA quantification without reverse transcription into cDNA, resulting in sensitive and reliable detection of mRNA expressed at low abundance. Since the effect of SHOX on validated genes differed between U2OS and NHDF cells, we concluded that the SHOX transcriptional regulation is strongly cell type-dependent (Fig. S2). Most strongly and significantly regulated was the chondrogenic matrix gene *CTGF*, which showed a reduced expression upon *SHOX*-transfections in NHDF and an increased expression in U2OS cells (Fig. 3B). *In situ* hybridization of *Ctgf* on wildtype and *Tg(Col2a1-SHOX)* embryonic limbs showed an increased and a more distal expression in the transgenic limbs (Fig. 3C).

The connective tissue growth factor gene *CTGF* represents a target of SHOX transactivating functions

Analyses of the *SHOX*-transgenic mouse and human cell lines overexpressing *SHOX* have demonstrated a regulatory effect of SHOX on *Ctgf/CTGF* expression. In addition, previous ChIP-Seq data on chicken micromass cultures transduced with RCAS-Shox [26] suggest *Ctgf* as a putative cell target of SHOX with several binding sites identified within the 5′ region of the gene. Computational analyses of the human *CTGF* upstream region (5 kb) identified more than 40 binding motifs of the ATTA/TAAT type which have been reported to be the target sites of SHOX [8,26]. Of these, eight motifs were arranged as palindromes. Furthermore, the region with the highest ChIP-Seq reads in the chicken *Ctgf* locus includes an ECR (evolutionary conserved region) that is also present in the human *CTGF* upstream sequence (Fig. 4A). To demonstrate that *CTGF* could be directly targeted by SHOX, we performed luciferase reporter gene assays in NHDF and U2OS cells. We used two constructs: the smaller one included the human ECR sequence (ECR) and the larger construct included the ECR as well as putative SHOX binding sites (ECR+) (Fig. 4B). As shown in Fig. 4C, significant regulatory effects of SHOX on the ECR+ reporter constructs were observed in both NHDF and U2OS cell lines, whereas for the ECR reporter construct a significant regulation could only be demonstrated in NHDF cells. To confirm a direct binding of SHOX to the *CTGF* upstream region, electrophoretic mobility shift assays (EMSA) were carried out using two oligonucleotide sequences of the ECR+ construct (Oligo 1 and Oligo 2) encompassing the ATTA/TAAT motifs (Fig. 4B). As controls, mutant SHOX proteins (p.Y141D, p.R153L and p.A170P; previously detected in patients with short stature) were used [10]. While the wildtype SHOX and p.R153L proteins bound to the tested sequences, p.Y141D and p.A170P did not (Fig. 4D). Further subdivision of oligonucleotides 1 and 2 narrowed down SHOX binding to all fragments where ATTA/TAAT sites were present (Fig. 4B and 4D). To demonstrate physiological relevance of these data, immunohistological staining on sections from human pubertal growth plate specimen were carried out. Using CTGF and SHOX specific antibodies, coexpression was detected in hypertrophic chondrocytes (Fig. 4E).

Discussion

Generation and expression studies of *Col2a1-SHOX*-transgenic mice

For a small number of human protein-coding genes, a mouse ortholog does not exist [35]. One approach to learn more about the biology of these human genes is to introduce them into mice. We have generated transgenic mice that express the human *SHOX* cDNA in embryonic limbs under the control of the murine *Col2a1* promoter/enhancer. Expression of the *SHOX*-transgene was detected between E12.5 and E14.5. Compared to *Col2a1*, a highly abundant major structural component of the extracellular matrix, the expression of the transcription factor *SHOX* was very weak and differed between animals. The generation of a transgenic mouse using a different promoter and/or enhancer may eventually yield in higher *SHOX* expression levels. However, low expression levels are characteristic for *SHOX* and have been found in all tissues and cell lines tested [22], suggesting that SHOX functions do not rely on high mRNA or protein abundance in the cell.

Analysis of skeletal parameters

Phenotypic analyses of the developing limbs in transgenic mice did not reveal significant differences compared to wildtype (with the exception of cortical thickness in female tibiae at 12 weeks and almost significant differences in female femora). Thus, there may be gender-specific effects in the transgenic mice during postnatal growth, however, to address this question, more detailed experiments would be necessary. Phenotypic clinical features have been previously assessed in patients with isolated SHOX

deficiency and LWD [6], but not much data on cortical bone structures, bone volume or mineral density is available. Patients with Turner syndrome (45,X) suffer from a high fracture risk and have reduced cortical bone structures and bone mineral density [36], but whether this is due to reduced *SHOX* expression is not known. Disorganization of the growth plate has been noted in some of our *SHOX*-transgenic mice, but is not a consistent feature. Disturbed growth plate morphology has been described in patients with LWD [17], but no data is available on patients with additional *SHOX* copies.

Gene expression and microarray analyses

To determine if the critical stages in endochondral ossification were altered in the transgenic mice, we carried out expression analysis of embryonic limb marker genes and could demonstrate that expression of these genes remained intact. A key question also concerned the extent to which the human gene is correctly read by the mouse transcriptional machinery. We therefore tested expression of all three known SHOX target genes [20,24,26] and obtained elevated mean expression levels for *Agc1* and *Nbbp* as expected, probably due to the conserved SHOX-sensitive binding sites in the *Agc1 and Nppb* enhancer and promoter regions, while the human SHOX-sensitive binding sites in the *Fgfr3* promoter do not exist in mouse.

To further search for effects of the *SHOX*-transgene, we carried out microarray analysis and identified many regulated genes belonging to the extracellular matrix and skeletal muscle pathways. It is interesting that several of these genes, including *Postn* and *Matn4*, have been previously also identified as targets in *Shox2*-deficient mice and thus may represent targets for both SHOX and Shox2 [37]. The mouse Shox2 protein is 79% identical to human SHOX and their 60 amino acid binding domains (the homeodomain) are identical [33]. *In situ* analysis have demonstrated a more proximal expression domain of the *SHOX* paralog *SHOX2* in human and also in chick embryonic limbs [33,34], and conditional deletion of *Shox2* in the developing mouse limbs dramatically impairs the formation of the proximal limb elements [38,39]. A substitution of the *Shox2* locus by human *SHOX* in mouse has demonstrated that *SHOX* is able to ameliorate but not to fully rescue *Shox2*-deficient limb anomalies, suggesting only partial functional redundancy [25].

We have selected eight putatively regulated genes (*Postn, Aspn, Ogn, Isl1, Ctgf, Efemp1, Matn4, Mef2c*) for further analysis and could demonstrate a significant deregulation in E12.5-E14.5 *SHOX*-transgenic limbs compared to wildtype in seven of the eight genes. To further validate these candidates, we also tested them in NHDF and U2OS cells and 5/8 (NHDF) and 4/8 (U2OS) were shown to be significantly regulated in these human cells. Taken together, our data demonstrate that the identified target genes of Tg(Col2a1-SHOX) are SHOX-specific and do not represent transgenic artifacts. It is also reassuring that the human *SHOX* is expressed in the appropriate stage- and cell-type specific manner in mouse and we confirm previous data that SHOX can act both as an activator and repressor of target genes in a cell-type specific fashion [25,26].

CTGF represents a direct SHOX target gene

Quantitative analyses in mouse and human cells identified *CTGF/Ctgf* as the most consistently regulated candidate target gene. Enhanced and slightly distalized expression of *Ctgf* was also seen at E12.5 (the stage of highest *SHOX* expression) in transgenic mouse limbs using WISH. Further evidence for *Ctgf* as a target of Shox was derived from ChIP-Seq data in chicken which identified several Shox binding sites in the *Ctgf* upstream region. Multiple

SHOX binding motifs and an ECR were identified, and by luciferase and EMSA experiments, we could show that the extended ECR region (ECR+) is responsive to SHOX in human cells. The finding that the *CTGF* mRNA was either down- (NHDF) or upregulated (U2OS) indicates a complex transcriptional regulation. The remarkable accumulation of Shox-mediated reads (respective binding sites) identified in the chicken *Ctgf* upstream region using ChIP-Seq (Fig. 4A) suggests that additional response elements outside the ECR may also be sensitive to SHOX, and these, together with a spatio-temporal composition of cofactors, may contribute to the fine regulation of *CTGF* expression in a given cellular environment. Physiological relevance of the SHOX-*Ctgf*/*CTGF* relationship is suggested by the coexpression of both, SHOX and CTGF proteins in hypertrophic chondrocytes of the human growth plates.

According to its expression pattern, *SHOX* deficiency results in shortening and deformation of radii/ulnae and tibiae/fibulae. Comparable to *SHOX* deficiency, the skeletal defects in *Ctgf* null mice are also specific for radii/ulnae and tibiae/fibulae and not for the proximal elements of the limbs [40]. Interestingly, the phenotypes of *SHOX*- as well as *Ctgf-transgenic* mice [41] are less severe than the loss-of-function phenotypes and strongly dependent on the expression level of the transgenes. Even though *Ctgf*-transgenic mice show more stigmata than *SHOX*-transgenic individuals, phenotypic differences were reported only at postnatal stages and also include cortical thickness and *Agc1* expression [41]. Since *Agc1* has been found to be reduced in *Ctgf* mutant mice [40] and to be regulated by SHOX in human cells [20], the demonstrated regulation may be indirect and mediated through *Ctgf*. This is also supported by our finding that the response of *CTGF* is an immediate consequence following *SHOX* overexpression, whereas the regulation of *AGC1* occurs at a later time point (Fig. S2). *Ctgf* null mice suffer from multiple defects, such as failure in growth plate chondrogenesis, angiogenesis, extracellular matrix production and bone formation/mineralization [40]. A role of SHOX during angiogenesis has been speculated, since *Shox* expression was detected in the vasculature of the developing chicken limbs [34]. However, a contribution of SHOX in other *CTGF*-associated conditions such as fibrotic disease, inflammation and cancer [42,43,44] is not known.

In summary, we have established a transgenic mouse model expressing *SHOX* under the control of the *Col2a1* promoter and enhancer. By combining data from mouse and chicken micromass cultures and human cell culture experiments, we could identify activating or repressing effects of SHOX on target genes, depending on spatio-temporal conditions and cell types. We have also demonstrated a direct regulatory effect on *CTGF* which may take place in the hypertrophic zone of the human growth plate. We have shown a direct binding of the SHOX protein to a highly conserved upstream region of the *CTGF* gene, identified by ChIP-Seq, resulting in regulatory effects in reporter gene assays in human cell lines. Since CTGF is involved in various biological processes, the effect of SHOX on *CTGF* expression in these different processes can now be investigated.

Supporting Information

Figure S1 Marker and target gene analysis during embryonic development. (A): WISH of limb marker genes from E11.5 to

E14.5. At E11.5, when *Tg(Col2a1-SHOX)* expression was first detected in the developing limb, limb buds in transgenic animals were indistinguishable from the wildtype. Expression of the *Shh* morphogen as a marker gene during limb initiation and outgrowth was normal. Also at E12.5 when *Tg(Col2a1-SHOX)* is most prominently expressed, chondrocyte proliferation in the transgenic animals appeared normal, as represented by *Col2a1* expression comparable to the wildtype. Also, the *SHOX*-homologue *Shox2* and its downstream gene *Runx2* were normally expressed in *SHOX*-transgenic animals at E12.5. *Runx2* is known to regulate chondrocyte maturation and *Ihh* expression, which was also unaffected in *Tg(Col2a1-SHOX)* limbs at E13.5. Following chondrocyte proliferation at E14.5 in both wildtype and transgenic embryos, a specific *Col10a1* pattern is detected which defines chondrocyte hypertrophy. (B): Quantitative RT-PCR on embryonic limb RNA of stages E12.5–E14.5 using primers for the SHOX target genes *Fgfr3*, *Agc1* and *Nppb*. cDNA of wildtype and transgenic littermates of each stage (N = 8–12) were measured individually and in duplicates. Measurements were normalized to *Adam9* and *Sdha*; values on y-axis represent relative normalized expression. The expression of *Fgfr3* was unaltered in transgenic limbs. Mean *Agc1* expression was increased during E12.5 and E13.5, a trend which did, however not reach significance (E12.5: 2.0-fold, $p = 0.068$; E13.5: 2.6-fold, $p = 0.092$; E14.5: 1.3-fold, $p = 0.377$). *Nppb* expression levels were weakly increased at E13.5 (1.7-fold, $p = 0.104$).

Figure S2 nCounter analysis of eight selected candidate genes in NHDF and U2OS cells. RNA was isolated 6 h, 12 h and 24 h after transfection of expression constructs for SHOX, SHOX Y141D (a defective SHOX variant (1)) and a control (pCDNA4). Measurements were carried out in triplicates and normalized to *ADAM9*, *HPRT1* and *SDHA*. As a control, *SHOX* expression upon its target gene *AGC1* was analyzed. Upon strong increase of *SHOX*, *AGC1* was significantly activated 12 hours after *SHOX*-tranfection. Values on y-axis represent absolute counts of mRNA. Significancies of the *SHOX*-transfected samples are indicated in each diagram by asterisks. *: $p \leq 0.05$, **: $p \leq 0.01$, ***: $p \leq 0.001$.

File S1 Contains Table S1, Genes, gene characterization, fold regulation and p-values of eight selected upregulated genes in the microarray. Table S2, Primers, Probes and Oligonucleotides.

Acknowledgments

We thank Andrea Vortkamp for providing p1757 vector and *Col2* and *Shh* probe vectors, Jochen Hecht for providing *Runx2* and *Runx3* probe vectors and Nenja Krüger for experimental support. We thank Roland Knopf for animal care. This publication is dedicated to Ruediger J. Blaschke.

Author Contributions

Conceived and designed the experiments: KUB GAR. Performed the experiments: KUB AG IS RR KK AM GM. Analyzed the data: KUB KK WR LL NG GAR. Contributed reagents/materials/analysis tools: MK. Wrote the paper: KUB GR.

References

1. Rao E, Weiss B, Fukami M, Rump A, Niesler B, et al. (1997) Pseudoautosomal deletions encompassing a novel homeobox gene cause growth failure in idiopathic short stature and Turner syndrome. Nat Genet 16: 54–63.

2. Shears DJ, Vassal HJ, Goodman FR, Palmer RW, Reardon W, et al. (1998) Mutation and deletion of the pseudoautosomal gene SHOX cause Leri-Weill dyschondrosteosis. Nat Genet 19: 70–73.

3. Ross JL, Scott C, Jr., Marttila P, Kowal K, Nass A, et al. (2001) Phenotypes Associated with SHOX Deficiency. J Clin Endocrinol Metab 86: 5674–5680.

4. Rappold GA, Ross JL, Blaschke RJ, Blum W (2002) Understanding *SHOX* deficiency and its role in growth disorders. A reference guide. Oxfordshire, UK: TMG Healthcare Communications Ltd.

5. Shears DJ, Guillen-Navarro E, Sempere-Miralles M, Domingo-Jimenez R, Scambler PJ, et al. (2002) Pseudodominant inheritance of Langer mesomelic dysplasia caused by a SHOX homeobox missense mutation. Am J Med Genet 110: 153–157.

6. Rappold G, Blum WF, Shavrikova EP, Crowe BJ, Roeth R, et al. (2007) Genotypes and phenotypes in children with short stature: clinical indicators of SHOX haploinsufficiency. J Med Genet 44: 306–313.

7. Belin V, Cusin V, Viot G, Girlich D, Toutain A, et al. (1998) SHOX mutations in dyschondrosteosis (Leri-Weill syndrome). Nat Genet 19: 67–69.

8. Rao E, Blaschke RJ, Marchini A, Niesler B, Burnett M, et al. (2001) The Leri-Weill and Turner syndrome homeobox gene SHOX encodes a cell-type specific transcriptional activator. Hum Mol Genet 10: 3083–3091.

9. Sabherwal N, Schneider KU, Blaschke RJ, Marchini A, Rappold G (2004) Impairment of SHOX nuclear localization as a cause for Leri-Weill syndrome. J Cell Sci 117: 3041–3048.

10. Schneider KU, Marchini A, Sabherwal N, Roth R, Niesler B, et al. (2005) Alteration of DNA binding, dimerization, and nuclear translocation of SHOX homeodomain mutations identified in idiopathic short stature and Leri-Weill dyschondrosteosis. Hum Mutat 26: 44–52.

11. Schneider KU, Sabherwal N, Jantz K, Roth R, Muncke N, et al. (2005) Identification of a major recombination hotspot in patients with short stature and SHOX deficiency. Am J Hum Genet 77: 89–96.

12. Chen J, Wildhardt G, Zhong Z, Roth R, Weiss B, et al. (2009) Enhancer deletions of the SHOX gene as a frequent cause of short stature: the essential role of a 250 kb downstream regulatory domain. Journal of Medical Genetics 46: 834–839.

13. Benito-Sanz S, Aza-Carmona M, Rodriguez-Estevez A, Rica-Etxebarria I, Gracia R, et al. (2012) Identification of the first PAR1 deletion encompassing upstream SHOX enhancers in a family with idiopathic short stature. Eur J Hum Genet 20: 125–127.

14. Rosilio M, Huber-Lequesne C, Sapin H, Carel JC, Blum WF, et al. (2012) Genotypes and phenotypes of children with SHOX deficiency in France. J Clin Endocrinol Metab 97: E1257–1265.

15. Kanaka-Gantenbein C, Kitsiou S, Mavrou A, Stamoyannou L, Kolialexi A, et al. (2004) Tall stature, insulin resistance, and disturbed behavior in a girl with the triple X syndrome harboring three SHOX genes: offspring of a father with mosaic Klinefelter syndrome but with two maternal X chromosomes. Horm Res 61: 205–210.

16. Jorge AA, Souza SC, Nishi MY, Billerbeck AE, Liborio DC, et al. (2007) SHOX mutations in idiopathic short stature and Leri-Weill dyschondrosteosis: frequency and phenotypic variability. Clin Endocrinol (Oxf) 66: 130–135.

17. Munns CF, Glass IA, LaBrom R, Hayes M, Flanagan S, et al. (2001) Histopathological analysis of Leri-Weill dyschondrosteosis: disordered growth plate. Hand Surg 6: 13–23.

18. Munns CJ, Haase HR, Crowther LM, Hayes MT, Blaschke R, et al. (2004) Expression of SHOX in human fetal and childhood growth plate. J Clin Endocrinol Metab 89: 4130–4135.

19. Marchini A, Marttila T, Winter A, Caldeira S, Malanchi I, et al. (2004) The short stature homeodomain protein SHOX induces cellular growth arrest and apoptosis and is expressed in human growth plate chondrocytes. J Biol Chem 279: 37103–37114.

20. Aza-Carmona M, Shears DJ, Yuste-Checa P, Barca-Tierno V, Hisado-Oliva A, et al. (2011) SHOX interacts with the chondrogenic transcription factors SOX5 and SOX6 to activate the aggrecan enhancer. Hum Mol Genet 20: 1547–1559.

21. Blum WF, Crowe BJ, Quigley CA, Jung H, Cao D, et al. (2007) Growth hormone is effective in treatment of short stature associated with short stature homeobox-containing gene deficiency: Two-year results of a randomized, controlled, multicenter trial. J Clin Endocrinol Metab 92: 219–228.

22. Durand C, Roeth R, Dweep H, Vlatkovic I, Decker E, et al. (2011) Alternative splicing and nonsense-mediated RNA decay contribute to the regulation of SHOX expression. PLoS One 6: e18115.

23. Waterston RH, Lindblad-Toh K, Birney E, Rogers J, Abril JF, et al. (2002) Initial sequencing and comparative analysis of the mouse genome. Nature 420: 520–562.

24. Marchini A, Haecker B, Marttila T, Hesse V, Emons J, et al. (2007) BNP is a transcriptional target of the short stature homeobox gene SHOX. Hum Mol Genet 16: 3081–3087.

25. Liu H, Chen CH, Espinoza-Lewis RA, Jiao Z, Sheu I, et al. (2011) Functional redundancy between human SHOX and mouse Shox2 genes in the regulation of sinoatrial node formation and pacemaking function. J Biol Chem 286: 17029–17038.

26. Decker E, Durand C, Bender S, Rodelsperger C, Glaser A, et al. (2011) FGFR3 is a target of the homeobox transcription factor SHOX in limb development. Hum Mol Genet 20: 1524–1535.

27. Long F, Schipani E, Asahara H, Kronenberg H, Montminy M (2001) The CREB family of activators is required for endochondral bone development. Development 128: 541–550.

28. Minina E, Wenzel HM, Kreschel C, Karp S, Gaffield W, et al. (2001) BMP and Ihh/PTHrP signaling interact to coordinate chondrocyte proliferation and differentiation. Development 128: 4523–4534.

29. Yang Y, Topol L, Lee H, Wu J (2003) Wnt5a and Wnt5b exhibit distinct activities in coordinating chondrocyte proliferation and differentiation. Development 130: 1003–1015.

30. Pfaffl MW (2001) A new mathematical model for relative quantification in real-time RT-PCR. Nucleic Acids Res 29: e45.

31. Blaschke RJ, Monaghan AP, Schiller S, Schechinger B, Rao E, et al. (1998) SHOT, a SHOX-related homeobox gene, is implicated in craniofacial, brain, heart, and limb development. Proc Natl Acad Sci U S A 95: 2406–2411.

32. Cheah KS, Lau ET, Au PK, Tam PP (1991) Expression of the mouse alpha 1(II) collagen gene is not restricted to cartilage during development. Development 111: 945–953.

33. Clement-Jones M, Schiller S, Rao E, Blaschke RJ, Zuniga A, et al. (2000) The short stature homeobox gene SHOX is involved in skeletal abnormalities in Turner syndrome. Hum Mol Genet 9: 695–702.

34. Tiecke E, Bangs F, Blaschke R, Farrell ER, Rappold G, et al. (2006) Expression of the short stature homeobox gene Shox is restricted by proximal and distal signals in chick limb buds and affects the length of skeletal elements. Dev Biol 298: 585–596.

35. Stahl PD, Wainszelbaum MJ (2009) Human-specific genes may offer a unique window into human cell signaling. Sci Signal 2: pe59.

36. Soucek O, Lebl J, Snajderova M, Kolouskova S, Rocek M, et al. (2011) Bone geometry and volumetric bone mineral density in girls with Turner syndrome of different pubertal stages. Clinical Endocrinology 74: 445–452.

37. Vickerman L, Neufeld S, Cobb J (2011) Shox2 function couples neural, muscular and skeletal development in the proximal forelimb. Dev Biol 350: 323–336.

38. Cobb J, Dierich A, Huss-Garcia Y, Duboule D (2006) A mouse model for human short-stature syndromes identifies Shox2 as an upstream regulator of Runx2 during long-bone development. Proc Natl Acad Sci U S A 103: 4511–4515.

39. Yu L, Liu H, Yan M, Yang J, Long F, et al. (2007) Shox2 is required for chondrocyte proliferation and maturation in proximal limb skeleton. Dev Biol 306: 549–559.

40. Ivkovic S, Yoon BS, Popoff SN, Safadi FF, Libuda DE, et al. (2003) Connective tissue growth factor coordinates chondrogenesis and angiogenesis during skeletal development. Development 130: 2779–2791.

41. Tomita N, Hattori T, Itoh S, Aoyama E, Yao M, et al. (2013) Cartilage-specific over-expression of CCN family member 2/connective tissue growth factor (CCN2/CTGF) stimulates insulin-like growth factor expression and bone growth. PLoS One 8: e59226.

42. Dhar A, Ray A (2010) The CCN family proteins in carcinogenesis. Exp Oncol 32: 2–9.

43. Cicha I, Goppelt-Struebe M (2009) Connective tissue growth factor: context-dependent functions and mechanisms of regulation. Biofactors 35: 200–208.

44. Leask A, Holmes A, Abraham DJ (2002) Connective tissue growth factor: a new and important player in the pathogenesis of fibrosis. Curr Rheumatol Rep 4: 136–142.

2

A Reliable Way to Detect Endogenous Murine β-Amyloid

Andrew F. Teich*, Mitesh Patel, Ottavio Arancio

Department of Pathology and Cell Biology, Taub Institute for Research on Alzheimer's Disease and the Aging Brain, Columbia University, New York, New York, United States of America

Abstract

Unraveling the normal physiologic role of β-amyloid is likely crucial to understanding the pathogenesis of Alzheimer's disease. However, progress on this question is currently limited by the high background of many ELISAs for murine β-amyloid. Here, we examine the background signal of several murine β-amyloid ELISAs, and conclude that the majority of the background is from non-APP derived proteins. Most importantly, we identify ELISAs that eliminate this background signal.

Editor: Colin Combs, University of North Dakota, United States of America

Funding: This work was supported by grants to OA by the National Institutes of Health (AG034248), and by grants to AFT from the Alzheimer's Association (NIRG-11-203583) and the Louis V. Gerstner, Jr. Scholars Program. The funders had no role in study design, data collection and analysis, decision to publish, or preparation of the manuscript.

Competing Interests: The authors have declared that no competing interests exist.

* E-mail: aft25@columbia.edu

Introduction

The Amyloid Hypothesis of Alzheimer's disease states that the accumulation of high levels of cerebral β-amyloid protein is a central event in this illness [1]. This hypothesis has subsequently fueled a massive research effort into understanding the physiology of β-amyloid protein. Although β-amyloid may reach high (nanomolar) concentrations in Alzheimer's disease [2], β-amyloid is also produced in the brain throughout life, and the normal *in vivo* concentration in the rodent brain has been estimated to be in the picomolar range [3,4]. Understanding how β-amyloid is normally regulated may help us understand how it accumulates to high levels in Alzheimer's disease, and this realization has inspired investigation into the normal physiologic function of β-amyloid [4,5,6,7,8]. In addition, others have recently started using wild-type rodents to study the function of γ-secretase inhibitors [9,10]. In all of these cases, studying β-amyloid in wild-type animals requires a reliable method of measuring low, picomolar concentrations of endogenous cerebral β-amyloid. However, it has been noted by several groups that ELISAs that use common β-amyloid antibodies (such as 4G8 and Signet 9153) give a very high background reading with wild-type rodent brain, presumably due to non-specific binding of various rodent brain proteins with these antibodies [10,11,12]. This has prompted some authors to claim that the ELISA method is flawed at measuring endogenous murine β-amyloid [11,12]. To correct this flaw, a solid-phase extraction technique has been proposed to chromatographically separate wild-type rodent β-amyloid from the proteins in rodent brain that lead to this non-specific binding [11,12]. However, some investigators who wish to measure endogenous murine β-amyloid may not have expertise with solid-phase extraction, which is a technique that is not commonly used in the Alzheimer's disease scientific community. In this study, we use hippocampal tissue from amyloid precursor protein knock-out (APP-KO) mice and wild-type (WT) littermates to investigate this issue further. We conclude that 1) The majority of the background signal seen in ELISAs for murine β-amyloid is from non-APP derived proteins, and most importantly, 2) We identify ELISAs that eliminate this background signal.

Materials and Methods

Animals

APP-KO mice [13] were bread against a B6 background; all mice were purchased from Jackson labs. Hemizygous transgenic (HuAPP695SWE)2576 mice expressing mutant human *APP* (K670N,M671L) [14] were used as positive controls in several experiments (these mice are from a colony that derives from a gift from Karen Hsiao-Ashe). Unless otherwise specified, mice were between 4 and 8 months (adults) when they were sacrificed for tissue analysis. This study was carried out in strict accordance with the recommendations in the Guide for the Care and Use of Laboratory Animals of the National Institutes of Health. The protocol was approved by the Institutional Animal Care and Use Committee of Columbia University (Protocol Number: AC-AAAD9255).

ELISA

Mouse hippocampi were homogenized in 880 µl of tissue lysate buffer (20 mM Tris-HCl (pH 7.4), 1 mM ethylenediaminetetraacetic acid, 1 mM ethyleneglycoltetraacetic acid, 250 mM sucrose) supplemented with protease inhibitors (Roche). The tissue homogenates were treated with diethanolamine to extract soluble β-amyloid. We did not perfuse the mice with heparin before-hand, as our goal is to compare the ability of various ELISAs to measure β-amyloid without this requirement. Although heparin perfusion is sometimes done to eliminate IgG from the brain, it is difficult to do this when measuring acute changes in β-amyloid, or when measuring β-amyloid in a hippocampal slice preparation. Thus, we sought to determine whether there is an ELISA technique that can measure β-amyloid in all circumstances without any requirements for prior heparin perfusion.

We measured β-amyloid in our hippocampal homogenate using the following ELISA kits: Covance Colorimetric BetaMark^{TM}

Beta-Amyloid x-42 ELISA Kit (Catalog Number: SIG-38956), Covance Colorimetric BetaMark™ Beta-Amyloid x-40 ELISA Kit, (Catalog Number: SIG-38954), Invitrogen Aβ 42 Mouse ELISA Kit (Catalog Number: KMB3441), Invitrogen Aβ 40 Mouse ELISA Kit (Catalog Number: KMB3481), Wako Human/Rat(Mouse) β-Amyloid (40) ELISA Kit (Catalog Number: 294-62501), and Wako Human/Rat(Mouse) β-Amyloid (42) ELISA High-Sensitive Kit (Catalog Number: 292-64501). All ELISA assays were performed according to the manufacturer's protocol. In addition, ELISAs using either 6E10 or M3.2 as a capture antibody and an HRP-conjugated 4G8 as a detection antibody were prepared as previously described [11] (all antibodies were purchased from Covance). Costar 96-well plates were incubated overnight at 4°C with capture antibody (in 0.1 M sodium bicarbonate, pH 8.2) at a dilution of 4 μg/mL. Plates were blocked the following morning in Block Ace (AbD Serotech) and then incubated overnight with 50 μl of brain lysate. The following morning the plates were washed with PBST and then incubated for 2 hours at room temperature with HRP-4G8 (Covance) at a dilution of 1 μg/mL. Plates were then washed and read at 620 nM wavelength 30 minutes after adding 100 μl of TMB substrate (Covance). The signal was normalized to the protein concentration for each sample. Between 5 and 6 mice were used in each group when comparing WT and APP-KO tissue for a given ELISA. All data analysis was done using Microsoft Excel; statistical significance was calculated using a 2-tailed T-test.

Western Blot

Western blotting was performed as previously described [15]. Hippocampal tissue was homogenized in RIPA buffer (Fisher scientific) with protease inhibitor (Roche) at 4°C, followed by centrifugation at 2,000 rpm for 1 min. The supernatant was electrophoresed on 4–12% Bis-Tris NuPAGE gels (Invitrogen) and then immunoblotted on nitrocellulose membrane. All primary antibodies were used at a 1:1,000 concentration for immunoblotting. HRP-conjugated secondary antibodies were purchased from Millipore. Although we do not quantify the bands in this paper, the same amount of protein (30 μg) was loaded for each sample for ease of comparison.

Results

When interpreting the signal from a β-amyloid ELISA, we assume that any signal with APP-KO tissue is pure noise, whereas signal from WT tissue is noise plus genuine signal from murine β-amyloid. We first compared the ability of six commercially available ELISAs from three different companies to distinguish WT from APP-KO hippocampal lysate (see Methods for details and catalog numbers). We first tried the Covance β-amyloid x-40 and x-42 ELISA kits. We found that both kits gave a strong signal for both WT and APP-KO mice (Figure 1A). In fact, there was no significant difference between WT and APP-KO mice for the x-40 kit (p-value = 0.3). For the x-42 kit the difference was significant (p-value = 0.03), but there was still a strong signal generated by APP-KO hippocampal tissue. For both of the Covance kits, the signal generated by the APP-KO tissue was statistically different from blank (p-value = 0.0001 for the x-40 kit; p-value = 0.015 for the x-42 kit). We next tried the Invitrogen mouse β-amyloid 40 and β-amyloid 42 kits. Both kits gave a substantial and statistically significant difference between WT and APP-KO mice (Figure 1B). However, the APP-KO signal was still statistically different from blank in both cases (p-value = 0.002 for the β-amyloid 42 kit; p-value = 0.0004 for the β-amyloid 40 kit). Interestingly, the total amount of β-amyloid detected by the Invitrogen kits is significantly

less than the amount detected by the Covance kits (when normalized by protein levels). This may be due to the fact that the Covance kits are detecting significant amounts of other proteins in addition to β-amyloid. Consistent with this second possibility, the APP-KO tissue has a strong signal with the Covance kits. Finally, we ran the same experiment with the β-Amyloid 40 and β-Amyloid 42 High-Sensitive kits from Wako chemicals. These kits both gave a large, statistically significant difference between WT and APP-KO tissue (Figure 1C). In fact, the β-Amyloid 42 High-Sensitive kit gave a signal with the APP-KO tissue that was effectively at our blank reading (it was actually slightly below). The Wako β-Amyloid 40 kit gave a positive background reading with APP-KO tissue, but it was not statistically significantly different from the blank reading (p-value = 0.49).

How much of the background noise in the above ELISAs is from fragments of APP other than β-amyloid and how much is from non-APP derived proteins? Since APP-KO mice give a robust signal with many of the above kits, we assume that a large portion of the noise in these kits is due to proteins not derived from APP. However, there remains the possibility that some of the signal in WT mice is coming from APP-derived fragments other than β-amyloid. Definitively answering this question is somewhat tricky, as APP has multiple proteolytic fragments. In addition, most commercially available ELISA kits do not disclose which antibodies are used in them; this is true both for the Covance and the Invitrogen kits that had significant background activity from APP-KO tissue. Thus, in order to further investigate this issue, we decided to design an ELISA that was specific for human β-amyloid, and measure the signal from WT and APP-KO tissue with this ELISA. This ELISA should not detect murine β-amyloid [11]. Thus, if the signal from APP-KO and WT mice is equal in this ELISA, then this argues that at least in this case, the noise is coming entirely from non-APP derived proteins. To do this, we made an ELISA using 6E10 (a human-specific β-amyloid antibody) as the capture antibody and HRP-conjugated 4G8 (which recognizes both rodent and human β-amyloid) as the detection antibody. As seen in Figure 1D, both APP-KO and WT mice gave a strong signal with this ELISA, and both signals are of comparable strength. To verify that our ELISA was working correctly, we first successfully measured step-wise dilutions of human synthetic β-amyloid (data not shown). As an additional control, we then asked whether this ELISA could have detected murine β-amyloid if the capture antibody had been for rodent β-amyloid. To do this, we ran the exact same ELISA with APP-KO and WT tissue, but changed the capture antibody from human-specific 6E10 to rodent-specific M3.2; the ELISA was identical in all other respects. This ELISA should now give a clear difference between WT and APP-KO tissue, although there may well be some background signal with the APP-KO tissue. As expected, this ELISA gave a strong signal in WT tissue and a smaller, but statistically significant signal in APP-KO tissue (Figure 1D). The very strong signal seen with the WT tissue in the M3.2 ELISA suggests that there is a significant contribution from proteins other than β-amyloid. Since the APP-KO tissue only gives a small signal, it is likely that this ELISA not only cross-reacts with other non-APP related proteins, but also cross-reacts with other proteins derived from APP other than β-amyloid. The M3.2 and 4G8 antibodies are known to cross-react with multiple APP fragments [16,17]. Thus, it is reasonable to assume that other APP-related fragments partially contribute to the WT signal in our M3.2 ELISA.

We next quantified the "signal to noise" ratio for each ELISA we had run (Figure 1E), which we define as the ratio of the WT

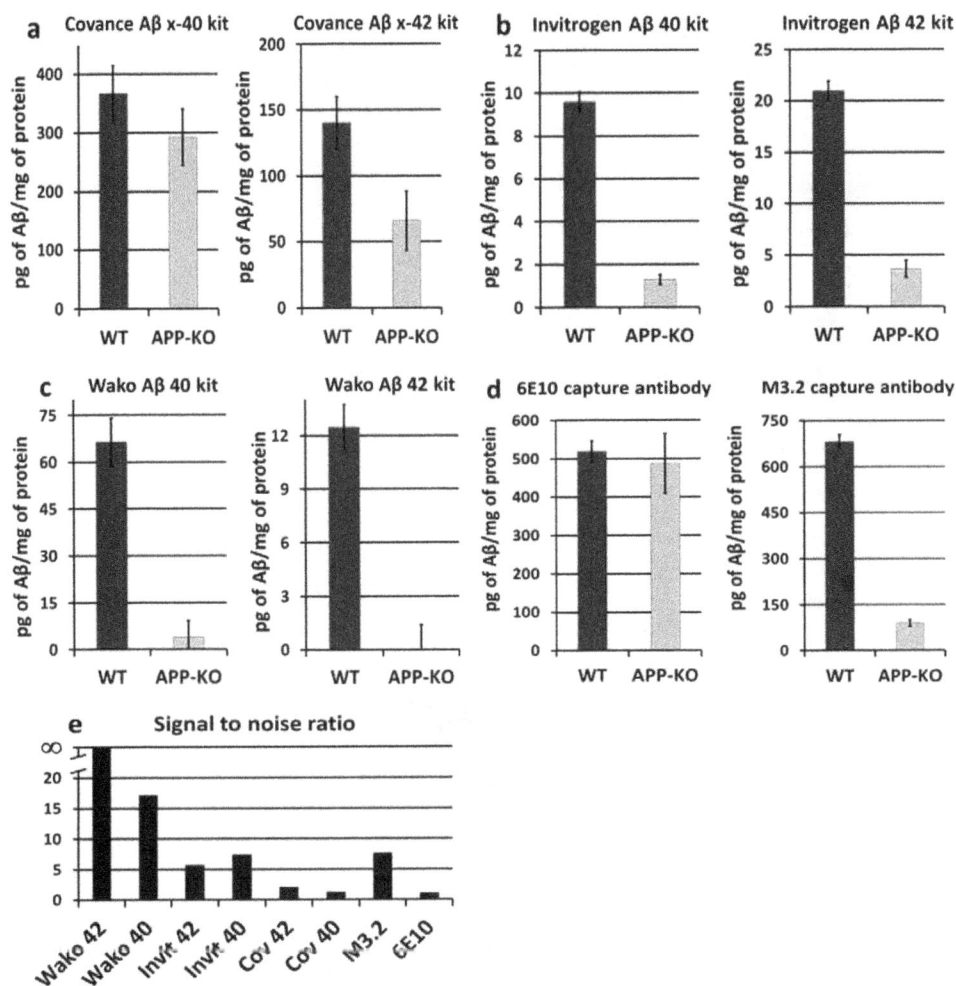

Figure 1. The amount of background signal with APP-KO tissue varies widely between different ELISAs. A) The Covance Colorimetric BetaMark™ Beta-Amyloid x-40 ELISA Kit (left) and x-42 Kit (right). Both kits give a signal with APP-KO tissue that is significantly above baseline. The x-40 kit gives a signal with APP-KO tissue that is not significantly different from WT (the difference between APP-KO and WT is significant for the x-42 kit). **B)** The Invitrogen Aβ 40 Mouse ELISA Kit (left) and Aβ 42 Mouse ELISA Kit (right). Both kits give a signal with APP-KO tissue that is significantly above baseline, but also significantly different from WT tissue. **C)** The Wako β-Amyloid (40) ELISA Kit (left) and β-Amyloid (42) ELISA High-Sensitive Kit (right). Both kits give a signal with APP-KO tissue that is not statistically significant from baseline. **D)** Both WT and APP-KO give a similar level of background signal with an ELISA for human β-amyloid (6E10 capture antibody – left), but show a clear difference when a rodent-specific capture antibody is used (M3.2 antibody – right). **E)** "Signal to noise" ratio of the WT signal divided by the APP-KO signal for each ELISA. Error bars in A–D are standard error.

signal to APP-KO signal for each ELISA. Note that we give the Wako β-Amyloid 42 High-Sensitive kit a ratio of "∞" because the denominator is zero. Also note that the Invitrogen kits and the M3.2 homemade ELISA all give a large difference between WT and APP-KO brain tissue. However, because the APP-KO signal is statistically significant in all of these ELISAs, the signal/noise ratio for these ELISAs is much smaller than the Wako ELISA kits.

Finally, we ran western blots using the three antibodies from our homemade ELISAs (M3.2, 4G8, and 6E10) on hippocampal tissue from WT, APP-KO, and APP transgenic mice expressing human APP (Hu*APP*695SWE) (Figure 2). Although antibodies may bind proteins in a western blot that they do not bind in an ELISA [2], we did this experiment as an additional test of antibody specificity for β-amyloid. Surprisingly, WT and APP-KO tissue showed a very similar band pattern for all three antibodies. 4–8 month old APP transgenic tissue also showed a similar band pattern, although 4G8 and 6E10 detected β-amyloid when brain tissue from 28 month old transgenic mice was used. In addition, 6E10 detected

an 87 kD band in APP transgenic tissue that is probably full-length human APP. In summary, the similar band pattern between WT and APP-KO tissue is consistent with our general conclusion that many β-amyloid antibodies show non-specific binding. Because we have used APP-KO tissue in this analysis, this also supports the conclusion that much of the non-specific binding is with proteins that are not derived from APP.

Discussion

In conclusion, we have confirmed prior reports of non-specificity of β-amyloid antibodies when performing ELISAs for murine β-amyloid with rodent brain tissue [10,11,12]. By performing our experiments with APP-KO mice, we have demonstrated that much of the observed cross-reactivity is not with other proteolytic fragments of APP, but with other non-APP related proteins found in rodent brain. Most importantly, we have shown that not all β-amyloid ELISAs show high non-specificity,

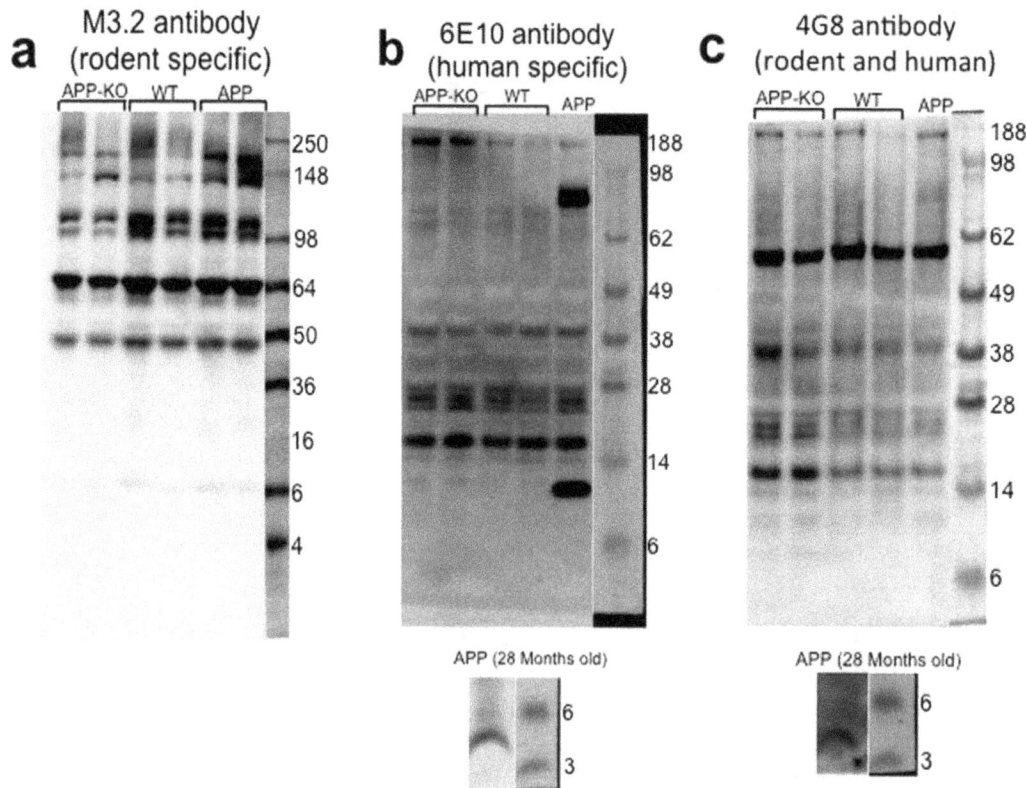

Figure 2. Several β-amyloid antibodies show high non-specificity for β-amyloid by western blot. A) Western blot with the antibody M3.2 (specific for rodent β-amyloid) using hippocampal tissue from APP-KO, WT, and APP transgenic mice expressing human APP (Hu*APP*695SWE). All three genotypes show a similar band pattern. **B)** Western blot with the antibody 6E10 (specific for human β-amyloid). All three genotypes show a similar band pattern. However, with 6E10, the APP transgenic mouse tissue shows an additional 8 kD band (presumably a human β-amyloid dimer) as well as a band near 87 kD (presumably full-length human APP). In addition, a 4.5 kDa band (β-amyloid) is detected when older mouse brain is used that has high levels of β-amyloid protein. **C)** Western blot with 4G8. All three genotypes also showed a similar band pattern with this antibody as well. In addition, a 4.5 kDa band (β-amyloid) is detected when older mouse brain is used that has high levels of β-amyloid protein.

and investigators may detect endogenous murine β-amyloid in rodent brain tissue using an ELISA without a prior solid-phase extraction step. Finally, we want to emphasize that our findings here pertain to the detection of *endogenous* murine β-amyloid. For example, the Covance kits will have less of an issue measuring fluctuations of human β-amyloid in transgenic animals. This is because the huge amount of β-amyloid that is produced in transgenic mice will be much higher than the background signal. For example, some authors have estimated that 35 times as much β-amyloid is being produced in transgenic mice than in wild-type rodents [10]. Thus, the Covance kits may still be quite useful in these kinds of experiments.

Measuring natural variations in murine endogenous β-amyloid is important for understanding the physiology of this protein, and

the work done here will hopefully make it easier for researchers to select the appropriate kit for their work. An interesting side issue is whether there are significant differences in endogenous β-amyloid levels in different strains of mice, and whether this affects the choice of kit for different mouse strains. In our own work we have used the Wako high-sensitive β-amyloid 42 kit on several different strains of mice (C57B6, FVB, C57B6/129 hybrids, and C57B6/ Swiss Webster hybrids), and we have found roughly comparable levels of β-amyloid 42. Thus, although the work in this paper was done with C57B6 mice, we believe that the results extrapolate to other mouse strains as well.

In table 1, we summarize the strengths and weaknesses of the six commercially available kits tested here. In summary, the ELISAs tested in this paper range from being rodent-specific (Invitrogen) to

Table 1. Summary of Commercially Available Kits We Tested.

	Covance Aβ x-40	Covance Aβ x-42	Invitrogen Aβ 40	Invitrogen Aβ 42	Wako Aβ 40	Wako Aβ 42
Species	Rodent and Human	Rodent and Human	Rodent	Rodent	Rodent and Human	Rodent and Human
Signal to Noise Ratio	Poor	Poor	Good	Good	Excellent	Excellent
Cost	Lowest	Lowest	Low-Intermediate	Low-Intermediate	Highest	Highest

working with both rodents and humans (Covance and Wako). The fact that the Invitrogen kit is rodent specific gives this kit a unique advantage over the Covance and Wako kits. Namely, the Invitrogen kit can be used to measure the physiology of endogenous rodent β-amyloid in transgenic mice that also express human β-amyloid. Any experiment that examines endogenous β-amyloid processing in transgenic mice must use this kit, as the Wako and Covance kits will not distinguish between endogenous and human β-amyloid. The signal-to-noise ratio ranges from poor (Covance) to excellent (Wako). Thus, in terms of signal-to-noise ratio, the Wako kits are the best. Finally, cost must be a consideration when choosing an ELISA kit. The Covance kits are cheaper than either the Invitrogen or Wako kits, so this may be a reason to choose these kits if they fit the experiment one is trying to do (for example, measure human β-amyloid in a transgenic mouse).

Author Contributions

Conceived and designed the experiments: AFT MP OA. Performed the experiments: AFT MP. Analyzed the data: AFT MP OA. Contributed reagents/materials/analysis tools: AFT OA. Wrote the paper: AFT OA.

References

1. Hardy J, Allsop D (1991) Amyloid deposition as the central event in the aetiology of Alzheimer's disease. Trends Pharmacol Sci 12: 383–388.
2. Wang J, Dickson DW, Trojanowski JQ, Lee VM (1999) The levels of soluble versus insoluble brain Abeta distinguish Alzheimer's disease from normal and pathologic aging. Exp Neurol 158: 328–337.
3. Cirrito JR, May PC, O'Dell MA, Taylor JW, Parsadanian M, et al. (2003) In vivo assessment of brain interstitial fluid with microdialysis reveals plaque-associated changes in amyloid-beta metabolism and half-life. J Neurosci 23: 8844–8853.
4. Puzzo D, Privitera L, Fa' M, Staniszewski A, Hashimoto G, et al. (2011) Endogenous amyloid-β is necessary for hippocampal synaptic plasticity and memory. Annals of Neurology 69: 819–830.
5. Puzzo D, Privitera L, Leznik E, Fa M, Staniszewski A, et al. (2008) Picomolar amyloid-beta positively modulates synaptic plasticity and memory in hippocampus. J Neurosci 28: 14537–14545.
6. Plant LD, Boyle JP, Smith IF, Peers C, Pearson HA (2003) The production of amyloid beta peptide is a critical requirement for the viability of central neurons. J Neurosci 23: 5531–5535.
7. Lopez-Toledano MA, Shelanski ML (2004) Neurogenic effect of beta-amyloid peptide in the development of neural stem cells. J Neurosci 24: 5439–5444.
8. Abramov E, Dolev I, Fogel H, Ciccotosto GD, Ruff E, et al. (2009) Amyloid-beta as a positive endogenous regulator of release probability at hippocampal synapses. Nat Neurosci 12: 1567–1576.
9. Anderson JJ, Holtz G, Baskin PP, Turner M, Rowe B, et al. (2005) Reductions in beta-amyloid concentrations in vivo by the gamma-secretase inhibitors BMS-289948 and BMS-299897. Biochem Pharmacol 69: 689–698.
10. Best JD, Jay MT, Otu F, Ma J, Nadin A, et al. (2005) Quantitative measurement of changes in amyloid-beta(40) in the rat brain and cerebrospinal fluid following treatment with the gamma-secretase inhibitor LY-411575 [N2-[(2S)-2-(3,5-difluorophenyl)-2-hydroxyethanoyl]-N1-[(7S)-5-methyl-6-oxo-6,7-d ihydro-5H-dibenzo[b,d]azepin-7-yl]-L-alaninamide]. J Pharmacol Exp Ther 313: 902–908.
11. Lanz TA, Schachter JB (2006) Demonstration of a common artifact in immunosorbent assays of brain extracts: development of a solid-phase extraction protocol to enable measurement of amyloid-beta from wild-type rodent brain. J Neurosci Methods 157: 71–81.
12. Lanz TA, Schachter JB (2008) Solid-phase extraction enhances detection of beta-amyloid peptides in plasma and enables Abeta quantification following passive immunization with Abeta antibodies. J Neurosci Methods 169: 16–22.
13. Zheng H, Jiang M, Trumbauer ME, Sirinathsinghji DJ, Hopkins R, et al. (1995) beta-Amyloid precursor protein-deficient mice show reactive gliosis and decreased locomotor activity. Cell 81: 525–531.
14. Hsiao K, Chapman P, Nilsen S, Eckman C, Harigaya Y, et al. (1996) Correlative memory deficits, Abeta elevation, and amyloid plaques in transgenic mice. Science 274: 99–102.
15. Gong B, Cao Z, Zheng P, Vitolo OV, Liu S, et al. (2006) Ubiquitin Hydrolase Uch-L1 Rescues beta-Amyloid-Induced Decreases in Synaptic Function and Contextual Memory. Cell 126: 775–788.
16. Morales-Corraliza J, Mazzella MJ, Berger JD, Diaz NS, Choi JH, et al. (2009) In vivo turnover of tau and APP metabolites in the brains of wild-type and Tg2576 mice: greater stability of sAPP in the beta-amyloid depositing mice. PLoS One 4: e7134.
17. Venezia V, Russo C, Repetto E, Salis S, Dolcini V, et al. (2004) Apoptotic cell death influences the signaling activity of the amyloid precursor protein through ShcA and Grb2 adaptor proteins in neuroblastoma SH-SY5Y cells. J Neurochem 90: 1359–1370.

3

Creation of Resveratrol-Enriched Rice for the Treatment of Metabolic Syndrome and Related Diseases

So-Hyeon Baek[1], Woon-Chul Shin[1], Hak-Seung Ryu[2,3], Dae-Woo Lee[2,3], Eunjung Moon[4], Chun-Sun Seo[1], Eunson Hwang[4], Hyun-Seo Lee[5], Mi-Hyun Ahn[5], Youngju Jeon[5], Hyeon-Jung Kang[1], Sang-Won Lee[3,6], Sun Yeou Kim[7], Roshan D'Souza[8], Hyeon-Jin Kim[8], Seong-Tshool Hong[5]*, Jong-Seong Jeon[2,3]*

1 National Institute of Crop Science, Rural Development Administration, Iksan, Chonbuk, Korea, 2 Graduate School of Biotechnology, Kyung Hee University, Yongin, Gyeonggi, Korea, 3 Crop Biotech Institute, Kyung Hee University, Yongin, Gyeonggi, Korea, 4 Graduate School of East-West Medical Science, Kyung Hee University, Yongin, Gyeonggi, Korea, 5 Laboratory of Genetics and Department of Microbiology, Chonbuk National University Medical School, Jeonju, Chonbuk, Korea, 6 Department of Plant Molecular Systems Biotechnology, Kyung Hee University, Yongin, Gyeonggi, Korea, 7 College of pharmacy, Gachon University, Incheon, Korea, 8 BDRD Research Institute, JINIS Biopharmaceuticals Inc., Wanju, Chonbuk, Korea

Abstract

Resveratrol has been clinically shown to possess a number of human health benefits. As a result, many attempts have been made to engineer resveratrol production in major cereal grains but have been largely unsuccessful. In this study, we report the creation of a transgenic rice plant that accumulates 1.9 µg resveratrol/g in its grain, surpassing the previously reported anti-metabolic syndrome activity of resveratrol through a synergistic interaction between the transgenic resveratrol and the endogenous properties of the rice. Consumption of our transgenic resveratrol-enriched rice significantly improved all aspects of metabolic syndrome and related diseases in animals fed a high-fat diet. Compared with the control animals, the resveratrol-enriched rice reduced body weight, blood glucose, triglycerides, total cholesterol, and LDL-cholesterol by 24.7%, 22%, 37.4%, 27%, and 59.6%, respectively. The resveratrol-enriched rice from our study may thus provide a safe and convenient means of preventing metabolic syndrome and related diseases without major lifestyle changes or the need for daily medications. These results also suggest that future transgenic plants could be improved if the synergistic interactions of the transgene with endogenous traits of the plant are considered in the experimental design.

Editor: M. Lucrecia Alvarez, TGen, United States of America

Funding: This work was supported by a grant from the Next-Generation BioGreen 21 Program (No. PJ0079732012) and RDA program (No. PJ0075392012), Rural Development Administration, Korea. The funders had no role in study design, data collection and analysis, decision to publish, or preparation of the manuscript.

Competing Interests: RD and H. Kim are employees of BDRD Research Institute, JINIS Biopharmaceuticals, Inc. BDRD Research Institute, JINIS Biopharmaceuticals, Inc. helped to carry out in vivo efficacy assay of transgenic grains using a mouse model. All other authors declare no competing interest.

* E-mail: seonghong@chonbuk.ac.kr (STH); jjeon@khu.ac.kr (JSJ)

Introduction

Resveratrol (3,5,4'-trihydroxy-trans-stilbene) is a non-flavonoid polyphenol-type stilbene compound found in several fruits and vegetables. Although resveratrol has various beneficial health effects, its effect on metabolic syndrome is the best characterized [1]. Since this finding, a major research objective has been to create transgenic plants that accumulate resveratrol. The transfer of stilbene synthase (STS) genes has been previously accomplished in a number of plants [2]. Of these transgenic plant studies, however, reasonable levels of resveratrol production were only observed in a few cases, including transgenic tobacco [3,4], tomato [5], and lettuce [6]. Notably, resveratrol production has not been successfully achieved in human-edible agronomically significant crops such as cereal grains. A grain crop plant with proven activity against metabolic syndrome is therefore an ideal target for resveratrol production, considering that metabolic syndrome and related diseases could be controlled by dietary intake.

The *Oryza sativa* japonica variety Dongjin (Dongjin rice), developed by the Rural Development Administration of Korea, yields a grain that is rich in fiber and in polyphenols that confer low levels of anti-metabolic syndrome activity [7]. It is thus reasonable to assume that a transgenic Dongjin rice strain that expresses resveratrol may prevent and treat metabolic syndrome and related diseases through a synergistic effect of its innate and transgenic properties. To test this hypothesis, we generated transgenic resveratrol-enriched rice and assessed its efficacy in controlling metabolic syndrome and related diseases in a mouse model.

Results and Discussion

Production of Transgenic Rice

We cloned the resveratrol biosynthesis gene, stilbene synthase (STS), from the peanut *Arachis hypogaea* variety Palkwang, a well-known plant species that contains high quantities of resveratrol [8]. Sequence analysis of the cloned cDNA, designated *AhSTS1* (GenBank accession no. DQ124938), showed a high similarity to previously identified STSs (Figure S1). In the peanut, STS appeared to be highly expressed in the early and middle stages of the developing pods after flowering but not in the leaves (Figure S2). To determine whether *AhSTS1* encodes a functional STS enzyme, we cloned the *4-coumaroyl-CoA ligase (4CL)* gene from

Arabidopsis thaliana (*At4CL2*). The product of this gene converts *p*-coumaric acid into *coumaroyl-CoA* by coupling it with a coenzyme. We reasoned that the coexpression of *AhSTS1* and *At4CL2* might lead to resveratrol production using *p*-coumaric acid and malonyl-CoA [9,10]. *AhSTS1* and *At4CL2* were cotransformed into *E. coli*, and the production of the recombinant AhSTS1 and At4CL2 proteins was confirmed using western blot analysis with anti-His and anti-MBP antibodies, respectively (Figure S3). GC-MS analysis of the culture grown in medium supplemented with *p*-coumaric acid demonstrated that one fraction eluted by HPLC was identical to the resveratrol standard (Figure S4). This finding establishes AhSTS1 as an active STS enzyme. In contrast, cells transformed with control vectors did not produce resveratrol.

Several transgenic cereal plants have been produced with the aim of accumulating an adequate quantity of resveratrol in the edible portion of cereal crops [11,12]. However, these transgenic cereal plants failed to accumulate resveratrol in the grain, likely because of unfavorable chimeric constructs or because the foreign gene was inserted into a chromosomal locus that was unfavorable for expression. In this study, we constructed a chimeric fusion between the maize *Ubiquitin1* (*Ubi1*) promoter, which produces high levels of activity in monocots [13] and *AhSTS1* to express *AhSTS1* in rice. Then, we conducted phenotypic expression analysis at each step before proceeding to the next step to confirm the proper expression of the transgene during the creation of our transgenic rice. We transferred the chimeric construct into embryonic calli induced from the mature embryo of Dongjin rice using the *Agrobacterium*-mediated transformation method to generate transgenic calli. Somatic embryos formed from the transgenic calli were germinated on N6 medium containing phosphinothricin (PPT) to regenerate small plantlets. HPLC analysis of the metabolite profiles of the T_1 seeds produced by 398 plants showed that 129 T_1 lines of Dongjin rice produced over 0.1 µg resveratrol per gram of seed.

We planted all of the 129 T_1 seeds in a paddy field to enable a thorough analysis of agricultural traits. Southern blot analysis demonstrated that these transgenic rice lines carried one to four copies of the transgene (Figure 1A), and RT-PCR analysis indicated that all of the transgenic lines exhibited high levels of *AhSTS1* expression (Figure 1B). In the rice paddy field, many of the transgenic plants displayed partial sterility at the flowering stage. However, several lines were completely fertile. A similar infertility phenotype was observed in tobacco expressing high levels of an *STS* gene [4], suggesting that transgenic overexpression of *STS* affects the fertility of plants.

Resveratrol Analysis of Transgenic Rice

To assess the biosynthetic profile of the transgene in Dongjin rice, we analyzed resveratrol and the related resveratrol glucoside piceid from all tissues of the transgenic rice plants using HPLC. The health benefits of piceid are less than resveratrol [14,15]. In the wild-type Dongjin rice, HPLC analysis failed to detect resveratrol or piceid (Figure 2B). In the leaves of the transgenic rice plant, however, we detected high levels of piceid ranging from 1.2–174.4 µg/g and low levels of resveratrol ranging from 0–8.9 µg/g (Figure 3A). On the other hand, the grains of the transgenic rice contained comparable levels of resveratrol (0.1–4.8 µg/g) but a relatively low quantity of piceid (0.1–10.4 µg/g) compared with the corresponding levels in the leaves (Figure 2C and 3B). These quantities in the grain of the transgenic rice are similar to the levels of resveratrol (0.8–5.8 µg/mL) reported in high-quality red wine [16]. Based on agricultural, biochemical, and genetic traits, we chose the homozygous transgenic line RS18 as a candidate strain for further experiments. The RS18 line

contained a single transgene copy and exhibited a relatively high expression level of the transgene among all of the transgenic lines produced; its performance with respect to agronomic traits was similar to that of the parental Dongjin rice (Table S1).

As mentioned above, the grains of the transgenic plants, including RS18, contained a relatively high quantity of resveratrol compared with piceid, whereas the reverse was observed in the leaves. This unequal distribution of the two metabolites could be due to glucosyltransferase activity. Glucosyltransferase activity is known to be involved in the formation of piceid from resveratrol [17,18]. The relatively high level of piceid in the leaves suggested that the leaves might have higher glucosyltransferase activity than the grains. Quantification of glucosyltransferase activity in wild type and RS18 rice showed that the RS18 leaves exhibited much higher resveratrol glucosyltransferase activity than the grains. The wild-type leaves and grains did not show any resveratrol glucosyltransferase activity (Figure 4). This observation suggests that resveratrol-specific glucosyltransferase activity is induced in response to resveratrol production in the transgenic plant. Further study is necessary to identify the gene(s) responsible for this resveratrol glucosyltransferase activity.

Rice seeds are milled to create polished white rice, which is then consumed. We compared the resveratrol levels of unpolished brown grains and polished white grains of the RS18 strain. The results indicated that the unpolished and polished grains contained similar levels of resveratrol, 1.9 and 1.7 µg/g, respectively (Table S2). This finding suggests that most of the resveratrol accumulates in the endosperm rather than in the other tissues, such as the aleurone layer and the embryo.

Assessment of Anti-metabolic Syndrome Activity of Transgenic Resveratrol-enriched Rice

We examined the efficacy of the transgenic resveratrol-enriched rice on metabolic syndrome and on related diseases associated with the levels of blood glucose, triacylglycerol, total cholesterol, LDL cholesterol and HDL cholesterol using an *in vivo* mouse model. We assessed in the mice whether the innate characteristics of the rice and the transgenic resveratrol had a synergistic effect that boosts anti-metabolic syndrome activity. C57BL inbred mice were fed a high-fat diet (HFD) for 12 weeks to induce metabolic syndrome and related diseases. The food consumption rate was the same among different mouse groups (each individual mouse consumed 4 g of food per day). The diet-induced obesity mice were fed the HFD in the control group or a modified HFD in the experimental group, in which the carbohydrate source was replaced with the resveratrol-enriched rice (Table S3). We periodically monitored the changes in the blood profiles and body weight in each mouse group under continuous HFD conditions (Table 1 and Figure 5). Compared with the HFD control, supplementation with resveratrol at the same level as that produced by RS18 resulted in modest improvement, consistent with previous reports on the effects of resveratrol on metabolic syndrome and related diseases [1,19,20]. The consumption of Dongjin rice also resulted in a similar improvement in lipid profile and blood glucose levels, as expected due to its endogenic nature. Notably, the consumption of the resveratrol-enriched Dongjin rice significantly improved all aspects of metabolic syndrome and related diseases, lowering the blood glucose by 22.0%, triacylglycerol by 37.4%, total cholesterol by 27.0%, and LDL cholesterol by 59.6%, whilst increasing the HDL cholesterol by 14.8% (RS18 compared with the HFD control). An RS18-half group with a modified HFD, in which only half the amount of corn starch was replaced by RS18 rice, failed to have an effect similar to that observed in the RS18 group, indicating a dose-dependent effect of

A

B

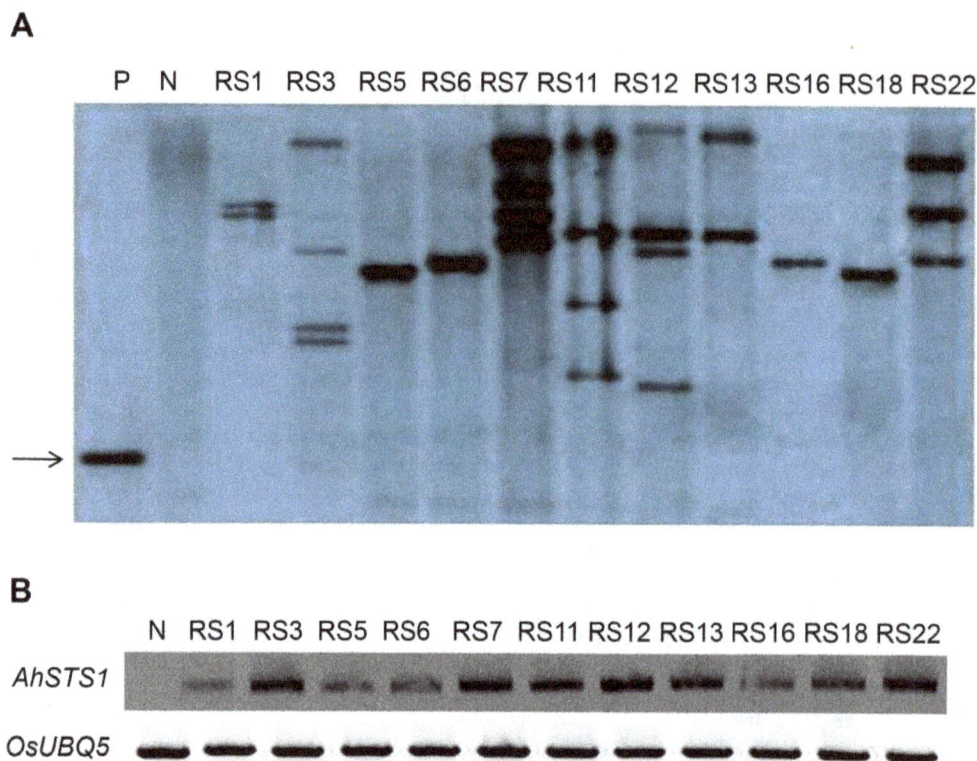

Figure 1. Molecular characterization of transgenic rice lines expressing *AhSTS1*. (A) Southern blot analysis. Genomic DNA in lanes P and RS1 to RS22 were digested with *Bam*HI (specific to the T-DNA region). The arrow indicates the fragment (1.2 kb) hybridized with the *AhSTS1* cDNA probe. P, pSB2220 vector; N, non-transgenic wild-type Dongjin; lanes RS1 - RS22, representative transgenic Dongjin lines out of 129 T_1 samples. (B) RT-PCR analysis. Total RNA from leaf samples of the same lines as in (A) was analyzed. *OsUBQ5* was included as a PCR control.

the resveratrol-enriched rice. As expected from the blood profiles, body weights were greatly reduced in mice fed the resveratrol-enriched rice (RS18 group; 24.7% compared with the control) and was different from the other treatments (the resveratrol supplementation group, Dongjin rice group, and RS18-half group) (Figure 5A). Micro-CT image analysis of abdominal fat deposition showed that the total, visceral, and subcutaneous fat volumes in the resveratrol-enriched rice group (RS18) were 21.55%, 16.33%, and 3.10%, respectively, which were significantly lower than the fat volumes from the HFD control (25.43%, 20.02%, and 3.83%, respectively) (Figure 5B). Representative images clearly indicated that the total, visceral and subcutaneous fat accumulation volumes were lowest in the RS18 group compared with the other treatments (Figure 5C).

The most important finding from this experiment was the synergistic effect of Dongjin rice and transgenic resveratrol in the RS18 group compared with treatment by resveratrol supplementation or Dongjin rice alone. The resveratrol-enriched Dongjin rice, RS18, was thus found to be as effective at treating metabolic syndrome and related diseases as typical pharmaceutical drugs for these disorders in reducing the blood glucose, LDL/total cholesterol, or body weight. Hence, resveratrol-enriched rice is a potentially feasible and viable choice to treat most, if not all, aspects of metabolic syndrome and related diseases.

The central nervous system controls nutrient levels in an effort to maintain metabolic homeostasis through the feedback and crosstalk of many organs [21]. In the brain, Sirt1, a nicotinamide adenine dinucleotide (NAD^+)-dependent deacetylase, is a key regulator of the energy homeostasis involved in glucose and lipid metabolism [22–24]. To examine the effect of transgenic rice

grains on the level of Sirt1 protein, we treated human neuroblastoma SH-SY5Y cells with ethanol extracts from the grains of RS18 (50 and 100 μg/mL). Western blot analysis indicated that the levels of Sirt1 protein were higher in the treated cells than in untreated cells. Similar increases in Sirt1 protein were observed in cells treated with 100 μM resveratrol (Figure 6A). Moreover, mice fed a HFD supplemented with transgenic grain (RS18) had higher *Sirt1* expression in the brain, liver, skeletal muscle and adipose tissues. Among these tissues, *Sirt1* expression in the liver of the RS18-fed mice was significantly increased in comparison to that observed in the control mice fed a HFD alone (Figure 6B). A previous study reported that glucose and blood cholesterol levels were reduced in *Sirt1* transgenic mice [25]. Thus, these results suggest that treatment with resveratrol-enriched transgenic grains may improve metabolic syndrome and related diseases associated with the disturbance of hepatic lipid metabolism and of glucose and lipid homeostasis by upregulating *Sirt1* expression.

Conclusions

After the etiological agent of the French Paradox was identified as resveratrol [26], the creation of transgenic cereal plants that accumulate resveratrol in their grains has been a major research objective. Although transgenic cereal plants have been produced with the aim of accumulating resveratrol in their grains, resveratrol was only detected at low levels in the leaves and stems of the previously created transgenic plants [19]. In this study, we report the first successful creation of rice with resveratrol-enriched grains, using the approach of validating the expression of the transgene at each step. Because the resveratrol-enriched rice was created using

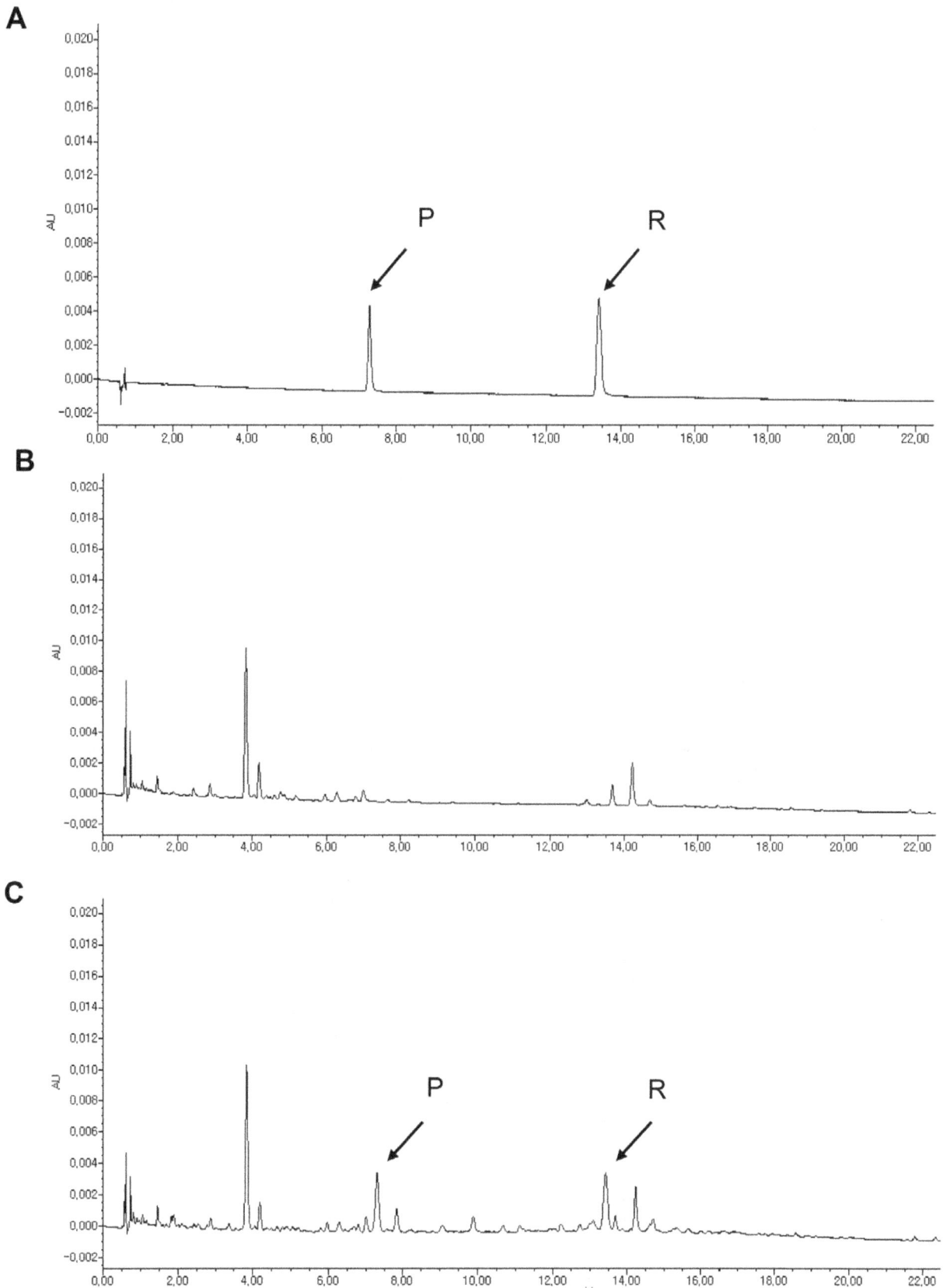

Figure 2. The identification of resveratrol and piceid in the grains of wild-type Dongjin and transgenic rice using HPLC. (A) A standard mixture of piceid (P) and resveratrol (R). (B) Wild-type Dongjin rice. (C) Transgenic Dongjin rice RS18. The arrows indicate the positions of piceid (P) and resveratrol (R).

A

B

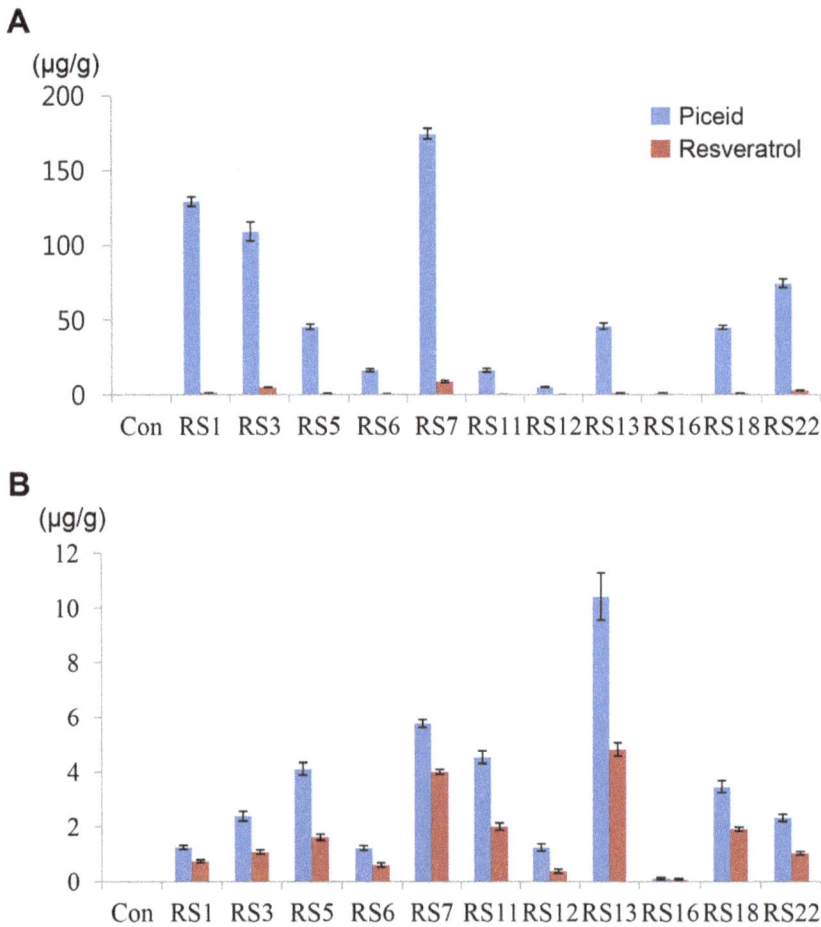

Figure 3. The quantification of the piceid and resveratrol levels in the leaves and grains of wild-type Dongjin rice and representative transgenic Dongjin rice lines out of 398 T₁ samples. (A) Leaves. (B) Grains.

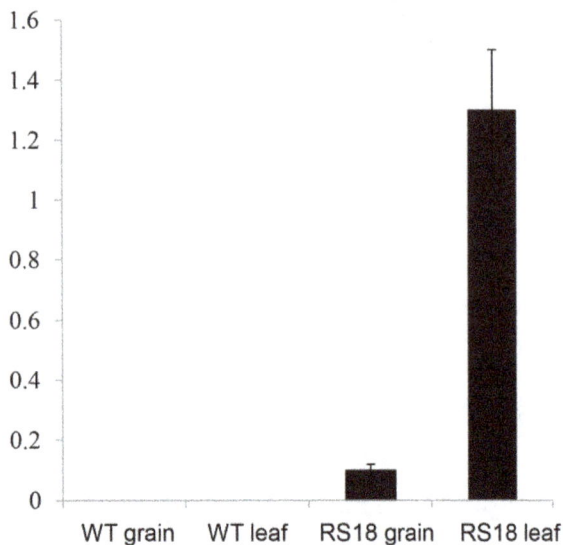

Figure 4. Measurement of resveratrol-specific glucosyltransferase activity in rice leaves and grains. WT, wild-type Dongjin rice; RS18, *AhSTS1* transgenic Dongjin rice line. Remarkably high resveratrol-specific glucosyltransferase activity was found only in the leaves of the transgenic rice line RS18.

a rice variety with endogenous anti-metabolic syndrome characteristics, it has more potent anti-metabolic syndrome activity than resveratrol itself due to synergistic effects and can be used to treat and prevent metabolic syndrome and related diseases. Both the severity and prevalence of metabolic syndrome and related diseases, such as obesity, cardiovascular diseases, and diabetes, among many others, are currently more serious in developing countries than in developed countries. Moreover, because access to medical care is more limited in developing countries, these disorders are a more serious problem. We believe that our resveratrol-enriched rice will be an excellent alternative for the management of metabolic syndrome and related diseases in both developed and developing countries.

Materials and Methods

Plant Materials

The leaves and developing pods of the peanut cultivar Palkwang were used for total RNA isolation and resveratrol measurements. Seven-week-old leaves of the wild-type cultivar Dongjin and transgenic Dongjin rice plants were used for molecular characterization. Eight-week-old leaves and mature grains of rice were used to determine the levels of resveratrol and piceid.

Table 1. The effects of the resveratrol-enriched rice on blood lipid and glucose levels.

	Week 0	Week 4	Week 8	Week 12
TG (mg/dL)				
CTL	84.5±17.6	86.4±19.7	81.8±6.2	81.5±22.7
Resv	84.3±20.7	78.9±28.9	64.4±13.9[b]	60.2±11.9[b]
DJ	85.5±19.3	70.2±25.7[a]	65.5±20.9[b]	64.7±19.3[b]
RS18-half	83.1±18.9	75.4±19.1	60.5±6.6[b]	64.3±14.7[b]
RS18	82.2±15.6	60.6±10.9[b]	60.5±14.6[b]	51.0±11.3[b,c]
TC (mg/dL)				
CTL	178.5±41.8	177.3±29.7	196.0±15.3	198.3±30.7
Resv	180.9±37.8	170.3±26.6	172.7±11.2[b]	157.3±4.2[b]
DJ	178.4±27.8	171.5±39.2	170.1±23.5[b]	167.2±21.3[b]
RS18-half	177.4±32.4	160.0±26.9	163.4±9.7[b]	159.7±5.9[b]
RS18	181.8±24.4	157.1±20.0[a]	159.4±14.4[b]	144.7±12.2[b,d]
HDL-C (mg/dL)				
CTL	82.7±5.6	85.9±19.6	85.0±8.9	83.4±6.9
Resv	84.0±4.1	89.4±24.1	99.5±3.4[b]	94.4±12.5[b]
DJ	82.3±8.8	86.5±17.9	86.7±8.7	89.2±12.1
RS18-half	84.5±8.0	78.8±14.3	84.7±7.0	89.3±11.3
RS18	78.4±7.2	85.4±5.2	95.3±6.6[b,c]	97.9±8.6[b,c]
LDL-C (mg/dL)				
CTL	78.4±43.1	74.2±31.6	93.8±11.2	90.8±15.1
Resv	85.9±36.1	65.1±35.8	60.3±15.1[b]	50.8±6.8[b]
DJ	79.1±29.4	70.9±20.1	70.3±18.9[b]	65.1±12.3[b]
RS18-half	76.3±30.2	66.1±29.9	66.6±10.1[b]	57.5±6.7[b,c]
RS18	87.0±23.7	59.6±19.9	51.9±20.1[b,c]	36.6±10.6[b,d]
Glucose (mg/dL)				
CTL	193.1±22.7	201.7±60.4	233.7±25.0	236.3±23.2
Resv	195.2±18.8	201.4±20.9	205.1±24.4[b]	203.3±18.5[b]
DJ	194.5±23.2	203.3±27.8	208.1±24.1[b]	207.4±19.1[b]
RS18-half	194.4±24.7	198.1±23.8	202.4±23.3[b]	200.4±19.4[b]
RS18	192.1±28.2	194.0±21.0	187.3±24.0[b,c]	184.3±18.3[b,d]

The values represent the means ± SEM (n = 16). CTL, mice fed a HFD (D12451); Resv, mice fed a HFD supplemented with resveratrol; DJ, mice fed a HFD in which the corn starch and sucrose were replaced with Dongjin rice; RS18-half, mice fed a HFD in which half of the corn starch and sucrose were replaced with the resveratrol-enriched rice; RS18, mice fed a HFD in which the corn starch and sucrose were replaced with the resveratrol-enriched rice. Values in a column with a superscripted letter indicate statistical significance as analyzed by an unpaired Student's t-test;
[a]p<0.05 compared with CTL;
[b]p<0.01 compared with CTL;
[c]p<0.05 compared with DJ;
[d]p<0.01 compared with DJ.

Cloning of AhSTS1 cDNA

Total RNA was isolated from developing peanut pods 40 days after flowering using TRIzol reagent (Invitrogen, Carlsbad, CA). The total RNA was reverse-transcribed with oligo-dT primers and the First Strand cDNA Synthesis Kit (Roche). An *STS* cDNA was cloned using RT-PCR of the first strand cDNA. Gene-specific primers (5′-ATGGTGTCTGTGAGTGGAATTC-3′ and 5′-CGTTATATGGCCACACTGC-3′) were designed based on the genomic DNA sequence of the *A. hypogaea STS* gene (GenBank

accession no. AF227963) to encompass the complete coding sequence.

Determination of AhSTS1 Activity

The 4CL enzyme converts *p*-coumaric acid into *coumaroyl-CoA* by coupling it with coenzyme A. Subsequently, three malonyl-CoA units are added to *coumaroyl-CoA* by STS with a loss of carbon dioxide, which results in the production of resveratrol [9,10]. *AhSTS1* was amplified from cDNA using the specific primers 5′-GGATCCATGGTGTCTGTGAGTG-3′ and 5′-CTCGAG-TATGGCCACACTGCGGAG-3′. The *At4CL2* gene (GenBank accession no. NM113019) was also amplified using RT-PCR from *Arabidopsis* leaf RNA using the gene-specific primers 5′-GGATC-CATGACGACACAAGATGTGATAG-3′ and 5′-CTCGAGGTTCATTAATCC ATTTGCTAGT-3′ (the substitutions required to create *Bam*HI and *Xho*I restriction sites are underlined). The amplified fragments of *AhSTS1* and *At4CL2* were cloned into pET28a, a plasmid carrying a kanamycin resistance marker, and the pMAL-c2x vector, which harbors an ampicillin marker. The *AhSTS1* and *At4CL2* coding sequences were inserted in frame with the His and MBP (maltose-binding protein) carboxyl terminal tags, respectively. The plasmids containing each *AhSTS1* and *At4CL2* gene were cotransformed into BL21 *E. coli* for the induction of protein expression [27]. Finally, *E. coli* cells carrying the resistance genes for kanamycin and ampicillin were selected. The cells were grown in LB supplemented with 100 µg/mL of kanamycin and ampicillin at 37°C. Protein expression was induced at $OD_{600} = 0.5$ by adding 1 mM isopropyl β-D-thiogalactopyranoside (IPTG). After 24 and 48 h, the cells were harvested by centrifugation and resuspended in lysis buffer (50 mM NaH_2PO_4, 300 mM NaCl, and 10 mM imidazole). After sonication, the samples were subjected to SDS-PAGE for western blot analysis.

To examine resveratrol production using the recombinant proteins, *E. coli* cells carrying both genes were grown in 2XYT medium (10 g/L yeast extract, 16 g/L tryptone, 5 g/L NaCl) supplemented with 5 mM *p*-coumaric acid (C9008, Sigma) and 0.1 mM IPTG at 28°C. After 48 h of incubation, 1 mL of the culture medium was centrifuged at 13,000 rpm for 15 min. The supernatant was transferred to a new tube, and 50 µL 1 N hydrochloric acid was added to adjust the pH to 9.0. These samples were stored overnight at −20°C. The tubes were thawed at room temperature, and resveratrol was isolated with two extractions of equal volumes of ethyl acetate, dried under nitrogen gas, and then resuspended in 100 µL of methanol. All of the samples were stored at −20°C until they were used for the resveratrol content analysis [10,28].

Binary Vector Construction and Rice Transformation

To overexpress *AhSTS1* in rice, the binary vector pSB22 was constructed by inserting an expression cassette encoding the maize *Ubi1* promoter [13], multiple cloning sites (*Bam*HI, *Sma*I, and *Sac*I), and the *nopaline synthase* (*Nos*) terminator into the *Hind*III and *Eco*RI sites of the pCAMBIA3300 vector carrying the herbicide resistance *bialaphos* (*bar*) gene. Subsequently, the *AhSTS1* cDNA was inserted between the *Bam*HI and *Sac*I site under the control of the *Ubi1* promoter. The resulting plasmid was designated pSB2220. This construct was introduced into rice plants using *Agrobacterium*-mediated transformation [29]. Three-week-old calli derived from the mature seeds of the Dongjin *japonica* rice variety were cocultivated with the *A. tumefaciens* strain LBA4404 carrying pSB2220. After 3–4 weeks, transgenic calli were selected on N6 medium containing 5 mg/L phosphinothricin (PPT) and 250 mg/L cefotaxime. The transgenic plants were regenerated on MS

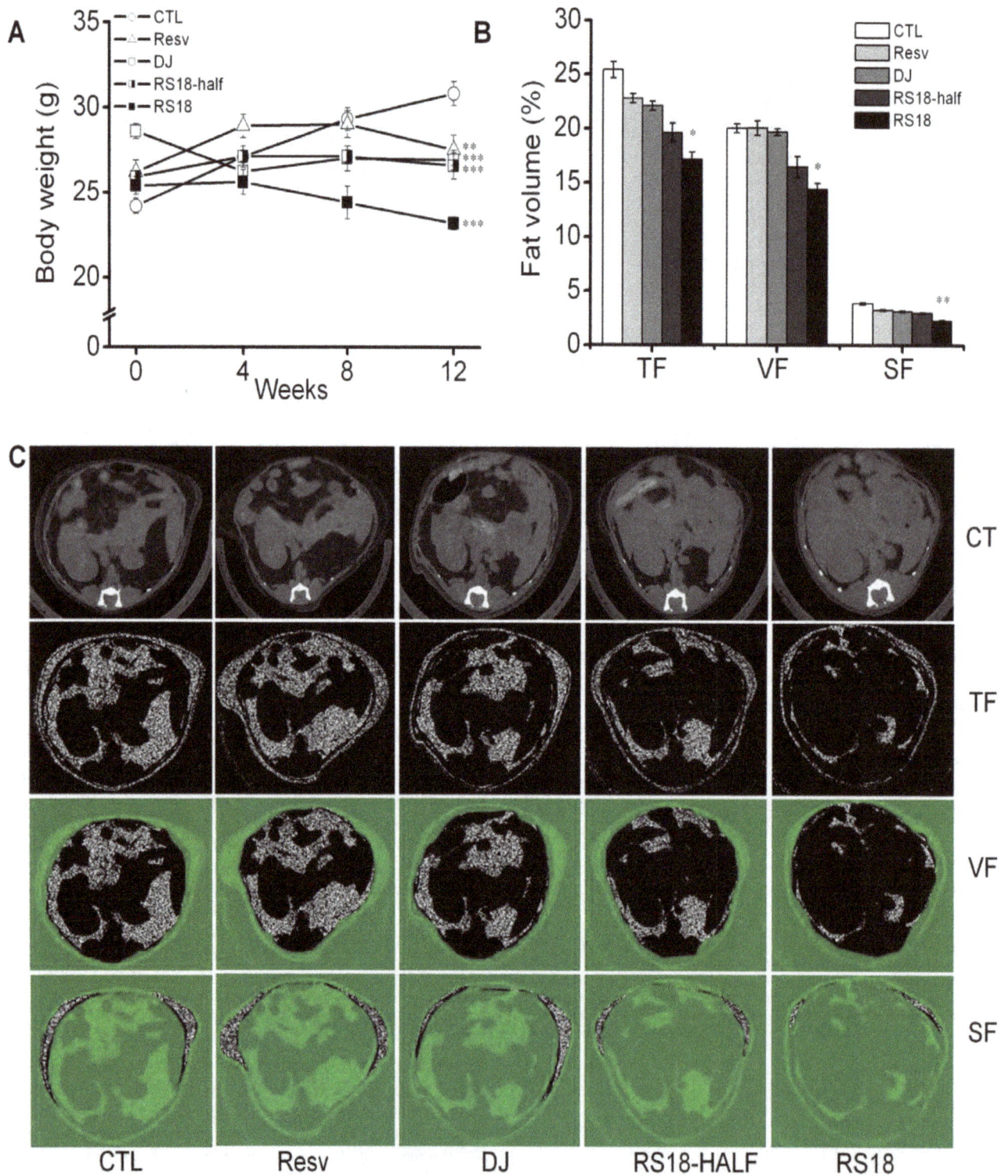

Figure 5. The effects of the resveratrol-enriched rice on body weight and body fat volume. (A) The body weight of mice during a 12-week period. The values represent the means ± SEM (n = 16). An unpaired Student's t-test was used for the statistical analysis; *p<0.05, **p<0.01, ***p<0.001 compared with CTL. (B) The fat volume of mice using *in vivo* micro-CT image analysis. (c) Representative images of micro-CT and fat area. The values represent the means ± SEM (n = 5). TF, total fat; VF, visceral fat; SF, subcutaneous fat (SF).

media supplemented with 0.1 mg/L NAA, 2 mg/L kinetin, 2% sorbitol, 3% sucrose, 1.6% phytagar, 5 mg/L PPT, and 250 mg/L cefotaxime. The plants were grown in a greenhouse with a 12 h photoperiod.

Southern Blot and RT-PCR Analysis

Approximately 3 μg of genomic DNA from the transgenic plants was digested with *Bam*HI and then subjected to electro-

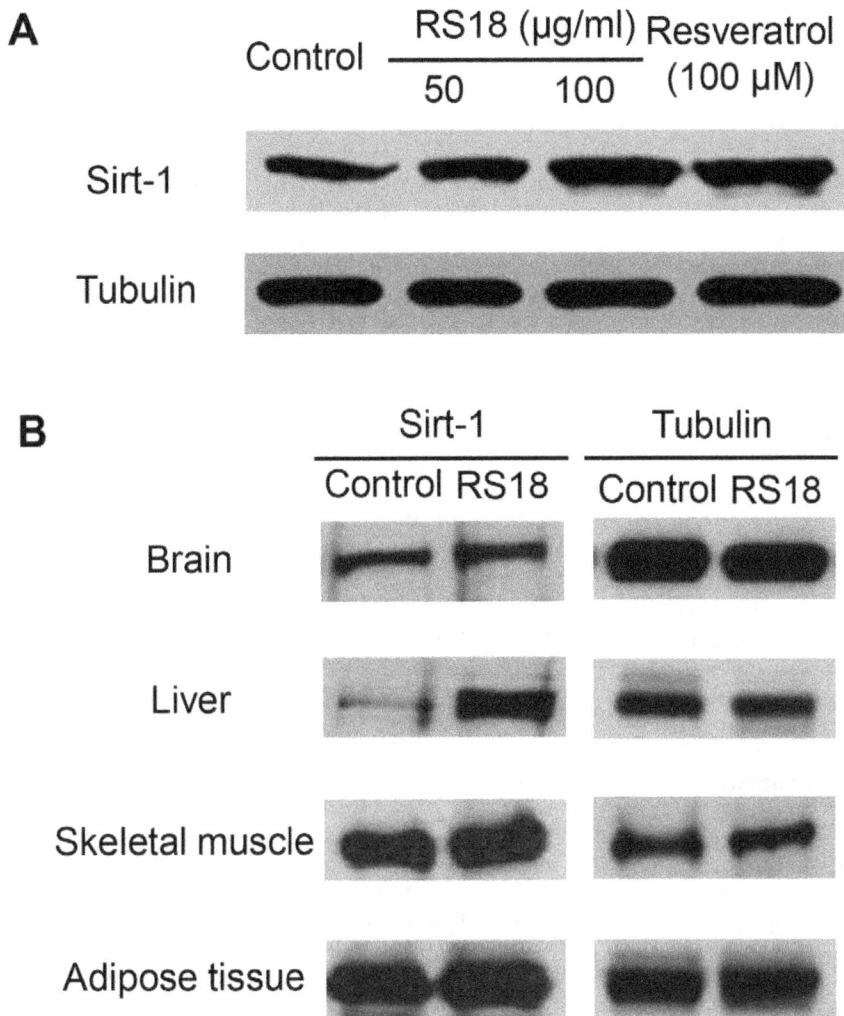

Figure 6. The effects of the resveratrol-enriched rice on Sirt1 protein levels in cells and in mice. (A) The level of Sirt1 protein in SH-SY5Y cells. The SH-SY5Y cells were treated with 70% ethanol extracts of RS18 transgenic grain (50 and 100 μg/mL) or resveratrol (100 μM) for 24 h. (B) Mice treated with RS18 transgenic grain. The organs, including the brain, liver, skeletal muscle and adipose tissues, were harvested from mice that had been fed RS18 transgenic grain. Subsequently, 30 μg of protein from each lysate was used for western blot analysis. Equal protein loading was confirmed using an anti-tubulin antibody.

phoresis on a 0.8% agarose gel. The DNA was transferred onto a Hybond N+ nylon membrane, and hybridization was performed using a [α-^{32}P] dCTP-labeled gene-specific probe according to the standard procedures for high-stringency hybridization conditions. The blot was hybridized in a solution containing 0.5 M sodium phosphate (pH 7.2), 1 mM EDTA, 1% (w/v) BSA, and 7% (w/v) SDS for 20 h at 60°C. First-strand cDNAs, prepared from harvested leaf samples, were used in the RT-PCR reactions with gene-specific primers and control primers for *OsUBQ5*. The *AhSTS1*-specific primers were 5′-ATGGTGTCTGTGAGTG-GAATTC-3′ and 5′-CGTTATATGGCCACACTGC-3′, and the *OsUBQ5*-specific primers were 5′-GACTACAACATCCA-GAAGGAGTC-3′ and 5′-TCATCTAATAACCAGTTC-GATTTC-3′.

Quantification of Resveratrol and Piceid

The resveratrol and piceid levels in the transgenic rice were determined by HPLC (ACQUITY UPLC, Waters, Milford, MA) using an instrument equipped with a UV-spectrophotometer at 308 nm (ACQUITY TUV, Waters). The results were calculated

using Empower software (Waters). Chromatographic separation was accomplished by injecting 1 μL of the samples onto an ACQUITY UPLC BEH-C18 1.7 μm column (2.1 mm×100 mm, Waters) at a ow rate of 0.4 mL/min. The mobile phase was 10 to 90% acetonitrile (ACN). A gradient elution was conducted as follows: 0 min, 10% ACN; 1.54 min, 10% ACN; 10 min, 15% ACN; 22 min, 25% ACN; 22.4 min, 90% ACN; and 25 min, 90% ACN. The column was then re-equilibrated with 10% ACN for 5 min prior to the next injection. The calibration curves were obtained using a weighted linear regression of the peak areas against known concentrations (0.5, 1, 2, 5 and 10 μg/mL of each) of resveratrol and piceid. The weights were obtained from a smoothed estimate of the within-triplicate standard deviation of each sample [30]. The correlations of each calibration ranged from 0.98922 to 0.99917.

The HPLC fraction was eluted and further verified using GC-MS analysis with the 6890/5973N GC/MS system (Agilent Technologies) equipped with an Rtx-5MS capillary column (30 mm×0.25 mm I.D., 0.25 μm film thickness, *Restek*, Germany). The fractions were dried, resolubilized in 10 μL of methoxyamine hydrochloride, resolved in pyridine (40 mg/mL), and incubated at 30°C for 90 min. Then, 90 μL of N-methyl-N-(trimethylsilyl)tri-fluoroacetamide was added, and the samples were incubated at 37°C for 30 min. The resveratrol standard was prepared by solubilizing 20 μg of the compound in the same way. The initial column temperature was 80°C for 5 min, followed by a 5°C/min ramp to 300°C. Sample volumes of 1 μL were injected with a split ratio of 25:1 using an autosampler system. The interface and ion source temperatures were set to 250°C. The resveratrol in the sample was identified by comparing the MS spectrum to the standard.

Glucosyltransferase Activity Assay

The enzymatic activity of glucosyltransferase was measured using a previously described method [18]. Briefly, to extract the total protein, 2 g of leaves or seeds were collected from transgenic Dongjin rice and wild-type Dongjin rice. The samples were ground to a fine powder in liquid nitrogen and suspended with extraction buffer [500 mM Tris-HCl, (pH 8.0), 5 mM sodium metabisulfite, 10% glycerol, 1% PVP-40 (polyvinyl polypyrroli-done), 1 mM phenylmethyl sulfonyl fluoride, 0.1% β-mercap-toethanol, and 10% insoluble PVP]. The slurries were filtered through two layers of nylon mesh (20 μm) followed by centrifu-gation at 13,000 rpm for 10 min at 4°C. The protein concentra-tion of the supernatant was determined using the Bradford reagent (BioRad, Hercules, CA). One milligram of total protein was used for the glucosyltransferase activity assay.

Each reaction mixture contained resveratrol (1 μg/mL) and rice protein extract (1 mg) in 140 μL of reaction buffer (100 mM Tris, pH 9.0). The enzyme reaction was initiated by adding 10 μL of 25 mM uridine diphosphate glucose (UDPG). Each reaction was incubated at 30°C for 30 min and terminated by the addition of 150 μL of absolute methanol. The products of the enzyme reaction were extracted twice with equal volumes of *trichloroacetic acid (*TCA) and dried under nitrogen gas. The dried residues were resuspended in 100 μL methanol. All of the samples were filtered through a 0.45 μm nylon filter after mixing with the same volume of 20% ACN for HPLC analysis. The control reactions without total protein extract or UDPG did not yield any detectable piceid.

Animal Care and Diets

All of the procedures performed with animals were in accordance with established guidelines and were reviewed and approved by the Ethics Committee of Chonbuk National University Laboratory Animal Center. C57BL/6 female mice were purchased from Joongang Experimental Animal Co. (Seoul, Korea) at six weeks of age. The mice were housed at 10 animals per cage, with food (10% kcal as fat; D12450B; Research Diets Inc., New Brunswick, NJ) and water available ad libitum unless otherwise stated. They were maintained under a 12 h light/12 h dark cycle at a temperature of 22°C and humidity of 55±5%. After one week of acclimation, the animals were provided with a high-fat diet (HFD) containing 45% kcal as fat (D12451, Research Diets Inc.) for 12 weeks to induce metabolic syndrome and related diseases. After 12 weeks on the HFD, a total of 100 mice were randomly divided into the following groups: HFD diet (CTL), HFD supplemented with resveratrol (Resv), HFD in which the corn starch and sucrose were replaced with Dongjin rice (DJ), HFD in which half of the corn starch and sucrose were replaced with resveratrol rice (RS18-half); and HFD in which the corn starch and sucrose were replaced with resveratrol rice (RS18) (Table S3).

In vivo Efficacy Assay

The blood glucose and lipid levels were measured at 0, 4, 8, and 12 weeks during treatment. The food consumption of each mouse group was regularly monitored. Blood samples were drawn from the tail after 5 h of fasting, and blood-glucose measurements were taken using an Accu-check Glucometer (Roche Diagnostic Corporation, Indianapolis, IN). The serum was separated by centrifuging at 13,000 rpm for 10 min and immediately stored at −20°C until assayed. The triglycerides, total cholesterol, and HDL cholesterol levels in the serum were quantitatively de-termined using an enzymatic colorimetric method (Asan Pharm., Yongin, Korea) [31]. The LDL cholesterol levels were calculated using the Friedewald equation [(LDL) = (T-CHO) − (HDL) − (TG)/5].

The body weights were measured at 0, 4, 8, and 12 weeks after treatment. For fat analysis, the total body fat was determined by high-resolution *in vivo* micro-CT (Skyscan 1076; SkyScan, Ko-nitch, Belgium) with a high resolution CCD/phosphor screen detector. Before CT, the mice were anesthetized with zoletil and rumpun (4:1) and placed on a radio-transparent mouse bed in a supine position with the caudal end closest to the micro-CT. The hind legs were extended and held in place with clear tape to ensure the correct anatomical position. Micro-CT images of the abdomen were captured at the level of the L1–L5 inter-vertebral disks, and the total fat, visceral fat and subcutaneous fat areas were analyzed using CTan Ver.1.10, Skyscan software (Skyscan).

Determination of the Sirt1 Protein Level

Transgenic rice grains were extracted with 70% EtOH under ultrasonic conditions for 1.5 h. After repeating this process three times, the extracts were evaporated and then freeze-dried with a yield of 8.9%. SH-SY5Y cells were seeded at approximately 1×10^6 cells in 60 mm culture dishes. After 24 h, the cells were treated with 70% ethanol extracts of transgenic grains (50 and 100 μg/mL) or resveratrol (100 μM) for 24 h. Six-week-old female C57BL/6 mice were randomly assigned to the control and transgenic rice groups. The control group was fed a HFD alone for 18 months. The transgenic rice group was fed a HFD with RS18 transgenic grain for 18 months. The organs assayed included the brain, liver, skeletal muscle and adipose tissues harvested from the mice. The cells and tissues were lysed in cold lysis buffer (0.1% SDS, 150 mM NaCl, 1% NP-40, 0.02% sodium azide, 0.5% sodium deoxycholate, 100 μg/mL PMSF, 1 μg/mL aprotinin, and phosphatase inhibitor in 50 mM Tris-HCl, pH 8.0). The levels of Sirt1 were determined by western blot analysis using an anti-Sirt1 antibody (Santa Cruz Biotechnology, Santa Cruz, CA). Briefly, 30 μg of protein was separated by SDS-PAGE (8% acrylamide gel) and transferred to a nitrocellulose membrane. The membrane was blocked with 5% non-fat skim milk in Tris-buffered saline with Tween-20 and incubated overnight with the primary antibody at 4°C. The membranes were then incubated with the secondary antibody for 1 h at room temperature. The membranes were developed using ECL reagents.

Supporting Information

Figure S1 Comparison of the deduced amino acid sequence of AhSTS1 and previously identified STS protein sequences. These proteins contain conserved domain regions, such as the malonyl-CoA binding sites, a dimer interface, and active sites, which are indicated by *, •, and ▲, respectively. The black boxes indicate identical or conserved residues.

Figure S2 Northern blot analysis of total RNA isolated from peanut leaves and pods. The pods were collected during the early (1), middle (2), and late (3) stages of development. The *AhSTS1* cDNA was used as a probe. Strong signals were only observed in the early and middle stages of the developing peanut pods. Ethidium bromide staining of the rRNAs demonstrated equal RNA loading.

Figure S3 Western blot analysis of the recombinant AhSTS1 and At4CL2 proteins. The *AhSTS1* and *At4CL2* genes were expressed to produce fusion proteins containing a His6-tag or an MBP-tag, respectively. Total proteins were prepared from *E. coli* cells carrying *AhSTS1* or *At4CL2* at 24 and 48 h after adding 1 mM isopropyl β-D-thiogalactopyranoside (IPTG) and hybridized with rabbit anti-His6 and anti-MBP serum. AhSTS1-His6, 60 kDa; 4CL2-MBP, 103 kDa.

Figure S4 GC-MS analysis of the eluted resveratrol fraction. The MS spectrum of the resveratrol standard (A) is identical to that of the HPLC peak fraction (B). The arrows indicate the position of resveratrol.

Table S1 The major agronomic characteristics of wild-type Dongjin rice and the *AhSTS1* transgenic rice line RS18.

Table S2 The resveratrol content in unpolished and polished grains of the transgenic rice line RS18.

Table S3 The formulation of the diets (g).

Author Contributions

Conceived and designed the experiments: SB SYK SH JJ. Performed the experiments: SB WS HR DL EM CS EH HL MA YJ H. Kang SL RD H. Kim. Analyzed the data: SB HR SL SYK SH JJ. Wrote the paper: SB HR SL SYK SH JJ.

References

1. Guarente L (2006) Sirtuins as potential targets for metabolic syndrome. Nature 444: 868–874.
2. Zhuang H, Kim YS, Koehler RC, Dore S (2003) Potential mechanism by which resveratrol, a red wine constituent, protects neurons. Ann N Y Acad Sci 993: 276–286.
3. Hain R, Reif HJ, Krause E, Langebartels R, Kindl H, et al. (1993) Disease resistance results from foreign phytoalexin expression in a novel plant. Nature 361: 153–156.
4. Fischer R, Budde I, Hain R (1997) Stilbene synthase gene expression causes changes in flower color and male sterility in tobacco. Plant J 11: 489–498.
5. Morelli R, Das S, Bertelli A, Bollini R, Lo Scalzo R, et al. (2006) The introduction of the stilbene synthase gene enhances the natural antiradical activity of *Lycopersicon esculentum* mill. Mol Cell Biochem 282: 65–73.
6. Liu S, Hu Y, Wang X, Zhong J, Lin Z (2006) High content of resveratrol in lettuce transformed with a stilbene synthase gene of *Parthenocissus henryana*. J Agric Food Chem 54: 8082–8085.
7. Choi H, Moon JK, Park BS, Park HW, Park SY, et al. (2012) Comparative nutritional analysis for genetically modified rice, Iksan483 and Milyang204, and nontransgenic counterparts. J Kor Soc Appl Biol Chem 55: 19–26.
8. Sobolev VS, Khan SI, Tabanca N, Wedge DE, Manly SP, et al. (2011) Biological activity of peanut (*Arachis hypogaea*) phytoalexins and selected natural and synthetic stilbenoids. J Agric Food Chem 59: 1673–1682.
9. Sparvoli F, Martin C, Scienza A, Gavazzi G, Tonelli C (1994) Cloning and molecular analysis of structural genes involved in flavonoid and stilbene biosynthesis in grape (*Vitis vinifera L.*). Plant Mol Biol 24: 743–755.
10. Beekwilder J, Wolswinkel R, Jonker H, Hall R, de Vos CHR, et al. (2006) Production of resveratrol in recombinant microorganisms. Appl Environ Microbiol 72: 5670–5672.
11. Leckband G, Lörz H (1998) Transformation and expression of a stilbene synthase gene of *Vitis vinifera* L. in barley and wheat for increased fungal resistance. Theor Appl Genet 96: 1004–1012.
12. Stark-Lorenzen P, Nelke B, Hänßler G, Mühlbach HP, Thomzik JE (1997) Transfer of a grapevine stilbene synthase gene to rice (*Oryza sativa* L.). Plant Cell Rep 16: 668–673.
13. Christensen AH, Sharrock RA, Quail PH (1992) Maize polyubiquitin genes: structure, thermal perturbation of expression and transcript splicing, and promoter activity following transfer to protoplasts by electroporation. Plant Mol Biol 18: 675–689.
14. Kimura Y, Okuda H, Arichi S (1985) Effects of stilbenes on arachidonate metabolism in leukocytes. Biochim Biophys Acta 834: 275–278.
15. Meng X, Maliakal P, Lu H, Lee MJ, Yang CS (2004) Urinary and plasma levels of resveratrol and quercetin in humans, mice, and rats after ingestion of pure compounds and grape juice. J Agric Food Chem 52: 935–942.
16. Souto AA, Carneiro MC, Seferin M, Senna MJH, Conz A, et al. (2001) Determination of *trans*-resveratrol concentrations in Brazilian red wines by HPLC. J Food Comp Anal 14: 441–445.
17. Lunkenbein S, Bellido M, Aharoni A, Salentijn EMJ, Kaldenhoff R, et al. (2006) Cinnamate metabolism in ripening fruit: Characterization of a UDP-glucose:-cinnamate glucosyltransferase from strawberry. Plant Physiol 140: 1047–1058.
18. Hall D, De Luca V (2007) Mesocarp localization of a bi-functional resveratrol/hydroxycinnamic acid glucosyltransferase of Concord grape (*Vitis labrusca*). Plant J 49: 579–591.
19. Delaunois B, Cordelier S, Conreux A, Clement C, Jeandet P (2009) Molecular engineering of resveratrol in plants. Plant Biotechnol J 7: 2–12.
20. Jang M, Cai L, Udeani GO, Slowing KV, Thomas CF, et al. (1997) Cancer chemopreventive activity of resveratrol, a natural product derived from grapes. Science 275: 218–220.
21. Morton GJ, Cummings DE, Baskin DG, Barsh GS, Schwartz MW (2006) Central nervous system control of food intake and body weight. Nature 7109: 289–295.
22. Bordone L, Guarente L (2005) Calorie restriction, SIRT1 and metabolism: understanding longevity. Nat Rev Mol Cell Biol 6: 298–305.
23. Liang F, Kume S, Koya D (2009) SIRT1 and insulin resistance. Nat Rev Endocrinol 5: 367–373.
24. Haigis MC, Sinclair DA (2010) Mammalian sirtuins: biological insights and disease relevance. Annu Rev Pathol 5: 253–295.
25. Bordone L, Cohen D, Robinson A, Motta MC, van Veen E, et al. (2007) SIRT1 transgenic mice show phenotypes resembling calorie restriction. Aging Cell 6: 759–767.
26. Liu BL, Zhang X, Zhang W, Zhen HN (2007) New enlightenment of French Paradox: resveratrol's potential for cancer chemoprevention and anti-cancer therapy. Cancer Biol Ther 6: 1833–1836.
27. Yang W, Zhang L, Lu Z, Tao W, Zhai Z (2001) A new method for protein coexpression in *Escherichia coli* using two incompatible plasmids. Protein Expr Purif 22: 472–478.
28. Sydor T, Schaffer S, Boles E (2010) Considerable increase in resveratrol production by recombinant industrial yeast strains with use of rich medium. Appl Environ Microbiol 76: 3361–3363.
29. Lee S, Jeon JS, Jung KH, An G (1999) Binary vectors for efficient transformation of rice. J Plant Biol 4: 310–316.
30. Normolle DP (1993) An algorithm for robust non-linear analysis of radioimmunoassays and other bioassays. Stat Med 12: 2025–2042.
31. Du H, You JS, Zhao X, Park JY, Kim SH, et al. (2010) Antiobesity and hypolipidemic effects of lotus leaf hot water extract with taurine supplementation in rats fed a high fat diet. J Biomed Sci 17 (Suppl 1): S42.

4

Levels of the Mahogunin Ring Finger 1 E3 Ubiquitin Ligase Do Not Influence Prion Disease

Derek Silvius[1], Rose Pitstick[1], Misol Ahn[2], Delisha Meishery[1], Abby Oehler[2], Gregory S. Barsh[3], Stephen J. DeArmond[2], George A. Carlson[1], Teresa M. Gunn[1]*

1 McLaughlin Research Institute, Great Falls, Montana, United States of America, 2 Institute for Neurodegenerative Diseases and Department of Pathology, University of California San Francisco, San Francisco, California, United States of America, 3 Departments of Genetics and Pediatrics, Stanford University, Stanford, California, United States of America

Abstract

Prion diseases are rare but invariably fatal neurodegenerative disorders. They are associated with spongiform encephalopathy, a histopathology characterized by the presence of large, membrane-bound vacuolar structures in the neuropil of the brain. While the primary cause is recognized as conversion of the normal form of prion protein (PrP^C) to a conformationally distinct, pathogenic form (PrP^{Sc}), the cellular pathways and mechanisms that lead to spongiform change, neuronal dysfunction and death are not known. Mice lacking the Mahogunin Ring Finger 1 (MGRN1) E3 ubiquitin ligase develop spongiform encephalopathy by 9 months of age but do not become ill. In cell culture, PrP aberrantly present in the cytosol was reported to interact with and sequester MGRN1. This caused endo-lysosomal trafficking defects similar to those observed when *Mgrn1* expression is knocked down, implicating disrupted MGRN1-dependent trafficking in the pathogenesis of prion disease. As these defects were rescued by over-expression of MGRN1, we investigated whether reduced or elevated *Mgrn1* expression influences the onset, progression or pathology of disease in mice inoculated with PrP^{Sc}. No differences were observed, indicating that disruption of MGRN1-dependent pathways does not play a significant role in the pathogenesis of transmissible spongiform encephalopathy.

Editor: Hyoung-gon Lee, Case Western Reserve University, United States of America

Funding: This work was funded by National Institutes of Health program project grant NS41997. The funders had no role in study design, data collection or analysis, decision to publish, or preparation of the manuscript.

Competing Interests: The authors have declared that no competing interests exist.

* E-mail: tmg@mri.montana.edu

Introduction

Transmissible spongiform encephalopathies, are rare but invariably fatal neurodegenerative disorders that affect humans and animals [1,2]. They are associated with misfolding and aggregation of the cellular prion protein, PrP^C, into a protease-resistant, pathogenic conformer referred to as PrP^{Sc}, with Sc referring to the prototypical Scrapie prion disease of sheep. PrP^{Sc} can be generated and propagated from endogenous PrP^C following infectious exposure to exogenous PrP^{Sc}, while rare inherited forms, such as familial Cruetzfeldt-Jakob disease, fatal familial insomnia and Gerstmann-Sträussler-Scheinker syndrome, result from autosomal dominant mutations in the prion protein gene (*PRNP*). Most prion diseases are characterized by spongiform changes, starting with the development of vacuoles in the neuropil and progressing to widespread vacuolation of the central nervous system (CNS). At advanced stages, there is typically neuronal loss, astrogliosis and cerebellar atrophy (predominantly affecting granular neurons), but no inflammatory response. Despite progress in understanding the primary cause of prion diseases, the cellular and molecular mechanisms that lead to neurodegeneration and death are still under investigation.

Mice lacking the E3 ubiquitin ligase, mahogunin ring finger-1 (MGRN1) or the type I transmembrane protein, attractin (ATRN) develop age-dependent CNS vacuolation that is histologically

similar to that associated with prion diseases, without the accumulation of protease-resistant PrP^{Sc} [3,4]. The cellular role of ATRN remains unknown, although it has been shown to be required for membrane homeostasis [5]. The only ubiquitination target of MGRN1 identified to date is tumor susceptibility gene 101 (TSG101), a component of the endocytic trafficking machinery that sorts membrane proteins into multivesicular bodies [6,7]. Loss of MGRN1-dependent ubiquitination disrupts endo-lysosomal trafficking, leading to accumulation of activated epidermal growth factor receptor (EGFR) and alterations in the morphology of early endosomes, late endosomes and lysosomes. PrP is normally secreted and tethered to the plasma membrane by a GPI anchor, but ER stress and some pathogenic mutations in *PRNP* can induce mislocalization of PrP to the cytosol and induce non-transmissible neurotoxicity [8]. A recent study demonstrated that cytosolically exposed forms of PrP can bind to and sequester MGRN1 in HeLa cells [9], resulting in similar abnormalities in endo-lysosomal trafficking to those observed in cells in which *Mgrn1* was knocked down by siRNA. Over-expressing MGRN1 rescued the trafficking defects. Reduced immunostaining for MGRN1 was observed in the brains of transgenic mice expressing a transmembrane form of PrP, along with an age-dependent increase in lysosome size/number (based on Cathepsin D staining) in Purkinje cells. These data suggested that disrupted MGRN1-dependent endo-lysosomal trafficking could be the cellular

Table 1. Brain *Mgrn1* expression.

Genotype	*Mgrn1* relative quantification value (range)	p value[a]
Tg−; Mgrn1$^{md-nc/+}$	0.42 (0.22–0.79)	0.04
Tg−; Mgrn1$^{+/+}$	1.00 (0.82–1. 21)	n/a
Tg+; Mgrn1$^{md-nc/+}$	1.41 (0.76–2.61)[b]	0.21
Tg+; Mgrn1$^{+/+}$	4.43 (1.59–12.32)	0.001

[a]Student's t-test against *Tg−; Mgrn1*$^{+/+}$ value, p<0.05 significant.
[b]Student's t-test against *Tg−; Mgrn1*$^{mdnc/+}$ yields p = 0.04.

mechanism underlying spongiform neurodegeneration in prion diseses.

Cytosolically-exposed PrP has been proposed to contribute to the pathogenesis of inherited and transmissible spongiform encephalopathies [8]. Mutations in the hydrophobic domain of the prion gene that lead to increased production of a transmembrane form with its N-terminal domain exposed to the cytosol cause neurodegeneration with pathology reminiscent of prion disease in transgenic mice [10]. Similar transmembrane forms of prion protein have been detected in both genetic and transmitted prion diseases [10–12]. The relationship between cytosolic PrP and CNS vacuolation is unclear. We tested whether functional sequestration of MGRN1 by cytosolic PrP contributes to transmissible prion disease by inoculating mice expressing reduced or elevated levels of *Mgrn1* with Rocky Mountain Laboratory (RML) prions. No differences were observed in the onset, progression, or histopathology of disease. This indicates that altered MGRN1 function has little or no role in the pathogenesis of transmissible prion diseases and indirectly supports a role for plasma membrane PrP^{Sc}.

Materials and Methods

Ethics statement

All animal procedures adhered to Association for Assessment and Accreditation of Laboratory Animal Care guidelines and were approved by the Institutional Animal Care and Use Committee of the McLaughlin Research Institute.

Mice

Mgrn1^{md-nc} (null) mutant mice and *Tg(Mgrn1^I)C3Tmg* transgenic (hereafter referred to as *Tg+*) mice, which express wild-type *Mgrn1* isoform I from the human β-actin promoter, were described previously [4,13]. The *Tg(Mgrn1^I)C3Tmg* transgenic line completely rescue all aspects of the *Mgrn1* null mutant phenotype, including spongiform degeneration of the CNS. *Mgrn1* null mutant (*Mgrn1*$^{md-nc}$) mice are maintained by breeding heterozygotes with their homozygous mutant siblings. A wild-type control line was established by inbreeding +/+ animals that were generated by intercrossing *Mgrn1*$^{md-nc/+}$ mice; this line is re-generated from *Mgrn1* heterozygotes every 3 years. *Tg+; Mgrn1* null mutant mice were backcrossed to wild-type mice to generate *Tg+* and *Tg− Mgrn1*$^{md-nc/+}$ and wild-type (*Mgrn1*$^{+/+}$) mice. Mice were genotyped for the *Mgrn1*$^{md-nc}$ mutation and the transgene as previously described [13].

Mgrn1 expression

Mgrn1 expression in the brains of *Tg+* and *Tg−* mice that were homozygous wild-type at the *Mgrn1* locus or heterozygous for the null allele (*Mgrn1*$^{md-nc/+}$) was determined by quantitative RT-PCR. *Tg+* mice were obligate heterozygotes for the *Tg(Mgrn1^I)C3Tmg* transgene since they were generated by mating *Tg+; Mgrn1*$^{md-nc/+}$ animals to *Tg−; Mgrn1*$^{+/+}$ mice. Brain RNA was extracted from 3 mice of each genotype using TriPure reagent (Roche), followed by DNaseI treatment (Invitrogen) and cDNA synthesis (High Capacity cDNA Kit, Applied Biosystems, Foster City, CA) prior to amplification using SYBR green Brilliant II PCR master mix (Agilent) on a BioRad Opticon II. All samples were analyzed in triplicate on a single plate. Expression was normalized against *glucose phosphate isomerase (Gpi)* levels using the comparative Ct method to determine the relative quantification value [14]. *Mgrn1* primers: GATCTACGGCATCGAGAACAA and AGTGTGTCCCGCAGGTC. *Gpi* primers: CAACTGC-TACGGCTGTGAGA and CTTTCCGTTGGACTCCATGT.

Prion inoculations

All procedures involving animals adhered to Association for Assessment and Accreditation of Laboratory Animal Care guidelines and were reviewed and approved by the Institutional Animal Care and Use Committee of the McLaughlin Research Institute. Female mice of the following genotypes were inoculated intracerebrally at 36–60 days of age with RML prions using 10 μl of 10% brain homogenate from clinically ill mice: *Tg−; Mgrn1*$^{md-nc/+}$ (n = 4), *Tg−; Mgrn1*$^{+/+}$ (n = 7), *Tg+; Mgrn1*$^{md-nc/+}$

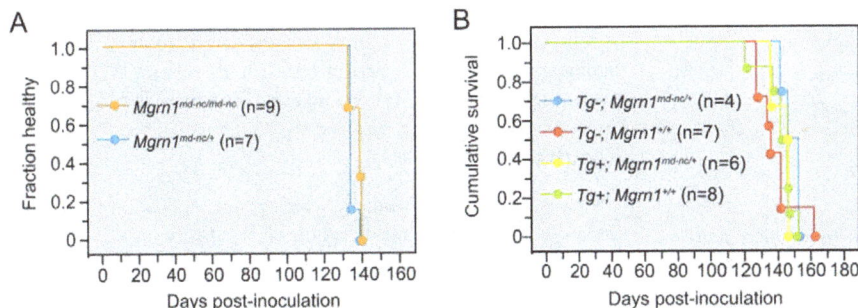

Figure 1. Kaplan-Meier plots for health status following RML prion inoculation. (A) Plot indicating proportion of healthy *Mgrn1*$^{md-nc/+}$ and *Mgrn1*$^{md-nc/md-nc}$ animals over time following RML prion inoculation. (B) Plot showing survival of *Mgrn1* transgenic (Tg+) and non-transgenic (Tg−) *Mgrn1*$^{md-nc/+}$ and *Mgrn1*$^{+/+}$ mice over time following inoculation with RML prions.

Table 2. Disease progression.

Genotype (n)	Average age at inoculation (range)	Average number of days post-inoculation to appearance of paresis (range) (SD)[a]	Average number of days post-inoculation to death or euthanasia (range) (SD)[a]
$Tg-$; $Mgrn1^{md-nc/+}$ (4)	53 (40–59)	103 (98–107) (6.4)	148 (141–152) (5.3)
$Tg-$; $Mgrn1^{+/+}$ (7)	47 (39–60)	103 (88–107) (11.6)	138 (127–160) (11.3)
$Tg+$; $Mgrn1^{md-nc/+}$ (6)	41 (36–60)	110 (94–115) (11.3)	142 (135–146) (5.6)
$Tg+$; $Mgrn1^{+/+}$ (8)	45 (37–59)	102 (94–111) (7.4)	141 (120–152) (9.5)

[a]No statistically significant differences by one-way ANOVA (all p>0.40).

(n = 6) and $Tg+$; $Mgrn1^{+/+}$ (n = 8). An independent cohort of 4 female and 3 male $Mgrn1^{md-nc/+}$ and 1 female and 8 male $Mgrn1^{md-nc/md-nc}$ mutant mice were also inoculated with RML prions; in these studies, $Mgrn1^{md-nc/+}$ mice were used as controls since they are phenotypically normal and a coisogenic $Mgrn1^{+/+}$ line was not available at the time. Animals were monitored daily for general health status, while neurologic status was assessed three times per week. At the first appearance of symptoms, mice were weighed prior to each neurological exam. Animals were euthanized when they showed signs of progressive neurological dysfunction or had lost 20% of their initial weight. One sagittal half of each brain was fixed in 10% formalin, while the other half was immediately frozen and stored at $-80°C$ until processed for immunoblotting (see below). One-way ANOVA was used to test whether there were any significant differences in average time to onset of paresis or survival following inoculation.

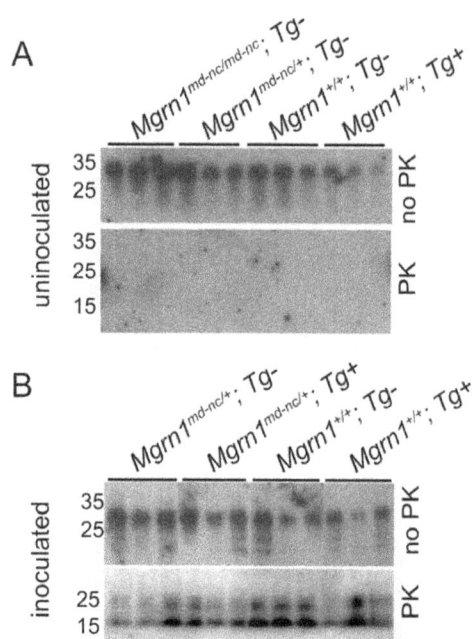

Figure 2. Brain PrP^C and PrP^{Sc} expression. (A-B) Brain protein lysates from uninoculated (A) and inoculated (B) transgenic and non-transgenic $Mgrn1^{+/+}$, $Mgrn1^{md-nc/+}$ and/or $Mgrn1^{md-nc/md-nc}$ mice were subjected to immunoblotting with an antibody against PrP. The 'no PK' panel shows all PrP species present. The protease-resistant (proteinase K treated, 'PK') panel shows that PrP^{Sc} is present in samples from all RML prion-inoculated animals but not in uninoculated animals, regardless of $Mgrn1$ genotype.

Immunoblotting

Mouse brain hemispheres were homogenized in 1.0 ml EMBO buffer (50 mM Tris-HCl pH 7.4, 150 mM NaCl, 1 mM EDTA 1% Triton X) per 0.1 g of brain weight. Whole protein was quantified using bicinchoninic acid assay (Pierce). For PrP^{Sc} detection, samples were digested with proteinase K (1 μg/50 μg total protein) for 1 h at 37 C, followed by addition of PMSF (2 mM final concentration). Samples (20 μg of total protein) were subjected to electrophoresis through 10 or 14% Novex Tris-Glycine gels (Invitrogen) and transferred to PVDF membranes. Blots were blocked with 5% non-fat dry milk in Tris-buffered saline with Tween-20 (TBST) for 1 h, incubated with anti-prion antibody D18 (0.5ug/mL) overnight at 4 C, then washed and incubated at room temperature for 2 hrs with HRP-conjugated goat anti-human (Bio-Rad) secondary antibody (1:5000 in 1% not-fat dry milk) before being developed using ECL Western Blotting Substrate (Thermo Scientific Pierce).

Histology

Formalin-fixed samples were processed, paraffin-embedded, and sectioned at 8 μm thickness. Sections of septum, hippocampus/thalamus, midbrain and cerebellum/brainstem were stained with hematoxylin and eosin (H&E, Fisher Scientific SH26–500D and 245–658) or processed by immunohistochemistry for PrP (R2 monoclonal antibody, a gift from Dr. Stanley Prusiner) or the astrocyte marker, GFAP (Dako #Z0334). The secondary antibody for R2 was biotinylated goat anti-human Kappa chain (Vector Laboratories #BA–3060) and for GFAP was biotinylated goat anti-rabbit IgG (Vector Laboratories #BA–1000). The Peroxidase substrate DAB kit (Vector Laboratories #SK–4100) was used for color development. The percentage area occupied by vacuoles in a high power field was measured on H&E stained sections from 2 animals of each genotype. Each brain region was scored on one section for each animal. A score of 0 indicates no vacuoles were present. A score of up to 5 could reflect a small amount of real vacuolation or artifact and is considered to represent 'no significant lesion' (NSL). Scores of 5–19, 20–69 and ≥70 indicate mild, moderate and severe vacuolation, respectively [15].

Results

One possible mechanism by which prion replication may cause disease is by inducing misdirection of PrP to the cytoplasm [11]. Since $Mgrn1$ over-expression rescued the endo-lysosomal trafficking defects associated with the presence of cytosolically exposed forms of PrP *in vitro* [9], we set out to test whether $Mgrn1$ levels influence PrP^{Sc}-mediated prion disease *in vivo* by inoculating mice that express no $Mgrn1$ and mice that over-express $Mgrn1$ with RML prions. A $Mgrn1$ isoform I transgene $(Tg(Mgrn1^I)C3Tmg)$ that rescues all aspects of the $Mgrn1^{md-nc/md-nc}$ phenotype, including

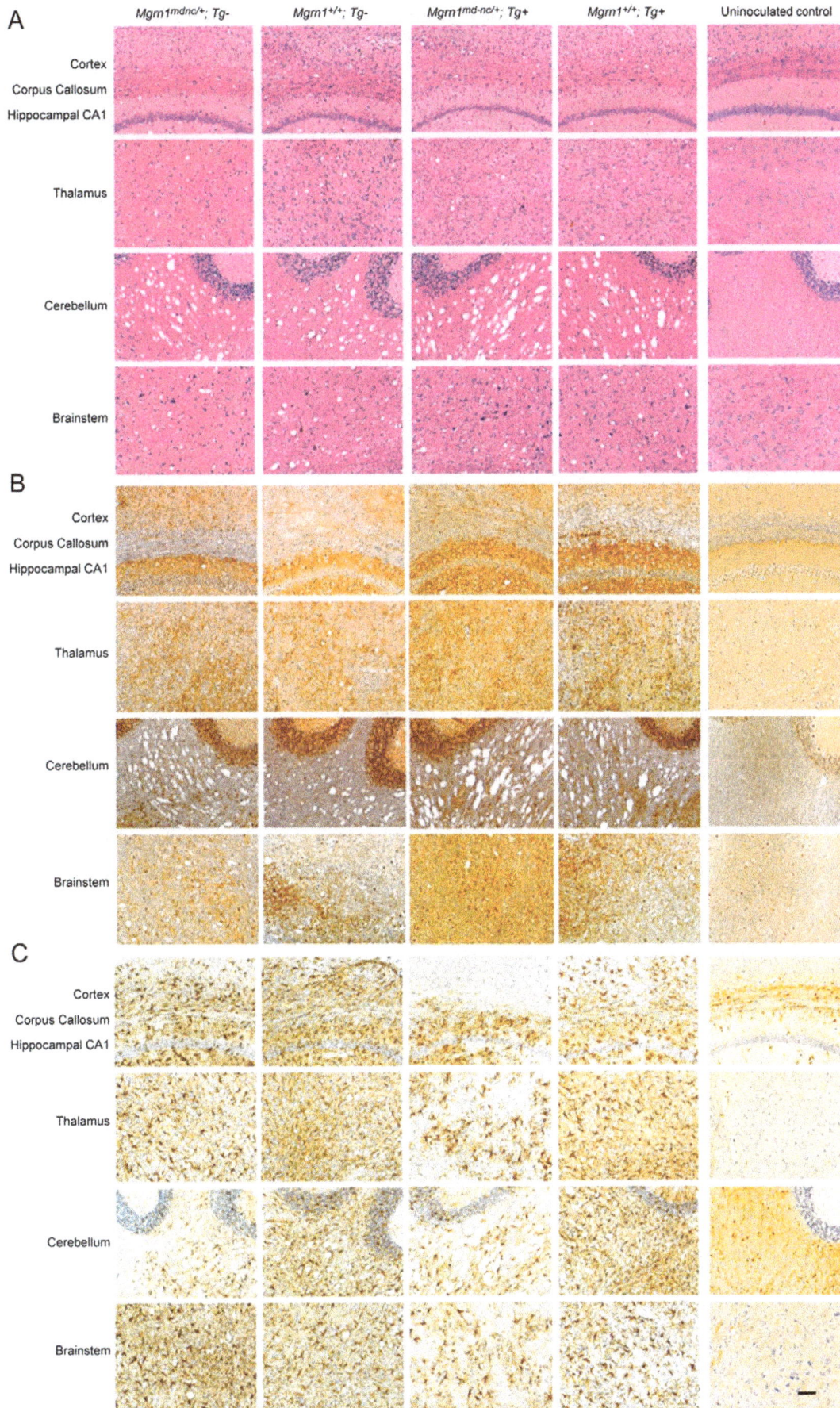

Figure 3. Histopathology and immunohistology of prion inoculated mice expressing normal or elevated levels of *Mgrn1*. (A) Hematoxylin and eosin-stained sections of indicated brain regions of non-transgenic and transgenic *Mgrn1*$^{md-nc/+}$ and *Mgrn1*$^{+/+}$ mice inoculated with RML prions and an uninoculated animal. Similar levels of vacuolation were observed in inoculated animals, regardless of genotype. As indicated in Table 2, the white matter of the cerebellum was most severely affected, followed by the brainstem and thalamus. (B) Immunohistochemistry against PrP on sections adjacent to those shown in A. The overall level and distribution of PrP was similar in inoculated mice regardless of their genotype. (C) Immunohistochemistry against GFAP on sections adjacent to those shown in A and B showing similar levels of astrocytosis in inoculated animals across genotypes. All images in–C were taken at the same magnification and are shown to same scale. Scale bar (in last panel): 100 μm.

CNS vacuolation, was previously shown to be expressed in the brain [13] but its expression level relative to endogenous *Mgrn1* was not assessed. Since an antibody that recognizes endogenous MGRN1 in mouse brain lysates is not available, we performed quantitative RT-PCR to assess *Mgrn1* expression in the brains of *Tg(Mgrn1^I)C3Tmg* transgenic and non-transgenic wild-type (*Mgrn1*$^{+/+}$) and *Mgrn1*$^{md-nc/+}$ mice. *Mgrn1* mRNA in the brain showed statistically significant differences consistent with genotype: expression in non-transgenic *Mgrn1*$^{md-nc/+}$ brains was significantly reduced relative to wild-type samples, transgenic (Tg+) *Mgrn1*$^{md-nc/+}$ and *Mgrn1*$^{+/+}$ mice had significantly higher levels (3-4-fold) than their non-transgenic counterparts, and *Tg+*; *Mgrn1*$^{md-nc/+}$ brains expressed similar levels to wild-type brains (Table 1).

In our colony, *Mgrn1*$^{md-nc}$ null mutant mice start to show spongiform encephalopathy with reactive astrocytosis between 7 and 9 months of age. Unlike prion-inoculated mice, however, they do not develop obvious neurological symptoms and can live to at least 24 months of age. To test whether loss of MGRN1 function can contribute to the pathogenesis of prion disease, male and female mice homozygous for the *Mgrn1*$^{md-nc}$ null mutation and heterozygous controls were inoculated with RML prions and carefully monitored for signs of illness and neurological symptoms associated with prion infection (one or more of the following: weakness in rear, paresis, wobble in rear, abnormal gait, abnormal posture, generalized tremor, tail rigidity, poor righting reflex). No significant differences were observed in disease incubation time (Figure 1A). This indicates that absence of MGRN1 does not accelerate the pathogenesis of scrapie, but does not distinguish whether RML prions cause disease by disrupting MGRN1 function or act independent of MGRN1.

As over-expression of MGRN1 in cell culture reversed the endosomal trafficking defects associated with the presence of cytosolic PrP, we tested whether *in vivo* over-expression of MGRN1 protected against RML prion-induced disease. Female non-transgenic and transgene-positive *Mgrn1*$^{+/+}$ and *Mgrn1*$^{md-nc/+}$ mice were inoculated with RML prions and carefully monitored for signs of illness and neurological symptoms. No statistically

significant differences were observed in the average time to onset of symptoms or survival time between mice expressing different levels of *Mgrn1* (Table 2 and Figure 1B). Protease-resistant PrP was detected in brain lysates from inoculated mice of every genotype, but not in samples from uninoculated animals (Figure 2), indicating that *Mgrn1* levels do not affect the conversion of PrP^C to PrP^{Sc}. Histological and immunohistological analyses revealed no obvious genotype-dependent differences in the distribution or degree of vacuolation (Figure 3A and Table 3), the level or distribution of PrP (using an antibody that recognizes PrP^C and PrP^{Sc}) (Figure 3B), or in reactive astrocytic gliosis (using an antibody against GFAP) (Figure 3C) in the brains of inoculated animals. Vacuoles were most abundant in the white matter of the cerebellum, followed by the brainstem, but even in these regions it was only mild to moderate (Table 3). Interestingly, cerebellar vacuoles in *Mgrn1* null mutant mice are most common in the granule layer [13], which was the least vacuolated region of the cerebellum in all RML prion-inoculated animals in this study.

Discussion

Mgrn1 null mutant mice develop CNS vacuolation that is similar both histologically and in its anatomical distribution to that observed in animals inoculated with RML prions. The possibility that the same pathogenic mechanism might cause spongiform change in *Mgrn1* null mutant mice and prion diseases was suggested by the observation that cytosolically exposed forms of PrP can bind to and sequester MGRN1 [9]. In that study, the presence of cytosolic PrP in HeLa cells caused abnormalities in endo-lysosomal trafficking similar to those observed when *Mgrn1* expression is knocked down by siRNA, and rescue of these defects by over-expression of MGRN1 was consistent with functional sequestration of MGRN1 by cytosolic PrP. Our results indicate that neither loss of *Mgrn1* nor its over-expression *in vivo* influences the onset, progression or outcome of disease caused by RML prion inoculation, including the distribution and severity of vacuolation. Since the *Tg(Mgrn1^I)C3Tmg* transgene has been shown to rescue vacuolation associated with loss of *Mgrn1* [13], this suggests that

Table 3. Severity of vacuolation.

Genotype	Average severity of vacuolation by brain region (scores)[a]										
	Ctx	Hipp	Thal	Hyp	Sep	Caud	Fbr	Mol	Gran	Wm	Bst
Tg−; Mgrn1$^{md-nc/+}$ (n = 2)	NSL (4, 5)	NSL (4, 0)	NSL (5, 4)	NS (0, 2)	NS (4, 2)	NSL (0, 1)	NSL (3, 0)	NSL (0, 0)	NSL (0, 4)	Mild (15, 10)	Mild (10, 7)
Tg−; Mgrn1$^{+/+}$ (n = 2)	NSL (0, 4)	NSL (3, 5)	Mild (6, 5)	NS (0,0)	NS (0, 4)	NSL (2, 0)	NSL (0, 0)	NSL (3, 0)	NSL (5, 0)	Mild (20, 15)	Mild (10, 10)
Tg+; Mgrn1$^{md-nc/+}$ (n = 2)	NSL (3, 0)	NSL (0, 3)	NSL (2, 5)	NS (2, 0)	NS (0, 0)	NSL (na, 0)	NSL (na, 0)	NSL (0, 0)	NSL (0, 0)	Mild (20, 30)	Mild (15, 10)
Tg+; Mgrn1$^{+/+}$ (n = 2)	NSL (0, 1)	NSL (3, 0)	NSL (3, 1)	NS (0, 0)	NS (0, 3)	NSL (0, 0)	NSL (0, 0)	NSL (0, 0)	NSL (3, 1)	Mild (30, 10)	Mild (8, 5)

[a]Actual vacuolation scores (percentage of area occupied by vacuoles) provided in parenthesis. Brain regions: Ctx: cortex; Hipp: hippocampus; Thal: thalamus; Hyp: hypothalamus; Sep: septum; Caud: caudate nucleus; Fbr: basal forebrain; Mol: molecular layer of cerebellum; Gran: granule layer of cerebellum; Wm: white matter of cerebellum; Bst: brainstem.
Severity scores: NSL (no significant lesions), average vacuolation score <5; Mild, average vacuolation score 5–19; Mod (moderate), average vacuolation score 20–69; na: brain region not examined.

loss of MGRN1-dependent cellular processes is not the underlying cause of spongiform encephalopathy caused by RML prions. Furthermore, our data suggest that either cytosolic PrP is not produced in this disease model or it does not play a significant role in the pathogenesis of transmissible prion diseases. We cannot, however, rule out the possibility that functional sequestration of MGRN1 may contribute to the neurotoxicity associated with cytosolic PrP.

Our results are consistent with other studies that have suggested cytosolic PrP does not make a significant contribution to prion disease, particularly the pathogenesis of CNS vacuolation. For example, transgenic mice expressing cytosolic PrP did not develop spongiform change, even when the transgene was expressed on a *Prnp* null background and the mice were inoculated with RML prions, nor did the inoculated mice get sick or accumulate protease-resistant PrP [16]. Spongiform change was not observed a patient carrying a truncation mutation in *PRNP* (Q160X) that has been shown to lead to significant cytoplasmic retention of PrP [17,18]. In several studies, accumulation of cytosolic PrP was shown to not be toxic to human or mouse neurons in primary culture or to N2a cells [19–21], and in two of those studies, the presence of cytosolic PrP was in fact associated with protection against apoptosis. Together, these data suggest that the presence of cytosolic prion protein is not sufficient to cause prion disease and is not functionally equivalent to loss of MGRN1.

An alternative mechanism for vacuolar degeneration of neurons and their synapses is the accumulation of PrP^{Sc} in neuronal cellular membranes. At least 80% of PrP^{Sc} formed accumulates in the neuronal plasma membranes, especially in synaptic regions [22], and most vacuoles occur in pre- and post-synaptic structures [23]. In experimental scrapie and sporadic CJD, PrP^{Sc} accumulation and vacuolation begin focally in the brain and progress by axonal transport of PrP^{Sc} to different regions of the central nervous system [22]. The brain regions affected in the terminal stages of prion disease are determined by the strain of prions (PrP^{Sc}) [24]. Neuronal dysfunction and morphological changes (vacuolation) appear to be caused directly by accumulation of PrP^{Sc} in plasma membranes [22,25] and are related to the great effect of PrP^{Sc} has on membrane functions [26–28]. Dendritic degeneration, which is an additional abnormal step in synapse pathobiology, is caused specifically by PrP^{Sc} activation of Notch-1 signaling in the neuronal plasma cell membrane [29,30]. Therefore the effects of PrP^{Sc} on membrane pathobiology cannot be ignored. PrP^{Sc} accumulates to a lesser degree by endocytosis into lysosomes and by phagocytosis into autophagosomes that release PrP^{Sc} into the extracellular space [25], and ingestion of PrP^{Sc} by activated microglia causes release of cytokines from microglia that cause nerve cell death [31].

The similar disease progression of *Mgrn1* null mutant mice, transgenic mice that over-express *Mgrn1*, and controls inoculated with RML prions indicates that MGRN1-dependent processes are not necessary for the pathogenesis of transmissible prion disease. Further studies, along with a better understanding of the origin of CNS vacuoles, will be needed to determine whether PrP^{Sc} and loss of MGRN1 act through the same downstream pathways to cause this intriguing phenotype.

Acknowledgments

We are grateful to Sarah Anderson, Richard Bennett and Julie Amato for technical assistance and Janet Peters and Anita Pecukonis for animal care.

Author Contributions

Conceived and designed the experiments: TMG GAC GSB. Performed the experiments: DS RP MA DM AO. Analyzed the data: TMG GAC MA SJD. Wrote the paper: TMG.

References

1. Colby DW, Prusiner SB (2011) Prions. Cold Spring Harb Perspect Biol 3: a006833.
2. Imran M, Mahmood S (2011) An overview of human prion diseases. Virol J 8: 559.
3. He L, Lu XY, Jolly AF, Eldridge AG, Watson SJ, et al. (2003) Spongiform degeneration in mahoganoid mutant mice. Science 299: 710–712.
4. Gunn TM, Inui T, Kitada K, Ito S, Wakamatsu K, et al. (2001) Molecular and phenotypic analysis of Attractin mutant mice. Genetics 158: 1683–1695.
5. Azouz A, Gunn TM, Duke-Cohan JS (2007) Juvenile-onset loss of lipid-raft domains in attractin-deficient mice. Exp Cell Res 313: 761–771.
6. Jiao J, Sun K, Walker WP, Bagher P, Cota CD, et al. (2009) Abnormal regulation of TSG101 in mice with spongiform neurodegeneration. Biochimica et Biophysica Acta 1792: 1027–1035.
7. Kim BY, Olzmann JA, Barsh GS, Chin LS, Li L (2007) Spongiform neurodegeneration-associated E3 ligase Mahogunin ubiquitylates TSG101 and regulates endosomal trafficking. Molecular Biology of the Cell 18: 1129–1142.
8. Miesbauer M, Rambold AS, Winklhofer KF, Tatzelt J (2010) Targeting of the prion protein to the cytosol: mechanisms and consequences. Curr Issues Mol Biol 12: 109–118.
9. Chakrabarti O, Hegde RS (2009) Functional depletion of mahogunin by cytosolically exposed prion protein contributes to neurodegeneration. Cell 137: 1136–1147.
10. Hegde RS, Mastrianni JA, Scott MR, DeFea KA, Tremblay P, et al. (1998) A transmembrane form of the prion protein in neurodegenerative disease. Science 279: 827–834.
11. Hegde RS, Tremblay P, Groth D, DeArmond SJ, Prusiner SB, et al. (1999) Transmissible and genetic prion diseases share a common pathway of neurodegeneration. Nature 402: 822–826.
12. Ma J, Wollmann R, Lindquist S (2002) Neurotoxicity and neurodegeneration when PrP accumulates in the cytosol. Science 298: 1781–1785.
13. Jiao J, Kim HY, Liu RR, Hogan CA, Sun K, et al. (2009) Transgenic analysis of the physiological functions of Mahogunin Ring Finger-1 isoforms. Genesis 47: 524–534.
14. Livak KJ, Schmittgen TD (2001) Analysis of relative gene expression data using real-time quantitative PCR and the 2(-Delta Delta C(T)) Method. Methods 25: 402–408.
15. Carlson GA, Ebeling C, Yang SL, Telling G, Torchia M, et al. (1994) Prion isolate specified allotypic interactions between the cellular and scrapie prion proteins in congenic and transgenic mice. Proc Natl Acad Sci U S A 91: 5690–5694.
16. Norstrom EM, Ciaccio MF, Rassbach B, Wollmann R, Mastrianni JA (2007) Cytosolic prion protein toxicity is independent of cellular prion protein expression and prion propagation. J Virol 81: 2831–2837.
17. Jayadev S, Nochlin D, Poorkaj P, Steinbart EJ, Mastrianni JA, et al. (2011) Familial prion disease with Alzheimer disease-like tau pathology and clinical phenotype. Ann Neurol 69: 712–720.
18. Heske J, Heller U, Winklhofer KF, Tatzelt J (2004) The C-terminal globular domain of the prion protein is necessary and sufficient for import into the endoplasmic reticulum. J Biol Chem 279: 5435–5443.
19. Roucou X, Guo Q, Zhang Y, Goodyer CG, LeBlanc AC (2003) Cytosolic prion protein is not toxic and protects against Bax-mediated cell death in human primary neurons. J Biol Chem 278: 40877–40881.
20. Fioriti L, Dossena S, Stewart LR, Stewart RS, Harris DA, et al. (2005) Cytosolic prion protein (PrP) is not toxic in N2a cells and primary neurons expressing pathogenic PrP mutations. J Biol Chem 280: 11320–11328.
21. Restelli E, Fioriti L, Mantovani S, Airaghi S, Forloni G, et al. (2010) Cell type-specific neuroprotective activity of untranslated prion protein. PLoS One 5: e13725.
22. Bouzamondo-Bernstein E, Hopkins SD, Spilman P, Uyehara-Lock J, Deering C, et al. (2004) The neurodegeneration sequence in prion diseases: evidence from functional, morphological and ultrastructural studies of the GABAergic system. J Neuropathol Exp Neurol 63: 882–899.
23. Lampert P, Hooks J, Gibbs CJ Jr, Gajdusek DC (1971) Altered plasma membranes in experimental scrapie. Acta Neuropathol (Berl) 19: 81–93.
24. DeArmond SJ, Yang S-L, Lee A, Bowler R, Taraboulos A, et al. (1993) Three scrapie prion isolates exhibit different accumulation patterns of the prion protein scrapie isoform. Proc Natl Acad Sci USA 90: 6449–6453.
25. Bajsarowicz K, Ahn M, Ackerman L, DeArmond BN, Carlson G, et al. (2012) A Brain Aggregate Model Gives New Insights into the Pathobiology and Treatment of Prion Diseases. J Neuropathol Exp Neurol 71: 449–466.
26. DeArmond SJ, Qiu Y, Wong K, Nixon R, Hyun W, et al. (1996) Abnormal plasma membrane properties and functions in prion-infected cell lines. Function

and Dysfunction in the Nervous System. Plainview, NY: Cold Spring Harbor Laboratory Press.pp. 531–540.

27. Kristensson K, Feuerstein B, Taraboulos A, Hyun WC, Prusiner SB, et al. (1993) Scrapie prions alter receptor-mediated calcium responses in cultured cells. Neurology 43: 2335–2341.

28. Wong K, Qiu Y, Hyun W, Nixon R, VanCleff J, et al. (1996) Decreased receptor-mediated calcium response in prion-infected cells correlates with decreased membrane fluidity and IP3 release. Neurology 47: 741–750.

29. Ishikura N, Clever J, Bouzamondo-Bernstein E, Samayoa E, Prusiner SB, et al. (2005) Notch-1 activation and dendritic atrophy in prion diseases. PNAS 102: 886–891.

30. Spilman P, Lessard P, Sattavat M, Bush C, Tousseyn T, et al. (2008) A γ-secretase inhibitor and quinacrine reduce prions and prevent dendritic degeneration in murine brains. PNAS 105: 10595–10600.

31. Giese A, Brown DR, Groschup MH, Feldmann C, Haist I, et al. (1998) Role of microglia in neuronal cell death in prion disease. Brain Pathol 8: 449–457.

5

Reversal Learning and Associative Memory Impairments in a BACHD Rat Model for Huntington Disease

Yah-se K. Abada[1,2]*, Huu Phuc Nguyen[3], Bart Ellenbroek[4], Rudy Schreiber[5]

1 Neuropharmacology, EVOTEC AG, Hamburg, Germany, **2** Brain Research Institute Dept. of Neuropharmacology, University of Bremen – FB 2, Bremen, Germany, **3** Institute of Medical Genetics and Applied Genomics, University of Tübingen, Tübingen, Germany, **4** School of Psychology, Victoria University of Wellington, Wellington, New Zealand, **5** Behavioral Physiology & Pharmacology, University of Groningen, Groningen, The Netherlands

Abstract

Chorea and psychiatric symptoms are hallmarks of Huntington disease (HD), a neurodegenerative disorder, genetically characterized by the presence of expanded CAG repeats (>35) in the HUNTINGTIN (HTT) gene. HD patients present psychiatric symptoms prior to the onset of motor symptoms and we recently found a similar emergence of non motor and motor deficits in BACHD rats carrying the human full length mutated HTT (97 CAG-CAA repeats). We evaluated cognitive performance in reversal learning and associative memory tests in different age cohorts of BACHD rats. Male wild type (WT) and transgenic (TG) rats between 2 and 12 months of age were tested. Learning and strategy shifting were assessed in a cross-maze test. Associative memory was evaluated in different fear conditioning paradigms (context, delay and trace). The possible confound of a fear conditioning phenotype by altered sensitivity to a 'painful' stimulus was assessed in a flinch-jump test. In the cross maze, 6 months old TG rats showed a mild impairment in reversal learning. In the fear conditioning tasks, 4, 6 and 12 months old TG rats showed a marked reduction in contextual fear conditioning. In addition, TG rats showed impaired delay conditioning (9 months) and trace fear conditioning (3 months). This phenotype was unlikely to be affected by a change in 'pain' sensitivity as WT and TG rats showed no difference in their threshold response in the flinch-jump test. Our results suggest that BACHD rats have a profound associative memory deficit and, possibly, a deficit in reversal learning as assessed in a cross maze task. The time course for the emergence of these symptoms (i.e., before the occurrence of motor symptoms) in this rat model for HD appears similar to the time course in patients. These data suggest that BACHD rats may be a useful model for preclinical drug discovery.

Editor: Yoshitaka Nagai, National Center of Neurology and Psychiatry, Japan

Funding: This work was supported by Neuromodel (grant no. 215618-2), an academic-industrial Initial Training Network on innovative treatment approaches of Huntington and Parkinson disease, funded through the people programme FP7 of the European Union. The funders had no role in study design, data collection and analysis, decision to publish, or preparation of the manuscript.

Competing Interests: The authors have the following interests: Dr. Rudy Schreiber and Bart Ellenbroek were both employees of EVOTEC AG, Germany. Currently, only Dr. Yah-se ABADA is working at EVOTEC AG. There are no patents, products in development or marketed products to declare. In addition, all co-authors have no financial interest to report.

* E-mail: yah-se.abada@evotec.com

Introduction

Huntington disease (HD) is one of the neurodegenerative disorders where the origin has been unequivocally identified, that is, an elongation of polyglutamine (>35 CAG repeats) in the *HUNTINGTIN* (*HTT*) gene on chromosome 4 [1]. Patients carrying the mutation present a combination of motor symptoms such as chorea, psychiatric symptoms, and cognitive changes [2–3]. The disease is associated with degeneration of neurons in the striatum (especially the Medium Spiny Neurons, MSN) and cortex [4–6]. Treatments to delay HD onset or inhibit the mechanisms by which neural loss occurs are still lacking [7], and therefore there is a continuing need for improved animal models to support drug discovery efforts.

During the last decades, many animal models for HD have been generated, from insects (*Drosophilae melanogaster*), to nonhuman primates (*Macaca mulatta*), including several rodents models [8–11]. The availability of such a wide range of models increases the potential opportunities to understand the disease progression and to find a cure. Besides selection of a reliable and valid animal model, the timing of drug treatment is of critical importance for

HD drug discovery studies. One plausible explanation for the recent failure of monoclonal antibodies against the beta-Amyloid protein to reverse symptoms in patients with advanced Alzheimer's disease in Phase III studies, has been that therapeutic intervention is needed at a time point when the disease has not yet caused too much neurodegeneration for treatments to be effective [12]. Accordingly, it is increasingly recognized that identification and validation of prodromal symptoms and biomarkers is critical. For HD, cognitive impairments may consist of prodromal symptoms that could be used as clinical endpoints in drug discovery. HD patients present several impairments in executive and visuospatial mnemonic functions [13–15]. Cognitive impairment appears to occur *before* the emergence of motor symptoms. For example, patients exhibited impairments in the California verbal learning test (CVLT) and the Wechsler memory scale (WMS)] in the absence of motor disturbances [16]. An important aim for future animal model development is to identify, characterize and validate cognitive symptoms that occur *before* the onset of motor symptoms. It is not yet clear to what extent the occurrence of cognitive and motor symptoms are adequately reflected in the current rodent

HD models. In fact, several studies reported cognitive impairments *after* the appearance of motors deficits. For example, in a tgHD rat model of HD that carry the human mutation with 51 CAG repeats [17], Fielding and colleagues [18] have not found significant impairment in object recognition, set shifting, and operant tests, although motor deficits were present at 13 months of age. Deficits were shown at 12 months of age in radial maze and at later ages (15 to 20 months old rats) in choice reaction time tasks, spatial, and location recognition memory tests, whereas in the R6/2 mouse model for HD, selective deficits in spatial, visual and reversal discrimination were observed before or during subtle motor deficits (3.5–5.5 weeks and 7–8 weeks respectively) [19–24].

Herein, we used BACHD rat, a novel model for HD that has been recently established [25]. Like the mouse BACHD model [26], the rat model carries the full length human mutant *HUNTINGTIN* (fl-*mHTT*) with 97 CAG-CAA mix repeats under control of the human HD promoter gene. An advantage of the rat model is that behavioral processes related to learning and memory and pharmacological validation have been well described for this species. We have previously found progressive motor deficits during rotarod testing, starting as early as 2 months of age; a decrease in spontaneous locomotor activity, as well as, gait deficits in a catwalk test. However, we were unable to show a significant cognitive impairment in an object recognition task or robust sensorimotor deficits in a prepulse inhibition test [27]. These results were somewhat unexpected in light of the cognitive deficits reported in patients. Therefore, we decided to perform a more profound evaluation of the cognitive phenotype of BACHD rats. The cognitive performance of different age cohorts of BACHD rats was assessed in reversal learning and associative memory tests. The cross-maze and the fear conditioning paradigms tests (contextual, delay and trace conditioning) were selected, because they have proven to be efficient for cognitive assessment in rodents. An important consideration for selection of the rat fear conditioning paradigms was that the neural circuitry for fear conditioning has been well described in rodents and humans and that these neural circuits are well conserved [28–29]. This offers a potentially powerful translational approach as fear conditioning studies in tandem with functional brain imaging studies in both species could be used for future drug discovery studies.

Materials and Methods

Ethics statement

The study was carried out in strict accordance with the German animal welfare act and the EU legislation (EU directive 2010/63/EU). The protocol was approved by the local ethics committee *Behörde für Gesundheit und Verbraucherschutz* (BGV, Hamburg).

Husbandry and genotyping

Wild type (WT) and transgenic (TG) BACHD rats, carrying the mutant human HTT gene, under the control of the human huntingtin promoter and its regulatory elements were used. The transgene contains 97 CAG-CAA mix repeats, which produces a particular stability of the repeat length, and additional 20 kb upstream and 50 kb downstream sequences that reduce its position effect [25]. Two transgenic males were supplied from the original BACHD colony of the Universitäts Klinikum Tübingen (UKT, Germany) and an in-house breeding colony was preserved and maintained at EVOTEC AG (Hamburg, Germany) by cross-breeding these males with wild type female rats. BACHD animals were maintained on a Sprague-Dawley background. All the animals at weaning were group-housed 2 to 4 per cage with wood shavings and a filter top. The environment

was enriched with a play tunnel and shredded paper. BACHD rats were maintained in climate controlled housing, with a 12-h reversed dark/light cycle (light from 19:00 to 07:00). Rats had free access to food and water except during experiments.

Ear punches were taken at weaning to determine their genotype. Genotyping was performed before and after all the studies using a validated protocol. Briefly, Genomic DNA was prepared from ear biopsy tissue using proteinase K digestion, followed by phenol/chloroform extraction (Qiagen DNeasy Tissue kit). Primers flanking the polyQ repeat in exon 1 were designed to recognize whether or not the rat carried at least one copy of the mutation, and were used to PCR amplify the polyQ regions [Q3: 5′ – AGG TCG GTG CAG AGG CTC CTC - 3′ and Q5: 5′ – ATG GCG ACC CTG GAA AAG CTG - 3′]. Gene status was confirmed in parallel by using designed primers from UKT [exon 1: FW 5′-ATG GCG ACC CTG GAA AAG CTG- 3′ and RV: 5′ -AGG TCG GTG CAG AGG CTC CTC- 3′; exon 67: FW 5′- TGT GAT TAA TTT GGT TGT CAA GTT TT- 3′ and RV: 5′ –AGC TGG AAA CAT CAC CTA CAT AGA CT- 3′]. The PCR product was run on an automated apparatus PTC-200 (Peltier Thermal Gradient Cycler) and the Agilent 2100 Bioanalyser (Agilent technologies) was used to determine the fragments size.

Our concern in this longitudinal study was to reduce as much as possible potential confounds that hamper the interpretation or extrapolation of the results. Therefore only male rats were used in the cognitive tests as the female estrus cycle may influence experimental outcomes [30–31].

Strategy and shifting (Cross-Maze)

The strategy shifting test is a standard dual-solution task which was used to assess the respective contributions of response (or egocentric) and place (or allocentric) learning strategies on memory [32]. It determines the relative involvement of these 2 strategies during the course of learning. We essentially used the same method as has been described for testing the BACHD mice [33].

Spatial alternation was assessed using a modified version of the standard cross-maze; the home made maze consists of 4 identical arms (50 cm×12 cm×20 cm) at 90 degrees to each other. The maze was made with clear Plexiglas, elevated 45 cm above the floor, and a T-maze was created by closing one arm (north, N) with a guillotine door. The T-maze configuration was as follow: 2 arms [east (E) and west (W)] are at 180 degrees to each other, and the last arm (south, S) was perpendicular to these arms. Two holes were present: one at the end of the E and W arms each, and spatial cues were placed on a black curtain which surrounded the maze. A home cage was put at one end of the arms (E or W) to motivate the animals to explore the maze and find the exit into this home cage (where they were additionally rewarded with food pellets). One week prior to the test, rats received small sucrose food pellets in addition to their normal diet. One day prior to the start of the experiments, all rats received a 5-min habituation session in the apparatus. During that period, food was not available.

The next day, the acquisition sessions started and a rat was placed in the S arm. The home cage containing sucrose food pellets was placed under the hole in the W arm. The rat had to guide itself in the maze and reach the home cage. During acquisition, rats received one trial per day for 7 days. During the first 2 days, the goal arm (i.e. arm giving access to the home cage) was baited with small sucrose food pellets. The same training procedure was run during reversal and extended reversal training sessions except that the rats had to reach this time the home cage placed underneath the E arm (opposite of the previously learned arm). When a rat made a wrong choice (entrance into the arm without home cage), it was allowed to trace back to the goal arm. If the rat failed to reach the home cage within 2 min, the rat was gently guided manually to the goal arm and the trial ended 20 s later. The maze was cleaned after each animal crossing with a 10% ethanol solution to avoid any bias related to odor.

Three probe trials were performed at days 8, 16 and 22, at the end of the acquisition, reversal and extended reversal training sessions, respectively. During the probe trials the S arm was closed and the N arm was used as the new start arm; the strategy (place or response) that the animal used to reach the goal arm was assessed. If the animal during the acquisition training had learned to use a place strategy, it would select the W arm. However, if the animal had used a response strategy (i.e. learned to turn left), it would select the E arm.

Fear conditioning

Classical fear conditioning (FC) is a form of associative learning in which subjects express fear responses to a neutral conditioned stimulus (CS) after it has been paired with an aversive, unconditioned stimulus (US). The tests were run in an apparatus (Med Associates Inc., Italy) consisting of a ventilated sound-attenuated box and a rectangular testing chamber (30×26×25 cm) with stainless steel rod floor. Measurements were accomplished through a front digital video recording camera, connected to a computer with video freeze software. All rats received 5 min acclimatization one day prior to training and testing days. The chambers were wiped with a 70% ethanol solution and were dried prior to each rat testing. Three different tasks were used: contextual, delay and trace fear conditioning.

Contextual fear conditioning. The training session consisted of a 5 min acclimatization followed by 6 pairings (1 min inter trial time) of a 0.6-mA, 1-s foot shock. Animals were returned to their home cage 3 min after receiving the last foot-shock. On the next day, conditioned freezing was assessed by placing rats in the conditioning chambers for 5 min, in the absence of foot shock. For the evaluation of long term memory (LTM), animals were re-exposed one and 2 months later during a 5-min sessions to the conditioning chambers.

Delay conditioning. The testing protocol is similar to the contextual fear paradigm, except that on the training day, after 3 min acclimatization, rats received 6 pairings (120 s inter trial time) of a 30-s tone (85 dB) with a 0.6-mA, 2-s foot shock. The foot shock terminated at the same time as the tone and rats were removed from the testing chambers 60 s after the last pairing. On the testing day, rats were tested for contextual freezing in the conditioning chambers for 3 min, in the absence of tone or foot shock. One hour later, an altered context was generated with white polyvinyl chloride materials that covered the shock-grid bars and the inside of the conditioning boxes. Freezing was assessed in the altered context without tone for 3min, followed by a 3-min tone presentation in the absence of foot shock.

Trace fear conditioning. The test was adapted from an existing protocol [34]. In this associative learning paradigm, rats received during the training day eight trials of a 85 dB, 10 s tone (CS), followed by 20 s trace period, after which a 1 s–0.6 mA foot shock (US) was delivered. Each CS-US pairing was separated by a random inter-trial interval (ITI) that varied between 60 and 120 s. The random ITI time was used to prevent time between foot-shocks to be used as a cue for the US. Rats were removed from the chamber 60 s after the last CS-US presentation.

Retention tests for contextual, auditory and trace fear memory were carried out 24 h after conditioning. Rats were first tested for tone and trace period in an altered context made with a white polyvinyl chloride insert to cover the shock-grid bars and the inside of the conditioning boxes. Each rat was given 2 min habituation, followed by four presentation of the CS with varied ITI, in the absence of US. Freezing behavior during the four CS presentations and trace periods were averaged for each animal. Following CS and trace retention testing periods, contextual retention test was measured by placing the animals back into the original context for 2 min during which freezing was scored, without exposure to the CS or US.

For all the paradigms, freezing behavior was defined as the lack of any movement, except respiration. The percent of time spent freezing was assessed using the linear methods of observation measures (video freeze software).

The 'flinch-jump' test

The method has been described by Lehner and colleagues [35]. Rats were placed individually into the fear conditioning boxes (Med Assoc. Italy). Shocks were delivered to the grid floor of the test box through a shock generator. After a 3-min period of habituation to the test box, shock titrations continued to increase in a stepwise manner (0.05 mA, 0.05–0.6 mA range). In this way, the 'flinch' and 'jump' thresholds in mA is defined for each rat. The interval between shocks was 2 min, and each animal was tested only once at each intensity. Behavior for each rat was recorded through a front digital video recording camera and analysis was done blind to the genotype. The 'flinch' threshold was defined as the lowest shock intensity that elicited a detectable response. The 'jump' threshold was defined as the lowest shock intensity that elicited simultaneous removal of at least three paws (including both hind paws) from the grid.

Statistical Analysis

All data were analyzed using *GraphPad* and *InVivoStat* software. Differences between groups were assessed with Student's t-test or mix ANOVA with repeated measures, with the factor GENO-TYPE as between subject and TIME or TEST as within subject variable. When significance was found, a Bonferroni – post hoc analysis was performed when appropriate. For the cross maze, the learning index is defined as the ratio of the mean number of correct choices over trials per animal. Therefore, we have generated a binary data set with 2 possible outcomes (correct choice vs. incorrect choice). The hypothetical value that results is "½" because each animal has 50% chance at every trial. The one sample t-test was used to evaluate the learning index in each population with a hypothetical value set at "½". Chi-square ($\chi2$) analyses were computed on animal's choice during Acquisition (A), Reversal (R), and Extend Reversal (ER) learning in the cross maze test, in order to determine discrepancies between groups and to determine potential changes in strategies between both probe trials. The Chi square test assesses whether an observed frequency distribution (i.e. the number of correct choices) differs from a theoretical distribution, and if this distribution is independent (i.e. the choice is genotype dependent). Finally, a Mann Whitney U-test was used to analyze 'flinch-jump' data. The significance level was set for all analysis at 0.05.

Results

We inspected each BACHD rat cohort animals prior to the experiments and all animals looked healthy. No global differences in phenotype were observed between wild type (WT) and transgenic (TG) rats. Only male rats were used during the study. No difference in weight was found between WT and TG (data not shown).

Acquisition, Reversal learning and Strategy shifting in a Cross-maze

The cross maze test was performed in independent cohorts of 2 months (n = 17 per genotype) and 6 months (n = 15 per genotype) old BACHD rats in order to avoid any bias related to recall or long term memory of the task. Data from acquisition and reversal training sessions at the first 2 days were not analyzed because the goal arm was baited with sucrose food pellets to guide BACHD rats to the home cage. A schematic representation of the cross-maze task is presented in Fig. 1a. One WT rat of 2 months was removed from the data analysis because it did not show any interest in the task and did not make a choice (turn left or right) during the experimental period.

Learning indices during acquisition, reversal and extended reversal learning

The learning index is calculated as the ratio of the mean number of correct choices over trials per animal (Fig. 1b). Both cohorts of 2 and 6 months old BACHD rats displayed improved learning during *acquisition* (A) (2 months: WT, t = 5.91 and P<0.0001; TG, t = 4.02 and P<0.001; 6 months: WT, t = 4.35 and P<0,001; TG, t = 5.776 and P<0.0001). During *reversal* (R), 2 months old WT and TG have a learning index above the chance level of 0.5, although both groups did only reach a statistical trend (p value between 0.05 and 0.1; 2 months: WT, t = 2.126 and P = 0.0532; TG, t = 1.884, and P = 0.0805). The 6 months old cohort (WT and TG) showed a (R) learning index below chance level (<0.5) and did not reach statistical significance. In the *extended reversal* training (ER), both WT and TG rats of 2 months presented a statistically significant learning index (WT, t = 3.809 and P<0.01; TG, t = 2.874 and P<0.05). Six Months old WT rats had a significant learning index during (ER) (t = 3.323, and P<0.01) whereas TG did not reach statistical significance (t = 0.743, and P>0.1); 6 months old TG learning index was around 0.5. A comparison between the (A) and (ER) learning index in WT and in TG rats showed only a difference for

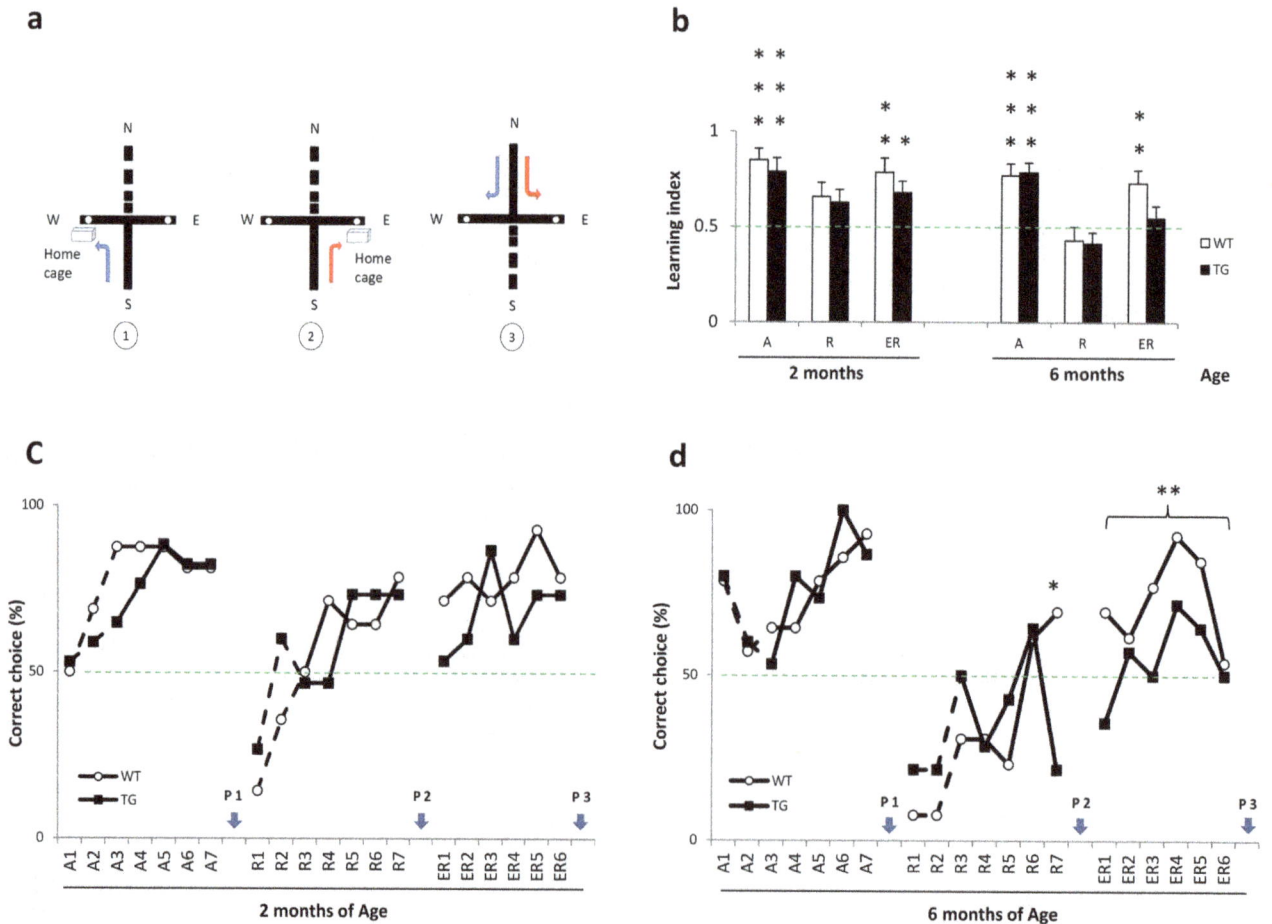

Figure 1. Cross-maze task. 2 months (WT n=17, TG n=17) and 6 months (WT n=15, TG n=15) BACHD rats were used. [**a**] Schematic representation of the cross-maze task. The north (N) arm is closed. The rat starts training in the south arm (S) and reaches the home cage through the hole located in the west arm (w, (1) acquisition) or east arm (E, (2) reversal). During (3) probe trial days 8 (P1), 16 (P2) and 23 (P3), the (S) arm is closed and the rat starts in the (N) arm. Rats reaching the home cage arm are Place learners, while those reaching the other arm are Response learners. [**b**] Learning index. Mean number of correct choices over acquisition (A), reversal (R) and extended reversal (ER) trials in BACHD rats. Both WT and TG rats of each age showed difficulties during (R); however, with (ER) training, 2 months old rats have improved learning whereas 6 months old TG rats have a learning index barely above chance level. Asterisks indicate significant difference from the hypothetical value (One sample test, *p<0.05, **p<0.01 and ***p<0.001). [**c** and **d**] Training trials. The percentage of correct choices made during acquisition (A), reversal (R) and extended reversal (ER) training are depicted. For the first 2 days, results where the goal arm was baited with sucrose food pellets are presented by dashed lines. There was no difference in acquisition training for both age cohorts. 6 months old TG rats (d) differed significantly from WT rats during reversal trial 7 and overall extended reversal trials (ER1 to ER6). Asterisks indicate significant difference (Chi square test, *p<0.05 and **p<0.01).

6 months old TG rats (A vs. ER, t = 2.846, P<0.01). A closer look at the 6 months rats (ER) bar graph (Fig. 1b) suggests a difference between WT and TG rats, and statistical analysis found a trend (t = 1.737, P = 0.0946).

Correct choices during acquisition, reversal and extended reversal learning

The progress of 2 and 6 months old BACHD rats in learning the task during training sessions for (A), (R) and (ER) is presented in Fig. 1 (c–d). We analyzed the number of correct choices that both groups made. All rats had one trial per day. There was no statistical difference during (A) in 2 and 6 months old BACHD rats. For (R) and (ER), only animals with learning index higher than 0.5 during acquisition were analyzed. That is, animals that actually learned the task. Four (WT, n = 2 and TG, n = 2) 2 months old rats and two (WT, n = 1 and TG, n = 1) 6 months old rats did not reach the criteria and therefore were not included in the analysis. Although 2 months old TG rats made correct choices during (R) and (ER) training, as did WT control rats, 6 months old TG rats made fewer correct choices. In fact, WT – but not TG – animals made more correct choices over the course of (ER) training. The difference between 6 months WT and TG rats was significant already on reversal trial 7 ($\chi2 = 6.23$, P<0.05) and persisted during (ER) trainings ($\chi2 = 8.594$, P<0.01).

Strategy shifting during probe trials

On trial days 8, 16, and 23 a probe trial was done to assess which strategy rats used to solve the task (Fig. 2). The north arm (N) was now the new start arm. Rats entering the same arm as during training sessions were designated place learners (allocentric learning) and rats entering the opposite arm were designated response learners (egocentric learning). Data were only analyzed for animals that made (1) correct arm choices with a learning index greater than 0.5 during each training session, and (2) were successful for the two last trials prior to the probe trial. Two months old rats exhibited a preference for response learning on P1 (*response:* WT = 73% and TG = 77%) and P2 (*response:* WT = 67% and TG = 57%). This preference for response learning was maintained on P3 (*response:* WT = 64% and TG = 67%). Six months old WT rats again exhibited a clear response learning during P1 (WT = 73%) while only half of TG rats were response learners. However, during P2 and P3, WT rats have adopted a place learning strategy (WT, *place:* P2 = 66% and P3 = 57%), whereas TG rats showed a response learning strategy (TG, *response:* P2 = 50% and P3 = 60%). Although WT rats results between both probe sessions (P1→P3) would suggest a shifting towards a place learning ($\Delta = 30\%$), this was not statistically significant ($\chi2 = 1.606$, P>0.05).

Contextual fear conditioning

Rats of 4 months (WT, n = 13 and TG, n = 13), 6 months (WT, n = 7 and TG, n = 9) and 12 months (WT, n = 16 and TG, n = 6) of age underwent a one day training session in conditioning chambers. Baseline activity was recorded 5 min before foot shocks were given and contextual memory was measured 24h later (Fig. 3a).

As shown in Fig. 3b, there was no significant difference between WT and TG baselines at all testing ages. However, TG rats expressed a significant lower freezing behaviour when re-exposed to the conditioning context (4 months: t = 5.757, P<0.0001; 6 months: t = 4.987, P<0.001 and 12 months: t = 5.147, P<0.0001). Visual inspection of fig. 3b indicates a decrease in percentage of freezing between 4 months, 6 and 12 months old

rats. In fact, a 2-way ANOVA analysis on Context results showed significant GENOTYPE (F (1, 79) = 76.53, P<0.001) and AGE (F (2, 79) = 13.42, P<0.001) effects. No interaction between GENOTYPE x AGE was found.

We next evaluated long term memory for contextual freezing in 4 months old rats by exposing them again to the conditioning chambers 1 and 2 months after the contextual test (retention tests, Fig. 3c). A progressive '*extinction*', characterized by a decrease in percentage freezing was observed in WT and TG rats (2-way ANOVA, GENOTYPE: F (1, 72) = 82.92, P<0.001 and AGE: F (2, 72) = 24.48, p<0.001). This trend is sustained as no interaction (GENOTYPE x AGE) was found.

Delay and Trace conditioning

The delay conditioning experiment evaluated the acquisition of a tone (85dB) fear conditioning when presented for 30 s before a 2 s foot-shock co-termination (Fig. 4a). Thirteen WT and fifteen TG rats of 9 months of age were given 6 trials training sessions. Baseline activity was recorded 3 min prior to the first trial and expressed as percentage freezing. WT and TG rats did not show differences in baseline freezing behavior (Fig. 4b). A 3-min retention test was performed after 24 h in the conditioning context and, for the tone, in an altered context. Both WT and TG rats expressed a trend for increased freezing to the context and the tone. TG rats showed a lower percentage freezing to the context and to the tone than WT rats. A 2-way ANOVA revealed significant effects for the main factors GENOTYPE and TEST (GENOTYPE, F (1,52) = 18.84, P<0.001; TEST, F (2,52) = 132.04, P<0.0001) as well as a significant interaction between both factors (F (2,52) = 5.04, P<0.01).

In the trace fear conditioning paradigm, the memory for context, tone and trace training is evaluated in 3 months old rats (Fig. 4c). The highest freezing responses were found during the trace period (Fig. 4d). No differences in baseline activity were found, but for all the stimuli (context, tone and trace), TG rats displayed a significantly lower freezing response than WT rats. A 2-way ANOVA analysis showed significant effects for the main factors GENOTYPE and TEST (GENOTYPE, F (1,78) = 39.34, P<0.0001; TEST, F (3,78) = 96.42, P< 0.0001), as well as a significant interaction between both factors (F (3,78) = 15.08, P<0.0001).

Flinch-Jump test

Thirty BACHD rats (n = 15 per genotype) of 6 months of age underwent the flinch-jump test. A flinch response was observed in all rats (Fig. 5a. WT, Mean = 0.23±0.048 SEM; TG, Mean = 0.25±0.042 SEM); whereas only 7 WT and 9 TG rats presented a jump response (Fig. 5b. Mean = 0.535±0.037 SEM and Mean = 0.538±0.048 SEM, respectively). In fact, statistical analysis did not reveal significant differences between WT and TG rats (Mann Whitney U-test: flinch, U = 85 and P=0.23; Jump, U = 29 and P=0.82).

Discussion

We investigated the cognitive phenotype of BACHD rats at ages 2 through 12 months. Learning deficits were found at 6 months of age in a cross-maze test. Pronounced associative memory deficits were found in context, delay (context and tone) and trace (context, tone and trace) fear conditioning. This fear conditioning phenotype is unlikely to be confounded by altered pain sensitivity, as WT and TG rats showed no differences in foot-shock intensity threshold as determined in a flinch-jump test. This is the first study to report robust and specific memory deficits in BACHD rats.

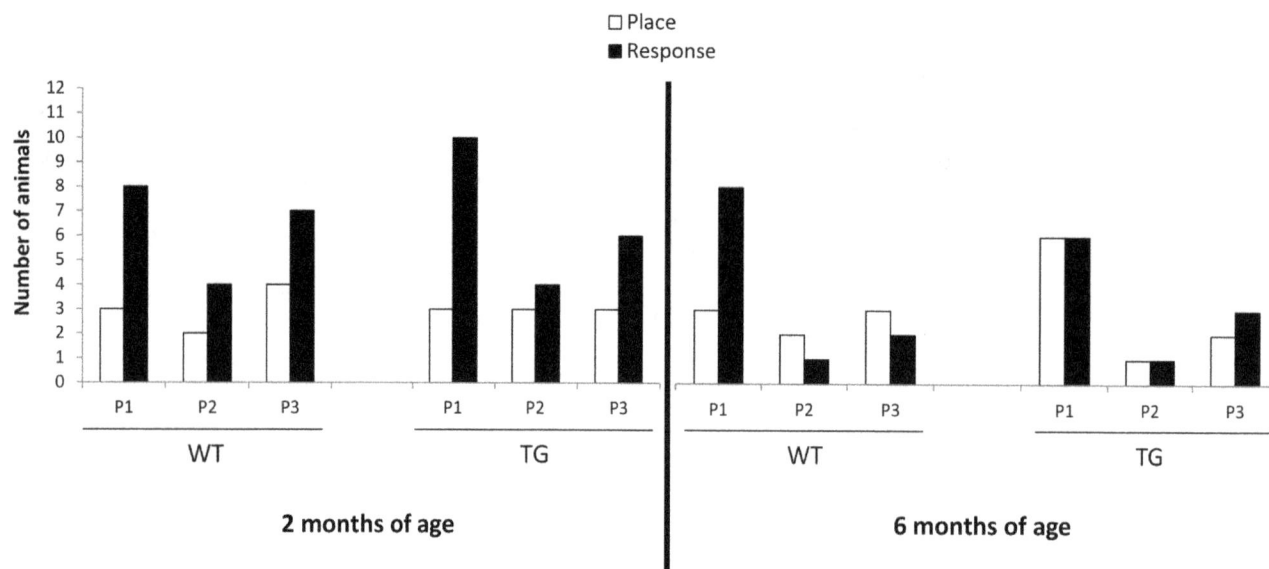

Figure 2. Strategy shifting. Number of rats that exhibited Place (P) or Response (R) learning strategy during each Probe trial P1, P2 and P3 (days 8, 16 and 23 respectively) are represented for WT and TG cohorts of 2 and 6 months of age. The size corresponds to animals that made (1) correct arm choices with a learning index greater than 0.5 during each training session, and (2) were successful for the two last trials prior to the probe trials in the cross maze.

Reversal learning deficit in cross maze test

We have investigated BACHD rats in a spatial memory paradigm with the cross maze test. A T-maze standard dual solution task [32] has proven to be useful in distinguishing between spatial and non spatial learning in animals. Rats were trained over several trials from the same start arm to consistently enter the arm where a baited home-cage was located. We used a home-cage baited with sucrose pellets as an alternative to traditional motivational procedures that use water or food deprivation. Since BACHD transgenic rats show reduced food intake [25], we felt that procedures that avoid food deprivation may be less liable to potential confounds and misinterpretation of behavioral data.

During Acquisition training (A) both TG and WT rats learned the task and showed a significant learning index (~0.8). This indicates that a 'return to home-cage' is an effective incentive and that a training protocol of only one trial per day is sufficient. These observations are in line with results from a study in adult $B6D2F_2$ mice (cross of C57BL/6J and DBA/2J strains) in a Lashley III maze. It was demonstrated that one training trial per day with a home-cage reward procedure led to a significant learning index (~0.7) after just 4 days [36].

During Reversal training (R), the home cage was located at the end of another arm, different from the initially trained arm. The same start arm was used. *All* rats of 2 and 6 months of age initially had difficulties finding the new location. This was confirmed by a comparison of the percentage correct choice of the last two acquisition trials (6 and 7) with the first two reversal trials (1 and 2). A difficulty in finding the new location is perhaps not surprising, as reversal learning is more challenging per se because rats have to disengage from a previous learned task in order to acquire a new task. We decided to extend the reversal training for 6 days (Extended Reversal training; ER) and all rats eventually learned the new task, although the 6 months old TG rats performed significantly worse than WT rats. We previously reported similar reversal learning deficits in adult transgenic BACHD mice of 10 months of age in a cross-maze task [33]. In 6 months old $Hdh^{(CAG)150}$ knock-in mice, cognitive impairments were shown in

compound reversal of an extra-dimensional shift task (EDS) [37]. Reversal learning difficulties were also reported in a spatial operant reversal test paradigm of 9 months old tgHD rats and in 27-week old YAC128 mice in a water T-maze task [38–40]. These data are in accordance with the present findings, suggesting a progressive cognitive decline between 2 and 6 months of age.

To discover which learning strategy rats have adopted, a test trial was performed after training was finished. The new start arm was now located opposite of the arm used during training. Accordingly, rats which used spatial cues to find the correct arm would enter the same baited arm as during trainings (place learners); whereas rats that used 'body turn response' learning (stimulus response, S-R) should enter the non baited arm (response learners). The first probe trial (P1) demonstrated that BACHD rats of 2 and 6 months of age were predominantly *response* learners. Using a similar cross maze task, this preference for the *response* strategy on (P1) was also observed in WT and TG BACHD *mice* [33]. Consistent with our findings, homozygote tgHD rats of 6 and 12 months of age were also mostly response learners in a Morris water maze task [41]. Interestingly, our results are in contrast with findings suggesting that during (A), *place* learning is typically adopted by rats in a cross-maze [32], [42–43]. The reason why 2 months old rats maintain their preference for response learning during P3 is unclear, but may involve a developmental time scale of spatial representation. In fact, spatial memory in the cross-maze involves association of the object (landmarks) to their spatial location (home cage); we assume that the network underlying the memory of spatial location in 2 months old WT rats is slower to develop [44]. The preference (shift) for *place* strategy was only observed in 6 months old WT during the 2nd and 3rd probe trial (P2 and P3), whereas TG rats maintained their *response* strategy. The same strategy was seen in adult BACHD mice during reversal probe trial [33]. The reason why BACHD rats maintain response learning may probably involve altered functioning of fronto-hippocampal (place learning) vs. fronto-striatal (response learning) circuitry [45–47]. Indeed, TG rats show already at 3 months of age abundant *htt* aggregates in the CA3 region of the hippocam-

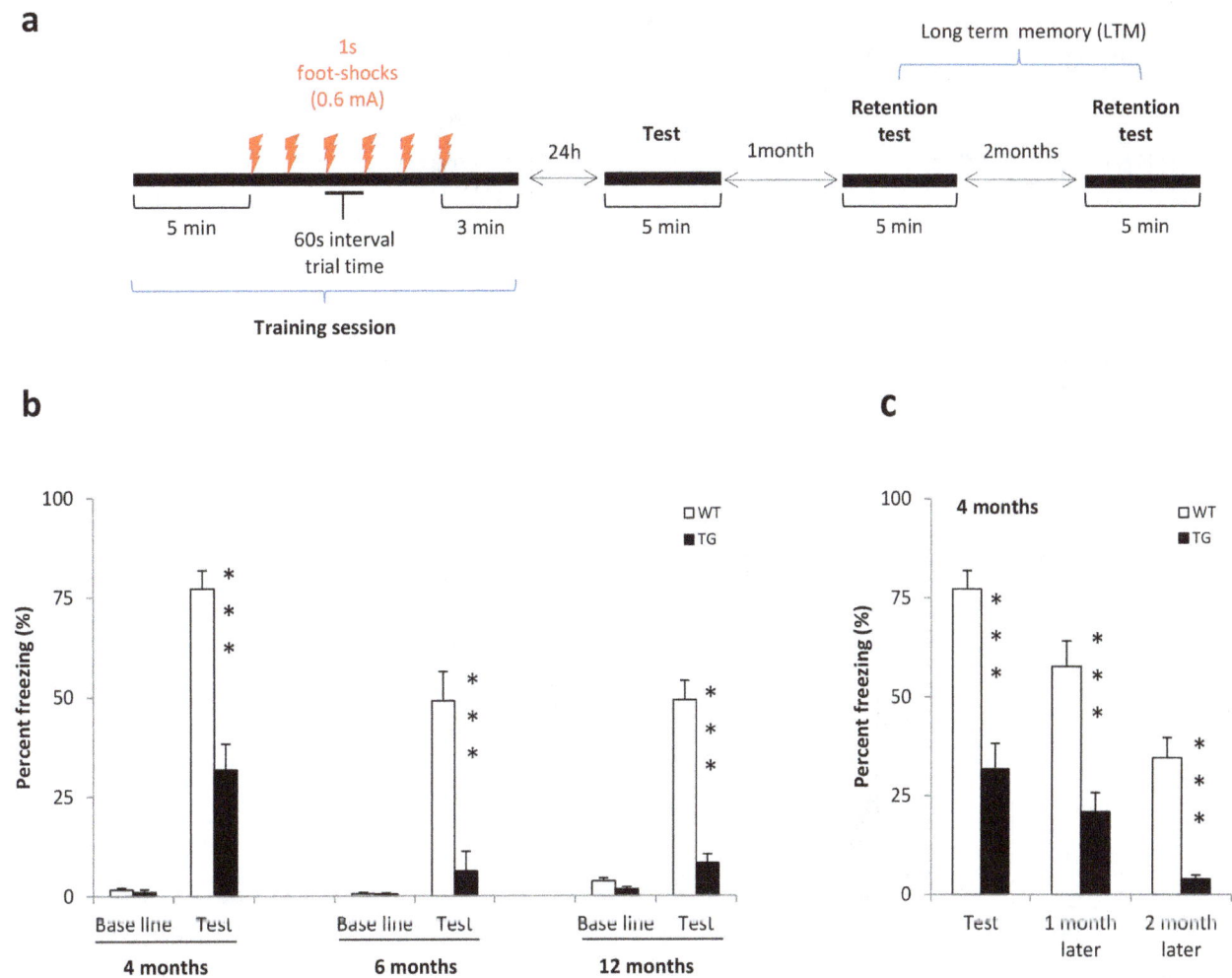

Figure 3. Contextual fear conditioning. [a] Schematic illustration of the contextual fear conditioning protocol. [b–c] Results are expressed as Mean ± SEM of percentage freezing. 4 months (WT n = 13, TG n = 13), 6 months (WT n = 7, TG n = 9) and 12 months (WT n = 16, TG, n = 6) BACHD rats were used. No difference in baseline responding to training was observed. TG rats showed less fear memory to the context as they freeze less in comparison with WT rats at 4, 6, and 12 months of age [b]. Long term memory was assessed 1 month and 2 months after retention testing were conducted in the 4 months old rats cohort (i.e. they were tested at 5 and 6 months of age respectively) [c]. TG compared to WT still had lower freezing to the context. A progressive freezing 'extinction' was observed. Asterisks indicate significant differences between WT and TG rats (***p<0.001).

pus, whereas only few aggregates were present in the caudate-putamen [25]. Therefore, the striatal-based 'body turn response' might prevail during learning and subsequent probe testing in TG rats.

The rodent data are consistent with human data. Similar impairments in reversal learning and strategy, when attention has to be shifted from one perceptual dimension to another, have been demonstrated in early and advanced-stage HD patients in an extradimensional shift (EDS) test and in a Wisconsin Card Sorting Test (WCST). These patients made perseverative errors suggesting memory inflexibility [13–15], [48]. Finally, cognitive set shifting ability in EDS and WCST involves cortical and basal ganglia circuitry system, especially the prefrontal cortex and the caudate nucleus [49].

Associative learning deficits in fear conditioning test

We performed an extensive characterization of BACHD rats in various fear conditioning paradigms and found very robust deficits across all age cohorts and under all experimental conditions. A

technical challenge that was successfully mastered was the selection of an appropriate current intensity. One that was not too high – high intensities would lead to a generalized freezing response – or too low – low intensities would lead to large variability in freezing and inconsistent fear conditioning [50]. A confound of the BACHD fear conditioning phenotype by motor deficits seems unlikely since rats did not show any difference in percent of time freezing during habituation. In order to address if altered sensitivity to foot shocks (US) may have confounded the fear conditioning phenotype, we employed a flinch-jump test and found no differences between WT and TG rats. We used a relatively low shock intensity (0.6 mA) which may explain that not all rats showed a 'jump' reflex. Our data are consistent with findings in Wistar rats where no correlation was found between pain sensitivity, conditioned and novelty-evoked fear responses in 'flinch-jump', 'tail flick' and 'contextual' fear tests [35]. Together, these data suggest that the deficit in conditioned fear responses in BACHD rats are not confounded by motor deficits or altered sensitivity to foot shocks.

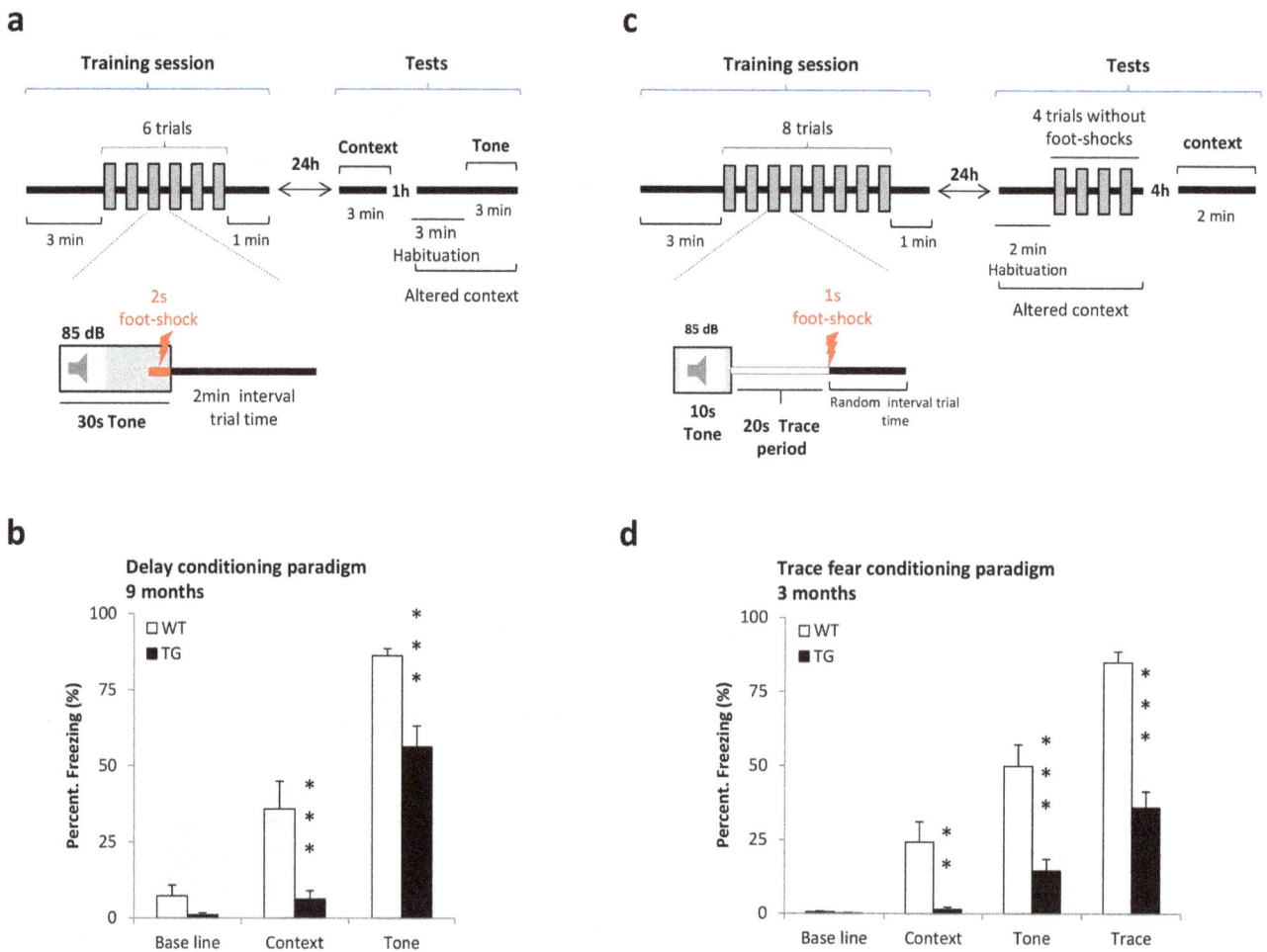

Figure 4. Delay and Trace fear conditioning. [a] Schematic illustration of the delay fear conditioning paradigm. [b] Results in 9 months old BACHD rats are expressed as Mean ± SEM (WT n = 13, TG n = 15). TG rats presented a significant lower freezing response to the context and to the tone. [c] Schematic illustration of trace fear conditioning paradigm. [d] Results in 3 months old BACHD rats are expressed as Mean ± SEM (WT n = 13, TG n = 15). TG rats, compared to WT rats, showed a significantly lower freezing response to the context, to the tone and trace period during retention tests. Asterisks indicate significant differences between WT and TG rats (**p<0.01 and ***p<0.001).

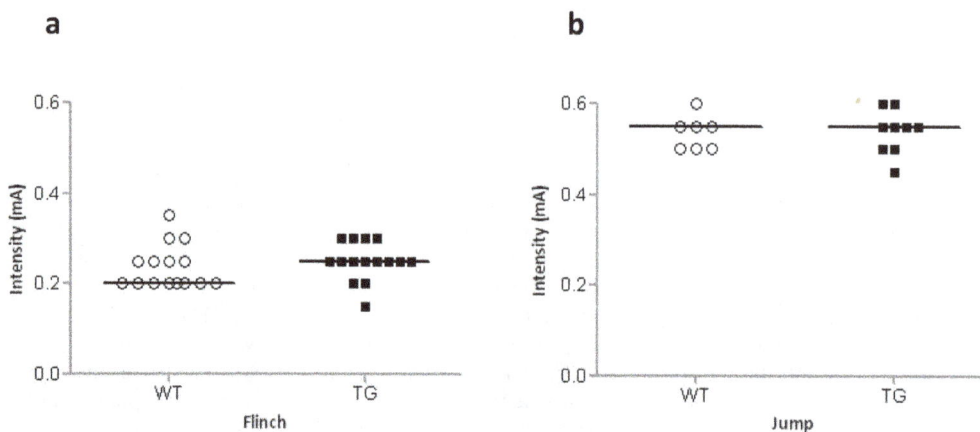

Figure 5. Flinch-jump test. Sensitivity of 6 months old BACHD rats to foot-shocks for [a] flinch and [b] jump (WT n = 15, TG n = 15). Individual intensity response is plotted and bars indicate median values for each genotype. There was no difference between WT and TG rats in current intensities that elicited a flinch or a jump response.

What are the mechanisms underlying the fear conditioning deficits in BACHD rats? Learning in contextual fear conditioning is thought to involve association of stimuli present in the conditioned chamber (texture, shape, dimensions) with the (US) itself. More complex stimuli may put a higher demand on effective learning and memory [51], and thus may affect subjects with impaired activity in fear conditioning circuitry to a larger extent than unaffected subjects. The amygdala and hippocampus are involved in complex stimuli learning [52] and BACHD rats show *htt* aggregates in both brain areas [25]. Therefore, TG rat's lower freezing response to the context could be associated to a hippocampal-amygdala dysfunction. Such a conclusion is consistent with our findings from delay and trace fear conditioning testing. As expected from a stimulus with a higher salience, all rats showed higher levels of fear conditioning to the tone than to the context. The BACHD rats showed again a fear conditioning impairment. During trace fear conditioning, the Tone and shock (US) are separated by a time interval and the Trace period becomes predictive of the (US). Using these more complex stimuli, BACHD rats showed a clear deficit in fear conditioning. Impairments in fear conditioning have also been reported in mouse models for HD. For example, 5 weeks old R6/2 mice showed less contextual freezing than their wild-type control, although no difference was observed in tone conditioning [53] (Bolivar et al. 2003). In addition, a reduced fear expression during extinction retrieval and a reinstatement of a fear conditioning in R6/2 mice was not associated with a weakness in CS-US, but with neuronal *hypoactivation* in the prelimbic cortex, a subregion region of the prefrontal cortex [54]. Four months old CAG140 Knock-In (KI) mice have shown an increased freezing response during training, but, again displayed no deficit in recall tone fear conditioning [55]. We reported that adult transgenic BACHD mice present *higher* freezing rates to the context and tone during retention testing, and attributed this impairment to emotional deficits [33]. The difference in fear conditioning phenotypes between BACHD mice and rats is surprising. However, in view of the robustness of the rat phenotype and the fact that we are eventually interested in the translation of these findings to humans, it would be more sensible to perform fear conditioning studies in a non human primate model for HD [9], rather than undertaking an effort to further characterize the mouse fear conditioning phenotype.

Interestingly, reversal learning impairments in a cross-maze appeared at 6 months of age, whereas associative learning and memory deficits in fear conditioning tasks were already present at 3 months of age. Matching the different onset of these deficits with the emergence of *htt* aggregates in brain areas involved in the circuitry underlying cross maze behavior and fear conditioning will be helpful to translate the findings from rodents to humans. Especially for fear conditioning the functional neuroanatomy has been well described [28]. Rodent data support a role for the amygdala in the acquisition of conditioned fear, whereas the hippocampus and the medial prefrontal cortex (mPFC) are required for consolidation of long-term memory [56–59]. Human

functional magnetic resonance imaging (fMRI) studies in delay and trace fear conditioning, have demonstrated a role of the hippocampus and other brain regions that support working memory processes in encoding temporal information and maintaining the associative representation CS-US during trace intervals [29]. Wide spread *htt* aggregates have been observed in brain areas involved in fear conditioning such as the neocortex, hippocampus, and the amygdala of BACHD rats [25]. However, the behavioral effects occurred at an earlier age than the *htt* aggregates (12 months). It is possible that more subtle molecular and cellular deficits in the cortex, hippocampus and amygdala contribute to the early deficits in fear conditioning. Further studies should address the developmental mechanisms underlying the disease progression in BACHD rats.

Conclusion

Our study is the first to provide evidence of progressive cognitive deficits in a transgenic BACHD rat model for HD. TG animals showed difficulties in associative learning at 3 months of age in a fear conditioning test, and impairments in spatial memory at 6 months of age, mainly in reversal training where attention has to be shifted from one set of learning to another. BACHD rats recapitulate some of the cognitive impairments seen in HD patients. The precise time course for development of the cognitive symptoms requires further studies in additional age cohorts. As fear conditioning deficits appeared already in the youngest cohort tested, animals of 1 and 2 months of age need to be tested to determine if the onset of the fear conditioning is similar to the onset of, for example, rotarod deficits that occur at 2 months of age [27]. Emergence of cognitive deficits before motor deficits might more closely mimic the time course in HD patients [16]. In conclusion: the robust fear conditioning phenotype offers a firm foundation for future studies aimed to further characterize the time course for the associative memory deficit and its underlying neural circuitry. In addition, this functional readout can be validated for drug discovery approaches that target *htt* aggregates, using, for example, adenovirus-based viral transfection methods against htt [60].

Acknowledgments

We would like to thank Drs. Susan Boyce and Frederic Machet for helpful discussions, Dr. Heinz von der Kammer for the animal breeding process management, and Prof. Dr. Michael Koch for proof reading the article and his insightful comments. We address our sincere thanks to Yu-Taeger Libo for supply of the non commercial primers sample and interesting discussions. We specially thank Dr. Pierre Ilouga and Insa Winzenborg for advices on statistical analysis. The authors are appreciative of the care and maintenance of the animal breeding by Mrs Brigitta Miethke.

Author Contributions

Conceived and designed the experiments: RS YSA BE. Performed the experiments: YSA. Analyzed the data: YSA. Contributed reagents/materials/analysis tools: HN. Wrote the paper: YSA RS.

References

1. Huntington's Disease Collaborative Research Group (1993) A novel gene containing a trinucleotide repeat that is unstable on Huntington's disease chromosomes. Cell 72: 971–983.

2. Huntington G (1872) "On Chorea". The Medical and Surgical Reporter: A weekly Journal, (Philadelphia: S. Butler), vo. 26, no 15, 317–321.

3. Myers RH, Vonsattel JP, Stevens TJ, Cupples LA, Richardson EP, et al. (1988) Clinical and neuropathologic assessment of severity in Huntington's disease. Neurology. 38(3): 341–7.

4. Vonsattel JP, Myers RH, Stevens TJ, Ferrante RJ, Bird ED, et al. (1985) Neuropathological classification of Huntington's disease J Neuropathol Exp Neurol. 44(6): 559–77.

5. Douaud G, Gaura V, Ribeiro MJ, Lethimonnier F, Maroy R, et al. (2006) Distribution of grey matter atrophy in Huntington's disease patients: a combined ROI-based and voxel-based morphometric study. Neuroimage. 32(4): 1562–75.

6. Jones L, Hughes A (2011) Pathogenic mechanisms in Huntington's disease. Int Rev Neurobiol. 98: 373–418.

7. Pidgeon C, Rickards H (2013) The pathophysiology and pharmacological treatment of Huntington disease. Behav Neurol. 26(4): 245–53. doi: 10.3233/BEN-2012-120267.

8. Jackson GR, Salecker I, Dong X, Yao X, Arnheim N, et al. (1998) Polyglutamine-expanded human huntingtin transgenes induce degeneration of Drosophila photoreceptor neurons. Neuron. 21(3): 633–42.

9. Yang SH, Cheng PH, Banta H, Piotrowska-Nitsche K, Yang JJ, et al. (2008) Towards a transgenic model of Huntington's disease in a non-human primate. Nature. 453(7197): 921–4. Epub 2008 May 18.

10. Crook ZR, Housman D (2011) Huntington's disease: can mice lead the way to treatment? Neuron. 69(3): 423–35.

11. Vlamings R, Zeef DH, Janssen ML, Oosterloo M, Schaper F, et al. (2012) Lessons learned from the transgenic Huntington's disease rats. Neural Plast. 2012: 682712. Epub 2012 Jul 18.

12. Mullard A (2012) Sting of Alzheimer's failures offset by upcoming prevention trials. [Review]. Nat Rev Drug Discov. 11(9): 657–60. doi: 10.1038/nrd3842.

13. Lawrence AD, Barbara JS, John RH, Anne ER, Klaus WL, et al. (1996) Executive and mnemonic functions in early Huntington's disease. Brain 119: 1633–1645.

14. Lawrence AD, John RH, Anne ER, Ann K, Charles C, et al. (1998) Evidence for specific cognitive deficits in preclinical Huntington's disease. Brain 121: 1329–1341.

15. Lawrence AD, Sahakian BJ, Rogers RD, Hodge JR, Robbins TW (1999) Discrimination, reversal, and shift learning in Huntington's disease: mechanisms of impaired response selection. Neuropsychologia. 37(12): 1359–74.

16. Hahn-Barma V, Deweer B, Dürr A, Dodé C, Feingold J, et al. (1998) Are cognitive changes the first symptoms of Huntington's disease? A study of gene carriers. J Neurol Neurosurg Psychiatry. 64(2): 172–7.

17. von Hörsten S, Schmitt I, Nguyen HP, Holzmann C, Schmidt T, et al. (2003) Transgenic rat model of Huntington's disease. Hum Mol Genet. 12(6): 617–24. Epub 2012 Mar 1.

18. Fielding SA, Brooks SP, Klein A, Bayram-Weston Z, Jones L, et al. (2012) Profiles of motor and cognitive impairment in the transgenic rat model of Huntington's disease. Brain Res Bull. 88(2–3): 223–36.

19. Cao C, Temel Y, Blokland A, Ozen H, Steinbusch HW, et al. (2006) Progressive deterioration of reaction time performance and choreiform symptoms in a new Huntington's disease transgenic ratmodel. Behav Brain Res. 170(2): 257–61.

20. Kántor O, Temel Y, Holzmann C, Raber K, Nguyen HP, et al. (2006) Selective striatal neuron loss and alterations in behavior correlate with impaired striatal function in Huntington's disease transgenic rats. Neurobiol Dis. 22(3): 538–47.

21. Nguyen HP, Kobbe P, Rahne H, Wörpel T, Jäger B, et al. (2006) Behavioral abnormalities precede neuropathological markers in rats transgenic for Huntington's disease. Hum Mol Genet. 15(21): 3177–94. Epub 2006 Sep 19.

22. Zeef DH, van Goethem NP, Vlamings R, Schaper F, Jahanshahi A, et al. (2012) Memory deficits in the transgenic rat model of Huntington's disease. Behav Brain Res. 227(1): 194–8.

23. Mangiarini L, Sathasivam K, Seller M, Cozens B, Harper A, et al. (1996) Exon 1 of the HD gene with an expanded CAG repeat is sufficient to cause a progressive neurological phenotype in transgenic mice. Cell. 87(3): 493–506.

24. Lione LA, Carter RJ, Hunt MJ, Bates GP, Morton AJ, et al. (1999) Selective discrimination learning impairments in mice expressing the human Huntington's disease mutation. J Neurosci. 19(23): 10428–37.

25. Yu-Taeger L, Petrasch-Parwez E, Osmand AP, Redensek A, Metzger S, et al. (2012) A Novel BACHD Transgenic Rat Exhibits Characteristic Neuropathological Features of Huntington Disease. J Neurosci. 32(44): 15426–15438.

26. Gray M, Shirasaki DI, Cepeda C, André VM, Wilburn B, et al. (2008) Full-length human mutant huntingtin with a stable polyglutamine repeat can elicit progressive and selective neuropathogenesis in BACHD mice. J Neurosci. 28(24): 6182–95.

27. Abada YS, Nguyen HP, Schreiber R, Ellenbroek B (2013) Assessment of Motor, sensory motor gating and Recognition memory in a novel full-length rat model for Huntington disease. doi: 10.1371/journal.pone.0068584.

28. Fendt M, Fanselow MS (1999) The neuroanatomical and neurochemical basis of conditioned fear. [Review]. Neurosci Biobehav Rev. 23(5): 743–60.

29. Knight DC, Cheng DT, Smith CN, Stein EA, Helmstetter FJ, et al. (2004) Neural substrates mediating human delay and trace fear conditioning. J Neurosci. 24(1): 218–28.

30. Farr SA, Flood JF, Scherrer JF, Kaiser FE, Taylor GT, et al. (1995) Effect of ovarian steroids on footshock avoidance learning and retention in female mice. Physiol Behav. 58(4): 715–23.

31. Pearson R, Lewis MB (2005) Fear recognition across the menstrual cycle. Horm Behav. 47(3): 267–71. Epub 2005 Jan 16.

32. Tolman EC, Ritchie BF, Kalish D (1946) Studies in spatial learning; place learning versus response learning. J Exp Psychol. 36: 221–9.

33. Abada YS, Schreiber R, Ellenbroek B (2013) Motor, emotional and cognitive deficits in adult BACHD mice: A model for Huntington's disease. Behav Brain Res. 238: 243–51. doi: 10.1016/j.bbr.2012.10.039. Epub 2012 Oct 30.

34. Blum S, Runyan JD, Dash PK (2006) Inhibition of prefrontal protein synthesis following recall does not disrupt memory for trace fear conditioning. BMC Neurosci. 7: 67.

35. Lehner M, Wisłowska-Stanek A, Maciejak P, Szyndler J, Sobolewska A, et al. (2010) The relationship between pain sensitivity and conditioned fear response in rats. Acta Neurobiol Exp (Wars). 70(1): 56–66.

36. Blizard DA, Weinheimer VK, Klein LC, Petrill SA, Cohen R, et al. (2006) 'Return to home cage' as a reward for maze learning in young and old genetically heterogeneous mice. Comp Med. 56(3): 196–201.

37. Brooks SP, Betteridge H, Trueman RC, Jones L, Dunnett SB (2006) Selective extra-dimensional set shifting deficit in a knock-in mouse model of Huntington's disease. Brain Res Bull. 69(4): 452–7. Epub 2006 Mar 10.

38. Brooks SP, Janghra N, Higgs GV, Bayram-Weston Z, Heuer A, et al. (2012) Selective cognitive impairment in the YAC128 Huntington's disease mouse. Brain Res Bull. 88(2–3): 121–9. Epub 2011 May 20.

39. Fink KD, Rossignol J, Crane AT, Davis KK, Bavar AM, et al. (2012) Early cognitive dysfunction in the HD 51 CAG transgenic rat model of Huntington's disease. Behav Neurosci. 126(3): 479–87. doi: 10.1037/a0028028.

40. Van Raamsdonk JM, Pearson J, Slow EJ, Hossain SM, Leavitt BR, et al. (2005) Cognitive dysfunction precedes neuropathology and motor abnormalities in the YAC128 mouse model of Huntington's disease. J Neurosci. 25(16): 4169–80.

41. Kirch RD, Meyer PT, Geisler S, Braun F, Gehrig S, et al. (2012) Early deficits in declarative and procedural memory dependent behavioral function in a transgenic rat model of Huntington's disease. Behav Brain Res. 239C: 15–26. doi: 10.1016/j.bbr.2012.10.048.

42. Packard MG, McGaugh JL (1996) Inactivation of hippocampus or caudate nucleus with lidocaine differentially affects expression of place and response learning. Neurobiol Learn Mem. 65(1): 65–72.

43. Packard MG (1999) Glutamate infused posttraining into the hippocampus or caudate-putamen differentially strengthens place and response learning. Proc Natl Acad Sci U S A. 96(22): 12881–6.

44. Ainge JA, Langston RF (2012) Ontogeny of neural circuits underlying spatial memory in the rat. Front Neural Circuits. 6: 8. doi: 10.3389/fncir.2012.00008. Epub 2012 Mar 1.

45. Restle F (1957) Discrimination of cues in mazes: a resolution of the place-vs.-response question. [Review]. Psychol Rev. 64(4): 217–28.

46. Pych JC, Chang Q, Colon-Rivera C, Haag R, Gold PE (2005) Acetylcholine release in the hippocampus and striatum during place and response training. Learn Mem. 12(6): 564–72.

47. Ciamei A, Morton AJ (2009) Progressive imbalance in the interaction between spatial and procedural memory systems in the R6/2 mouse model of Huntington's disease. Neurobiol Learn Mem. 92(3): 417–28.

48. Josiassen RC, Curry LM, Mancall EL (1983) Development of neuropsychological deficits in Huntington's disease. Arch Neurol. 40(13): 791–6.

49. Rogers RD, Andrews TC, Grasby PM, Brooks DJ, Robbins TW (2000) Contrasting cortical and subcortical activations produced by attentional-set shifting and reversal learning in humans. J Cogn Neurosci. 12(1): 142–62.

50. Baldi E, Lorenzini CA, Bucherelli C (2004) Footshock intensity and generalization in contextual and auditory-cued fear conditioning in the rat. Neurobiol Learn Mem. 81(3): 162–6.

51. Rescorla RA (1972) "Configural" conditioning in discrete-trial bar pressing. J Comp Physiol Psychol. 79(2): 307–17.

52. Phillips RG, LeDoux JE (1992) Differential contribution of amygdala and hippocampus to cued and contextual fear conditioning. Behav Neurosci. 106(2): 274–85.

53. Bolivar VJ, Manley K, Messer A (2003) Exploratory activity and fear conditioning abnormalities develop early in R6/2 Huntington's disease transgenic mice. Behav Neurosci. 117(6): 1233–42.

54. Walker AG, Ummel JR, Rebec GV (2011) Reduced expression of conditioned fear in the R6/2 mouse model of Huntington's disease is related to abnormal activity in prelimbic cortex. Neurobiol Dis. 43(2): 379–87. doi: 10.1016/j.nbd.2011.04.009. Epub 2011 Apr 16.

55. Hickey MA, Kosmalska A, Enayati J, Cohen R, Zeitlin S, et al. (2008) Extensive early motor and non-motor behavioral deficits are followed by striatal neuronal loss in knock-in Huntington's disease mice. Neuroscience. 157(1): 280–95. doi: 10.1016/j.neuroscience.2008.08.041.

56. Faure A, Höhn S, Von Hörsten S, Delatour B, Raber K, et al. (2011) Altered emotional and motivational processing in the transgenic rat model for Huntington's disease. Neurobiol Learn Mem. 95(1): 92–101. doi: 10.1016/j.nlm.2010.11.010. Epub 2010 Nov 25.

57. LeDoux JE (2000) Emotion circuits in the brain. [Review]. Annu Rev Neurosci. 23: 155–84.

58. Gilmartin MR, Helmstetter FJ (2010) Trace and contextual fear conditioning require neural activity and NMDA receptor-dependent transmission in the medial prefrontal cortex. Learn Mem. 17(6): 289–96. doi: 10.1101/lm.1597410.

59. Gilmartin MR, Kwapis JL, Helmstetter FJ (2012) Trace and contextual fear conditioning are impaired following unilateral microinjection of muscimol in the ventral hippocampus or amygdala, but not the medial prefrontal cortex. Neurobiol Learn Mem. 97(4): 452–64. doi: 10.1016/j.nlm.2012.03.009. Epub 2012 Mar 14.

60. Ramaswamy S, Kordower JH (2012) Gene therapy for Huntington's disease. [Review]. Neurobiol Dis. 48(2): 243–54. doi: 10.1016/j.nbd.2011.12.030. Epub 2011 Dec 24.

6

Intact Olfaction in a Mouse Model of Multiple System Atrophy

Florian Krismer[1], Gregor K. Wenning[1], Yuntao Li[1,2], Werner Poewe[1], Nadia Stefanova[1]*

1 Division of Neurobiology, Department of Neurology, Innsbruck Medical University, Innsbruck, Austria, **2** The Second School of Clinical Medicine, The Second Affiliated Hospital, Nanjing Medical University, Nanjing, China

Abstract

Background: Increasing evidence suggests that olfaction is largely preserved in multiple system atrophy while most patients with Parkinson's disease are hyposmic. Consistent with these observations, recent experimental studies demonstrated olfactory deficits in transgenic Parkinson's disease mouse models, but corresponding data are lacking for MSA models.

Methods: Olfactory function and underlying neuropathological changes were investigated in a transgenic multiple system atrophy mouse model based on targeted oligodendroglial overexpression of α-synuclein as well as wild-type controls. The study was divided into (1) a pilot study investigating olfactory preference testing and (2) a long-term study characterizing changes in the olfactory bulb of aging transgenic multiple system atrophy mice.

Results: In our pilot behavioral study, we observed no significant differences in investigation time in the olfactory preference test comparing transgenic with wild-type animals. These findings were accompanied by unaffected tyrosine hydroxylase-positive cell numbers in the olfactory bulb. Similarly, although a significant age-related increase in the amount of α-synuclein within the olfactory bulb was detected in the long-term study, progressive degeneration of the olfactory bulb could not be verified.

Conclusions: Our experimental data show preserved olfaction in a transgenic multiple system atrophy mouse model despite α-synucleinopathy in the olfactory bulb. These findings are in line with the human disorder supporting the concept of a primary oligodendrogliopathy with variable neuronal involvement.

Editor: John Duda, Philadelphia VA Medical Center, United States of America

Funding: This study was supported by grants of the Austrian Science Fund (FWF): F04404-B19 and P25161-B24. The funders had no role in study design, data collection and analysis, decision to publish, or preparation of the manuscript.

Competing Interests: The authors have declared that no competing interests exist.

* E-mail: nadia.stefanova@i-med.ac.at

Introduction

Multiple system atrophy (MSA) is a rapidly progressive neurodegenerative disorder of unknown etiopathogenesis. It is characterized clinically by autonomic failure accompanied by parkinsonism and cerebellar ataxia [1]. The distinction of early stage MSA from related parkinsonian syndromes including Parkinson's disease (PD) can be challenging [2]. However, previous reports suggested that assessment of olfactory function is an important pointer in the differential diagnosis. MSA patients show intact or mildly impaired olfaction whereas most PD patients are hyposmic or sometimes anosmic [3–8]. Even more interestingly, olfactory disturbances may predate the onset of classic motor features in PD [9,10]. Deficits in PD patients include impairment of odor detection, discrimination and identification [10,11].

α-synuclein (αSYN) is a key protein in the pathogenesis of MSA and PD with the former being characterized by glial cytoplasmic inclusions (GCIs, Papp-Lantos bodies) and the latter by neuronal Lewy bodies as their subcellular hallmark feature. These αSYN-positive inclusions are also observed in the olfactory tract, predominantly affecting the anterior olfactory nucleus [12,13].

In preclinical research, αSYN pathology may be replicated by transgenic (tg) ovexpression of αSYN under oligodendroglial [14–16] or neuronal promoters [17] mimicking MSA- or PD-like inclusion pathology, respectively.

Recently, olfactory disturbances have been studied in tg mouse models of PD. Behavioral alterations and olfactory bulb pathology in these models are reminiscent of the human disorder with age-related impairment in odor detection and discrimination [18–20] as well as extensive olfactory bulb pathology [21–23]. In contrast, smell disturbances in MSA models were only studied once in the context of glial derived neurotrophic factor (GDNF) replacement therapy [24]. This study reported olfactory impairment in tg versus wild-type (wt) animals in the saline-treated study arm; however, olfactory bulb pathology was not investigated [24].

In the present study, we investigated olfactory behavior and assessed neuropathological changes within the olfactory bulb (OB) and their age-related evolution in an established tg MSA mouse model featuring overexpression of αSYN in oligodendrocytes [14].

Methods

The study was split into two parts: (1) a pilot study determining behavioral olfactory deficits and immunohistochemical differences in 9-months old animals and (2) a confirmatory long-term study (LTS) focusing on the analysis of OB aging. In the LTS, mice with 2, 6 and 18 months of age were studied. Both subprotocols compared homozygous tg MSA mice to age- and strain-matched non-littermate wt controls of the inbred C57BL/6 strain.

Animals

The generation and characterization of tg mice with targeted overexpression of human αSYN (hαSYN) under the oligoden-droglial proteolipid protein promotor (PLP-hαSYN) were described previously [14]. Tg and wt mice were originally obtained from P. Kahle (University of Tübingen, Tübingen, Germany) and Charles River Laboratories (Charles River Laboratories, Sulzfeld, Germany), respectively. Mice were bred and maintained in a temperature-controlled specific pathogen free room with a 12-h light/dark cycle and free access to food and water at the Animal Facility of Innsbruck Medical University. Genotyping was performed by tail clip polymerase chain reaction (PCR) using the following primers: Forward: 5′-ATG GAT GTA TTC ATG AAA GG-3′; reverse: 5′-TTA GGC TTC AGG TTC GTA G-3′.

This study was carried out in strict accordance with the Austrian guidelines for the care and use of laboratory animals and all in vivo protocols were approved by the Austrian Federal Ministry of Science and Research (do. Zi. 6001). All efforts were made to minimize the number of animals used and their suffering.

Behavioral testing

We performed olfactory preference testing in 9 month old mice according to a previously published protocol [25,26]. This test is designed to identify specific odor detection deficiencies, based on the inability to sense attractive scents [25,26]. Briefly, four home cages sized 26 cm×45 cm×20 cm (width × length × height) were lined up next to each other separated by opaque filter paper. Animals were habituated to the unknown surrounding of an empty cage for a period of 1 hour with transfer steps to the next cage every 15 minutes. Moreover, a video camera was placed such that the entire arena cage (= last habituation cage) was in focus. Thereafter, 5 cm×5 cm squares of scented filter paper were placed on the opposite end of the mice's current position. The animal's behavior was recorded on video for a period of 3 minutes. Different scents were randomly presented at an interval of at least one minute. The following odorants were used: distilled water (control, inherent smell of the filter paper), peanut butter (Skippy, Unilever) dissolved in mineral oil (Sigma-Aldrich, Vienna, Austria), vanilla (Oetker GmbH, Bielefeld, Germany) and cinnamon (Invero, Wiener Neudorf, Austria) dissolved in mineral oil. Test cages were extensively cleaned after each mouse. Finally, to avoid confounding of data owing to task learning, mice performed the test once only.

Tissue preparation and immunohistochemistry

Mice were sacrificed at the designated time points by transcardial perfusion with 10 ml of 0.1 M phosphate buffered saline (PBS; Sigma-Aldrich, Vienna, Austria) followed by ice-cold 4% paraformaldehyde (PFA; Merck, Darmstadt, Germany) in PBS under deep thiopental (Sandoz, Kundl, Austria) anesthesia (i.e. 120 mg/kg body-weight thiopental). Brains were quickly removed and post-fixed in 4% PFA dissolved in PBS at 4°C over night. After cryoprotection in 25% PBS-sucrose solution (Sigma-Aldrich, Vienna, Austria), brains were slowly frozen in 2-methylbutane

(Merck, Darmstadt, Germany) and stored at -80°C until further processing.

Serial coronal sections (40 μm) were cut on a freezing microtome (Leica, Nussloch, Germany). One series of sections per animal was mounted on slides and underwent cresyl violet (CV) staining. Free floating sections were stained according to a standard immunoperoxidase protocol [27,28]. The following primary antibodies were used: mouse anti-tyrosine hydroxylase (TH, 1:1000; Sigma, St. Louis, Missouri, USA), rat anti-mouse CD11b (1:150; Serotec, Oxford, UK), rat anti-human-α-synuclein (15G7, 1:200; Enzo Life Sciences, Exeter, UK). For immunohis-tochemistry, secondary antibodies were biotinylated anti-mouse or anti-rat IgG (Vector Laboratories, Burlingame, California, USA) as appropriate. Following incubation with avidin-biotin complex (ABC) reagent (Vectastain ABC kit, Vector Laboratories, Burlingame, California, USA), immunohistochemical reactions were visualized by 3,3′-diamino-benzidine-tetrahydrochloride (DAB; Sigma, St. Louis, Missouri, USA). For immunofluorescence, Alexa 488- or Alexa 594-conjugated anti-rat or anti-mouse IgG (Molecular Probes, Life Technologies, Paisley, UK), as appropriate, were applied as secondary antibodies.

Image analysis

Image analyses were performed by a blinded investigator. Nikon E-800 (Nikon, Vienna, Austria) microscope equipped with a digital camera (Nikon DXM 1200, Nikon, Vienna, Austria) connected to a computer-assisted analysis system (Stereo Investigator Software, MicroBrightField Europe, Magdeburg, Germany) was used. Regions of interest were outlined manually according to the Paxinos and Franklin Mouse Brain Atlas (1997, Academic Press, San Diego). The optical fractionator workflow was exploited to generate an unbiased estimate of TH- and 15G7-immunoreactive cell numbers in the granular layer and glomerular layer of the OB. Microglial activation was determined by measuring optical densitiy (OD) of CD11b immunoreactivity. Briefly, staining brightness was measured in the glomerular and granular layer of the OB (OD_{ROI}) and a blank area ($OD_{Background}$). Next, the OD ratio was calculated according to the following formula: OD ratio $= -\log (OD_{ROI}/OD_{Background})$ as previously described.[27] OB atrophy was evaluated by outlining the OB bilaterally on 3 adjacent CV stained sections and measuring the respective area using Stereo Investigator Software.

For immunofluorescence, imaging was performed using a DMI 4000B Leica microscope equipped with Digital Fire Wire Color Camera DFC300 FX and Application Suite V3.1 software by Leica (Leica, Nussloch, Germany).

Statistical analysis

Statistical analysis was performed using SPSS 20.0 (SPSS Inc., Chicago, Illinois, USA). If not stated otherwise, data are expressed as mean ± standard error of mean (SEM). Group differences were analyzed by Student's T-test, Kruskal-Wallis-Test, one-way analysis of variance (ANOVA) or two-way ANOVA and Bonferroni correction of multiple comparisons as appropriate. The significance level was set at $p < 0.05$; all tests were two-sided.

Results

Pilot study: olfactory preference test in aged MSA mice

In the olfactory preference testing paradigm, there was no significant difference in investigation time between 9-month old tg and wt animals ($F_{1, 16} = 0.000$, $p = 0.989$, ANOVA). Although the different scents were a non-significant term in our ANOVA model ($F_{3, 48} = 2.189$, $p = 0.101$, ANOVA), pairwise comparisons re-

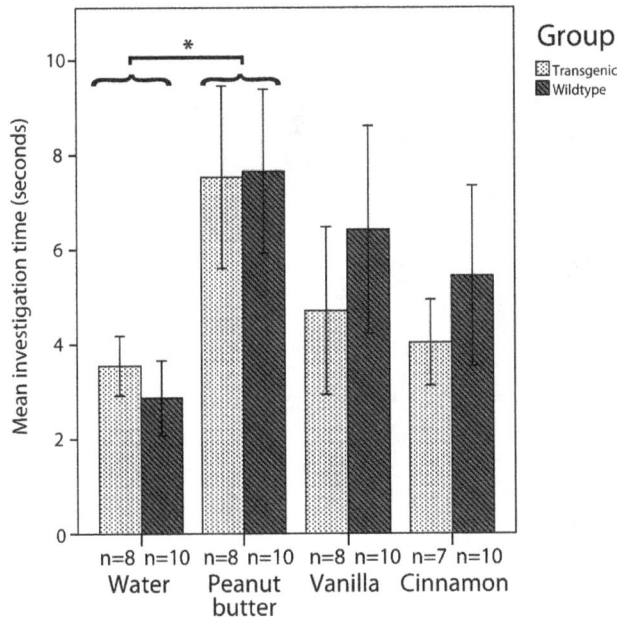

Figure 1. Olfactory behavior. Mean investigation time in seconds in olfactory preference testing of 9 months old mice. Data are expressed as mean; error bars indicate the standard error of mean. Sample sizes are reported below the X-axis. * P<0.05.

vealed that the inherent smell of the filter paper (water sample) was less attractive than peanut butter to the animals irrespective of the underlying genotype (p = 0.041, post-hoc analysis of ANOVA model with Bonferroni correction for multiple comparisons) (Figure 1)

Immunohistochemical analysis of TH Immunoreactive cells within the glomerular layer of the OB yielded no difference in the number of dopaminergic neurons in 9 months old mice (p = 0.129, Student's T-test, Table 1). In addition, stereological analysis of 15G7-immunoreactive cells revealed αSYN pathology of the OB in 9 months old mice already (Table 1).

Long-term study: OB ageing

As indicated above, the LTS involved assessment of neuropathology in 2, 6 and 18 month-old animals to detect age- and αSYN-related neurodegeneration in the OB.

We observed an age-dependent accumulation of hαSYN immunoreactive inclusions in the OB of tg mice (p = 0.036, Kruskal-Wallis; Table 1). However, the highest number of MSA-like cytoplasmic inclusion patholgoy was observed in 9 months old animals (Table 1) with a statistically significant difference to 2 months (p = 0.048, Mann-Whitney-U-Test with Bonferroni correction for multiple comparisons) and 6 months (p = 0.024, Mann-Whitney-U-Test with Bonferroni correction for multiple comparisons) old animals, but not compared to 18 months old mice (p = 0.558, Mann-Whitney-U-Test with Bonferroni correction for multiple comparisons). Transgenic hαSYN driven by the PLP promoter was detected in CNPase-positive glial cells of the OB (Figure 2b) similar to other CNS regions [29,30]. The increase in αSYN load did not convert into OB degeneration, in particular, there were no significant differences in OB volume between the two genotypes ($F_{1, 39} = 3.263$, p = 0.079, ANOVA; Figure 2a). Likewise, the number of TH-ir cells remained stable over time ($F_{3, 39} = 1.248$, p = 0.305, ANOVA; Table 1) and genotype was a non-significant term in the ANOVA model ($F_{1, 39} = 0.377$, p = 0.543, ANOVA; Table 1).

In contrast, tg animals featured early OB microglial activation with significantly increased CD11b immunoreactivity in the glomerular layer at 2 months compared to wt animals (p<0.01, ANOVA model with Bonferroni correction). However, wt animals showed an age-related increase in microglial activation catching up the difference which was present at 2 months (Figure 2c). In the granular layer, we observed age-related enhancement of CD11b immunoreactivity in tg as well as wt animals ($F_{3, 38} = 21.5$, p<0.001, ANOVA) with differences between the two genotypes failing to reach statistical significance (Figure 2d).

Discussion

The presence of olfactory deficits is an important diagnostic pointer in patients presenting with parkinsonism. MSA patients show largely preserved olfactory function whereas most PD patients are hyposmic [3–8]. In preclinical research, olfaction has been extensively studied in PD αSYN mouse models revealing deficits of odor detection and discrimination [18–20]. In addition,

Table 1. Stereological cell counts of 15G7 and TH immunohistochemistry.

Age	Genotype	15G7, mean ± SEM (n)	TH, mean ± SEM (n)	p-value[1]
2 months	Transgenic	77532.1±7258.3 (n=5)	67893.9±8327.7 (n=5)	0.226
2 months	Wildtype	Not applicable	53583.1±5948.7 (n=4)	
6 months	Transgenic	108711.4±11531.2 (n=6)	73244,8±10118.7 (n=6)	0.619
6 months	Wildtype	Not applicable	82077.8±13940.0 (n=6)	
9 months	Transgenic	160520.2±10012.3 (n=5)[2,3]	69613,5±17201.7 (n=6)	0.129
9 months	Wildtype	Not applicable	111041.0±18096.1 (n=7)	
18 months	Transgenic	124320.8±10888.0 (n=8)[2]	75983.0±7448.5 (n=8)	0.197
18 months	Wildtype	Not applicable	62436.0±5622.8 (n=6)	

π... Data from the pilot study;
[1]... Comparing TG vs. WT animals in TH IHC by Student's T Test;
[2]... p<0.05 compared to 2 month old mice (Mann-Whitney U Test with Bonferroni correction for multiple comparisons);
[3]... p<0.05 compared to 6 month old mice (Mann-Whitney U Test with Bonferroni correction for multiple comparisons). 15G7... human αSYN, TH... tyrosine hydroxylase, TG... transgenic, WT... wildtype, IHC... immunohistochemistry, SEM... standard error of mean.

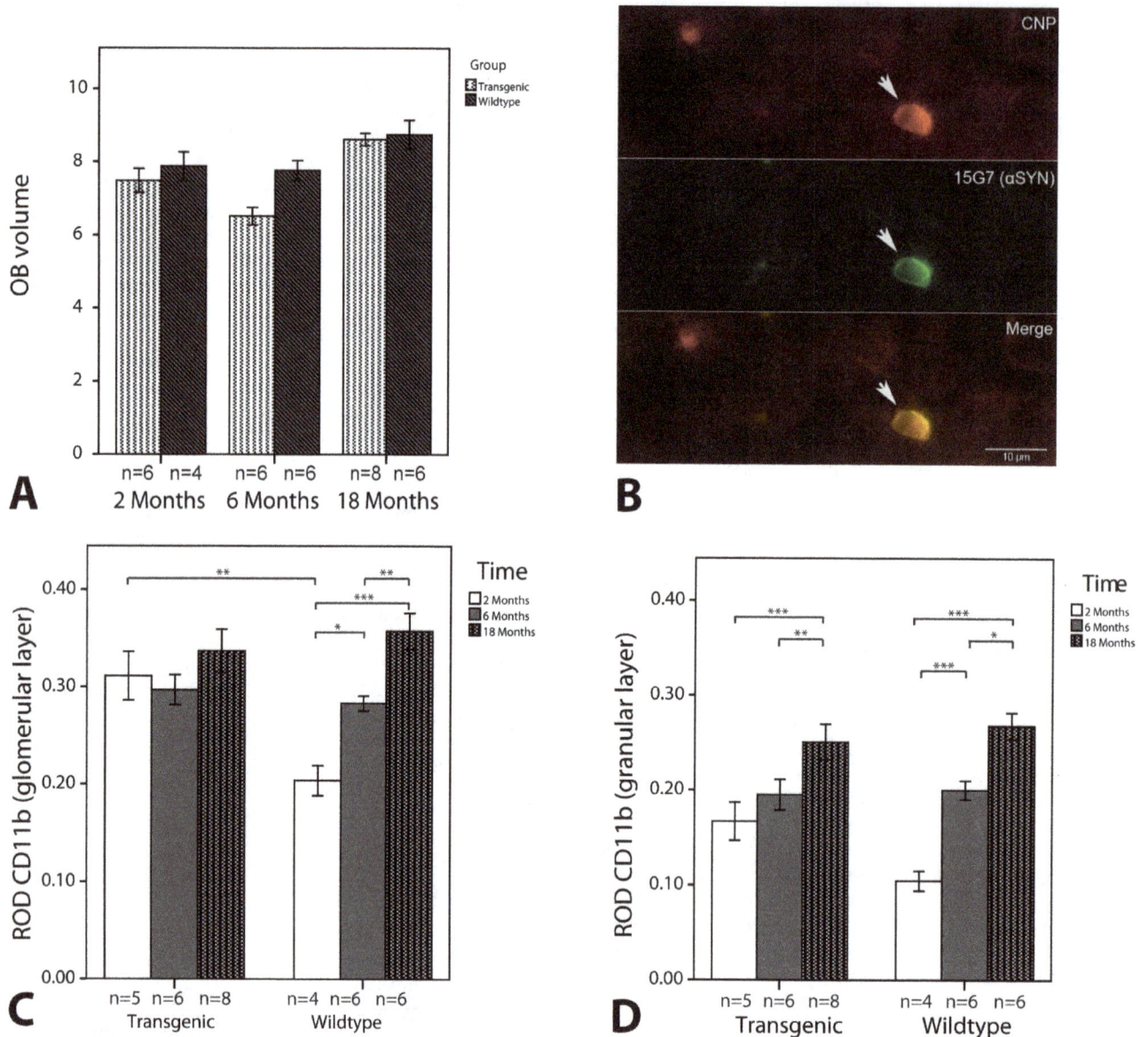

Figure 2. Immunohistochemistry of confirmatory long-term study. A…OB volume; B… Immunofluorescence for CNPase (red) and 15G7 (green) confirmed αSYN expression in glial cells (arrow), scale bar: 10 μm; ROD of CD11b immunohistochemistry in the (C) glomerular layer and the (D) granular layer of the OB; All data are expressed as mean; error bars indicate the standard error of mean. Sample sizes are reported below the X-axis; * P<0.05, ** P<0.01, *** P<0.001.

extensive olfactory bulb pathology has been reported in PD αSYN mouse models [21–23].

In the present study, we explored olfactory function and OB pathology in a transgenic mouse model with targeted oligodendroglial overexpression of hαSYN driven by the PLP promoter [14] featuring MSA-like inclusion pathology. Our data clearly show that olfactory preference testing, a widely used behavioral paradigm to identify specific olfactory deficiencies (i.e. the ability to sense attractive scents) [25,26], was not impaired in the PLP-hαSYN mouse model. This finding is in contrast to a previous study in a related mouse model overexpressing αSYN under the myelin basic protein (MBP) promoter which reported increased pellet retrieval latencies in tg compared to wt animals [24]. The discrepancy might be due to differences between the models used - including myelination with PLP-hαSYN mice lacking obvious

demyelination at ages up to 18 months [14] and MBP-hαSYN mice showing myelin damage at young ages already [15]. Other methodological issues (i.e. olfactory preference testing versus buried pellet test) might also have contributed to the observed differences. To verify that the absence of obvious smell deficits in PLP-hαSYN mice is a reliable observation, we performed neuropathological work-up demonstrating lack of genotype-specific OB atrophy as well as lack of accelerated neuronal loss in the transgenic OB, despite a trend towards a lower TH-immunoreactive cell number within the OB of 9 months old tg animals. The PLP-hαSYN transgenic model reproduces MSA-like selective vulnerability of the different subpopulations of dopaminergic neurons in SNc and OB. Putting these considerations into a clinical perspective, it has to be emphasized that previous clinical studies in human MSA found varying degrees of olfactory deficits with impairment being less pronounced in MSA compared to PD [3,7,31–33]. In addition, αSYN inclusion pathology has been demonstrated in the human OB [12], however, a recent clinicopathological case series could not identify any smell deficit in MSA [5].

To exclude oversights due to age-related neurodegeneration, we subsequently conducted a long-term confirmatory study focusing on neuropathological read-outs. Age-dependent accumulation of hαSYN immunoreactive inclusions with a significant increase at 9 and 18 months compared 2 months was observed in transgenic OB. This finding is partly in agreement with the human disorder showing GCIs in the OB [12], however, the temporal evolution of GCI pathology in the OB has not been studied in MSA patients so far. Surprisingly, the inclusion pathology did not convert into neurodegeneration. Neither olfactory bulb volume nor TH-ir cell numbers within the glomerular layer of the OB were significantly different between wt and tg animals. In the glomerular layer of the OB, microglial activation was more pronounced in tg compared to wt animals at 2 months of age. However, there was no age-related effect on CD11b immunoreactivity in tg animals, whereas microglial activation continuously increased in wt animals. In the granular layer, age-related enhancement of CD11b immuno-reactivity was observed in both, tg and wt animals. These findings are in line with a previous study reporting early and sustained microglial activation affecting the striatum and the SNc of PLP-hαSYN mice [34]. Finally, it has to be acknowledged that additional olfactory tests may be helpful to study independent functional domains associated with olfaction in 9 months old mice and further longitudinal studies are required to exclude late-onset

olfactory deficits paralleling the progressive OB α-synucleinopathy that has been observed in the present study. However, preserved olfaction at 9 months of age clearly separates the PLP-hαSYN MSA mouse model from corresponding PD mouse models [18,19]. Semi-quantitative analysis of microglial activation by OD measurements of CD11b immunostainings may be affected by various factors; therefore, we applied counter-measures including (1) the calculation of relative OD values to account for different labeling intensities, and (2) the acquisition of all images during a single microscopy session at uniform microscopy settings to vigorously control confounding factors.

To the best of our knowledge this is the first analysis of olfactory behavior as well as candidate neuropathology in the context of a transgenic MSA mouse model. Our experimental data suggest preserved olfactory function providing further support to a recent notion claiming that olfactory deficits are unlikely in MSA [5] reflecting the unique oligodendrogliopathy.

Acknowledgments

We are grateful to Monika Hainzer and Martina Flatscher for their excellent technical assistance in genotyping and immunohistochemistry.

Author Contributions

Conceived and designed the experiments: FK GKW YL WP NS. Performed the experiments: FK YL NS. Analyzed the data: FK NS. Wrote the paper: FK NS. Reviewed/critiqued statistical analysis: GKW YL WP. Reviewed/critiqued the manuscript: GKW YL WP. Approved the final manuscript version: FK GKW YL WP NS.

References

1. Wenning GK, Colosimo C, Geser F, Poewe W (2004) Multiple system atrophy. Lancet Neurol 3: 93–103.
2. Hughes AJ, Daniel SE, Ben-Shlomo Y, Lees AJ (2002) The accuracy of diagnosis of parkinsonian syndromes in a specialist movement disorder service. Brain 125: 861–870.
3. Wenning GK, Shephard B, Hawkes C, Petruckevitch A, Lees A, et al. (1995) Olfactory function in atypical parkinsonian syndromes. Acta Neurol Scand 91: 247–250.
4. Suzuki M, Hashimoto M, Yoshioka M, Murakami M, Kawasaki K, et al. (2011) The odor stick identification test for Japanese differentiates Parkinson's disease from multiple system atrophy and progressive supra nuclear palsy. BMC Neurol 11: 157.
5. Glass PG, Lees AJ, Mathias C, Mason L, Best C, et al. (2012) Olfaction in pathologically proven patients with multiple system atrophy. Mov Disord 27: 327–328.
6. Kikuchi A, Baba T, Hasegawa T, Sugeno N, Konno M, et al. (2011) Differentiating Parkinson's disease from multiple system atrophy by [123I] meta-iodobenzylguanidine myocardial scintigraphy and olfactory test. Parkinsonism Relat Disord 17: 698–700.
7. Garland EM, Raj SR, Peltier AC, Robertson D, Biaggioni I (2011) A cross-sectional study contrasting olfactory function in autonomic disorders. Neurology 76: 456–460.
8. Doty RL (2012) Olfactory dysfunction in Parkinson disease. Nat Rev Neurol 8: 329–339.
9. Ross GW, Petrovitch H, Abbott RD, Tanner CM, Popper J, et al. (2008) Association of olfactory dysfunction with risk for future Parkinson's disease. Ann Neurol 63: 167–173.
10. Doty RL, Stern MB, Pfeiffer C, Gollomp SM, Hurtig HI (1992) Bilateral olfactory dysfunction in early stage treated and untreated idiopathic Parkinson's disease. J Neurol Neurosurg Psychiatry 55: 138–142.
11. Doty RL, Deems DA, Stellar S (1988) Olfactory dysfunction in parkinsonism: a general deficit unrelated to neurologic signs, disease stage, or disease duration. Neurology 38: 1237–1244.
12. Kovacs T, Papp MI, Cairns NJ, Khan MN, Lantos PL (2003) Olfactory bulb in multiple system atrophy. Mov Disord 18: 938–942.
13. Pearce RK, Hawkes CH, Daniel SE (1995) The anterior olfactory nucleus in Parkinson's disease. Mov Disord 10: 283–287.
14. Kahle PJ, Neumann M, Ozmen L, Muller V, Jacobsen H, et al. (2002) Hyperphosphorylation and insolubility of alpha-synuclein in transgenic mouse oligodendrocytes. EMBO Rep 3: 583–588.
15. Shults CW, Rockenstein E, Crews L, Adame A, Mante M, et al. (2005) Neurological and neurodegenerative alterations in a transgenic mouse model

expressing human alpha-synuclein under oligodendrocyte promoter: implications for multiple system atrophy. J Neurosci 25: 10689–10699.
16. Yazawa I, Giasson BI, Sasaki R, Zhang B, Joyce S, et al. (2005) Mouse model of multiple system atrophy alpha-synuclein expression in oligodendrocytes causes glial and neuronal degeneration. Neuron 45: 847–859.
17. Rockenstein E, Mallory M, Hashimoto M, Song D, Shults CW, et al. (2002) Differential neuropathological alterations in transgenic mice expressing alpha-synuclein from the platelet-derived growth factor and Thy-1 promoters. J Neurosci Res 68: 568–578.
18. Kim YH, Lussier S, Rane A, Choi SW, Andersen JK (2011) Inducible dopaminergic glutathione depletion in an alpha-synuclein transgenic mouse model results in age-related olfactory dysfunction. Neuroscience 172: 379–386.
19. Fleming SM, Tetreault NA, Mulligan CK, Hutson CB, Masliah E, et al. (2008) Olfactory deficits in mice overexpressing human wildtype alpha-synuclein. Eur J Neurosci 28: 247–256.
20. Taylor TN, Caudle WM, Shepherd KR, Noorian A, Jackson CR, et al. (2009) Nonmotor symptoms of Parkinson's disease revealed in an animal model with reduced monoamine storage capacity. J Neurosci 29: 8103–8113.
21. Ubeda-Banon I, Saiz-Sanchez D, de la Rosa-Prieto C, Mohedano-Moriano A, Fradejas N, et al. (2010) Staging of alpha-synuclein in the olfactory bulb in a model of Parkinson's disease: cell types involved. Mov Disord 25: 1701–1707.
22. Nuber S, Petrasch-Parwez E, Arias-Carrion O, Koch L, Kohl Z, et al. (2011) Olfactory neuron-specific expression of A30P alpha-synuclein exacerbates dopamine deficiency and hyperactivity in a novel conditional model of early Parkinson's disease stages. Neurobiol Dis 44: 192–204.
23. Tofaris GK, Garcia Reitbock P, Humby T, Lambourne SL, O'Connell M, et al. (2006) Pathological changes in dopaminergic nerve cells of the substantia nigra and olfactory bulb in mice transgenic for truncated human alpha-synuclein(1–120): implications for Lewy body disorders. J Neurosci 26: 3942–3950.
24. Ubhi K, Rockenstein E, Mante M, Inglis C, Adame A, et al. (2010) Neurodegeneration in a transgenic mouse model of multiple system atrophy is associated with altered expression of oligodendroglial-derived neurotrophic factors. J Neurosci 30: 6236–6246.
25. Witt RM, Galligan MM, Despinoy JR, Segal R (2009) Olfactory behavioral testing in the adult mouse. J Vis Exp 23: 949.
26. Kobayakawa K, Kobayakawa R, Matsumoto H, Oka Y, Imai T, et al. (2007) Innate versus learned odour processing in the mouse olfactory bulb. Nature 450: 503–508.
27. Stefanova N, Kaufmann WA, Humpel C, Poewe W, Wenning GK (2012) Systemic proteasome inhibition triggers neurodegeneration in a transgenic mouse model expressing human alpha-synuclein under oligodendrocyte

promoter: implications for multiple system atrophy. Acta Neuropathol 124: 51–65.

28. Stefanova N, Georgievska B, Eriksson H, Poewe W, Wenning GK (2012) Myeloperoxidase inhibition ameliorates multiple system atrophy-like degeneration in a transgenic mouse model. Neurotox Res 21: 393–404.

29. Stefanova N, Hainzer M, Stemberger S, Couillard-Despres S, Aigner L, et al. (2009) Striatal transplantation for multiple system atrophy–are grafts affected by alpha-synucleinopathy? Exp Neurol 219: 368–371.

30. Stemberger S, Poewe W, Wenning GK, Stefanova N (2010) Targeted overexpression of human alpha-synuclein in oligodendroglia induces lesions linked to MSA-like progressive autonomic failure. Exp Neurol 224: 459–464.

31. Nee LE, Scott J, Polinsky RJ (1993) Olfactory dysfunction in the Shy-Drager syndrome. Clin Auton Res 3: 281–282.

32. Muller A, Mungersdorf M, Reichmann H, Strehle G, Hummel T (2002) Olfactory function in Parkinsonian syndromes. J Clin Neurosci 9: 521–524.

33. Goldstein DS, Sewell L (2009) Olfactory dysfunction in pure autonomic failure: Implications for the pathogenesis of Lewy body diseases. Parkinsonism Relat Disord 15: 516–520.

34. Stefanova N, Reindl M, Neumann M, Kahle PJ, Poewe W, et al. (2007) Microglial activation mediates neurodegeneration related to oligodendroglial alpha-synucleinopathy: implications for multiple system atrophy. Mov Disord 22: 2196–2203.

7

Age at Death of Creutzfeldt-Jakob Disease in Subsequent Family Generation Carrying the E200K Mutation of the Prion Protein Gene

Maurizio Pocchiari*, Anna Poleggi, Maria Puopolo, Marco D'Alessandro, Dorina Tiple, Anna Ladogana

Department of Cell Biology and Neurosciences, Istituto Superiore di Sanità, Rome, Italy

Abstract

Background: The E200K mutation of the prion protein gene (*PRNP*) is the most frequent amino acid substitution in genetic Creutzfeldt-Jakob disease and is the only one responsible for the appearance of clustered cases in the world. In the Israel and Slovakian clusters, age of disease onset was reduced in successive generations but the absence of a clear molecular basis raised the possibility that this event was an observational bias. The aim of the present study was to investigate possible selection biases or confounding factors related to anticipation in E200K CJD patients belonging to a cluster in Southern Italy.

Methods: Clinical and demographical data of 41 parent-offspring pairs from 19 pedigrees of the Italian cluster of E200K patients were collected. Age at death of parents was compared with age at death of E200K CJD offspring. Subgroup analyses were performed for controlling possible selection biases, confounding factors, or both.

Results: The mean age at death/last follow-up of the parent generation was 71.4 years while that of CJD offspring was 59.3 years with an estimated anticipation of 12.1 years. When the same analysis was performed including only parents with CJD or carrying the E200K mutation (n = 26), the difference between offspring and parents increased to 14.8 years.

Conclusions: These results show that early age at death occurs in offspring of families carrying the E200K *PRNP* mutation and that this event is not linked to observational biases. Although molecular or environmental bases for this occurrence remain unsettled, this information is important for improving the accuracy of information to give to mutated carriers.

Editor: Corinne Ida Lasmezas, The Scripps Research Institute Scripps Florida, United States of America

Funding: This study was supported by the Istituto Superiore di Sanità (ISS)-NIH research program "Rare Diseases 2006" and the Italian Ministry of Health (CJD Registry). The funders had no role in study design, data collection and analysis, decision to publish, or preparation of the manuscript.

Competing Interests: The authors have declared that no competing interests exist.

* E-mail: maurizio.pocchiari@iss.it

Introduction

The glutamine to lysine change at codon 200 (E200K) is the most common mutation of the prion protein (PrP) gene (*PRNP*) accounting for more than 70% of genetic prion diseases worldwide [1–3]. This mutation is responsible for all known clustered cases in the world [4–7], including that described in the Italo-Greek villages of the Southeast area of the Calabria region in Italy [6]. In this cluster, mutated carriers either develop Creutzfeldt-Jakob disease (CJD) with great age variability (from the third to the eighth decade) or do not develop disease at all resulting in a cumulative penetrance (67%) [8] similar to the Slovakian cluster (60%) [9] but lower than that reported in Libyan Jews in Israel (96%) [10]. In the past 20 years of CJD surveillance in the Calabrian cluster we observed a consistent early age of CJD onset in offspring compared to those of their relatives suggesting anticipation of disease onset similarly to those reported in E200K CJD Libyan Jews leaving in Israel [11] and in the Slovakian cluster [12]. The aim of this study was to formally analyze data on age at death of E200K CJD patients in the Calabrian cluster to account for possible selection biases or confounding factors.

Methods

Protocol Approvals, Registrations, and Patient Consents

Clinical data on patients, including family history and genetic analyses, were obtained following the criteria of the CJD surveillance program (approved by the Ethical Committee of the Istituto Superiore di Sanità) and data are stored in a database, which is registered at the Italian data protection Agency. Written informed consents for clinical data collection, blood sampling, and genetic analysis were obtained from patients (or their next-on-kin) and healthy subject involved in the study.

Study Population and Data Collection

We studied 41 parent-offspring pairs from 19 apparently unrelated pedigrees of the E200K CJD Calabrian cluster. Parent generation consisted of 34 subjects because there were two (n = 5) or three (n = 1) CJD-affected siblings in the same family. Diagnosis and classification of genetic CJD patients were done by internationally established criteria [13].

Information on medical history and age at death (or age at the last follow-up for unaffected parents) were obtained as described in table 1. In 15 parent-offspring pairs we were unable to infer the CJD-transmitting parent. We then took the conservative approach of using the age at death (or last follow-up) of the young parent.

PRNP Analysis

We performed the direct complete sequencing of the *PRNP* gene in 22 patients and 2 healthy carriers included in this study. Genomic DNA was extracted from whole blood using the QIAamp DNA KIT (QIAGEN), according to the manufacture's recommendations. The *PRNP* open reading frame (ORF) was amplified by the PCR as previously described [8]. Referral clinicians provided genetic data on the other 9 CJD patients included in the study.

Statistical Analysis

We assessed anticipation as differences between the age at death (or age at last follow-up) of parents and those of their offspring by one-tailed paired t-test to account for within pair data dependency. The paired differences for each pair were used to estimate anticipation. In the calculation of overall statistics, parents' data were repeated for each offspring included in the study. We used age at death rather than age at onset because the latter is more difficult to ascertain in parental generation where data are mainly collected retrospectively. Moreover, the short disease duration of E200K patients (median 5 months) makes age at onset similar to age at death.

The influence of possible selection biases in data collection, confounding factors, or both was investigated by re-analyzing anticipation data by subgroups of pairs defined by different variables: gender, to control for sex-specific differences in mortality rates for genetic CJD in Italy [8]; polymorphism at codon 129 of the *PRNP* gene, to control for differences in the presentation of disease [1] and in disease duration [14]; birth cohort, to control for differences in the length of available follow-up or possible different exposure to causative or environmental agents; death cohort, to control for improvement in surveillance performance, diagnostic accuracy, or alertness to CJD; age at death, to investigate differences in anticipation between old cases - whose parents likely had enough time to develop CJD - and young cases - whose parents might develop the disease after their descendants; and parent's age at death, to investigate the statistical phenomenon of regression to the mean [15], which states that anticipation is evident only in offspring of old parents. Subgroups for continuous

covariates were obtained by splitting samples according to the observed median values (year 1937 for birth cohort; year 2000 for death cohort, age 61 years for age at death; and age 70 years for parents' age at death or age at last evaluation).

Data are summarized by mean, standard deviation (SD) of the mean, and ranges. Observed p-values of paired t-test are reported. Comparisons were carried out at a significance level of 0.05 with no adjustment for multiple comparisons because of the explorative approach of the subgroups analyses. Data are graphed by box-plots.

Results

Demographic and clinical characteristics of E200K CJD patients included in the study are reported in table 2. Prospective and retrospective CJD patients had similar sex distribution, age at death, and disease duration suggesting that data taken retrospectively were accurate. As expected because codon 200 mutation in the Calabrian cluster co-segregates with methionine at codon 129, more than 75% of patients were methionine homozygous.

Figure 1 shows that in 36 of 41 pairs (88.0%) offspring died of CJD earlier than their parent. Overall, mean age at death of CJD offspring was 59.3 years (SD = 10.6; Range: 39–87) while that of the parent generation was 71.4 years (SD = 11.8; Range: 48–92) with an anticipation of 12.1 years (SD = 13.1; paired t-test, p<0.0001) (Figure 2a). This value, however, underestimates the lag of anticipation because when both parents did not show signs of CJD (n = 15) we selected that with the younger age at death or at last evaluation. When the same analysis was performed with the more relevant group, i.e., including only parents with CJD, carrying the E200K mutation, or obligate carriers for the E200K mutation (n = 26), the difference between offspring and parents increased to 14.8±9.93 (p<0.0001) (Figure 2b).

In this restricted group we also analyzed the role of possible selection biases, confounding in data collection, or both (Table 3). Anticipation was statistically significant in all subgroups, but it was more marked in the group of pairs where offspring was born after 1938, where parents died at ≥70 years, and where offspring died at <61 years. The finding that anticipation was statistically significant in the group of pairs where the age at death (or at last evaluation) was relative early (<70 years) excludes the possibility that the observed anticipation is related to the so-called phenomenon of regression to the mean [15].

Finally, the influence of father's age at conception, as an indirect indicator of *de novo* genetic mutations, and mother's age at conception, as internal control, did not show any significant correlation with age at onset of disease (Figure 3).

Table 1. Source of diagnostic information in offspring and parent subjects.

Diagnosis/Source	Offspring	Parent
Probable and Definite CJD/CJD Registry	35	7
Probable and Definite CJD/Medical records	4	2
Possible CJD/Family recollection or Municipal Registry Offices	2	3
E200K carriers with no CJD/CJD Registry	–	2
Obligate E200K carriers with no CJD/CJD Registry, Family recollection, or Municipal Registry Offices	–	7
Unknown parent carriers/Family recollection or Municipal Registry Offices	–	13
TOTAL[a]	41	34

[a]The discrepancy between the total number of offspring and parent is because two (n = 5) or three (n = 1) offspring had the same parent.

Table 2. Demographic and clinical characteristics of E200K CJD patients included in the study.

CJD cases	Sex			*PRNP* polymorphism at codon 129		Age at death			Disease duration (months)		
	M	F	p[a]	MM	MV	Mean, SD	range	p[b]	Median	range	p[b]
Prospective	14	22	1	23	7	59.6, 11.1	39–87	0.58	4.5	2–24	0.97
Retrospective	4	5		–	–	61.8, 7.30	49–73		6 (n = 6)	2–16	

[a]Fisher's exact test.
[b]t-test.

Discussion

In affected E200K CJD families belonging to the Calabrian cluster, offspring develop disease about 12 years earlier than their parents similarly to what was previously reported in other E200K CJD clusters in Israel [11] and Slovakia [12]. All together, these data reveal that anticipation of disease onset occurs in families affected with the E200K CJD. The term anticipation was set aside by Lionel Penrose, who claimed that anticipation was a statistical artifact arising as result of ascertainment biases [16], particularly in diseases, such as CJD, where the molecular basis of the phenomenon is still unexplained [17,18]. Several ascertainment biases might contribute to the observation of anticipation, including the selection of parents with late onset of disease, the selection of offspring with early onset of disease, the selection of cases with simultaneous onset in parents and offspring, or any bias that would cause a truncation in reporting cases within a family

[19–20]. An optimal study design should consider only prospective cases, but this is impractical in CJD and other rare diseases with onset in adult age [21]. Retrospective studies, on the other hand, suffer from possible ascertainment, confounding, or both biases, but these can be minimized and adjusted for by statistical analyses. In our study, ascertainment biases in selection of cases [16] are highly unlikely because we systematically included all E200K cases belonging to the Calabrian cluster and referred to the Italian CJD Registry without any selection on positive or negative family history for TSE, age, or status as offspring or parent. Selection biases or confounding factors in data collection were investigated in an explorative approach by subgroups analyses in the restricted group where parents developed CJD or carried the E200K mutation. These cases were stratified according to gender, polymorphism at codon 129 of *PRNP*, birth cohort, death cohort, age at death, and parent's age at death. Although with some minor differences, anticipation of disease remained statistically significant

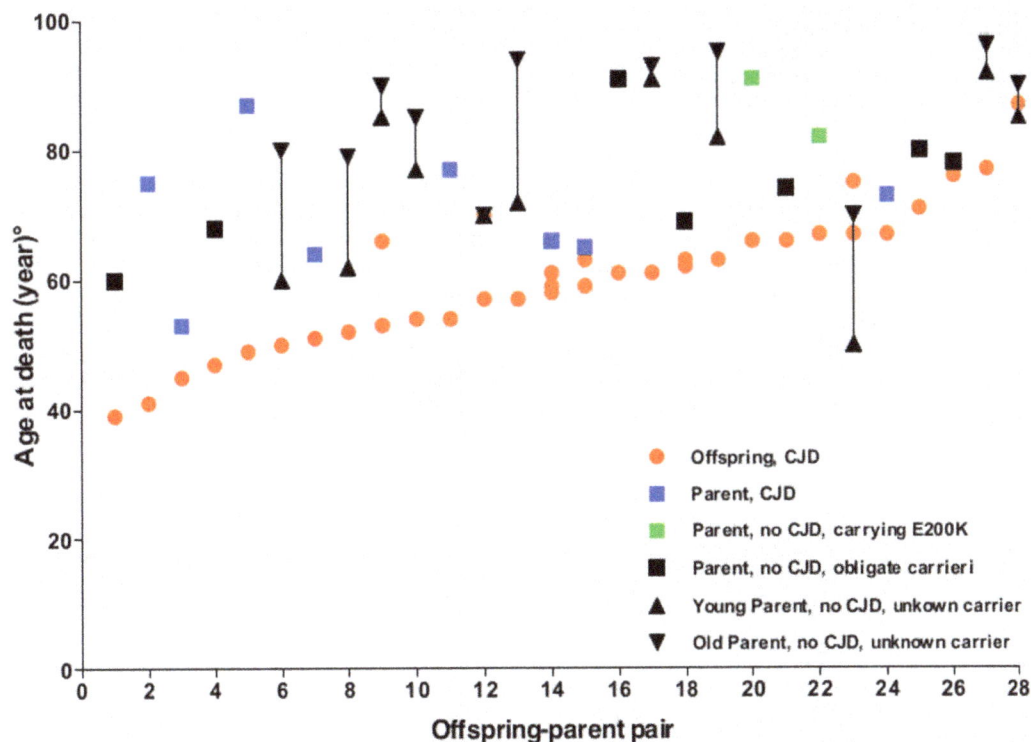

Figure 1. Parent-offspring pairs in the study population. Black circles, offspring with CJD; black squares, parents with CJD; black and white squares, unaffected parents carrying the E200K *PRNP* mutation; white squares, unaffected parents who are obligate carriers for the E200K *PRNP* mutation; white triangles, the oldest (up) and youngest (down) unaffected parents where it was unfeasible to discriminate the carrier. Vertical bars link the age between the oldest and youngest unaffected parents where it was unfeasible to discriminate the carrier.

A

B

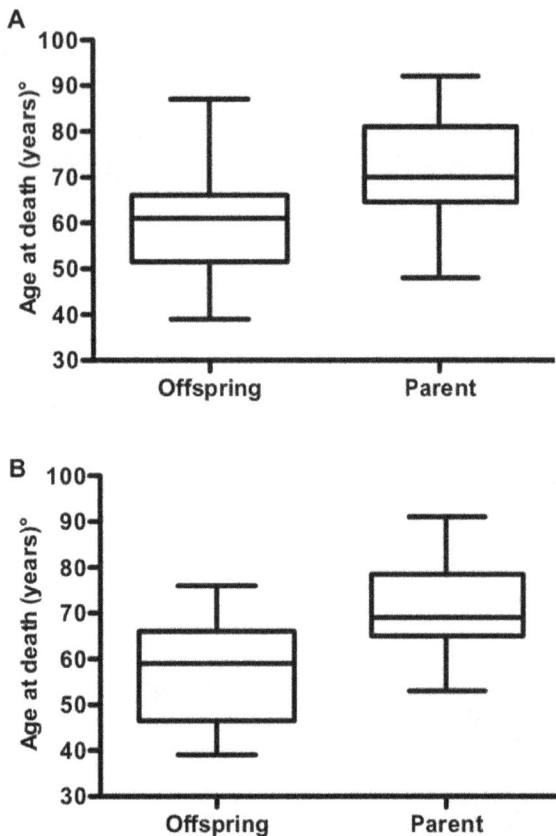

Figure 2. Age at death of offspring and parent generation. Box plots representing the age at death (or at the last clinical evaluation) for all offspring/parent pairs (a) or for pairs where parents had CJD, carried the E200K mutation, or were obligate carrier (b). The boxes extends from the 25th to the 75th percentile, bars extends to upper and lower adjacent values, and the lines in the middle of boxes represent the median value.

Table 3. Analyses of possible selection biases or confounding factors by offspring covariates.

	Number of pairs	Anticipation (years) Mean, SD	p-value[a]
Overall	26	14.7, 9.90	<0.0001
Gender			
Male	11	14.1, 10.6	0.0006
Female	15	15.3, 9.27	<0.0001
Codon 129 polymorphism[b]			
Met/Met	17	13.9, 10.4	<0.0001
Birth cohort			
<1939	10	9.70, 7.86	0.0018
≥1939	16	18.0, 9.89	<0.0001
Death cohort			
<2000	11	15.6, 12.6	0.0011
≥2000	15	15.4, 8.95	<0.0001
Age at death			
<61 years	14	17.7, 9.80	<0.0001
≥61 years	12	11.2, 9.20	0.0007
Parent's age			
<70 years	14	11.4, 6.96	<0.0001
≥70 years	12	18.6, 11.6	<0.0001

[a]one tailed paired t-test.
[b]Four patients were Met/Val at codon 129.

and relatively stables (between 9.7 and 18.6 years) in all subgroups, suggesting that the most common confounding factors do not influence the observed results and that anticipation is likely a biological phenomenon to be considered in E200K carriers.

What remain unknown are the pathogenic mechanisms of anticipation, which are unlikely related to the coding region of the *PRNP* gene but rather to the upstream regulatory regions of *PRNP*, other host genes, or environmental factors. In the *PRNP* coding region, the polymorphic codon 129 of the non-mutated allele does not influence age at onset of disease in E200K CJD patients [1] and other rare polymorphisms [22] would unlikely influence age at onset and be responsible for the anticipation phenomenon. On the other hand, polymorphisms in the upstream regulatory regions of *PRNP* may influence susceptibility and age at onset in genetic CJD as observed in sporadic CJD [23] and might therefore be a confounding factor in our study. Although it is unlikely that in all offspring/parent pairs the deleterious polymorphism is systematically present in the offspring generation, it would be of interest to search for such polymorphisms in E200K CJD patients and healthy carriers belonging to the Calabrian and other CJD clusters worldwide. Other genes may be involved in the anticipation phenomenon, but the few genome-wide association studies (GWAs) in human prion diseases have only identified risk loci with modest effects in determining susceptibility to one form of prion disease, i.e. variant or sporadic CJD, but not in others.

Overall, these studies have been so far disappointing, but further genetic association studies in homogenous population, such as that of the Italo-Greek minority of the Calabrian cluster with high inbreeding coefficient [24], would need to be performed for investigating whether there are genes, other than *PRNP*, that may account for differences in age at onset in E200K CJD patients. The lack of correlation between the father's age at conception and age at onset suggests that *de novo* genetic mutations do not play an important role in modulating the development of E200K CJD as reported for autism or other disorders [25].

An increased exposure to possible risk factors in the offspring generation may be also responsible for early age at onset of disease. In the last decades, the improvement of medical procedures in the cluster area may have exposed offspring carriers to possible iatrogenic factors. However, we could not find any known iatrogenic procedures, i.e., treatment with human-derived pituitary hormones, *dura mater* grafts, neurosurgery, or corneal transplants [26] in the medical history of E200K CJD patients and other potential medical procedures, such as general surgery or blood transfusion, recently associated to sporadic CJD, were not confirmed in genetic CJD patients [27]. It is also unlikely that the anticipation observed in the offspring generation results from an early diagnosis as we used age at death rather than age at onset for our analysis and because of the short median disease duration (less than 1 year).

Finally, changes of environmental factors in soil, drinking water, or food that have occurred in the last decades might have contributed to an increased susceptibility, thus an early age at onset, in the offspring generation. The increased concentration of some metal microcrystal pollutants, particularly barium stronzium and silver, in the soil of the Calabrain cluster area with respect to

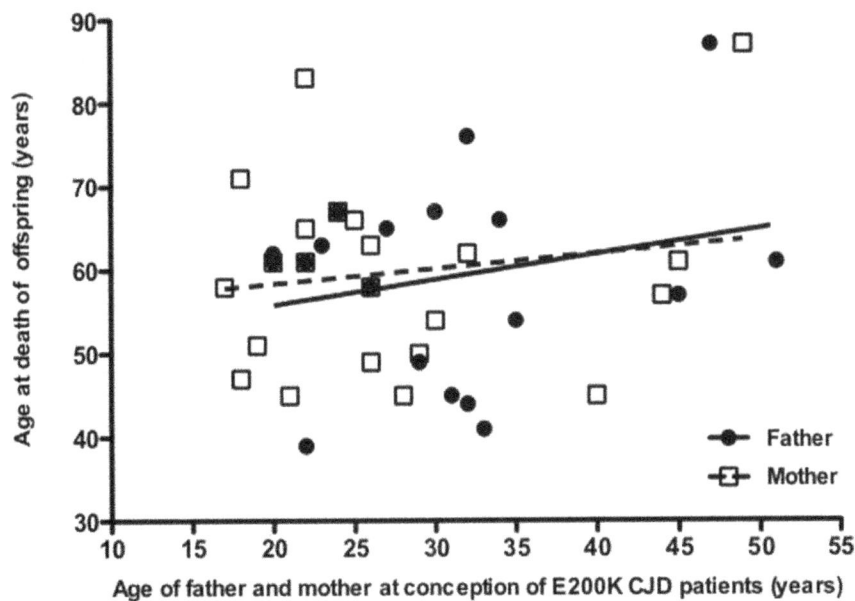

Figure 3. Father's and mother's age at conception versus age at death of offspring. The solid (father) and dashed (mother) lines denote the linear fit of data.

adjacent neighborhood was proposed as a responsible factor for the high incidence of familial CJD cases [28]. However, soil samples were only collected at the end of 2004 and no data on the concentration of these metals in the decades before 2004 are available to support this highly speculative hypothesis. Moreover, the finding that the anticipation phenomenon was also observed in pairs where parents lived in the cluster area and offspring emigrated in Northern Italy (14.1 years, n = 9) and that an anticipation (12.8 years) similar to that observed in Israel [11] is also present in Libyan Jews living in Italy (data not shown), weaken this possibility. Increased oxidative insults in offspring of E200K carriers with respect to their parents might also be responsible for the observed anticipation phenomenon. The substitution of glutamic acid to lysine makes the mutant PrPc more susceptible to oxidation and, once oxidized, likely more inclined to change conformation into PrP^{TSE} [29]. A recent study done in a new transgenic (Tg) mouse line expressing the E200K mutation and spontaneously developing prion disease [30] has shown that the exposure of these mice to copper, a redox active metal, in drinking water induces disease earlier than untreated mice [31]. These data suggest that the E200K mutant PrPc is likely less competent to protect cells from copper induced oxidation, which may finally lead to an acceleration of the misfolding process, hence to an early age at onset of disease. Assuming that humans would behave like mice, it remains to determine which life events have increased

oxidative insults in offspring that did not in their parents. Moreover, a recent study in the E200K CJD cluster of Slovakia reported an unbalance of manganese/copper concentration in brain tissues of genetic CJD cases in comparison to controls suggesting that metal disequilibrium might act as exogenous cofactor for the development of disease [32], but great caution is needed in the interpretation of these data.

The anticipation phenomenon is worth of further investigation because, if confirmed, will improve the accuracy of information to pass to mutated carriers and would possibly assist physicians in taking the difficult decision to start a preventive therapy, when would be available, at the right time.

Acknowledgments

We are very grateful to the families who generously contributed their time and materials to this research study and to the neurologists throughout Italy for their collaboration. We also thank Mrs. Cinzia Gasparrini, Mrs. Viviana Renzi, and Dr. Alessandra Garozzo for editorial and administrative assistance.

Author Contributions

Conceived and designed the experiments: M. Pocchiari AL. Performed the experiments: AP DT MD. Analyzed the data: M. Puopolo. Wrote the paper: M. Pocchiari AL.

References

1. Kovacs GG, Puopolo M, Ladogana A, Pocchiari M, Budka H, et al. (2005) Genetic prion disease: the EUROCJD experience. Human Genet 118: 166–174.

2. Mead S (2006) Prion disease genetics. Eur J Human Genet 14: 273–281.

3. Kong Q, Kong Q, Surewicz WK, Petersen RB, Zou W, et al. (2004) Inherited prion diseases. In Prusiner SB, editor. Prion Biology and Diseases. Cold Spring Harbor: Cold Spring Harbor Laboratory Press. 673–775.

4. Goldfarb LG, Mitrova E, Brown P, Toh BH, Gadjusek DC (1990) Mutation in codon 200 of scrapie amyloid protein gene in two clusters of Creutzfeldt-Jakob disease in Slovakia. Lancet 336: 514–515.

5. Goldfarb LG, Korczyn AD, Brown P, Chapman J, Gadjusek DC (1990) Mutation in codon 200 of scrapie amyloid precursor gene linked to Creutzfeldt-Jakob disease in Sephardic Jews of Libyan and non-Libyan origin. Lancet 336: 637–638.

6. D'Alessandro M, Petraroli R, Ladogana A, Pocchiari M (1998) High incidence of Creutzfeldt-Jakob disease in rural Calabria, Italy. Lancet 352: 1989–1990.

7. Miyakawa T, Inoue K, Iseki E, Kawanishi C, Sugiyama N, et al. (1998) Japanese Creutzfeldt-Jakob disease patients exhibiting high incidence of the E200K PRNP mutation and located in the basin of a river. Neurol Res 20: 684–688.

8. Ladogana A, Puopolo M, Poleggi A, Almonti S, Mellina V, et al. (2005) High incidence of genetic human transmissible spongiform encephalopathies in Italy. Neurology 64: 1592–1597.

9. Mitrova E, Belay G (2002) Creutzfeldt-Jakob disease with E200K mutation in Slovakia: characterization and development. Acta Virol 46: 31–39.

10. Spudich S, Mastrianni JA, Wrensch M, Gabizon R, Meiner Z, et al. (1995) Complete penetrance of Creutzfeldt-Jakob disease in Libyan Jews carrying the E200K mutation in the prion protein gene. Mol Med 1: 607–613.

11. Rosenmann H, Kahana E, Korczyn AD, Kahana I, Chapman J, et al. (1999) Preliminary evidence for anticipation in genetic E200K Creutzfeldt-Jakob disease. Neurology 53: 1328–1329.

12. Stelzer M, Wsólová L, Skripcáková I, Mitrová E (2012) Age at death and duration of gCJDE200K in relation to the affected generations and polymorphism M129V [abstract]. Prion 6(suppl): P107.

13. World Health Organization. Creutzfeldt-Jakob disease (CJD) and variant CJD (vCJD), excerpt from "WHO recommended standards and strategies for surveillance, prevention and control of communicable diseases". Available at: http://www.who.int/entity/zoonoses/diseases/Creutzfeldt.pdf [Accessed December 23, 2012].

14. Pocchiari M, Puopolo M, Croes EA, Budka H, Gelpi E, et al. (2004) Predictors of survival in sporadic Creutzfeldt-Jakob disease and other human transmissible spongiform encephalopathies. Brain 127: 2348–2359.

15. Bradley M, Bradley L, de Belleroche J, Orrell RW (2005) Patterns of inheritance in familial ALS. Neurology 64: 1628–1631.

16. Penrose LS (1948) The problem of anticipation in pedigrees of dystrophia myotonica. Ann Eugen 14: 125–132.

17. McInnis MG (1996) Anticipation: an old idea in new genes. Am J Hum Genet 59: 973–979.

18. Fraser FC (1997) Trinucleotide repeats not the only cause of anticipation. Lancet 350: 459–460.

19. Picco MF, Goodman S, Reed J, Bayless TM (2001) Methodological pitfalls in the determination of genetic anticipation: the case of Crohn disease. Ann Intern Med 134: 1124–1129.

20. Boonstra PS, Gruber SB, Raymond VM, Huang SC, Timshel S, et al. (2010) A review of statistical methods for testing genetic anticipation: looking for an answer in Lynch syndrome. Genet Epidemiol 34: 756–768.

21. Horwitz M, Goode EL, Jarvik GP (1996) Anticipation in familial leukemia. Am J Hum Genet 59: 990–998.

22. Bishop MT, Pennington C, Heath CA, Will RG, Knight RS (2009) PRNP variation in UK sporadic and variant Creutzfeldt Jakob disease highlights genetic risk factors and a novel non-synonymous polymorphism. BMC Med Genet 10: 146.

23. Sanchez-Juan P, Bishop MT, Croes EA, Knight RS, Will RG, et al. (2011) A polymorphism in the regulatory region of PRNP is associated with increased risk of sporadic Creutzfeldt-Jakob disease. BMC Med Genet 12: 73.

24. Biondi G, Perrotti E, Mascie-Taylor GC, Lasker GW (1990) Inbreeding coefficients from isonymy in the Italian-Greek villages. Ann Hum Biol 17: 543–546.

25. Kong A, Frigge ML, Masson G, Besenbacher S, Sulem P, et al. (2012). Rate of de novo mutations and the importance of father's age to disease risk. Nature 488: 471–475.

26. Brown P, Brandel JP, Sato T, Nakamura Y, MacKenzie J, et al. (2012) Iatrogenic Creutzfeldt-Jakob disease, final assessment. Emerg Infect Dis 18: 901–907.

27. Puopolo M, Ladogana A, Vetrugno V, Pocchiari M (2011) Transmission of sporadic Creutzfeldt-Jakob disease by blood transfusion: risk factor or possible biases. Transfusion 51: 1556–1566.

28. Purdey M (2005) Metal microcrystal pollutants; the heat resistant, transmissible nucleating agents that initiate the pathogenesis of TSEs? Med Hypotheses 65: 448–477.

29. Canello T, Frid K, Gabizon R, Lisa S, Friedler A, et al. (2010) Oxidation of helix-3 methionines precedes the formation of PK resistant PrPSc. PLoS Pathog 6: e1000977.

30. Friedman-Levi Y, Meiner Z, Canello T, Frid K, Kovacs GG, et al. (2011) Fatal prion disease in a mouse model of genetic E200K Creutzfeldt-Jakob disease. PLoS Pathog 7: e1002350.

31. Canello T, Friedman-Levi Y, Mizrahi M, Binyamin O, Cohen E, et al. (2012) Copper is toxic to PrP-ablated mice and exacerbates disease in a mouse model of E200K genetic prion disease. Neurobiol Dis 45: 1010–1117.

32. Slivarichová D, Mitrová E, Ursínyová M, Uhnáková I, Koscová S, et al. (2011) Geographic accumulation of Creutzfeldt-Jakob disease in Slovakia–environmental metal imbalance as a possible cofactor. Cent Eur J Public Health 19: 158–164.

8

TNNI3K, a Cardiac-Specific Kinase, Promotes Physiological Cardiac Hypertrophy in Transgenic Mice

Xiaojian Wang[1][9]**, Jizheng Wang**[1][9]**, Ming Su**[1]**, Changxin Wang**[1]**, Jingzhou Chen**[1]**, Hu Wang**[1]**, Lei Song**[2]**, Yubao Zou**[2]**, Lianfeng Zhang**[3]**, Youyi Zhang**[4]**, Rutai Hui**[1]*

1 Sino-German Laboratory for Molecular Medicine, State Key Laboratory of Cardiovascular Disease, FuWai Hospital & Cardiovascular Institute, Chinese Academy of Medical Sciences, Peking Union Medical College, Beijing, People's Republic of China, 2 Department of Cardiology, State Key Laboratory of Cardiovascular Disease, FuWai Hospital & Cardiovascular Institute, Chinese Academy of Medical Sciences, Peking Union Medical College, Beijing, People's Republic of China, 3 Key Laboratory of Human Disease Comparative Medicine, Ministry of Health, Institute of Laboratory Animal Science, Chinese Academy of Medical Sciences and Comparative Medical Center, Peking Union Medical College, Beijing, People's Republic of China, 4 Institute of Vascular Medicine, Peking University Third Hospital, Beijing, People's Republic of China

Abstract

Purpose: Protein kinase plays an essential role in controlling cardiac growth and hypertrophic remodeling. The cardiac troponin I-interacting kinase (TNNI3K), a novel cardiac specific kinase, is associated with cardiomyocyte hypertrophy. However, the precise function of TNNI3K in regulating cardiac remodeling has remained controversial.

Methods and Results: In a rat model of cardiac hypertrophy generated by transverse aortic constriction, myocardial TNNI3K expression was significantly increased by 1.62 folds ($P<0.05$) after constriction for 15 days. To investigate the role of TNNI3K in cardiac hypertrophy, we generated transgenic mouse lines with overexpression of human TNNI3K specifically in the heart. At the age of 3 months, the high-copy-number TNNI3K transgenic mice demonstrated a phenotype of concentric hypertrophy with increased heart weight normalized to body weight (1.31 fold, $P<0.01$). Echocardiography and non-invasive hemodynamic assessments showed enhanced cardiac function. No necrosis or myocyte disarray was observed in the heart of TNNI3K transgenic mice. This concentric hypertrophy maintained up to 12 months of age without cardiac dysfunction. The phospho amino acid analysis revealed that TNNI3K is a protein-tyrosine kinase. The yeast two-hybrid screen and co-immunoprecipitation assay identified cTnI as a target for TNNI3K. Moreover, TNNI3K overexpression induced cTnI phosphorylation at Ser22/Ser23 *in vivo* and *in vitro*, suggesting that TNNI3K is a novel upstream regulator for cTnI phosphorylation.

Conclusion: TNNI3K promotes a concentric hypertrophy with enhancement of cardiac function via regulating the phosphorylation of cTnI. TNNI3K could be a potential therapeutic target for preventing from heart failure.

Editor: Tohru Fukai, University of Illinois at Chicago, United States of America

Funding: This work was supported by Ministry of Science and Technology of China (2007DFC30340 to RTH); National Science and Technology Major Projects (2009ZX09501-026 to RTH); and the National Natural Science Foundation of China (30840041 to XJW, 30700322 to JZW). The funders had no role in study design, data collection and analysis, decision to publish, or preparation of the manuscript.

Competing Interests: The authors have declared that no competing interests exist.

* E-mail: huirutai@sglab.org

[9] These authors contributed equally to this work.

Introduction

In response to increased workload, the heart undergoes hypertrophic enlargement, which is characterized by an increase in the size of individual cardiac myocyte.[1] This hypertrophic response can be traditionally classified as either physiological or pathological. Physiological stimuli such as exercise lead to compensatory growth of the cardiomyocyte, accompanied by normal cardiac structure, preserved or improved cardiac function, and minimal alteration in cardiac gene expression pattern.[2] In contrast, the pathological hypertrophy, which is induced by persistent pressure or volume overload at various disease conditions, is associated with reactivation of fetal gene program, interstitial fibrosis, cardiac dysfunction and eventual heart failure.[3] As heart failure is almost invariably associated with cardiac hypertrophy, the elucidation of signaling cascades involved in these two forms of hypertrophy will be of critical importance for the design of specific therapy against heart failure.[4,5]

Protein kinase plays an essential role in regulating cardiac growth and hypertrophic response. Various kinases transmit hypertrophic signals from membrane bound receptors and change the phosphorylation status of functionally significant proteins.[6] Cardiac myofilament, the ultimate determinant in the control of cardiac contractility, is a central feature of kinase signal transduction. Levels of contractile protein phosphorylation are associated with stretch of the myocardium, the myofilament response to Ca^{2+} and the progression of cardiac remodeling.[7–9] Despite considerable progress has been made in elucidating the roles of various kinases in regulating myofilament during the past decades, understanding the molecular mechanism underlying myofilament phosphorylation and cardiac hypertrophy remains limited. This is due to, at least in part, lack of knowledge for the function of novel protein kinases in the heart. In this regard, it is crucial to identify novel genes potential involved in cardiac hypertrophy.

The cardiac troponin I-interacting kinase (TNNI3K), also known as CARK, is a novel cardiac-specific kinase. It contains a central kinase domain, flanking by an ankyrin repeat domain in the amino terminus and a serine-rich domain in the carboxyl terminus. TNNI3K is a functional kinase and directly interacts with cardiac troponin I (cTnI). [10] It has been suggested as a factor that moderates electrocardiographic parameters and the susceptibility for viral myocarditis. [11,12] Our group has verified that Mef2c, an important determinant for cardiac hypertrophy, play a critical role in regulating basal TNNI3K transcription activity.[13] The precise function of TNNI3K in regulating cardiac remodeling, however, has remained elusive and controversial. Some studies have shown that TNNI3K induces cardiomyocyte hypertrophy *in vitro* [14] and enhances cardiac performance and protects the myocardium from ischemic injury *in vivo*, [15] while others have shown that overexpression of TNNI3K can accelerate disease progression in mouse models of heart failure. [16]

To better understand the role of TNNI3K in cardiac remodeling, we generated transgenic mice overexpressing TNNI3K specifically in the heart. Our data demonstrated that increasing basal TNNI3K expression resulted in enhancement of cardiac function and adaptive hypertrophy. Furthermore, TNNI3K directly interacts with cTnI and induced cTnI phosphorylation at Ser22/Ser23 *in vivo* and *in vitro*. These data suggest that TNNI3K promotes cardiac remodeling via regulating the phosphorylation of cTnI.

Materials and Methods

2.1. Ethics Statement

All animal experiments were approved by the FuWai Administrative Panel for Laboratory Animal Care and were consistent with the Guide for the Care and Use of Laboratory animals published by the United States National Institutes of Health. Rats and Mice were housed in an AAALAC-accredited facility with a 12-hour light-dark cycle and allowed water and food ad libitum. All surgery was performed under anesthesia, and all efforts were made to minimize suffering.

2.2. Transverse aortic constriction (TAC) surgery

Male Sprague-Dawley rats (200–250 g) were anesthetized with 2% isoflurane. Adequacy of anesthesia was assessed by monitoring the respiratory rate as well as the loss of response to toe pinch. The rats were then intubated and ventilated using a rodent ventilator (Model 683, Harvard Apparatus, South Natick, MA, USA). Midline sternotomy was performed, and the transverse aorta was exposed. The aorta was ligated between the innominate and left common carotid arteries by tying a 7–0 silk suture around a tapered 22-gauge needle placed on top of the aorta. Sham-operated controls underwent an identical surgical procedure including isolation of the aorta, only without placement of the suture. At different time points (day 1 to day 15) after surgery, animals (n = 4 to 5 for each time point) were euthanized by overdose anesthesia (pentobarbital sodium 150 mg/kg, i.p.) and cervical dislocation. The hearts were removed; left ventricles were weighed and quickly frozen in liquid nitrogen for total RNA extraction.

2.3. Quantitative real-time PCR analysis

Total RNA was isolated from left ventricular tissue using Trizol and reverse transcribed with Superscript III transcript kit (Invitrogen, Carlsbad, CA, USA). SYBR green-based quantitative real-time PCR was carried out with the DNA Engine Opticon 2 real-time PCR Detector (BIO-RAD, Richmond, CA, USA) as previous described.[13] Melting curve analysis was used to confirm amplification specificity. GAPDH gene was used as internal control. The primers are listed in Table S1. All experiments were repeated at least twice in triplicate.

2.4. Plasmid Constructs and Generation of Transgenic Mice

A human wild-type TNNI3K cDNA (2508 bp, NM_015978) was subcloned into the *SalI/HindIII* site of between the 5.5-kb murine α-myosin heavy chain promoter (α−MHC) and the 0.6-kb human growth hormone (hGH) polyadenylation sequence, carried in the pBluescriptII-SK+ vector (Stratagene). The transgenic mice were generated in the key laboratory of Human Disease Comparative Medicines as previously described[17]. Briefly, an 8.7-kb DNA fragment was isolated, purified from transgenic vector after digestion with *NotI*, and microinjected into fertilized oocytes from C57BL/6J mice. The surviving eggs were surgically transferred into pseudopregnant females. The resulting pups were screened by diagnostic PCR using the primers for *TNNI3K*, 5′-ATG GCA AGA GCA TTG ACC TAG TC-3′ and 5′-GGA TGA TTG AGC TGG CAG AGA-3′. The *Fabpi* gene was amplified as internal control using the primers, 5′-TGG ACA GGA CTG GAC CTC TGC TTT CCT AGA-3′ and 5′-TAG AGC TTT GCC ACA TCA CAG GTC ATT CAG-3′. To determine the transgene copy number, Southern blot analysis was performed on tail genomic DNA digested with *EcoRI* and probed with a ^{32}P-labeled 0.6 kb *HindIII/NotI* hGH fragment. The purified transgene insert DNA was added into the *EcoRI* digested wild-type mouse DNA to yield the equivalent of 1, 5, and 10 copies of the gene per haploid genome (based on 3×10^9 base pairs per haploid genome). The signals were quantified using ImageJ, and the copy number was determined from the standard curve. Three independent founder lines were identified and mated to C57BL/6J wild-type mice. Transgenic hemizygous mice were born, studied, and compared with their wild-type counterparts.

2.5. Northern Blot Analysis

Transgenic mice and their wild-type counterparts were sacrificed by cervical dislocation at the age of 3 months. Total RNAs were isolated using Trizol (Invitrogen, Carlsbad, CA, USA) from multiple organs, including heart, liver, spleen, lung, and kidney. Aliquots (20 μg) of total RNA were separated on 1% agarose gels containing 2.2 M formaldehyde and were blotted on Hybond N+ membrane (Amersham Pharmacia, Piscataway, NJ, USA). The probe was a 611 bp TNNI3K cDNA fragment amplified from the transgenic vector using the following primers, 5′-AAA GAT TAG AAG ATG ACC TGC-3′ and 5′-ATC TTG AGC ATT CAC ATC TG-3′. The probe was labeled with ^{32}P using a Random Primer DNA Labeling Kit (TaKaRa, Dalian, China) based on supplier's protocol. After hybridization, the membranes were washed, and exposed to films (Kodak). The signal was detected using ImageJ software.

2.6. Echocardiography

Mice were weighted and anesthetized with 2.5% avertin (0.018 mL/g) given i.p. Adequacy of anesthesia was monitored by lack of reflex response to toe pinch. Two-dimensional short-

and long-axis views of the left ventricule (LV) were obtained by transthoracic echocardiography with the Vevo 770 Imaging System and a 30-MHz probe (VisualSonics, Toronto, Canada). M-mode tracings were recorded and used to determine LV end-diastolic diameter (LVEDD), LV end-systolic diameter (LVESD), and LV posterior wall thickness (LVPWT) and interventricular septum (IVS) in diastole over three cardiac cycles. LV fractional shortening (FS) was calculated with the formula %FS = (LVEDD−LVESD)/LVEDD. After echocardiography examination, the mice were sacrificed by cervical dislocation. The hearts were excised, rinsed in ice-cold saline, weighed, dissected into left and right ventricles, frozen in liquid nitrogen and stored at −80 °C.

2.7. In Vivo Hemodynamics Analysis in Transgenic Mouse

Non-invasive hemodynamic analysis was performed in 3-month-old TNNI3K transgenic mice and age-matched littermate controls as previous described.[18] Mice were anesthetized with an intraperitoneal injection of 2.5% avertin (0.018 mL/g). Adequacy of anesthesia was monitored by lack of reflex response to toe pinch. A 1.4 French Millar catheter-tip micromanometer catheter (SPR-719, Millar Instruments Inc, Houston, Texas) was inserted through the right carotid artery into the left ventricular. After stabilization for 10 min, the pressure signal was continuously recorded on a computer. The peak LV systolic pressure and LV end-diastolic pressure were measured, and the maximal slopes of systolic pressure increment (dP/dt_{max}) and diastolic pressure decrement (dP/dt_{min}), indexes of contractility and relaxation, respectively, were analyzed.

2.8. Histological and Morphometric Analysis

Hearts from transgenic mice and nontransgenic littermate controls were collected and fixed in 4% paraformaldehyde buffered with PBS, routinely dehydrated, and paraffin embedded. Hearts were sectioned at 4 μm and stained with hematoxylin and eosin, and Masson's Trichrome. Mean myocyte size was calculated by measuring 150 cells from sections stained with hematoxylin and eosin.

2.9. Cell Culture and Recombinant Adenovirus

Adenovirus encoding full-length human TNNI3K (Ad-TNNI3K) was constructed using AdEasy Adenoviral Vector System. The 1- to 2-day-old neonates were sacrificed by cervical dislocation and the primary ventricular cardiomyocyte was isolated by enzyme digestion as we previously described.[19] The cardiomyocytes were planted onto 35-mm-diameter wells (six well plates) at a density of $\approx 2 \times 10^5$ cells per square centimeter and cultured in DMEM supplemented with 10% fetal bovine serum, 100 units/ml penicillin/streptomycin and 0.1 mM bromodeoxyuridine (BrDu). The following day the cells were washed in phosphate-buffer saline and cultured in serum-free DMEM, containing penicillin/streptomycin (100 units/ml). For adenoviral infection, cardiomyocytes were incubated for 2 hours with Ad-TNNI3K and Ad-GFP at an approximate multiplicity of infection of 50. Forty-eight hours following infection, >95% of the cells were GFP positive. The cells were harvested at basal level or after isoproterenol-stimulation for 10 min (10 nmol/l). The protein kinase activity was assessed by Western Blot.

2.10. Western Blots

Western blots were carried out on extracts from left ventricular or cultured cardiomyocytes. Proteinase inhibitor and phosphatase inhibitor cocktail (Roche, Basel, Switzerland) were added accord-

ing to the supplier's protocol. Protein concentrations were determined by the method of Bradford[20] and thirty micrograms of proteins were loaded on gels. Antibodies to Akt, phospho-Thr-308-Akt, phospho-Ser-473-Akt, ERK, phospho-Thr-202/204-ERK, GAPDH were purchased from Cell Signaling Technology (Beverly, MA), Antibodies to total and phosphorylated cTnI were generous gift from Prof. Xianmin Meng. Proteins were detected by Western blotting and stained with NBT/BCIP (Promega, Madison, WI, USA) or ECL detection reagents (Amersham Pharmacia Biotech).

2.11. Yeast two-hybrid Screen and co-immunoprecipitation

A full-length human TNNI3K cDNA, fused to the GAL4 DNA binding domain, was used as bait in a yeast two-hybrid screen of approximately 3×10^6 clones of a human heart cDNA library (Clontech, California, USA). The interaction domain of TNNI3K to cTnI was determined by protein truncation test using immunoprecipitation techniques. The full-length and truncated TNNI3K cDNA sequence encoding amino acids 1–421, 422–726, 727–835, 1–726, and 422–835 of TNNI3K were cloned into pCMV-Myc vector between EcoRI and XhoI sites to generate pCMV-Myc-TNNI3K, pCMV-Myc-TNNI3K-ANK, pCMV-Myc-TNNI3K-PK, pCMV-Myc-TNNI3K-SR, pCMV-Myc-TNNI3K-ANK+PK, and pCMV-Myc-TNNI3K-PK+SR, respectively. Full-length human cTnI cDNA was cloned into pCMV-HA vector between EcoRI and KpnI sites to generate pCMV-HA-cTnI. The integrity of all of the constructs was verified by sequencing.

H9C2 cells were transiently transfected with 2 μg of plasmid DNA using Lipofectamine 2000 (Invitrogen) according to the manufacturer's protocol. Double transfections were carried out with the pCMV-HA-cTnI and empty pCMV-Myc or each of the TNNI3K plasmids. Forty-eight hours after transfection, cells were washed with cold PBS and harvested. Protein to protein interaction was assayed by co-immunoprecipitation using Pro-Found c-Myc-Tag IP/Co-IP Kits (Pierce Biotechnology, Rockford, IL) according to the manufacturer.

2.12. Immunoprecipitation of TNNI3K and Phospho Amino Acid Analysis.

A full-length myc-tagged human TNNI3K DNA was expressed in H9C2 cells. The myc-tagged TNNI3K protein was immunoprecipitated using ProFound Mammalian c-Myc Tag IP/Co-IP kit (Pierce, Rockford, IL) and was analyzed by SDS-PAGE followed by immunoblotting with rabbit anti-phosphoamino acid antibody, anti-phosphotyrosine antibody, anti-phosphothreonine antibody, and anti-phosphoserine antibody, respectively.

2.13. Statistics

All measurement data are expressed as mean±SE. The statistical significance of differences between groups was analyzed by Student's t-test. Differences were considered significant at a P-value<0.05.

Results

3.1. TNNI3K was involved in Cardiac Hypertrophy

To investigate whether TNNI3K is involved in cardiac hypertrophy, the expression of TNNI3K was examined in a rat model of cardiac hypertrophy generated by transverse aortic constriction (TAC). After constriction for 15 days, the heart weight/body weight ratio (HW/BW) and left ventricular weight/body weight ratio (LVW/BW) were increased by 59.2% and

64.3% in TAC rats compared with those of sham operated controls (both p<0.001) (Figure 1A and B). Atrial natriuretic peptide (ANP), a cardiac hypertrophic marker, was upregulated by 40 folds in TAC rats (Figure 1C). TNNI3K was significantly downregulated on day 1 (0.66 fold, P<0.05), return to basal levels on day 7, and then increased on day 15 (1.62 folds, P<0.05) (Figure 1D). This unique expression pattern indicates that TNNI3K is involved in cardiac remodeling.

3.2. Generation of Cardiac-Specific TNNI3K Transgenic Mice

To investigate the role of TNNI3K as a potential regulator of cardiac hypertrophy *in vivo*, transgenic mice overexpressing the human TNNI3K specifically in heart using the mouse αMHC promoter were generated (Figure 2A). By PCR genotyping, three independent transgenic founders were identified from 33 F0 mice (Figure 2B). These founders carried 2, 8, and 44 copies of transgene, and were designated as TG-L (low copy number), TG-M (medium copy number), and TG-H (high copy number), respectively (Figure 2C). The expression of the transgene was restricted to the heart. No expression was detected in the kidneys,

spleen, liver and lung (Fig. 2D). Furthermore, the transgene expression was positively associated with the transgene copy number. TG-H displayed 20-fold higher transgene expression level than TG-L (Figure 2D).

3.3. Overexpression of TNNI3K Induced Cardiac Hypertrophy *in vivo*

In all three lines of transgenic mice, no premature death or sign of heart failure were found after 1 year of observation. Both the TG-L and TG-M lines had no demonstrable cardiac phenotype when compared with their respective littermate controls (Table 1). In contrast, TG-H mice showed a unique profile of increased HW/BW ratio of 31.3% and 43.1% at 3 months and 12 months of age, respectively (P<0.01 and P<0.05). Therefore, further characterization studies of the transgenic phenotype were carried out using male mice from TG-H.

Cardiac size was significantly larger in TG-H mice than in non-transgenic littermates at the age of 3 month (Figure 3A). Macroscopic sections showed a phenotype of concentric ventricular hypertrophy in the transgenic mice, which was characterized with smaller chamber size and thicker ventricular wall (Figure 3B).

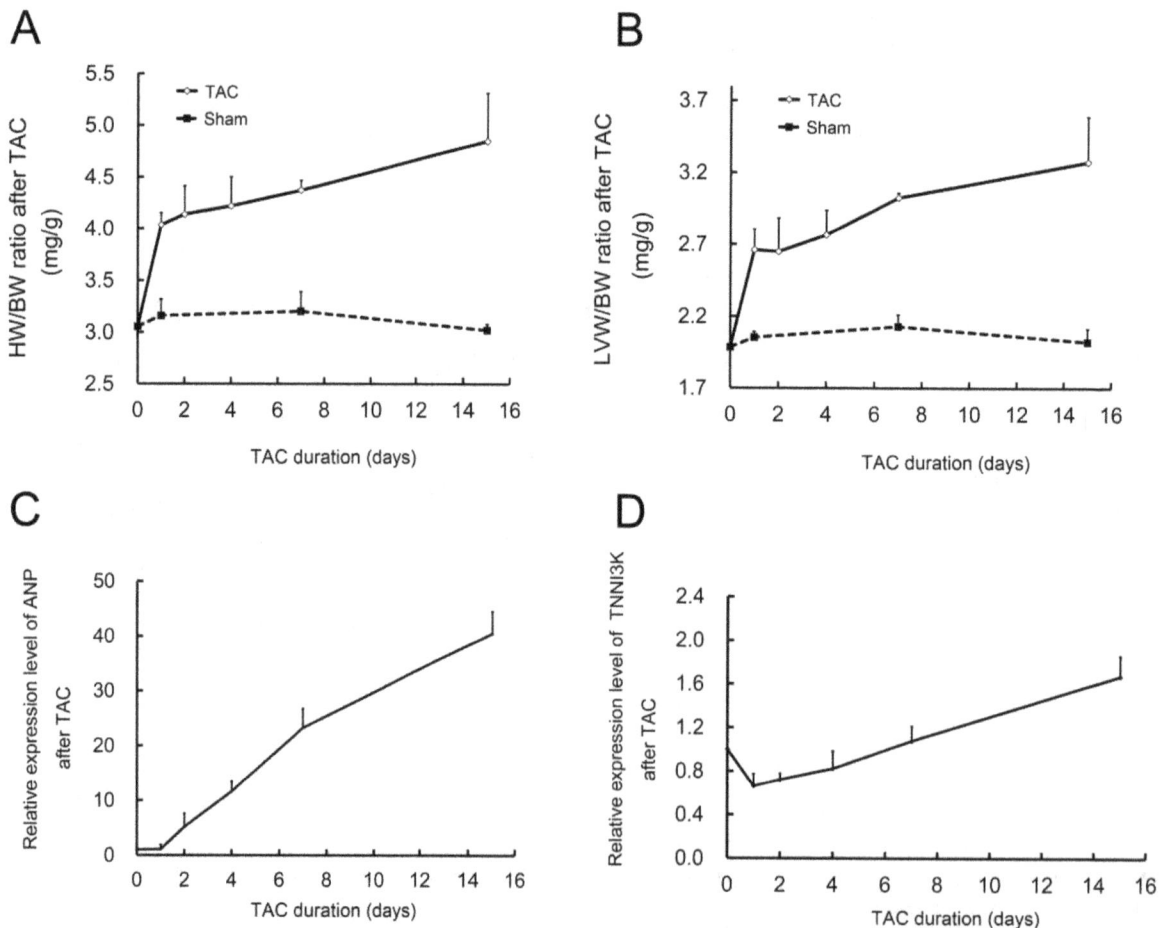

Figure 1. TNNI3K expression was dynamically regulated in the hypertrophic hearts in rats. A and B: the heart weight/body weight (HW/BW) and left ventricular weight/body weight (LVW/BW) were continuously increased after TAC. Five different time points were analyzed for TAC group and 3 different time points for sham operation group. Each time point included 4–5 animals. C and D: The expression of ANP and TNNI3K were detected by real-time PCR analysis. *Gapdh* was used for internal control and data were presented as fold-change compared with that of sham operation controls. Five different time points were analyzed (1, 2, 4, 7, 15 days post-operation, respectively). Each time point contained 4–5 animals.

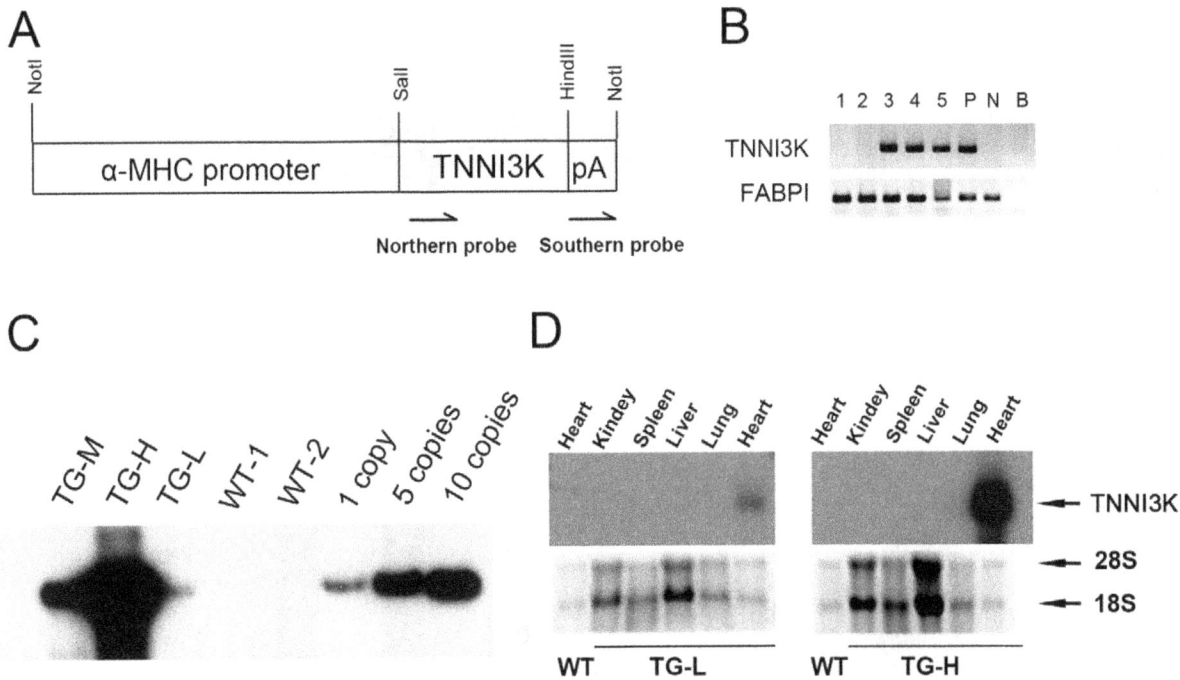

Figure 2. Generation of cardiac-specific *TNNI3K* transgenic mice. (A), Schematic of the TNNI3K transgene that was constructed with the α-MHC mouse promoter. pA: human growth hormone polyA sequences The positions of the Southern probe and northern probe were shown below the construct; (B); PCR genotyping of TNNI3K transgenic mice. 1–5: transgenic mice. P: positive control, wild-type mouse genomic DNA mixed with linearized transgenic fragment. N: negative control, wild-type mouse genomic DNA. B: blank, none DNA template. *FABPI* gene was amplified as internal control. (C), Southern blot analysis of wild-type and *TNNI3K* transgenic mice. Tail genomic DNA was digested with *EcoRI* and probed with *hGH* polyA sequence. Hybridization signals were present only in transgenic positive mice. Transgenic copy number was determined from the gray density against standard curve. 1 copy -10 copies: transgenic copy standards. (D). Northern blot analysis of RNA isolated from multiple tissues of the transgenic TG-L and TG-H lines. Hybridization signals were present only in the heart of transgenic mice. The RNA isolated from the heart of wild-type mouse was used as a negative control.

Upon microscopic observation, necrosis or myocyte disarray was not observed in TG-H mice (Figure 3C, upper panels). Masson-trichrome stain showed no interstitial fibrosis in TG-H mice (Figure 3C, lower panels). To examine whether the increase in cardiac size was due to cellular growth, the cross-section area of myocytes was quantitatively measured on the hematoxylin–eosin-stained LV myocardium of two representative transgenic mice and two controls, respectively. TG-H myocytes were substantially larger with a mean surface area 1.8-fold greater than those seen in littermate controls (Figure 3D).

Table 1. Gravimetric Data for the TNNI3K Transgenic Mouse Heart.

	3 months				12 months	
	NTG	TG-L	TG-M	TG-H	NTG	TG-H
N	11	6	6	6	6	6
BW (g)	24.5±0.5	24.5±0.6	26.2±1.2	26.4±1.0	38.0±0.7	38.1±0.7
HW/BW(mg/g)	5.1±0.2	5.4±0.4	5.5±0.3	6.7±0.2**	5.8±0.5	8.3±0.9*
LVW/BW(mg/g)	3.6±0.4	3.5±0.4	3.6±0.2	4.4±0.3*	3.8±0.5	5.8±0.9*

Heart weight/body weight ratios and left ventricular weight/body weight ratio were calculated from TNNI3K transgenic or non-transgenic mice at the indicated time points. Values given are Mean±SE and P values were calculated by Student's t-test. *: compared with non-transgenic mice, P<0.05, **: compared with non-transgenic mice, P<0.01.

To assess *in vivo* cardiac morphology and function, echocardiography were performed on TG-H and littermate control mice at the age of 3 months and 12 months, respectively. In agreement with the histological analysis, the 3-month-old TG-H mice showed a concentric cardiac hypertrophy. The left ventricular wall thickness was significantly increased (36.6% and 46.9% in IVS and LVPWT, respectively, both P<0.01), accompanied by decreased chamber dimension (19.0% and 10.8% in LVESD and LVESD, respectively, both P<0.01), and increased fraction shortening (Table 2 and Figure S1). Surprisingly, this phenotype of concentric hypertrophy persisted up to the age of 12 months (Table 2). Echocardiography showed increased ventricular wall thickness and decreased chamber dimension in the transgenic heart. More importantly, there is no loss of systolic functional performance between 3 and 12 months of age, indicating that cardiac hypertrophy in TNNI3K transgenic mice was compensated hypertrophy.

To evaluate the effect of overexpressing TNNI3K on cardiac function more accurately, noninvasive hemodynamic assessment was performed using LV catheterization. After anesthesia, there was no significant difference in heart rate and blood pressure between the 3-month-old TG-H mice and non-transgenic littermates. LV pressure was mildly increased both at systolic and diastolic in transgenic heart. In consistent with the echocardiography analysis, LV dP/dt_{max} and LV dP/dt_{min} were increased by 9% and 20% in transgenic mice compared with those in wild-type mice, respectively (Table 3), indicating enhanced contractility and diastolic function for the TG-H mice.

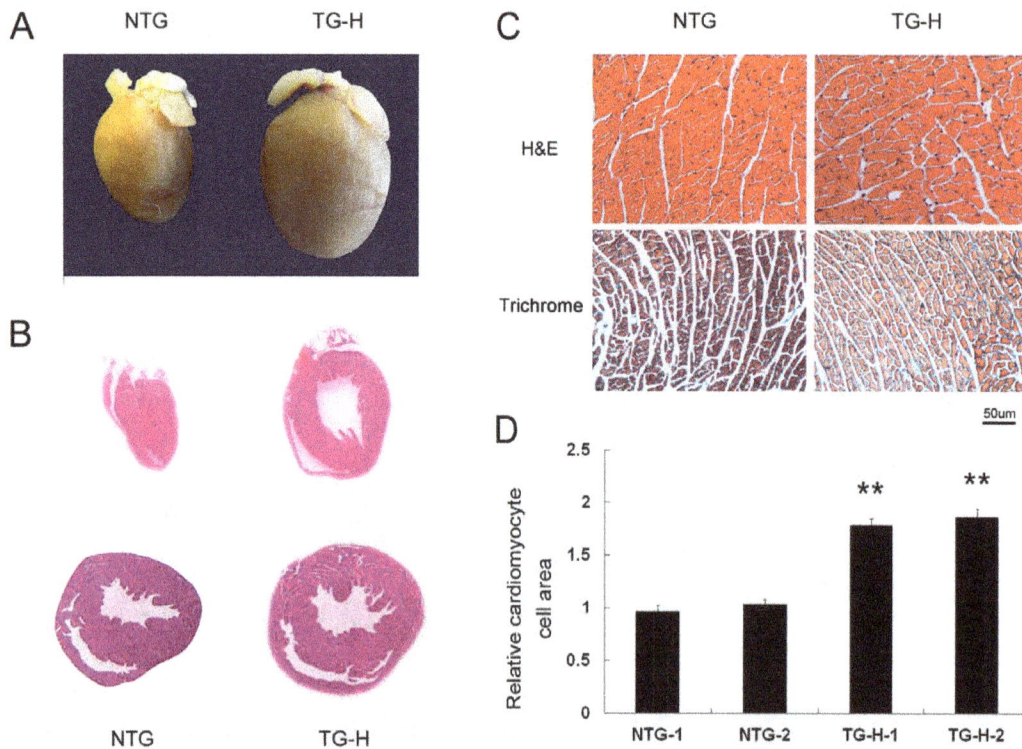

Figure 3. Cardiac histological analysis of TNNI3K transgenic mice at the age of 3 Months. (A) Whole heart, (B) Macroscopic view after hematoxylin–eosin-stained hearts revealed a concentric hypertrophy in TG-H mice. Upper was longitudinally sectioned. Lower was transversely sectioned. (C) Microscopic histological analysis of demonstrated cardiomyocyte hypertrophy without an increase in interstitial fibrosis in TG-H hearts. Upper panel was H&E stained sections. Lower panel was trichrome stained sections. Blue staining represents collagen deposition. Original magnification x200 (D) Cross-sectional areas of cardiomyocyte were quantified from hematoxylin–eosin-stained histological sections. At least 150 myocytes were measured each in two non-transgenic (NTG) hearts and two TG-H transgenic heart. **P<0.01 compared with wild-type.

Cardiac hypertrophy is accompanied by reprogramming of cardiac gene expression. To characterize the molecular phenotype of TNNI3K-induced cardiac hypertrophy, we examined the transcriptional levels of a set of hypertrophic markers, including atrial natriuretic peptide (*ANP*), brain natriuretic peptide (*BNP*), skeletal muscle α-actin (*Actc1*), α- and β-myosin heavy chain (*Myh6* and *Myh7*, respectively), phospholamban (*PLN*) and sarcoplasmic reticulum Ca^{2+} ATPase (*SERCA2a*). In 3-month-old TG-H transgenic hearts, the expression of ANP and BNP was increased

(P<0.01), whereas Actc1 was decreased (P<0.05). On the other hand, the expression of β-MHC was mildly increased, whereas α-MHC and SERCA2a mRNA were markedly increased (Figure 4). These discordant changes in gene expression indicate that the fetal gene program is differentially regulated in the TNNI3K transgenic heart.

Table 2. M-mode echocardiograms.

	3 months		12 months	
	NTG	**TG-H**	**NTG**	**TG-H**
N	13	7	6	5
IVS (mm)	0.71±0.03	0.97±0.03**	0.95±0.03	1.40±0.07**
LVPWT (mm)	0.66±0.02	0.97±0.05**	0.93±0.03	1.16±0.07*
LVESD (mm)	2.32±0.06	1.88±0.10**	2.75±0.25	1.83±0.21*
LVEDD (mm)	3.79±0.05	3.38±0.14**	4.23±0.22	3.67±0.15
%FS	38.7±1.4	46.5±2.8**	35.5±3.0	50.4±4.3*

LVEDD, LV end-diastolic diameter; LVESD, LV end-systolic diameter; LVPWT, LV posterior wall thickness, IVS, interventricular septum; %FS, fraction shorting. Values given are Mean±SE and P values were calculated by Student's t-test. *: compared with non-transgenic mice, P<0.05, **: compared with non-transgenic mice, P<0.01.

Table 3. Hemodynamic Analysis of TNNI3K Transgenic Mice.

	NTG	**TG-H**
N	8	6
HR (beats/min)	388.5±4.8	380.5±5.7
AP systolic (mmHg)	93.0±4.6	98.0±4.2
AP diastolic (mmHg)	62.7±3.4	61.2±2.5
LVP systolic (mmHg)	104.0±3.2	125.4±11.3
LVP diastolic (mmHg)	−17.3±1.0	−13.1±1.8
LVEDP (mmHg)	−3.3±0.5	−2.2±0.3
dP/dt$_{max}$ (mmHg/s)	6311.6±143.6	6881.8±514.6
dP/dt$_{min}$ (mmHg/s)	−6325.3±169.6	−7567.8±720.4

HR: heart rate; AP: arterial pressure; LVP: left ventricular pressure; LVEDP: left ventricular end-diastolic pressure, dP/dt$_{max}$ and dP/dt$_{min}$: average maximum and minimum values, respectively, of first derivative of ventricular pressure wave.

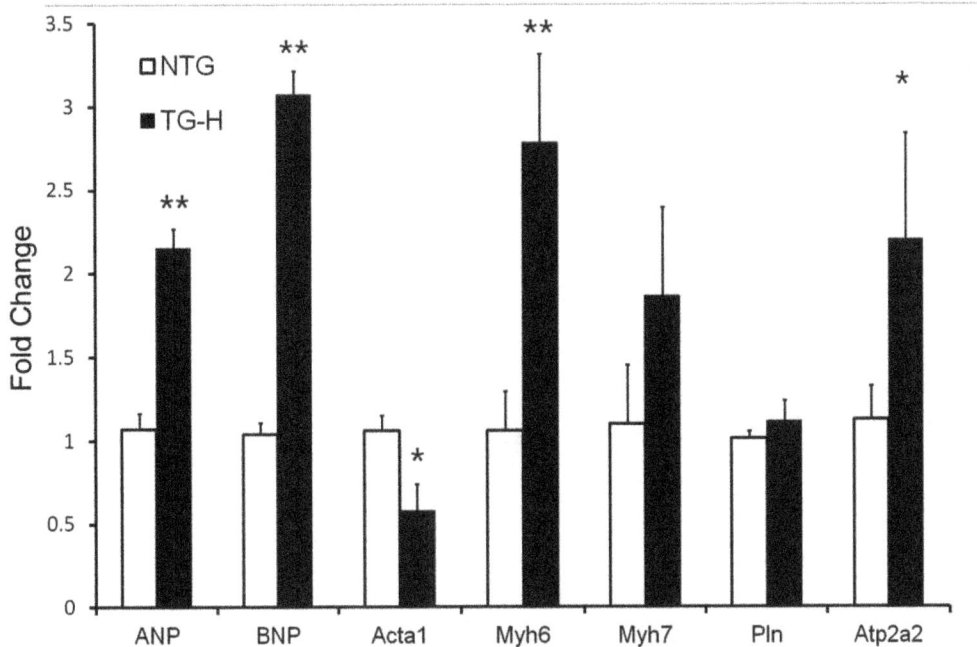

Figure 4. Assessment of hypertrophy marker genes. The mRNA levels were measured in left ventricle by RT-PCR analysis (N = 4 hearts in each group). *Gapdh* was used for internal control. **: $P<0.01$, *: $P<0.05$ vs nontransgenic littermates.

3.4. TNNI3K does not induce the activation of ERK and Akt

Akt and ERK were among the best characterized signaling cascades that induce cardiac hypertrophy. In the left ventricular of TG-H mice at the age of 3 months, however, no significant difference was identified in the amount of total or phosphorylated Akt and ERK (Figure 5A). Consistent with the *in vivo* results, overexpression of TNNI3K in cardiomyocyte with a recombinant adenovirus does not induced activated of Akt or ERK (Figure 5B). Therefore, TNNI3K might not be involved in these two signal pathway.

3.5. TNNI3K interacts with cTnI and induced cTnI phosphorylation at Ser22/Ser23

To gain insight into the proteins those physiologically interact with TNNI3K in the heart, we performed the yeast two-hybrid study using full-length TNNI3K as bait. Screening of human heart cDNA library resulted in the identification of thirteen TNNI3K-interacting factors, one of which was cTnI. Previous data suggest there is a physical interaction between TNNI3K and cTnI[10]. To map the region of TNNI3K that bind cTnI, amino- and carboxyl-terminal truncations of TNNI3K were co-expressed with full-length cTnI, and the TNNI3K protein was immunoprecipitated. As expected, the full-length TNNI3K efficiently immunoprecipitated cTnI (Figure 6B, lane 1). cTnI also binds the truncations of TNNI3K lacking the amino-terminal ANK-repeat domain and/or the central protein kinase domain. However, once the carboxyl-terminal serine-rich domain was truncated, all detectable cTnI binding was lost (Figure 6B). These results indicate that TNNI3K binds to cTnI in its carboxyl-terminal region bound by amino acids 727 and 835, outside the protein kinase domain.

TNNI3K is a functional kinase. To identify the specificity of phosphoamino-acid, TNNI3K was overexpressed in H9C2 cells, immunoprecipitated, and immunoblotted with anti-phosphoamino acid antibodies. As shown in Figure 7, only phosphotyrosine,

but no phosphoserine or phosphothreonine was detectable in TNNI3K, suggesting it is a protein-tyrosine kinase.

As TNNI3K is a functional kinase and it directly interacts of cTnI, we then performed immunoblot analysis to examine the effect of TNNI3K on cTnI phosphorylation. In the TNNI3K transgenic heart, the overexpression of TNNI3K was accompanied by increased cardiac troponin I phosphorylation at Ser22/Ser23 (Figure 8A). In cultured cardiomyocytes, infection with Ad-TNNI3K significantly induced cTnI phosphorylation at Ser22/Ser23 on the basal and isoproterenol-stimulated level, relative to cells infected with Ad-GFP (Figure 8B). Since the phosphorylation of troponin is a highly significant element in the control of cardiac contractility, these data suggested the role of TNNI3K in regulating cTnI phosphorylation and contractile function in the heart.

Discussion

In this study, we found that TNNI3K is associated with cardiac remodeling induced by hemodynamic overload. Overexpression of human TNNI3K specifically in the heart of transgenic mouse lines caused long-standing concentric hypertrophy associated with enhanced cardiac pump function. Furthermore, TNNI3K directly interacts with cTnI and induced cTnI phosphorylation at Ser22/Ser23 *in vivo* and *in vitro*. These data suggest that TNNI3K promotes cardiac remodeling via regulating the phosphorylation of cTnI.

In consistent with previous studies *in vitro*,[14,15] we found that overexpression TNNI3K could induce significantly enlargement of cardiomyocyte *in vivo*. Moreover, the TNNI3K transgene resulted in a concentric hypertrophy that is associated with enhanced cardiac function at the age of 3 months. There was no sign of pathological phenotype such as interstitial fibrosis. More importantly, the concentric hypertrophy remained up to 12 months of age and the left ventricular function was still normal. Therefore,

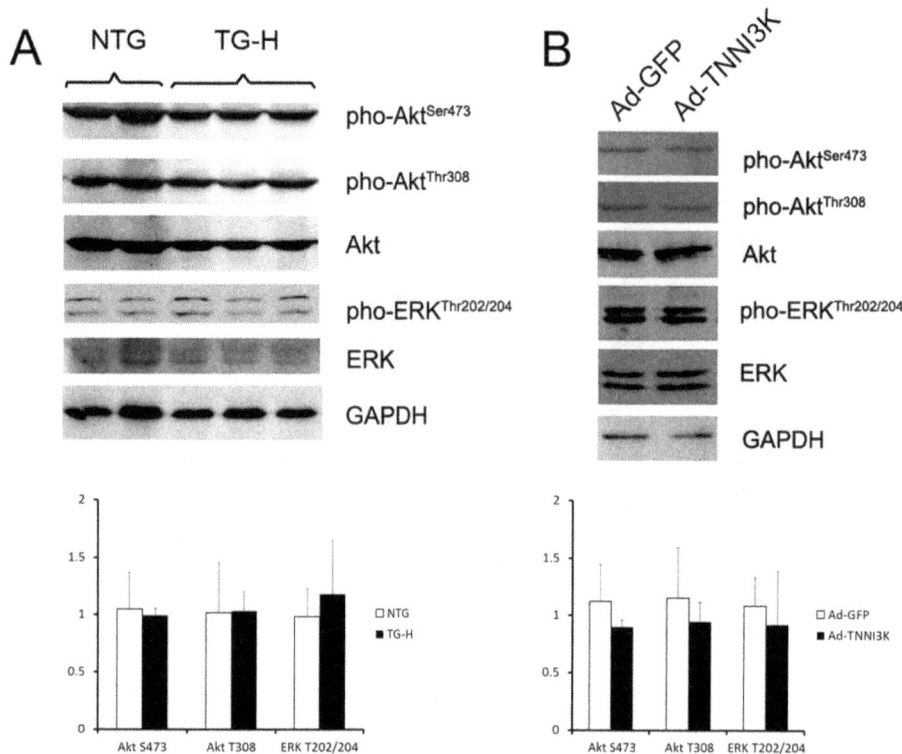

Figure 5. TNNI3K did not induce activation of Akt and ERK *in vivo* and *in vitro*. A: Western blot assessment for phosphorylation of Akt and ERK from transgenic hearts and nontransgenic control hearts at the age of 3 month. B: Cardiomyocytes were infected with Ad-TNNI3K and Ad-GFP for 48 hours and phosphorylation of Akt and ERK were analyzed by western blot. No significant phosphorylation increase was detected for the two effectors *in vivo* and *in vitro*. Each data point is shown as mean±SD.

we believe that the TNNI3K overexpression leads to an adaptive cardiac hypertrophy rather than maladaptive hypertrophy.

Cardiac hypertrophy is associated with alternation of cardiac gene expression.[21,22] For example, physiologic hypertrophy of the heart is generally associated with the induction of α-MHC expression; however, during pathologic hypertrophy, β-MHC is increased at the expense of α-MHC. In TNNI3K transgenic hearts, although β-MHC was mildly up-regulated, the expression of α-MHC was significantly increased. This result is in agreement with the previous report, showing that the adenovirus-mediated TNNI3K overexpression increases the content of α-MHC in cardiomyocytes,[14] Moreover, the expression of SERCA2a is also increased in transgenic heart. As α-MHC-to-β-MHC ratio and SERCA2a-to-phosholamban ratio are positively correlated with left ventricular contractility,[23,24] these molecular changes provide an explanation for the enhanced cardiac function and hyperhemodynamic state of TNNI3K transgenic heart.

To study the function of TNNI3K, we used "gain-of-function" strategy to generate transgenic mice in the C57BL/6J strain. Among the three transgenic lines, only the highest-copy-number transgenic mice developed a significantly cardiac remodeling. We believe the most likely explanation for this phenotype differences between the three lines of mouse is the endogenous counterbalancing mechanism. Firstly, C57BL/6J strain shows robust expression of TNNI3K in the heart at baseline.[16] Secondly, we generated transgenic mice expressing a wild-type CARK rather than a continuous-active CARK mutant. Therefore, lower expression level of exogenous CARK might be well tolerated by an endogenous counterbalancing mechanism.

Our observations with mice expressing TNNI3K are different from the previous report that described an accelerated disease progression in TNNI3K transgenic mice in response to pressure overload and in Csq transgenic mice.[16] This apparent discrepancy may be explained by the difference in the level of activation of TNNI3K signaling in the two mouse models. In Wheeler's study, the expression of TNNI3K in high-copy-number is comparable with that of low-copy-number transgenic lines. In our study, however, the difference in expression level is more than 20 folds. In addition, Wheeler et. al. found no major phenotype in the transgenic at baseline, which is consistent with our findings with the two lines that expresses low and moderate levels of TNNI3K. Hence, it is possible that because of differences in levels of activity of TNNI3K, pathways are less activated in Wheeler's mice than in our mice.

Up to date, little is known about the downstream targets of TNNI3K that are involved in the regulation cardiac hypertrophy. Given the facts that the IGF-1-phosphoinositide 3-kinase (PI3K)-Akt pathway and MEK1-ERK1/2 pathway are mainly advocated to mediate physiological cardiac growth,[2,25] and the physiological hypertrophy demonstrated by TNNI3K transgenic mice is reminiscent of phenotypes displayed in transgenic mice expressing a caPI3K mutant[26] and caMEK1 mutant,[27] we hypothesized that the functional role of TNNI3K may be accomplished through the two signaling pathways. However, the activities of Akt and ERK were not changed by TNNI3K overexpression either *in vivo* or *in vitro*, suggesting TNNI3K promotes cardiac hypertrophy through other signaling pathway.

Previous study reported that TNNI3K directly binds to cTnI.[10] In our study, this interaction was confirmed indepen-

A

B

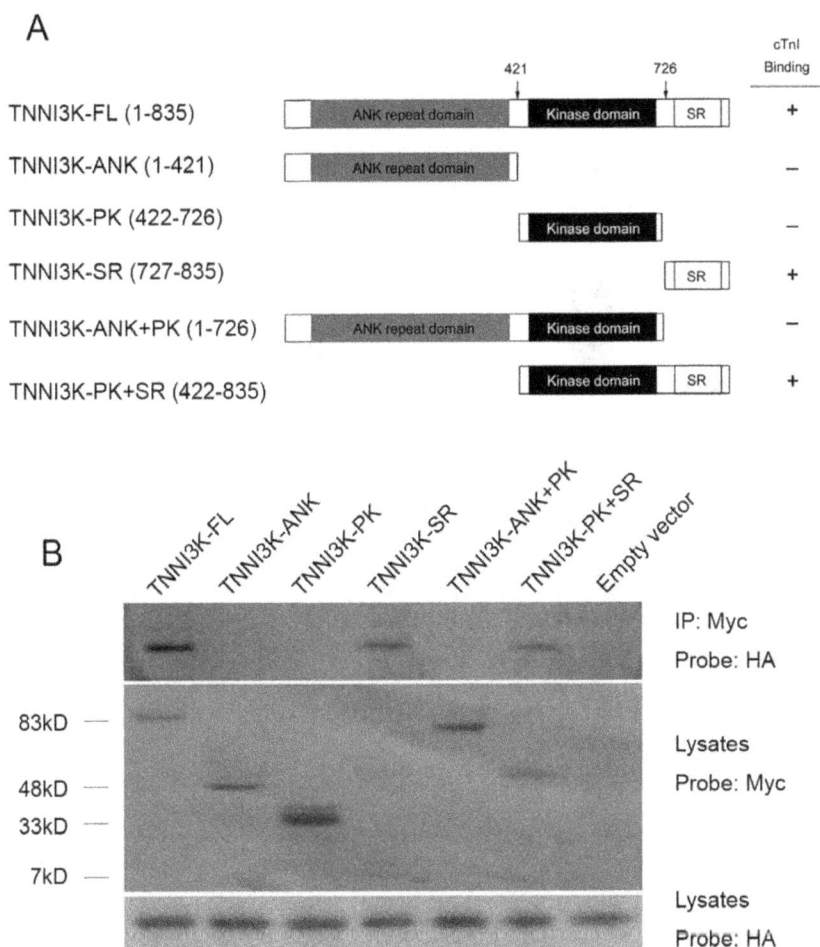

Figure 6. cTnI interacts with the serine-rich domain in the carboxyl terminus of TNNI3K. (A) Schematic representation of cTnI TNNI3K binding results. The selected domains on TNNI3K were indicated by labeled boxes. FL: full length, ANK: ANK repeat domain, PK: protein kinase domain, SR: serine rich domain. (B) H9C2 cells were transiently transfected with plasmids expressing full-length HA-tagged cTnI and the full-length Myc-tagged TNNI3K or the indicated truncations of Myc-tagged TNNI3K. TNNI3K was immunoprecipitated (IP) with anti-Myc antibodies and the presence of cTnI in the immunoprecipitate was assessed by immunoblotting with anti-HA antibodies. The expression of myc-TNNI3K from transfected constructs in the lysates is shown on the middle panel. The abundant of HA-cTnI in the lysates probed with anti-HA was used as an input.

dently both in the two-hybrid screen and in *in vitro* binding experiments. As the unique isoform expressed in heart muscle, cTnI plays a key role in the modulation of cardiac myofilament response to protein phosphorylation. cTnI can be phosphorylated at multiple amino acid residues by various kinases.[28] The phosphorylation level of cTnI at different residues has been proved to be a highly significant element in the control of cardiac contractility and a potential point of vulnerability in the network of

Figure 7. Detection of specificity of phosphoproteins in TNNI3K. Myc-tagged TNNI3K was overexpressed in H9C2 cells. TNNI3K proteins were immunoprecipitated with anti-myc antibody and Western blotted with anti-phosphoamino acid antibody (P-total), anti-phospho-tyrosine antibody (P-Tyr), anti-phosphothreonine antibody (P-Thr), anti-phosphoserine antibody (P-Ser) (top panel) or anti-myc (bottom panel).

signals by which hypertrophy and failure evolve.[29,30] Ser22/Ser23 locates in the cardiac-specific N terminus of cTnI. Phosphorylation at the Ser22/Ser23 sites alters the shape of the cTnI, resulting in accelerated relaxation and augmented contractility[31]. In the present study, we found that TNNI3K overexpression induced cTnI phosphorylation at Ser22/Ser23 *in vivo* and *in vitro*, suggesting that TNNI3K is a novel upstream regulator for cTnI phosphorylation. Moreover, the TNNI3K transgenic mice demonstrate a unique hypertrophic phenotype with enhanced cardiac function and hyperhemodynamic state, which is consistent with the lusitropic effect of cTnI phosphorylation on Ser22/Ser23. Therefore, TNNI3K may promote cardiac remodeling via regulating the phosphorylation of cTnI.

In conclusion, our study shows that upregulation of TNNI3K was characterized by long-standing concentric hypertrophy with enhancement of cardiac function. This prohypertrophic effect of TNNI3K was associated with the increased cTnI phosphorylation on Ser22/Ser23. As the phosphorylation state of Ser22 and/or Ser23 is significantly reduced in end-stage failing hearts[32,33], future studies investigating whether TNNI3K could be a potential therapeutic target for heart failure should be pursued.

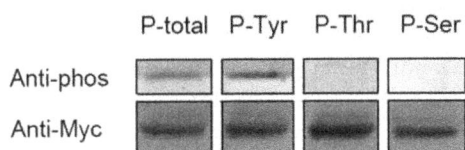

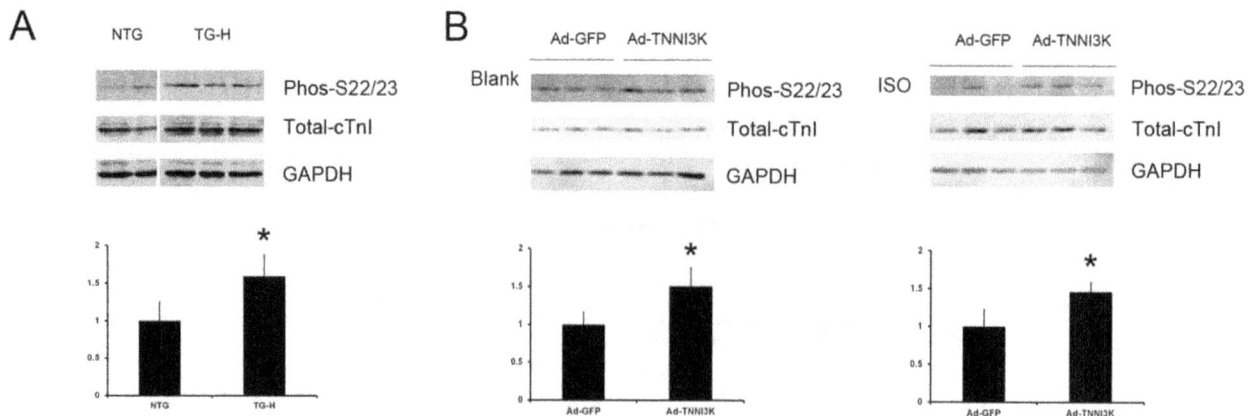

Figure 8. TNNI3K induced cTnI phosphorylation at Ser22/Ser23 *in vivo* and *in vitro*. (A), Heart lysates of TG-H and NTG mice were immunoblotted with antibodies against total and phosphorylated cTnI. TNNI3K induced cardiac troponin I phosphorylation at Ser22/Ser23 *in vivo*. (B), TNNI3K-induced cTnI phosphorylation at Ser22/Ser23 in cardiomyocytes on the basal and isoproterenol-stimulated level. Immunoblots are representative of 3 independent experiments. *: P<0.05 vs control group.

Supporting Information

Figure S1 Schematic M-mode echocardiographic tracings of TG-H and non-transgenic littermates at the age of 3 month. LVEDD: LV end-diastolic diameter, LVPWD: LV posterior wall thickness in diastole, IVSD: interventricular septum thickness in diastole.

Acknowledgments

We gratefully thank Shan Gao and Wei Dong for their superb assistance in generating and phenotyping transgenic mice. Thanks a lot to Dr. Yusheng Wei and Dr. Han Xiao for their great support in performing experiments.

Author Contributions

Conceived and designed the experiments: XJW JZW RTH. Performed the experiments: XJW JZW MS CXW JZC HW. Analyzed the data: XJW MS LS YBZ. Contributed reagents/materials/analysis tools: LFZ YYZ. Wrote the paper: XJW RTH.

References

1. Frey N, Olson EN (2003) Cardiac hypertrophy: the good, the bad, and the ugly. Annu Rev Physiol 65: 45–79.
2. Dorn GW, 2nd (2007) The fuzzy logic of physiological cardiac hypertrophy. Hypertension 49: 962–970.
3. Chien KR (1999) Stress pathways and heart failure. Cell 98: 555–558.
4. Frey N, Katus HA, Olson EN, Hill JA (2004) Hypertrophy of the heart: a new therapeutic target? Circulation 109: 1580–1589.
5. Hill JA, Olson EN (2008) Cardiac plasticity. N Engl J Med 358: 1370–1380.
6. Dorn GW, 2nd, Force T (2005) Protein kinase cascades in the regulation of cardiac hypertrophy. J Clin Invest 115: 527–537.
7. Copeland O, Sadayappan S, Messer AE, Steinen GJ, van der Velden J, et al. (2010) Analysis of cardiac myosin binding protein-C phosphorylation in human heart muscle. J Mol Cell Cardiol 49: 1003–1011.
8. Jacques AM, Copeland O, Messer AE, Gallon CE, King K, et al. (2008) Myosin binding protein C phosphorylation in normal, hypertrophic and failing human heart muscle. J Mol Cell Cardiol 45: 209–216.
9. Vorotnikov AV, Risnik VV, Gusev NB (1988) [Phosphorylation of troponin in the heart and skeletal muscle by Ca2+−phospholipid-dependent protein kinase]. Biokhimiia 53: 31–40.
10. Zhao Y, Meng XM, Wei YJ, Zhao XW, Liu DQ, et al. (2003) Cloning and characterization of a novel cardiac-specific kinase that interacts specifically with cardiac troponin I. J Mol Med (Berl) 81: 297–304.
11. Wiltshire SA, Leiva-Torres GA, Vidal SM (2011) Quantitative trait locus analysis, pathway analysis, and consomic mapping show genetic variants of Tnni3k, Fpgt, or H28 control susceptibility to viral myocarditis. J Immunol 186: 6398–6405.
12. Milano A, Lodder EM, Scicluna BP, Sun AY, Tang H, et al. (2011) Tnni3k is a Novel Modulator of Cardiac Conduction. Circulation 124: A16031.
13. Wang H, Chen C, Song X, Chen J, Zhen Y, et al. (2008) Mef2c is an essential regulatory element required for unique expression of the cardiac-specific CARK gene. J Cell Mol Med 12: 304–315.
14. Wang L, Wang H, Ye J, Xu RX, Song L, et al. (2011) Adenovirus-mediated overexpression of cardiac troponin I-interacting kinase promotes cardiomyocyte hypertrophy. Clin Exp Pharmacol Physiol 38: 278–284.
15. Lai ZF, Chen YZ, Feng LP, Meng XM, Ding JF, et al. (2008) Overexpression of TNNI3K, a cardiac-specific MAP kinase, promotes P19CL6-derived cardiac myogenesis and prevents myocardial infarction-induced injury. Am J Physiol Heart Circ Physiol 295: H708–716.

16. Wheeler FC, Tang H, Marks OA, Hadnott TN, Chu PL, et al. (2009) Tnni3k modifies disease progression in murine models of cardiomyopathy. PLoS Genet 5: e1000647.
17. Juan F, Wei D, Xiongzhi Q, Ran D, Chunmei M, et al. (2008) The changes of the cardiac structure and function in cTnTR141W transgenic mice. Int J Cardiol 128: 83–90.
18. Wang J, Xu N, Feng X, Hou N, Zhang J, et al. (2005) Targeted disruption of Smad4 in cardiomyocytes results in cardiac hypertrophy and heart failure. Circ Res 97: 821–828.
19. Wang X, Yang X, Sun K, Chen J, Song X, et al. (2010) The haplotype of the growth-differentiation factor 15 gene is associated with left ventricular hypertrophy in human essential hypertension. Clin Sci (Lond) 118: 137–145.
20. Bradford MM (1976) A rapid and sensitive method for the quantitation of microgram quantities of protein utilizing the principle of protein-dye binding. Anal Biochem 72: 248–254.
21. Izumo S, Lompre AM, Matsuoka R, Koren G, Schwartz K, et al. (1987) Myosin heavy chain messenger RNA and protein isoform transitions during cardiac hypertrophy. Interaction between hemodynamic and thyroid hormone-induced signals. J Clin Invest 79: 970–977.
22. Izumo S, Nadal-Ginard B, Mahdavi V (1988) Protooncogene induction and reprogramming of cardiac gene expression produced by pressure overload. Proc Natl Acad Sci U S A 85: 339–343.
23. Gupta MP (2007) Factors controlling cardiac myosin-isoform shift during hypertrophy and heart failure. J Mol Cell Cardiol 43: 388–403.
24. Catalucci D, Latronico MV, Ceci M, Rusconi F, Young HS, et al. (2009) Akt increases sarcoplasmic reticulum Ca2+ cycling by direct phosphorylation of phospholamban at Thr17. J Biol Chem 284: 28180–28187.
25. Weeks KL, McMullen JR (2011) The athlete's heart vs. the failing heart: can signaling explain the two distinct outcomes? Physiology (Bethesda) 26: 97–105.
26. Shioi T, Kang PM, Douglas PS, Hampe J, Yballe CM, et al. (2000) The conserved phosphoinositide 3-kinase pathway determines heart size in mice. EMBO J 19: 2537–2548.
27. Bueno OF, De Windt LJ, Tymitz KM, Witt SA, Kimball TR, et al. (2000) The MEK1-ERK1/2 signaling pathway promotes compensated cardiac hypertrophy in transgenic mice. EMBO J 19: 6341–6350.
28. Sumandea MP, Burkart EM, Kobayashi T, De Tombe PP, Solaro RJ (2004) Molecular and integrated biology of thin filament protein phosphorylation in heart muscle. Ann N Y Acad Sci 1015: 39–52.

29. Layland J, Solaro RJ, Shah AM (2005) Regulation of cardiac contractile function by troponin I phosphorylation. Cardiovasc Res 66: 12–21.

30. Scruggs SB, Solaro RJ (2011) The significance of regulatory light chain phosphorylation in cardiac physiology. Arch Biochem Biophys 510: 129–134.

31. van der Velden J, de Jong JW, Owen VJ, Burton PB, Stienen GJ (2000) Effect of protein kinase A on calcium sensitivity of force and its sarcomere length dependence in human cardiomyocytes. Cardiovasc Res 46: 487–495.

32. Bodor GS, Oakeley AE, Allen PD, Crimmins DL, Ladenson JH, et al. (1997) Troponin I phosphorylation in the normal and failing adult human heart. Circulation 96: 1495–1500.

33. Kooij V, Saes M, Jaquet K, Zaremba R, Foster DB, et al. (2010) Effect of troponin I Ser23/24 phosphorylation on Ca2+−sensitivity in human myocardium depends on the phosphorylation background. J Mol Cell Cardiol 48: 954–963.

9

Epigenetic Changes of Lentiviral Transgenes in Porcine Stem Cells Derived from Embryonic Origin

Kwang-Hwan Choi, Jin-Kyu Park, Hye-Sun Kim, Kyung-Jun Uh, Dong-Chan Son, Chang-Kyu Lee*

Department of Agricultural Biotechnology, Animal Biotechnology Major, and Research Institute for Agriculture and Life Science, Seoul National University, Seoul, Korea

Abstract

Because of the physiological and immunological similarities that exist between pigs and humans, porcine pluripotent cell lines have been identified as important candidates for preliminary studies on human disease as well as a source for generating transgenic animals. Therefore, the establishment and characterization of porcine embryonic stem cells (pESCs), along with the generation of stable transgenic cell lines, is essential. In this study, we attempted to efficiently introduce transgenes into Epiblast stem cell (EpiSC)-like pESCs. Consequently, a pluripotent cell line could be derived from a porcine-hatched blastocyst. Enhanced green fluorescent protein (EGFP) was successfully introduced into the cells via lentiviral vectors under various multiplicities of infection, with pluripotency and differentiation potential unaffected after transfection. However, EGFP expression gradually declined during extended culture. This silencing effect was recovered by *in vitro* differentiation and treatment with 5-azadeoxycytidine. This phenomenon was related to DNA methylation as determined by bisulfite sequencing. In conclusion, we were able to successfully derive EpiSC-like pESCs and introduce transgenes into these cells using lentiviral vectors. This cell line could potentially be used as a donor cell source for transgenic pigs and may be a useful tool for studies involving EpiSC-like pESCs as well as aid in the understanding of the epigenetic regulation of transgenes.

Editor: Aditya Bhushan Pant, Indian Institute of Toxicology Reserach, India

Funding: This work was supported by the Next-generation BioGreen 21 Program (PJ0094932013), Rural Development Administration, Republic of Korea. The funders had no role in study design, data collection and analysis, decision to publish, or preparation of the manuscript.

Competing Interests: The authors have declared that no competing interests exist.

* E-mail: leeck@snu.ac.kr

Introduction

Over the last three decades, the establishment of pluripotent cell lines from preimplantation mouse embryos has been considered to be one of the biggest events in developmental biology [1,2]. These cells, known as embryonic stem cells, have *in vivo* and *in vitro* differentiation potentials into three germ layers and can proliferate infinitely. Recently, mouse epiblast stem cells (EpiSCs) and induced pluripotent stem cells (iPSCs) were derived from postimplantation embryos and somatic cells, respectively [3,4]. These pluripotent cells are divided into "naïve" and "primed" states by their pluripotent status [5]. In permissive lines, pluripotent cells can be derived from embryos in both states. However, in nonpermissive lines such as human and pig, cells are only derived into the "primed" state, such as epiblast stem cells, if no additional treatment such as genetic manipulation and chemicals are performed [6–8].

Pluripotent cells are expected to be used as cell therapeutic material in degenerative disorders, and in domestic animals, as cell sources for generating transgenic animals and xenotransplantation [9]. In particular, in transgenic animal and xenotransplantation applications, pigs have been identified as an ideal animal model because of similarities between humans and pigs in physiological and immunological features, as well as organ size [10,11]. Therefore, many research groups have attempted to create transgenic pigs to produce pharmaceutical proteins and in xenotransplantation [12,13]. In addition, although authentic porcine embryonic stem cells (pESCs) have not yet been established, the characterization of pESCs, along with the generation of stable transgenic cell lines, has been studied for a long time [8,14–18].

To achieve these goals, genetic manipulation via transgenic technologies has been required in stem cell research. Transgenic stem cells using the homologous recombination technique were first reported in mouse embryonic stem cells by Thomas & Capecchi in 1987 [19]. Subsequently, researchers have successfully delivered transgenes into pluripotent stem cells using several methods, including electroporation [20], liposomal [21] and viral vectors [22,23], and nucleofection [24]. However, stably introducing transgenes in these cells has proven difficult because of the low efficiency and cytotoxic side effects. The delivery of transgenes using viral vectors, which are stably expressed, is considered the most useful tool for inducing low cytotoxicity and inserting transgenes into the host genome [25]. Moreover, lentiviral vectors belonging to retroviral families are able to infect several types of cells, as well as nondividing cells [26,27].

Transgenesis in porcine embryonic stem cells was first reported by Yang *et al.* (2009) [17]. In contrast to other reports using somatic cell nuclear transfer (SCNT) with transgenic donor cells [28,29], the transgene [humanized renilla green fluorescent protein (hrGFP)] was directly delivered into pESCs via electroporation. Stably hrGFP-expressing porcine pluripotent cell lines were successfully established by introducing plasmid vectors via electroporation. Transgenic porcine embryonic germ cell lines were reported by Rui *et al.* (2006) [30]. In this study enhanced green fluorescent protein (EGFP) transgenes were introduced into cells with a liposomal vector. In other studies involving mouse embryonic stem cells, GFP-expressing lines were successfully established using viral vectors [31,32]. However, GFP expression gradually decreased during passaging in mouse embryonic stem cells due to DNA methylation. In a similar case of transgenic animal production by lentiviral transduction, transgenes were silenced by DNA methylation in specific cell types [33,34].

Recently, we derived an epiblast stem cell-like pESC lines (EpiSC-like pESCs) from hatched blastocysts, which showed EpiSC-like features [8]. As we established these porcine pluripotent stem cell lines, we attempted to generate a transgenic pluripotent cell line. As mentioned above, transgenesis in porcine pluripotent cells is essential for applications such as the production of transgenic pigs and analysis of gene functions. Moreover it is important to characterize and optimize an efficient transfection system. The main purpose of this study was to successfully introduce transgenes into EpiSC-like pESCs using lentiviral vectors, and to optimize these viral infection conditions. Additionally, we evaluated the relationship between transgene expression and changes in the DNA methylation status of the inserted lentiviral transgene, particularly in the promoter regions of undifferentiated and differentiated EpiSC-like pESCs. Consequently, transgenes were successfully introduced into the cells via lentiviral vectors under various multiplicities of infection. Furthermore, it was confirmed that the expression of inserted lentiviral transgenes was controlled by DNA methylation. This cell line could potentially be used as a donor cell source for transgenic pigs and may be a useful tool for studies of gene functions involving EpiSC-like pESCs.

Materials and Methods

Animal Welfare

The care and experimental use of pigs and mice was approved by the Institute of Laboratory Animal Resources, Seoul National University (SNU-110509-1).

Isolation and Culture of Epiblast Stem Cell-like Porcine Embryonic Stem Cells

EpiSC-like pESCs were derived from in vitro-produced blastocysts. In vitro-produced and hatched blastocysts were seeded on feeder cells composed of mitotically inactivated mouse embryonic fibroblasts (MEFs) according to our previously studies [8,16]. After 5–7 days, primary colonies of embryonic stem cells were observed and cultured for approximately 7–10 days longer. Fully expanded colonies were mechanically dissociated using pulled-glass pipettes and transferred onto new feeder cells for subculture.

EpiSC-like pESCs were cultured in porcine embryonic stem cell media (PESM). PESM consisted of 1:1 mixture of Dulbecco's modified Eagle's medium (DMEM, low glucose) and Ham's F10 media containing 15% fetal bovine serum (FBS; collected and processed in the USA), 2 mM glutamax, 0.1 mM β-mercaptoethanol, 1× MEM nonessential amino acids, and 1× antibiotic–antimycotic (all from Gibco, USA). To support pluripotency and self-renewal, the embryonic stem cells were cultured in PESM with the following cytokines: 40 ng/ml human recombinant stem cell factor (hrSCF; R&D Systems, USA), 20 ng/ml human recombinant basic fibroblast growth factor (hrbFGF; R&D Systems), and 100 ng/ml heparin sodium salt (Sigma-Aldrich, USA). Media were changed every 24 h and all cells were cultured in humidified conditions with 5% CO_2 at 37°C. EpiSC-like pESCs were subcultured every 5–7 days using pulled glass pipettes. Expanded colonies were detached from the feeder cells and dissociated into

small clumps. These clumps were transferred into new feeder cells containing mitomycin-C-treated (Roche, Germany) MEFs.

Reverse Transcription-polymerase Chain Reaction (RT-PCR)

Total RNA was extracted from the cells using the TRIzol® Reagent (Invitrogen, USA) according to the manufacturer's instructions. cDNA was synthesized using the High Capacity RNA-to-cDNA Kit (Applied Biosystems, USA) at 37°C for 1 h. Derived cDNA samples were amplified with 2× PCR master mix solution (iNtRON, Korea) and 2 pmol primers as shown in Table 1. PCR reactions were performed in a thermocycler under the following conditions: 94°C for 5 min, 35 cycles of denaturation at 95°C for 30 s, annealing for 30 s (annealing temperatures depended on each primer set), extension at 72°C for 30 s, and a final extension at 72°C for 7 min. Amplified PCR products were visualized using electrophoresis on 1% agarose gel stained with ethidium bromide.

Immunocytochemistry (ICC) and Alkaline Phosphatase (AP) Staining

ICC and AP staining were performed to evaluate expression of genes related to pluripotency and AP activity. Before staining, all cell samples were preincubated for 10 min at 4°C and fixed with 4% paraformaldehyde for 30 min. After washing twice with Dulbecco's phosphate-buffered saline (DPBS; Welgene), samples were treated for 1 h with 10% goat serum in DPBS to blocking nonspecific binding. Serum-treated cells were incubated overnight at 4°C with the primary antibodies. The primary antibodies used were as follows: Oct4 (1:100; Santa Cruz Biotechnology, USA), Sox2 (1:100; Millipore, USA), Nanog (1:100; Santa Cruz Biotechnology), SSEA4 (1:100; Millipore), Tra-1-60 (1:100; Millipore), and Tra-1-81 (1:100, Millipore). When we used the antibodies for intracellular proteins such as Oct4, Sox2, and Nanog, fixed cells were treated for 5 min with 0.2% Triton-X100 (Sigma-Aldrich) before the serum blocking. After incubation with the primary antibody, the cells were treated for 3 h at room temperature with Alexa Fluor-conjugated secondary antibodies. Nuclei were stained with Hoescht 33342 (Molecular Probes, USA) or propidium iodide (PI; Sigma-Aldrich). Images of stained cells were captured using a LSM 700 Laser Scanning Microscope (Carl Zeiss, Germany) and processed with the ZEN 2009 Light Edition program (Carl Zeiss).

For AP staining, fixed EpiSC-like pESCs were incubated for 30 min at room temperature in the dark with 2% nitro blue tetrazolium chloride (NBT)/5-bromo-4-chloro-3-indolylphosphate toluidine salt (BCIP) stock solution (Roche) diluted in buffer solution (0.1 M Tris-HCl, 0.1 M NaCl, pH 9.5). Cells were then examined under an inverted microscope.

Embryoid Body (EB) Formation and in vitro Differentiation

To evaluate the ability of in vitro differentiation, embryoid bodies were generated from EpiSC-like pESCs. Cultured embryonic stem cell colonies were detached from feeder cells, and colonies were mechanically dissociated into small clumps. Suspension cultures of these clumps were obtained using the hanging-drop method for 5–6 days with PESM in the absence of cytokines. After hanging-drop culture, small clumps were aggregated and formed embryoid bodies. Cultured embryoid bodies were seeded on 0.1% gelatin-coated plates and cultured for 2–3 weeks with DMEM containing 15% FBS. After 2–3 weeks, differentiated cells were fixed in 4% paraformaldehyde and

Table 1. Primer sets for RT-PCR.

Gene	Primer sequence	Annealing temperature (°C)	Product size (bp)	Accession number
OCT4	5'-AACGATCAAGCAGTGACTATTCG-3'	60	153	AF074419
	5'-GAGTACAGGGTGGTGAAGTGAGG-3'			
NANOG	5'-AATCTTCACCAATGCCTGAG-3'	60	141	DQ447201
	5'-GGCTGTCCTGAATAAGCAGA-3'			
SOX2	5'-CAACTCTACTGCTGCGGCG-3'	56	317	EU519824
	5'-CGGGCAGTGTGTACTTATCCTTC-3'			
CRABP2	5'-CTGACCATGACGGCAGATGA-3'	60	185	NM001164509
	5'-CCCCAGAAGTGACCGAAGTG-3'			
DES	5'-CCTCAACTTCCGAGAAACAAGC-3'	60	108	NM001001535
	5'-TCACTGACGACCTCCCCATC-3'			
AFP	5'-CGCGTTTCTGGTTGCTTACAC-3'	60	483	NM214317
	5'-ACTTCTTGCTCTTGGCCTTGG-3'			
EGFP	5'-GCGACGTAAACGGCCACAAGTTC-3'	60	599	YP003162718
	5'-GACCATGTGATCGCGCTTCTCG-3'			
PSIP1	5'-CTCCTCCCTGGGCTTCGGAC-3'	60	114	XM001927571
	5'-CTCGAGCTGGCCAATGAGGAT-3'			
ß-ACTIN	5'-GTGGACATCAGGAAGGACCTCTA-3'	60	137	U07786
	5'-ATGATCTTGATCTTCATGGTGCT-3'			

analyzed using RT-PCR and immunostaining with differentiation-specific antibodies: neurofilament (ectoderm; 1:100; Millipore), vimentin (mesoderm; 1:100; Millipore), and cytokeratin 17 (endoderm; 1:100; Millipore) as described above.

Karyotype Analyses

Standard G-banding chromosome and cytogenetic analyses were used to karyotype the cell lines. Karyotyping was performed at Samkwang Medical Laboratories (Korea, http://www.smlab.co.kr/).

Lentiviral Vector Production

Lentiviral vectors containing enhanced green fluorescent protein (EGFP) were produced as previously described [35] with some modifications. HEK 293 LTV cells (Cell Biolabs, USA) were used as the packaging cell line and cultured according to the manufacturer's instructions. Four plasmids were used for the production of lentiviral vectors: self-inactivating lentiviral vector plasmid, pLL3.7; packaging plasmids, pLP1 and pLP2; and envelope plasmid, pLP/VSVG(Invitrogen). These plasmids were transfected into HEK 293 LTV cells using the calcium phosphate precipitation method. Two hours before transfection, the cells were incubated with 25 μM chloroquine (Sigma-Aldrich). After 12 h of transfection, transfected cells were treated with 15% glycerol solution for 90 s and cultured for another 24 h. Culture supernatants were harvested four times (every 12 h) and stored at 4°C. Harvested supernatants were filtered using 0.45-μm pore filters (Nalgene, USA) and concentrated by centrifugation at 18,000×g for 5 h at 4°C. The virus pellet was dissolved in PESM and stored at −76°C until use. The viral titer was calculated using the serial dilution method.

Lentiviral Transgene Transduction and Flow Cytometric Analyses

Three to four days after passaging into new feeder cells, EpiSC-like pESCs were transduced with lentiviral vectors under various multiplicities of infection (MOIs) of 1–100. Transductions were performed for 24 h in PESM containing 8 μg/ml polybrene (Sigma-Aldrich) and concentrated virus. These transduced EpiSC-like pESCs were cultured in PESM without virus for another 4–5 days and then passaged. The parts of them were analyzed by flow cytometry. To analyse the EGFP expression level in transduced EpiSC-like pESCs under each MOI, EpiSC-like pESC colonies were detached from feeder layers and dissociated into single cells using TrypLE™ Express (Gibco) and fixed with paraformaldehyde. Fixed cells were analyzed using flow cytometry (FACSCalibur) and Cell Quest software (Becton Dickinson, USA).The data were processed using the software FlowJo (Tree Star Inc., USA).

Genome Methylation Assay

To analyze methylation patterns in CMV promoter regions of lentiviral transgenes, genomic DNA of transduced cells was analyzed by bisulfite sequencing. First, genomic DNA was extracted using the G-spin™ Genomic DNA Extraction Kit for Cell/Tissue (iNtRON) and bisulfite treatment was performed using the EZ DNA Methylation-Gold™ Kit (Zymo Research, USA). Bisulfite-treated DNA samples were PCR-amplified with specific primers for the CMV promoter region (Table 2). Amplifications were performed using 2× PCR master mix solution and 2 pmol of primers designed using the Methprimer program (http://www.urogene.org/methprimer/index1.html). The resulting PCR products were separated by electrophoresis and purified from agarose gels using the MEGA-spin™ Agarose Gel Extraction Kit (iNtRON). Purified amplicons were cloned into the pGEMT-Easy Vector (Promega, USA) and transformed into *Escherichia coli* (DH5-α; Novagen, USA). Positive colonies were selected and plasmids were extracted from the selected colonies using the DNA-spin™ Plasmid DNA Purification Kit (iNtRON). The extracted plasmids were sequenced using an ABI PRISM 3730 automated sequencer (Applied Biosystems). Finally, the conversion rate of cytosine to thymine was calculated to be more than 99%, and converted sequences were processed using the BIQ Analyzer Program (http://biq-analyzer.bioinf.mpi-inf.mpg.de/) and the original sequences to analyze the methylation patterns.

5-Aza-2'-deoxycytidine (5-AzadC) and Trichostatin A (TSA) Treatments

Late-passage EGFP-transduced EpiSC-like pESCs were treated with inhibitors of repressive epigenetic markers to evaluate the relationship between viral transgene expression and epigenetic modifications. 5-Aza-2'-deoxycytidine (5-AzadC; Sigma-Aldrich) and trichostatin A (TSA; Sigma-Aldrich) (DNA methyl transferase and histone deacetylase inhibitors, respectively) were used based on previous reports [33,36], with some modifications. Three to four days after passage into new feeder cells, EpiSC-like pESCs were treated with 100 nM 5-AzadC (for 48 h), 10 nM TSA [for the first 24 h and dimethyl sulfoxide (DMSO) only for the second 24 h], or DMSO (for 48 h; Edwards Life sciences, USA) as a vehicle-only control. After treatments, cells were cultured for 2 more days without inhibitors or DMSO. Finally, treated cells were analyzed using flow cytometry. Bisulfite sequencing was performed with genomic DNA samples extracted as previously described.

Statistical Analyses

All efficiency data from flow cytometric analyses were statistically analyzed using the "R" program (http://www.r-project.org). Statistical significance between data was determined by one-way analysis of variance (ANOVA) and Tukey's honestly significant difference (HSD) test. Differences were considered significant when the P-value was less than 0.05.

Results

Establishment and Characterization of Epiblast Stem Cell-like Porcine Embryonic Stem Cells

First, the pluripotent cell line used in this study was generated from porcine preimplantation embryos. In vitro produced blastocysts were used for the derivation of EpiSC-like pESCs as previously described [8,37]. Two EpiSC-like pESC lines were established from 42 hatched blastocysts (Derivation efficiency: 4.78% (2/42)). Of these two established cell lines, only one was used for further studies. Established EpiSC-like pESCs are represented by typical flattened morphologies as previously reported [8], similar to mEpiSCs and human embryonic stem cells. Additionally, they possess AP activity (Fig. 1A) and are stably maintained over long periods (>50 passages in 1 year) with a normal karyotype (Fig. 1B; 36+ XX). These cells were analyzed for

pluripotent marker expression and their differentiation ability in vitro to verify their pluripotency, according to previously reported standards [8]. Expression of pluripotency-related transcription factors such as OCT4, SOX2, and NANOG were detected at the mRNA level (Fig. 1C). These factors, as well as EpiSC-like pESC surface markers such as SSEA4, TRA-1-60, and TRA-1-81, were also identified at the protein level using immunocytochemistry (Fig. 1D). When these cells were detached from feeder cells and cultured in suspension, they aggregated and subsequently formed embryoid bodies (EBs; Fig. 1E). The generated EBs spontaneously differentiated into three germ layers upon being placed on gelatin-coated plates. In the differentiated cells, the expression of differentiation marker genes on the three germ layers was detected by RT-PCR and immunostaining (ectoderm: CRABP, neurofilament, mesoderm: DES, vimentin, endoderm: AFP, cytokeratin 17; Fig. 1F, G). Thus, we confirmed that the established cell line was a pluripotent cell line with the differentiation potential to generate the three germ layers.

Lentiviral Transgene Introduction into Epiblast Stem Cell-like Porcine Embryonic Stem Cells

To deliver an exogenous gene into the EpiSC-like pESCs, lentiviral vectors were employed as a carrier. A lentiviral vector containing EGFP as a reporter was used to investigate the introduction of lentiviral transgenes into EpiSC-like pESCs. In a previous report, transgenic mouse and mouse ESCs were successfully produced using this lentiviral construct [38]. The estimated titer of produced lentiviral vectors in 293 LTV cells was approximately 9×10^9 viral particles/ml. Using these vectors, EpiSC-like pESCs were transduced for 24 h under several MOIs (1, 5, 10, 25, 50, 75, and 100). Unlike somatic cells, EpiSC-like pESCs lose their characteristics, including typical morphology and pluripotency, when cultured without feeder cells. Therefore, EpiSC-like pESCs were transfected in culture on feeder cells and MOIs were calculated based on the number of MEFs and EpiSC-like pESCs.

EGFP was successfully introduced into the cells via lentiviral vectors under various MOIs. EGFP expression was detected in the cells under an inverted microscope, although expression differences existed depending on the MOI (Fig. 2A). The EGFP expression levels quantified by flow cytometry significantly increased up to a MOI of 75 in a dose-dependent manner but decreased by approximately 5% at a MOI of 100 (Fig. 2B). EGFP expression was heterogeneous in the colony of cultured EpiSC-like pESCs, particularly concentrated on part of the boundary. Therefore, it is possible that expression levels could be increased up to 70–80% in a single colony by selecting the EGFP-expressing part of colonies during subculture (Fig. 3B). Cell characteristics such as viability and proliferation were rarely affected, except at a MOI of 100 in which the cells exhibited cellular toxicity post-transduction.

Table 2. Primer sets for bisulfite sequencing.

	Primer sequence	CpG	Annealing temperature (°C)	Product size (bp)
CMV	5'-ATGATTTTATGGGATTTTTTTATTTG-3'	14	56	279
	5'-ATTCACTAAACCAACTCTACTTATATAAAC-3'			
EGFP	5'- TGGGGTATAAGTTGGAGTATAATTATAATA-3'	18	54	259
	5'- AACTCCAACAAAACCATATAATC-3'			

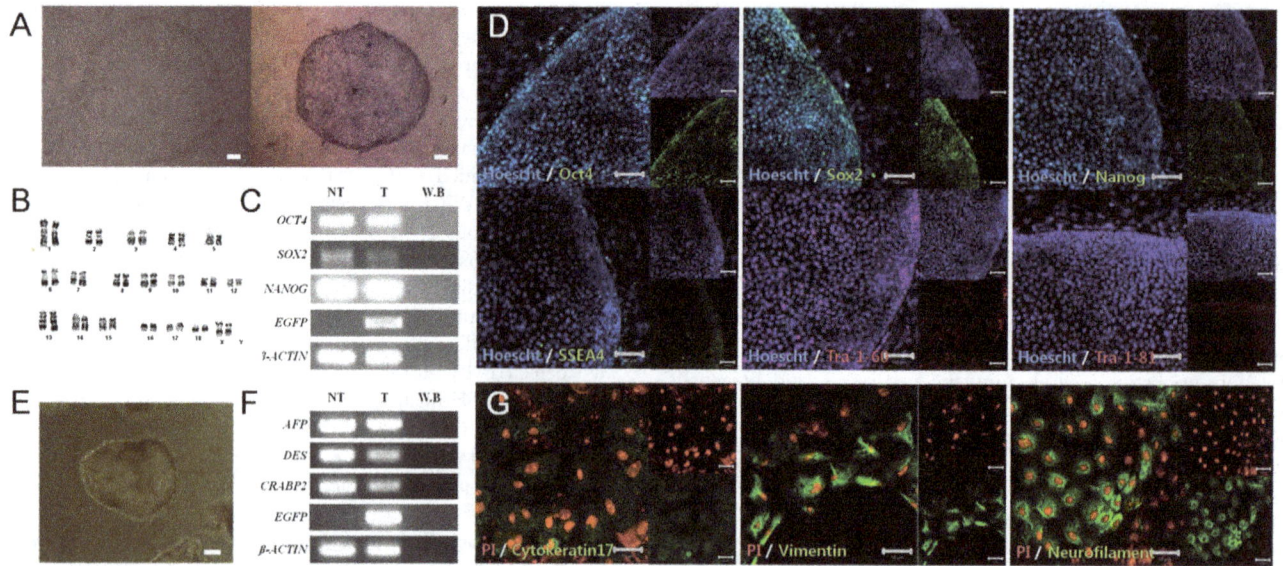

Figure 1. Derivation and characterization of EpiSC-like pESCs. (A) EpiSC-like pESCs derived from *in vitro*-produced embryos represented typical morphologies of mouse epiblast stem cells and human embryonic stem cells, and have alkaline phosphatase activity (left panel: no-stained colony; right panel: AP stained colony). (B) EpiSC-like pESCs have a normal karyotype (36+ XX) and (C) expressed genes related to pluripotency, as determined by RT-PCR (NT: non-transfected EpiSC-like pESCs passage 14, T: transfected EpiSC-like pESCs passage 14, W.B.: water blank). (D) Expression of pluripotent markers was detected using immunocytochemistry (passage number: 13). (E) Embryoid bodies were generated in suspension culture and spontaneously differentiated onto culture dishes. (F, G) The differentiated cells expressed differentiation marker genes at the mRNA and protein levels (Ectoderm: *CRABP2*, neurofilament, Mesoderm: *DES*, vimentin, Endoderm: *AFP*, cytokeratin 17; passage number: 17). Scale bars = 100 μm, except for A and G (scale bar = A: 200 μm, G: 50 μm).

EGFP-transduced EpiSC-like pESCs were characterized to assess whether pluripotency was affected by transduction. Transduced EpiSC-like pESCs under a MOI of 75 possessing the highest expression levels were used for characterization. Transduced EpiSC-like pESCs could be stably maintained over an extended time period (>50 passages) using the same general EpiSC-like pESC culture methods and had an identical karyotype (36+ XX) as before transfection (Fig. 3A). EGFP expression, as well as transcription factors related to pluripotency, was detected in transfected cells at the mRNA and protein levels as measured by RT-PCR and immunostaining, respectively (Figs. 1C and 3D). These cells could be differentiated into three germ layers *in vitro* as determined by EB formation (Fig. 3C, E). In brief, transduction-induced abnormalities, such as physiological features and pluripotency post-transfection, were not detected in EGFP-transduced EpiSC-like pESCs.

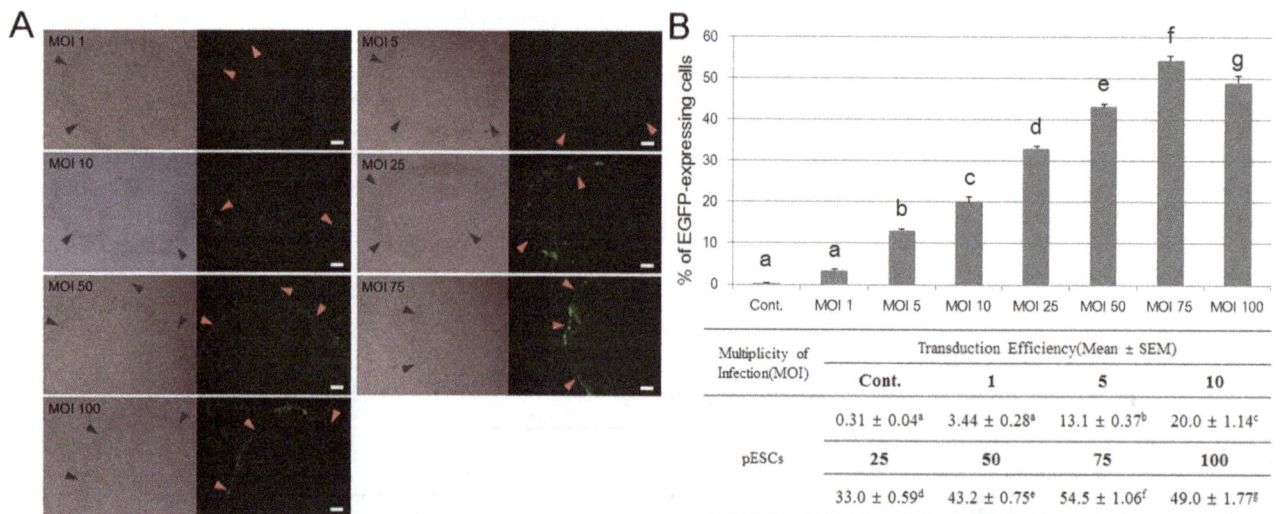

Figure 2. Lentiviral transduction of EpiSC-like pESCs. (A) Lentiviral-transduced EpiSC-like pESCs with several MOIs were cultured stably and passaged (left panels: bright field, right panels: EGFP, gray arrows: boundary of colonies, red arrows: EGFP-expressing cells; scale bar = 100 μm). (B) The transduced cells have different expression levels depending on MOI. Efficiency was measured using flow cytometry (mean ± S.E.M, n = 3). Values noted by a–h indicate they are significantly different.

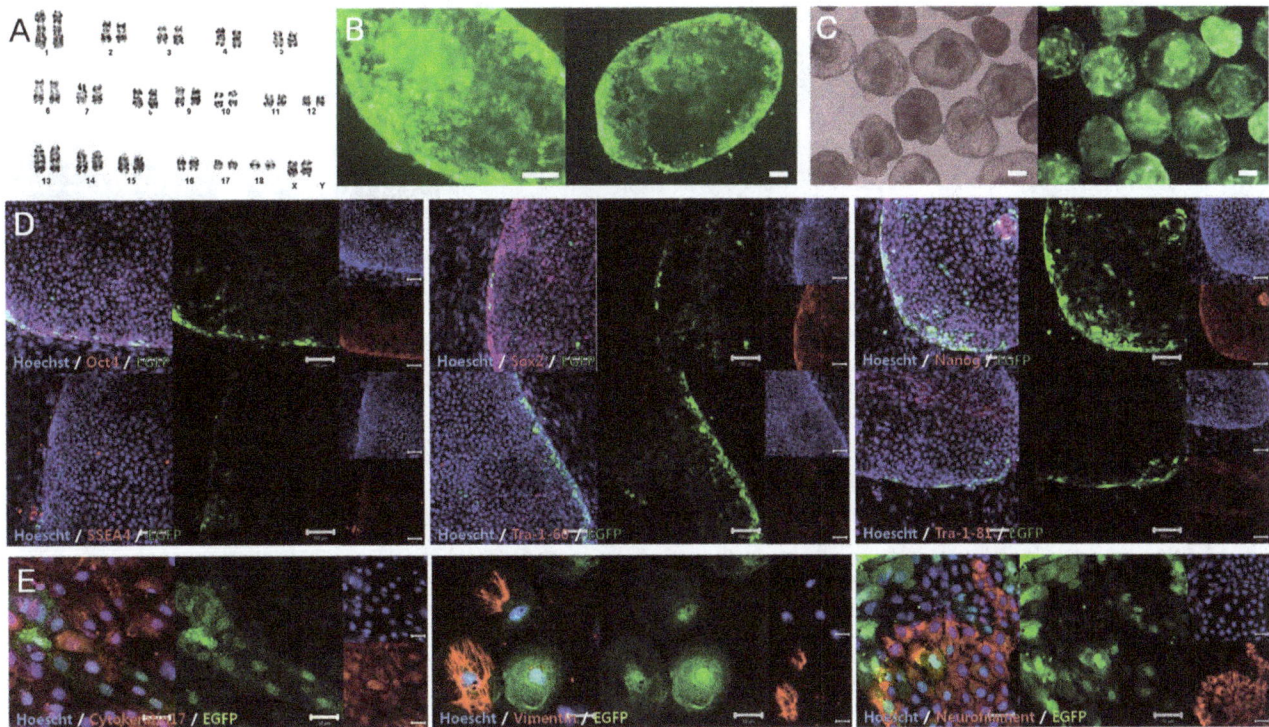

Figure 3. Characterization of EGFP-transduced EpiSC-like pESCs. (A) EGFP-transduced EpiSC-like pESCs have a normal karyotype (36+ XX). (B) The proportion of EGFP-expressing cells increased to 70–80% via selection for the EGFP-expressing part of colonies. (C) Embryoid bodies formed from EGFP-expressing EpiSC-like pESCs and expressed EGFP after aggregation (left panel: bright field, right panel: EGFP; scale bar = 100 μm). (D) EGFP-expressing EpiSC-like pESCs expressed pluripotent genes at the protein level (scale bar = 100 μm; passage number: 13) and (E) could be differentiated into three germ layers (scale bar = 50 μm; passage number: 23).

Decreased EGFP Expression with Extended Culture and Recovery of EGFP Expression by Differentiation

The proportion of cells expressing EGFP decreases when cultured for an extended time period without selection. Therefore, additional transfections were attempted to investigate this phenomenon. Thirteen days after transduction, using EpiSC-like pESCs (passage 12) under a MOI of 75, the proportion of cells expressing EGFP as measured by flow cytometry was 44.1±1.28% (Fig. 4A). Decreased EGFP expression was observed to be approximately 10% compared with the results of previous transduction (54.5±1.84% of cells expressed EGFP) (Fig. 2B). Furthermore, when cultured for longer than 13 days without selection, EGFP expression level decreased to 27.7±2.42% after 46 days. Unexpectedly, however, the decreased EGFP expression in transduced EpiSC-like pESCs was recovered by spontaneous *in vitro* differentiation. EGFP expression increased from 54.5% and 27.7% to 71.3% and 64.3% in the respective samples as measured by flow cytometry analyses. Note that EGFP expression recovered during differentiation, and the recovered EGFP expression levels of differentiated cells were higher than that of EpiSC-like pESCs.

Therefore, we hypothesized that these phenomena, including changes in EGFP expression due to extended culture and differentiation, were caused by epigenetic modifications and cellular status. First, to identify the effect of the cellular state on the expression of transgenes, MEFs and PEFs as somatic cell controls were transduced. Compared to transduced EpiSC-like pESCs, the rate of cells expressing EGFP in MEFs or PEFs was 81.2±0.70% and 74.8±5.71%, respectively, at a MOI of 75, which is similar to the expression level of differentiated cells from EpiSC-like pESCs (Fig. S2).

To analyze the effects of the epigenetic state among cell types on transgene expression, the DNA methylation patterns of the CMV promoter region in the transgene of each sample (Fig. 4A) were evaluated via bisulfite sequencing. During extended culture, methylation levels in the CMV region increased as the EGFP expression decreased (Fig. 4B). Cells from early passages have much lower methylation levels compared to those from late passage cells. Additionally, early passage cells had irregularly scattered patterns in terms of methylation sites. However, when spontaneously differentiated, the degree of DNA methylation did not change, but GFP expression increased during differentiation. The CMV promoters of differentiated cells were methylated similar to those of non-differentiated cells (Fig. 4B). When transfected somatic cells such as MEFs and PEFs were analyzed, the methylation pattern of the CMV promoter was completely unmethylated (Fig. S3).

Decreased EGFP Expression during Extended Culture due to DNA Methylation of Promoter Regions

To evaluate the effect of methylation on EGFP expression during extended culture, EpiSC-like pESCs were treated with inhibitors of repressive epigenetic marks. Control and experimental groups consisted of four: non-treated, DMSO-, 5-AzadC-, and TSA-treated samples. Late passage (passage 46) EpiSC-like pESCs expressed low levels of EGFP (4.11±0.28%). Samples treated with DMSO and TSA did not show a recovery of EGFP expression and showed similar expression levels as nontreated samples. Although these two groups did not affect expression of the transgenes, the silenced EGFP expression recovered to 20.0±1.6% upon treatment with 5-AzadC (Fig. 5A, B). The effect of 5-AzadC on the

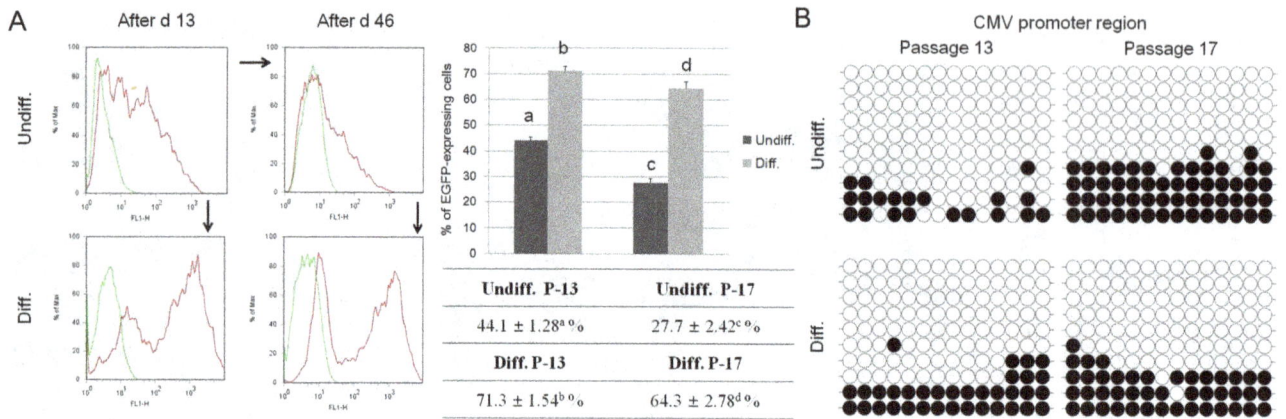

Figure 4. Change in EGFP expression during extended culture and *in vitro* differentiation. (A) EGFP expression levels of transduced EpiSC-like pESCs declined during extended culture. The decreased expression recovered during *in vitro* differentiation. (B) To analyze the effects of the epigenetic state on transgene expression, the DNA methylation patterns of the CMV promoter region in the transgene of each sample were evaluated via bisulfite sequencing. Along with decreases in EGFP expression, the DNA methylation levels of the CMV promoter region increased. Although DNA methylation levels increased during extended culture, they were not altered by *in vitro* differentiation. (Each circle indicates individual CpG dinucleotides. White and dark circles represent unmethylated and methylated CpGs, respectively. Each row represents one individual clone of amplified PCR products). Values noted by a–h indicate they are significantly different. Data are the mean± S.E.M. (*n* = 3).

expression of transgenes was determined via bisulfite sequencing. Among the four groups, CMV promoter region demethylation occurred after 5-AzadC treatment (Fig. 5C).

Discussion

Because pluripotent cells have great potential as a cell source, research in this area has focused on embryonic carcinoma cells to iPSCs [4,39]. Embryonic stem cell research began in 1981 by the establishment of mouse embryonic stem cells and was accelerated

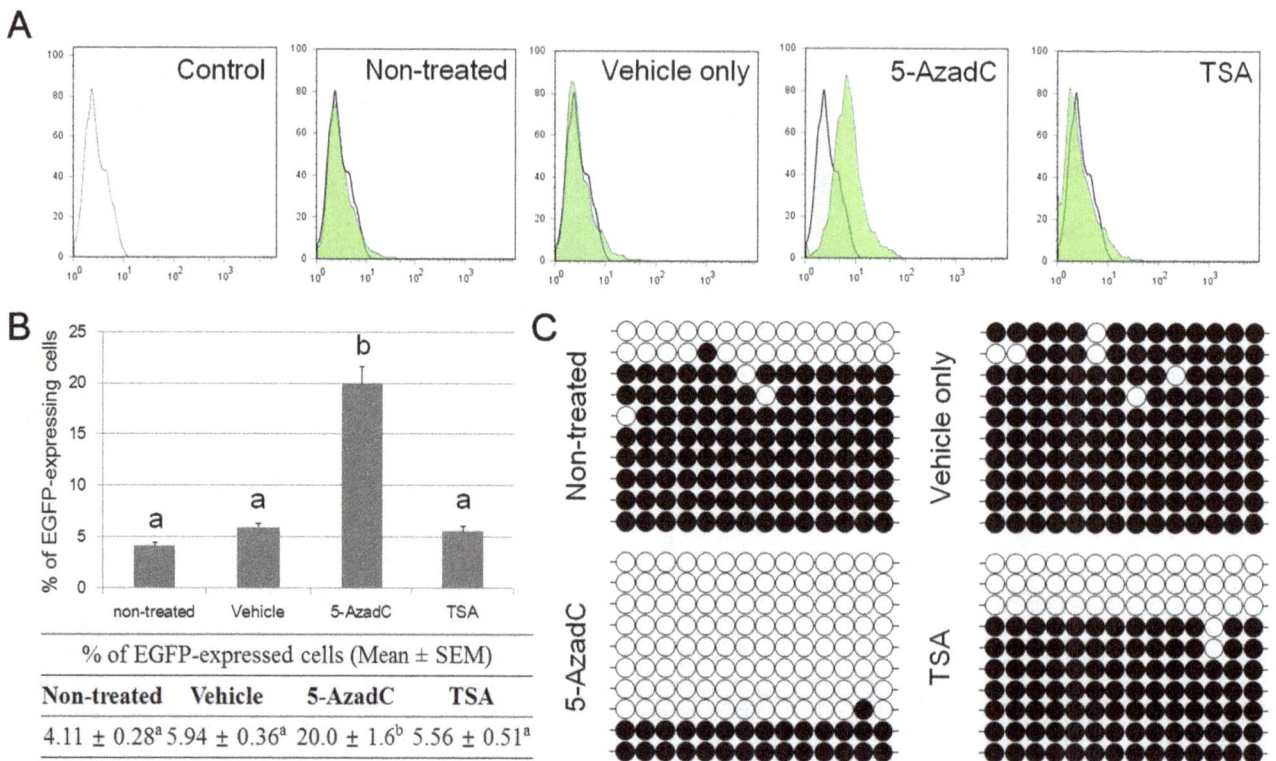

Figure 5. Silenced transgenes can be reactivated by treatment with the DNA methylase inhibitor, 5′-aza-2′-deoxycitidine (5-AzadC). (A) 5-AzadC treatment reactivates the silenced transgene, EGFP. (B) The recovered expression level was measured by flow cytometry. (C) 5-AzadC demethylated the methylated CMV promoter region. Values noted by a–h indicate they are significantly different. Data are the means± S.E.M. (*n* = 3).

by the establishment of human embryonic stem cells and iPSCs in 1998 and 2006, respectively [1,4,40]. The purpose of pluripotent stem cell research in humans and mice includes the elucidation of basic cellular mechanisms contributing to the maintenance of pluripotency. In humans, applications exist in cell therapies. However, in domestic animals, the research aim is to create an indefinite cell source for transgenic animals used as bioreactors and tissue engineering materials as well as preliminary studies for human research [9]. Because of the physiological and immunological similarities that exist between pigs and humans, porcine pluripotent cell lines have been identified as important candidates for preliminary studies on human disease as well as a source for generating transgenic animals [10–12]. Therefore, the establishment and characterization of pESCs, along with the generation of stable transgenic cell lines, are vitally important. In our previous studies, we developed a reproducible method for the establishment of EpiSC-like pESCs and iPSCs and determined that similar to humans, pig is a nonpermissive species and pig pluripotent stem cells show a primed pluripotent state regarding marker gene expression patterns and X chromosome inactivation [8,16].

To achieve such goals, stem cell engineering involving the introduction of transgenes into cells has been developed. In mice and humans, many transfection methods, including viral vectors, liposome-mediated gene delivery, and electroporation, have been studied [20–23]. In the porcine model, however, much fewer studies have been reported. In the first reported study, EGFP-transduced embryonic germ cells (EGCs) were established by the introduction of plasmid vectors using liposomal vectors [30]. In a second study, plasmid vectors containing humanized renilla green fluorescent protein (hrGFP) were introduced into pESCs was via electroporation [17]. Although GFP-expressing pESC lines were established via electroporation, transfection efficiency was very low (only three stably GFP-expression lines from 12 trials), and a GFP-expressing line was not obtained by retroviral and liposome-mediated transfection. Therefore, we undertook our study to develop a system for efficient introduction of transgenes into embryo-derived pluripotent stem cells. Our previous studies showed that EpiSC-like pESCs have very high cytotoxicity when treated with liposome-mediated and electroporation methods, as well as low colony-forming rates from dissociated single cells, similar to humans [8,41]. Therefore, viral vectors were selected because plasmid vectors are not suitable for generating a stable transgene expression cell line. And because lentiviral vectors can infect dividing cells as well as nondividing cells, lentiviral vectors were chosen for transfection of EpiSC-like pESCs having long doubling times of approximately 36 h (Fig. S1), similar to human ESCs [7,27].

Historically, although retro- and lentiviral vectors have been widely used for the production of transgenic animals and the establishment of transgenic pluripotent stem cells, silencing of the viral transgenes caused by epigenetic modifications and *trans*-acting factors remains an obstacle to be resolved [31–34,42–44]. To overcome these silencing problems in cells, particularly embryos and pluripotent cells, various approaches have been developed. The major difficulties post-transfection, the repression of long terminal repeats (LTRs) in the viral genome known as transcriptional regulator by *de novo* DNA methylation and repressive *trans*-acting factors [45,46], were resolved using bidirectional or internal promoters such as the CMV promoter, which is strongly expressed in various tissues, [47,48] and the deletion or modification of LTR sequences to prevent recruitment of repressive *trans*-acting factors [43,49]. Using regulatory elements, including woodchuck hepatitis virus response element (WRE) [50], HIV FLAP [51], and matrix attachment region (MAR) [52],

which are responsible for transcript stabilization and the translocation of provirus into nuclear and DNA loop formation, respectively, also improve transgene expression.

Prior to transfection, we established a new EpiSC-like pESC line using the whole seeding method, as previously reported [8]. The established line exhibited the general characteristics of a primed pluripotent state in terms of marker gene expression, ability to differentiate *in vitro*, and normal karyotype (Fig. 1). The lentiviral construct used in this study is pLL3.7 consisted of EGFP as a marker gene with the CMV promoter and *cis*-acting elements including FLAP, WRE, 3′SIN-LTR and multiple cloning site for shRNA. The reason this construct was chosen is as follows: 1) This vector has been proven to work efficiently in mouse embryos, primary cells and embryonic stem cell lines, and transgenic mice have been generated using mouse ESCs and embryos transfected with this vector [38]. 2) In previous transfection studies involving porcine pluripotent cells, the CMV promoter was used for the establishment of transgene-expressed cell lines, these resulting transduced cell lines stably expressed the transgene for more than 90 months [17,30]. When EpiSC-like pESCs were transfected with this vector under various MOIs, cytotoxicity was not detected up to a MOI of 75. At a MOI of 100, however, cytotoxicity occurred and reduced the number of EGFP-expressing cells (Fig. 2B). Moreover, EGFP expression was stronger at the edge of the colonies (Fig. 2A), likely because of metabolic upregulation and high cell density due to proliferation of dividing cells at the edge of the colonies.

To examine the decline in EGFP expression during extended culture, we assessed whether transgene expression was affected by DNA methylation or histone acetylation, which are known to epigenetically regulate gene expression. Methylated cytosine in DNA represses gene expression via recruitment of proteins associated with heterochromatin such as MeCP2. Acetylation on histone tails activates gene expression by increasing the negative charge of histones [44]. Bisulfite sequencing data indicated that the DNA methylation level of the promoter region and expression of the transgene were negatively correlated because DNA methylation in the CMV promoter region increased with a concomitant decrease in EGFP expression (Fig. 4). However, methylation levels of the EGFP region were not related to transgene expression (data not shown), in contrast to a previous report involving transgenic animals [33]. The EGFP region was hypomethylated regardless of EGFP expression.

The expression level of transgenes is dependent upon the vector construct, transfection methods used and cell types. It is therefore important that the characterization of vector activities in various cell types is given attention. In the case of the CMV promoters used in this study, differences in expression patterns have been reported in several papers via the characterization of the transcriptional activities in ESCs. Some studies showed that the CMV promoter is active in human or mouse ESCs during long term culture [38,53–55]. Conversely, other studies have reported that CMV-driven transgene expressions are rapidly downregulated or inactive in human or mouse ESCs. Silenced transgenes were not reactivated during differentiation [56–59]. In porcine studies, CMV-driven expression of GFP was stably expressed during the propagation of pESCs (over 20 months) [17], and transgenic porcine embryonic germ cell lines were successfully established using a CMV-EGFP construct [30]. Our data has shown that the CMV-driven expression of GFP was progressively downregulated by DNA methylations and reactivated during differentiation by transacting factors without changes of DNA methylation levels. This result is consistent with previous findings of the latter cases in humans and mice with the exception of the upregulation of transgenes during differentiations. The differences of our results with previous porcine studies may be as a result of the different vector construct and transfection method used. The reactivation of silenced transgenes during differentiation that we observed in this study indicate that trans-acting factors, as well as DNA methylations, affect transgene silencing in undifferentiated EpiSC-like pESCs.

When transduced cells were treated with 5-AzadC and TSA, inhibitors of DNA methylation and histone deacetylases, respectively, 5-AzadC allowed reactivation of silenced expression, but TSA did not (Fig. 5A, B). Bisulfite sequencing of the promoter regions showed that 5-AzadC-treated samples were hypomethylated in CMV promoter regions compared with the other treated groups, which had hypermethylated promoters. These results clearly demonstrate that decreased transgene expression is related to DNA methylation in promoter regions, not histone modifications.

In addition to silencing of gene expression during extended culture, the transfection efficiency of EpiSC-like pESCs was lower than that of other embryo-originated somatic cells (Fig. S2). However, PSIP1, a protein that participates in lentiviral provirus integration into host genomes [60], was examined by RT-PCR and no difference was observed among EpiSC-like pESCs, PEFs, and MEFs (Fig. S4). This result suggests that the low efficiency of transfection is not related to the low integrity of the provirus. In addition, because the copy number of an inserted lentiviral construct could affect EGFP expression, a correlation between EGFP expression level and transgene copy number was verified. However, because of heterogeneity in the transduced cells due to the random insertion of multiple copies, it is hard to quantitatively measure the copy number of inserted transgenes. So, transgene copy number was relatively quantified by real-time RCR. Real-time PCR results showed that changes in EGFP expression and transgene copy number are not correlated (Fig. S5). And also reactivation of silenced transgenes during differentiation clearly demonstrated that low expression levels are not due to low integrity (Fig. 4A). Note that DNA methylation levels of the CMV promoter region did not change during differentiation although transgene expression increased. Undifferentiated and differentiated cells at the same passage have similar levels of DNA methylation in the CMV promoter (Fig. 4B). This suggests that the recruitment of transcription factors in differentiated cells or the downregulation of repressive trans-acting factors, rather than demethylation of the promoter region, is involved in the unmethylated open chromatin for reactivation of silenced transgenes. In fact, many binding sites for putative transacting factors such as cyclic AMP-response elements (CRE), NF-Kappa B, AP-1, serum response elements exist in the CMV promoter [61]. These sites could be predicted by the program TFSEARCH ver. 1.3 (http://www.cbrc.jp/research/db/TFSEARCH.html). This mechanism is supported by the phenomenon that more methylated cells in late passages showed lower reactivation of transgenes (Fig. 4A). Therefore, low transgene expression in pluripotent cells is likely because of trans-acting factors as well as DNA methylation.

In conclusion, we were able to successfully derive EpiSC-like pESCs and introduced an EGFP transgene into these cells using lentiviral vectors. Transgene expression in EpiSC-like pESCs was altered during maintenance and differentiation due to epigenetic changes. Although we used modified lentiviral vectors containing regulatory elements, we could not prevent the transgene silencing that occurred due to DNA methylation and trans-acting factors.

Nonetheless the silenced transgene expression was reactivated by differentiation and treatment of 5-AzadC. Therefore, this system could be applied for the short term analysis of gene function in EpiSC-like pESCs, induction of differentiation, tracking of transplanted cells and the production of chimeras. Finally, this cell line could potentially be used as a donor cell source for transgenic pigs and serve as a useful tool for studies involving pESCs and gene therapy in humans, as well as aid in the understanding of epigenetic regulation of transgenes.

Supporting Information

Figure S1 Doubling time of epiblast stem cell-like porcine embryonic stem cells. Because EpiSC-like pESCs have long doubling times of approximately 36 h, similar to human ESCs, lentiviral vectors were chosen for transfection of EpiSC-like pESCs.

Figure S2 Lentiviral transduction efficiency of MEFs and PEFs at various MOIs. MEFs and PEFs were transfected at various MOIs for comparing with EpiSC-like pESCs. The transfection efficiency of embryo-originated somatic cells was higher than that of EpiSC-like pESCs. The EGFP expression levels quantified by flow cytometry significantly increased up to a MOI of 50 in a dose-dependent manner and were reached a plateau from MOI of 75.

Figure S3 Methylation level of the CMV promoter region in MEFs and PEFs. To compare the DNA methylation patterns of the CMV promoter region between transfected porcine pluripotent cells and somatic cells such as MEFs and PEFs, bisulfite sequencing was performed. The CpGs were completely unmethylated in CMV promoter of MEFs and PEFs.

Figure S4 Expression level of *PSIP1* in MEFs, PEFs, and EpiSC-like pESCs. PSIP1, a protein that participates in lentiviral provirus integration into host genomes, was examined by RT-PCR and no difference was observed among EpiSC-like pESCs, PEFs, and MEFs.

Figure S5 Correlation between the copy number of inserted transgenes and the EGFP expression level. To analyze the effects of transgene copy number on changes of transgene expression, the transgene copy number was relatively quantified by real-time RCR. Real-time PCR results showed that changes in EGFP expression and transgene copy number are not correlated.

Author Contributions

Conceived and designed the experiments: KHC CKL. Performed the experiments: KHC JKP HSK KJU DCS. Analyzed the data: KHC JKP CKL. Contributed reagents/materials/analysis tools: KHC. Wrote the paper: KHC CKL.

References

1. Evans MJ, Kaufman MH (1981) Establishment in culture of pluripotential cells from mouse embryos. Nature 292: 154–156.

2. Martin GR (1981) Isolation of a pluripotent cell line from early mouse embryos cultured in medium conditioned by teratocarcinoma stem cells. Proc Natl Acad Sci U S A 78: 7634–7638.

3. Tesar PJ, Chenoweth JG, Brook FA, Davies TJ, Evans EP, et al. (2007) New cell lines from mouse epiblast share defining features with human embryonic stem cells. Nature 448: 196–199.

4. Takahashi K, Yamanaka S (2006) Induction of pluripotent stem cells from mouse embryonic and adult fibroblast cultures by defined factors. Cell 126: 663–676.

5. Nichols J, Smith A (2009) Naive and primed pluripotent states. Cell Stem Cell 4: 487–492.

6. Hanna J, Markoulaki S, Mitalipova M, Cheng AW, Cassady JP, et al. (2009) Metastable pluripotent states in NOD-mouse-derived ESCs. Cell Stem Cell 4: 513–524.

7. Buecker C, Chen HH, Polo JM, Daheron L, Bu L, et al. (2010) A murine ESC-like state facilitates transgenesis and homologous recombination in human pluripotent stem cells. Cell Stem Cell 6: 535–546.

8. Park JK, Kim HS, Uh KJ, Choi KH, Kim HM, et al. (2013) Primed pluripotent cell lines derived from various embryonic origins and somatic cells in pig. PLoS One 8: e52481.

9. Keefer CL, Pant D, Blomberg L, Talbot NC (2007) Challenges and prospects for the establishment of embryonic stem cell lines of domesticated ungulates. Anim Reprod Sci 98: 147–168.

10. Hall V (2008) Porcine embryonic stem cells: a possible source for cell replacement therapy. Stem Cell Rev 4: 275–282.

11. Brevini TA, Antonini S, Cillo F, Crestan M, Gandolfi F (2007) Porcine embryonic stem cells: Facts, challenges and hopes. Theriogenology 68 Suppl 1: S206–213.

12. Houdebine LM (2009) Production of pharmaceutical proteins by transgenic animals. Comp Immunol Microbiol Infect Dis 32: 107–121.

13. Pierson RN, . (2009) Current status of xenotransplantation and prospects for clinical application. Xenotransplantation 16: 263–280.

14. Piedrahita JA, Anderson GB, Bondurant RH (1990) On the isolation of embryonic stem cells: Comparative behavior of murine, porcine and ovine embryos. Theriogenology 34: 879–901.

15. Ezashi T, Telugu BP, Alexenko AP, Sachdev S, Sinha S, et al. (2009) Derivation of induced pluripotent stem cells from pig somatic cells. Proc Natl Acad Sci U S A 106: 10993–10998.

16. Son HY, Kim JE, Lee SG, Kim HS, Lee E, et al. (2009) Efficient Derivation and Long Term Maintenance of Pluripotent Porcine Embryonic Stem-like Cells. Asian-australasian journal of animal sciences 22: 26–34.

17. Yang JR, Shiue YL, Liao CH, Lin SZ, Chen LR (2009) Establishment and characterization of novel porcine embryonic stem cell lines expressing hrGFP. Cloning Stem Cells 11: 235–244.

18. Kues WA, Herrmann D, Barg-Kues B, Haridoss S, Nowak-Imialek M, et al. (2013) Derivation and characterization of sleeping beauty transposon-mediated porcine induced pluripotent stem cells. Stem Cells Dev 22: 124–135.

19. Thomas KR, Capecchi MR (1987) Site-directed mutagenesis by gene targeting in mouse embryo-derived stem cells. Cell 51: 503–512.

20. Eiges R, Schuldiner M, Drukker M, Yanuka O, Itskovitz-Eldor J, et al. (2001) Establishment of human embryonic stem-transfected clones carrying a marker for undifferentiated cells. Curr Biol 11: 514–518.

21. Ko BS, Chang TC, Shyue SK, Chen YC, Liou JY (2009) An efficient transfection method for mouse embryonic stem cells. Gene Ther 16: 154–158.

22. Ma Y, Ramezani A, Lewis R, Hawley RG, Thomson JA (2003) High-level sustained transgene expression in human embryonic stem cells using lentiviral vectors. Stem Cells 21: 111–117.

23. Pfeifer A, Ikawa M, Dayn Y, Verma IM (2002) Transgenesis by lentiviral vectors: lack of gene silencing in mammalian embryonic stem cells and preimplantation embryos. Proc Natl Acad Sci U S A 99: 2140–2145.

24. Hohenstein KA, Pyle AD, Chern JY, Lock LF, Donovan PJ (2008) Nucleofection mediates high-efficiency stable gene knockdown and transgene expression in human embryonic stem cells. Stem Cells 26: 1436–1443.

25. Zhang X, Godbey WT (2006) Viral vectors for gene delivery in tissue engineering. Adv Drug Deliv Rev 58: 515–534.

26. Bukrinsky MI, Haggerty S, Dempsey MP, Sharova N, Adzhubel A, et al. (1993) A nuclear localization signal within HIV-1 matrix protein that governs infection of non-dividing cells. Nature 365: 666–669.

27. Naldini L, Blomer U, Gallay P, Ory D, Mulligan R, et al. (1996) In vivo gene delivery and stable transduction of nondividing cells by a lentiviral vector. Science 272: 263–267.

28. Tan G, Ren L, Huang Y, Tang X, Zhou Y, et al. (2011) Isolation and culture of embryonic stem-like cells from pig nuclear transfer blastocysts of different days. Zygote: 1–6.

29. Huang L, Fan N, Cai J, Yang D, Zhao B, et al. (2011) Establishment of a porcine Oct-4 promoter-driven EGFP reporter system for monitoring pluripotency of porcine stem cells. Cell Reprogram 13: 93–98.

30. Rui R, Qiu Y, Hu Y, Fan B (2006) Establishment of porcine transgenic embryonic germ cell lines expressing enhanced green fluorescent protein. Theriogenology 65: 713–720.

31. Cherry SR, Biniszkiewicz D, van Parijs L, Baltimore D, Jaenisch R (2000) Retroviral expression in embryonic stem cells and hematopoietic stem cells. Mol Cell Biol 20: 7419–7426.

32. Kosaka Y, Kobayashi N, Fukazawa T, Totsugawa T, Maruyama M, et al. (2004) Lentivirus-based gene delivery in mouse embryonic stem cells. Artif Organs 28: 271–277.

33. Hofmann A, Kessler B, Ewerling S, Kabermann A, Brem G, et al. (2006) Epigenetic regulation of lentiviral transgene vectors in a large animal model. Mol Ther 13: 59–66.

34. Park SH, Kim JN, Park TS, Lee SD, Kim TH, et al. (2010) CpG methylation modulates tissue-specific expression of a transgene in chickens. Theriogenology 74: 805–816 e801.

35. Nagano M, Watson DJ, Ryu BY, Wolfe JH, Brinster RL (2002) Lentiviral vector transduction of male germ line stem cells in mice. FEBS Lett 524: 111–115.

36. Kong Q, Wu M, Wang Z, Zhang X, Li L, et al. (2011) Effect of trichostatin A and 5-Aza-2'-deoxycytidine on transgene reactivation and epigenetic modification in transgenic pig fibroblast cells. Mol Cell Biochem 355: 157–165.

37. Lee SG, Park CH, Choi DH, Kim HS, Ka HH, et al. (2007) In vitro development and cell allocation of porcine blastocysts derived by aggregation of in vitro fertilized embryos. Mol Reprod Dev 74: 1436–1445.

38. Rubinson DA, Dillon CP, Kwiatkowski AV, Sievers C, Yang L, et al. (2003) A lentivirus-based system to functionally silence genes in primary mammalian cells, stem cells and transgenic mice by RNA interference. Nat Genet 33: 401–406.

39. Martin GR, Evans MJ (1974) The morphology and growth of a pluripotent teratocarcinoma cell line and its derivatives in tissue culture. Cell 2: 163–172.

40. Thomson JA, Itskovitz-Eldor J, Shapiro SS, Waknitz MA, Swiergiel JJ, et al. (1998) Embryonic stem cell lines derived from human blastocysts. Science 282: 1145–1147.

41. Amit M, Carpenter MK, Inokuma MS, Chiu CP, Harris CP, et al. (2000) Clonally derived human embryonic stem cell lines maintain pluripotency and proliferative potential for prolonged periods of culture. Dev Biol 227: 271–278.

42. Whitelaw CB, Lillico SG, King T (2008) Production of transgenic farm animals by viral vector-mediated gene transfer. Reprod Domest Anim 43 Suppl 2: 355–358.

43. Laker C, Meyer J, Schopen A, Friel J, Heberlein C, et al. (1998) Host cis-mediated extinction of a retrovirus permissive for expression in embryonal stem cells during differentiation. J Virol 72: 339–348.

44. Hotta A, Ellis J (2008) Retroviral vector silencing during iPS cell induction; an epigenetic beacon that signals distinct pluripotent states. J Cell Biochem 105: 940–948.

45. Hoeben RC, Migchielsen AA, van der Jagt RC, van Ormondt H, van der Eb AJ (1991) Inactivation of the Moloney murine leukemia virus long terminal repeat in murine fibroblast cell lines is associated with methylation and dependent on its chromosomal position. J Virol 65: 904–912.

46. Loh TP, Sievert LL, Scott RW (1990) Evidence for a stem cell-specific repressor of Moloney murine leukemia virus expression in embryonal carcinoma cells. Mol Cell Biol 10: 4045–4057.

47. Hamaguchi I, Woods NB, Panagopoulos I, Andersson E, Mikkola H, et al. (2000) Lentivirus vector gene expression during ES cell-derived hematopoietic development in vitro. J Virol 74: 10778–10784.

48. Golding MC, Mann MR (2011) A bidirectional promoter architecture enhances lentiviral transgenesis in embryonic and extraembryonic stem cells. Gene Ther 18: 817–826.

49. Miyoshi H, Blomer U, Takahashi M, Gage FH, Verma IM (1998) Development of a self-inactivating lentivirus vector. J Virol 72: 8150–8157.

50. Zufferey R, Donello JE, Trono D, Hope TJ (1999) Woodchuck hepatitis virus posttranscriptional regulatory element enhances expression of transgenes delivered by retroviral vectors. J Virol 73: 2886–2892.

51. Arhel NJ, Souquere-Besse S, Munier S, Souque P, Guadagnini S, et al. (2007) HIV-1 DNA Flap formation promotes uncoating of the pre-integration complex at the nuclear pore. EMBO J 26: 3025–3037.

52. Bode J, Benham C, Knopp A, Mielke C (2000) Transcriptional augmentation: modulation of gene expression by scaffold/matrix-attached regions (S/MAR elements). Crit Rev Eukaryot Gene Expr 10: 73–90.

53. Ward CM, Stern PL (2002) The human cytomegalovirus immediate-early promoter is transcriptionally active in undifferentiated mouse embryonic stem cells. Stem Cells 20: 472–475.

54. Zeng X, Chen J, Sanchez JF, Coggiano M, Dillon-Carter O, et al. (2003) Stable expression of hrGFP by mouse embryonic stem cells: promoter activity in the undifferentiated state and during dopaminergic neural differentiation. Stem Cells 21: 647–653.

55. Bagchi B, Kumar M, Mani S (2006) CMV promotor activity during ES cell differentiation: potential insight into embryonic stem cell differentiation. Cell Biol Int 30: 505–513.

56. Liew CG, Draper JS, Walsh J, Moore H, Andrews PW (2007) Transient and stable transgene expression in human embryonic stem cells. Stem Cells 25: 1521–1528.

57. Wang R, Liang J, Jiang H, Qin LJ, Yang HT (2008) Promoter-dependent EGFP expression during embryonic stem cell propagation and differentiation. Stem Cells Dev 17: 279–289.

58. Liu J, Jones KL, Sumer H, Verma PJ (2009) Stable transgene expression in human embryonic stem cells after simple chemical transfection. Mol Reprod Dev 76: 580–586.

59. Norrman K, Fischer Y, Bonnamy B, Wolfhagen Sand F, Ravassard P, et al. (2010) Quantitative comparison of constitutive promoters in human ES cells. PLoS One 5: e12413.

60. Engelman A, Cherepanov P (2008) The lentiviral integrase binding protein LEDGF/p75 and HIV-1 replication. PLoS Pathog 4: e1000046.

61. Meier JL, Stinski MF (1996) Regulation of human cytomegalovirus immediate-early gene expression. Intervirology 39: 331–342.

10

Icariin Ameliorates Neuropathological Changes, TGF-β1 Accumulation and Behavioral Deficits in a Mouse Model of Cerebral Amyloidosis

Zhi-Yuan Zhang*, Chaoyun Li, Caroline Zug, Hermann J. Schluesener

Division of Immunopathology of the Nervous System, Institute of Pathology and Neuropathology, University of Tuebingen, Tuebingen, Germany

Abstract

Icariin, a major constituent of flavonoids from the Chinese medicinal herb Epimedium brevicornum, exhibits multiple biological properties, including anti-inflammatory, neuroregulatory and neuroprotective activities. Therefore, Icariin might be applied in treatment of neurodegenerative disorders, including Alzheimer's disease (AD), which is neuropathologically characterized by β-amyloid aggregation, hyperphosphorylated tau and neuroinflammation. Potential therapeutic effects of Icariin were investigated in an animal model of cerebral amyloidosis for AD, transgenic APP/PS1 mouse. Icariin was suspended in carboxymethylcellulose and given orally to APP/PS1 mice. Therapeutic effects were monitored by behavioral tests, namely nesting assay, before and during the experimental treatment. Following an oral treatment of 10 days, Icariin significantly attenuated Aβ deposition, microglial activation and TGF-β1 immunoreactivity at amyloid plaques in cortex and hippocampus of transgenic mice 5 months of age, and restored impaired nesting ability. Our results suggest that Icariin might be considered a promising therapeutic option for human AD.

Editor: Jaya Padmanabhan, University of S. Florida College of Medicine, United States of America

Funding: Chaoyun Li is supported and financed by the Chinese government scholarship from the China Scholarship Council (CSC). The authors also acknowledge support by Deutsche Forschungsgemeinschaft (German Research Foundation) and Open Access Publishing Fund of Tuebingen University, for the open access publishing. The funders had no role in study design, data collection and analysis, decision to publish, or preparation of the manuscript.

Competing Interests: The authors have declared that no competing interests exist.

* Email: zhiyuan.zhang@medizin.uni-tuebingen.de

Introduction

Icariin is a natural flavonoid extracted from the Chinese tonic herb Epimedium and is considered the major pharmacologically active compound. Multi-functional Icariin possesses anti-tumor, anti-oxidant, vasorelaxant, anti-bacterial and anti-inflammatory activities [1] (Figure 1). Previous studies reported that Icariin ameliorated brain dysfunction induced by LPS [2], inhibited corticosterone-induced apoptosis in neurons [3], attenuated the ischemia/reperfusion damage to neurons [4], stimulated neurite growth [5] and thereby showed anti-neuroinflammatory and neuroprotective activities. Further, Icariin shows antidepressant-like activity [6]; protects against aluminium-induced learning and memory deficits due to its antioxidant activities; decreases Aβ1-40 content in the hippocampus of aluminium-intoxicated rodents [7]; and its metabolite Icaritin has neuroprotective effects on β amyloid-induced neurotoxicity to neuronal cells [8]. Therefore, Icariin is considered as a potential therapy against neurodegenerative diseases such as Alzheimer's disease (AD).

AD is the most common form of neurodegeneration and the major cause of dementia. AD is clinically defined by distinct behavioral and cognitive deficits, and the most characteristic neuropathological feature is extracellular deposit of aggregated β-amyloid (Aβ) peptide (amyloid plaques). It is proposed that Aβ peptides are toxic and causative in AD, contributing to memory loss, behavioral impairment and neurodegenerative pathology [9]. Beside the well known Aβ aggregation, neuroinflammation

also plays an important role in the pathophysiology of this multifactorial disorder [10]. Neuroinflammation is characterized by release of numerous inflammatory mediators, microglial and astroglial activation, in particular around senile plaques [11,12]. Neuroinflammation may contribute to neural dysfunction and cell death, establishing a self-perpetuating vicious cycle by which inflammation induces further neurodegeneration [13]. Therefore, anti-inflammatory drugs may have beneficial effects on neurodegenerative disorders, including AD [13]. In addition, up-regulated TGF-β1 has been observed in the brain of AD patients and AD animal models [14,15]. As a potent immunomodulator and with a central role in the response to neuroinflammation, TGF-β1 is considered to be involved in pathological progression of AD.

Our aim was to study the potential therapeutic effect of Icariin in a transgenic mouse model of AD, focusing on its effect on Aβ deposition, neuroinflammation, TGF-β1 expression and behavioral deficits. We used the APP/PS1-21 double transgenic mouse model of cerebral amyloidosis, co-expressing the KM670/671NL mutated human amyloid precursor protein and the L166P mutated human presenilin 1 (APP/PS1-21 mice) under the neuron-specific Thy1 promoter element. In this transgenic line, amyloid deposition occurred as early as 2 months in the cortex and 4 months in the hippocampus, accompanied by inflammatory responses and robust impairment of cognitive function [16–18].

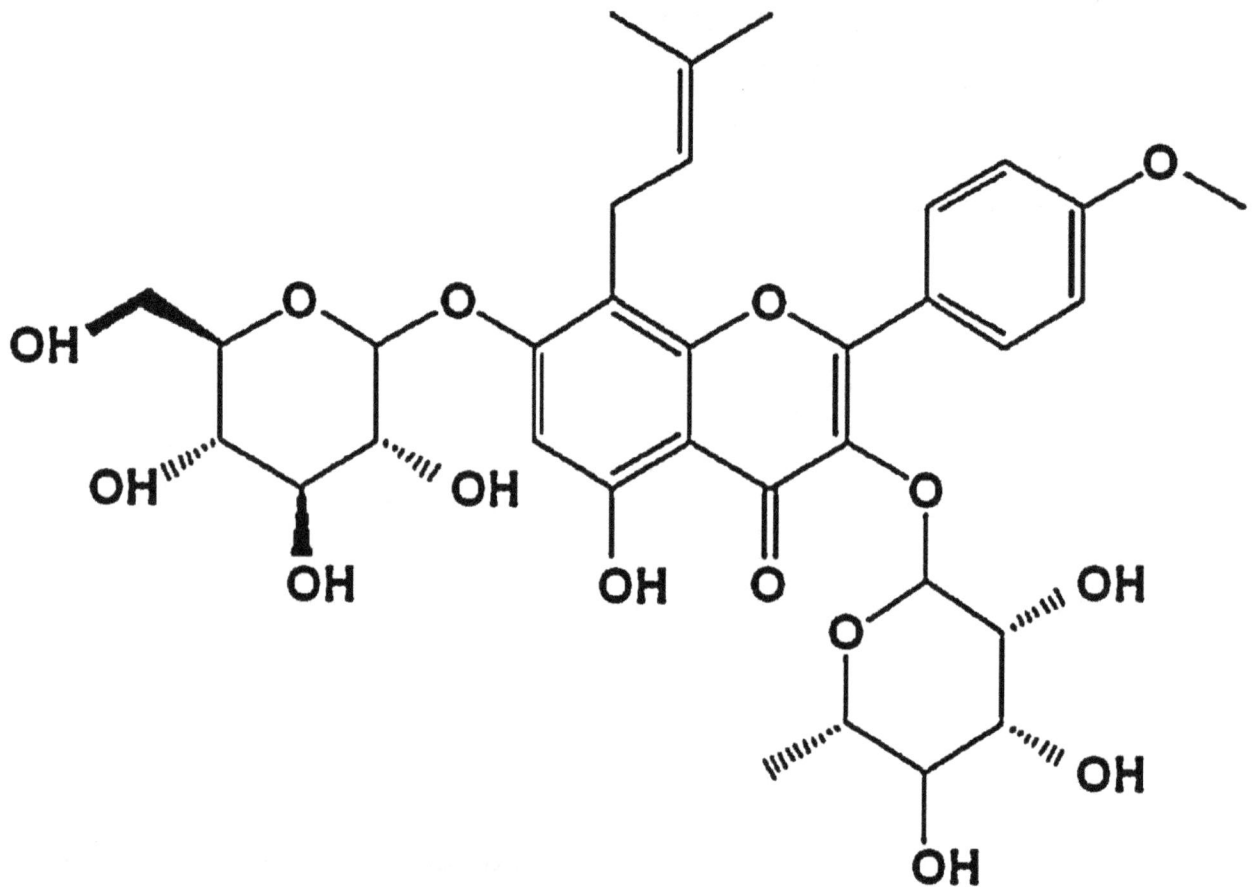

Figure 1. Molecular structure of Icariin.

Materials and Methods

Animals

Male APP/PS1-21 mice with a C57BL/6J background were obtained from Prof. M. Jucker. Heterozygous male APP/PS1-21 mice were bred with wild-type C57BL/6J females (Charles River Germany, Sulzfeld Germany). Offsprings were tail snipped and genotyped using PCR. Animals were housed under a 12 h light-12 h dark cycle with free access to food and water. Mouse diets were provided by SSNIFF Spezialdiaeten GmbH (Soest, Germany), diet number V1124-703 was for breeding pairs and diet number V2534-703 was for all the other mice. All experiments and protocols were licensed and approved by regional Administrative Council (Regierungspräsidium) Tuebingen according to The German Animal Welfare Act (TierSchG) of 2006.

Materials

Icariin (>98%) was purchased from MR Natural Product Co., Ltd. (Xi'an, China). For oral treatment, Icariin was suspended in 1% carboxymethylcellulose (CMC, Hercules-Aqualon, Düsseldorf, Germany) at a concentration of 12.5 mg/ml (Icariin/CMC solution).

Treatment with Icariin

14 transgenic APP/PS1-21 mice 5 months of age, 6 males and 8 females, were separated into two groups: group 1 received a 10-days Icariin treatment (100 mg/kg by daily gavage); group 2, as control, received the same volume of 1% CMC dissolved in water (approx. 200 ul).

Design and evaluation of nest construction assay

A nest construction assay [19] was modified to determine the deficits in affiliative/social behavior of APP/PS1 mice and potential changes following treatment.

Mice were individually housed for at least 24 hours in clean plastic cages with approximately 1 cm of wood chip bedding lining the floor and identification cards coded to render the experimenter blind to gender, age, and genotype of mice. Two hours prior to the onset of the dark phase of the light cycle, individual cages were supplied a 20×20 cm piece of paper towel torn into approximately 5×5 cm squared pieces. The next morning (approximately 16 hours later), cages were inspected for nest construction. Pictures were taken prior to evaluation for documentation. Paper towel nest construction was scored by a 3 point system: 1 = no biting or tears on the paper, 2 = moderate biting and/or tears on the paper but no coherent nest (not grouped into a corner of the cage) and 3 = the vast majority of paper torn into pieces and grouped into a corner of the cage [19].

Immunohistochemistry (IHC) and image evaluation/ analysis

Icariin-treated and control mice were sacrificed after 10-days-treatment. According to our previous data, Aβ deposition starts as early as 2 months of age, but develops very fast (exponential phase) around the age of 5 months. We presumed that a potential treatment may affect Aβ deposition most effectively during the exponential phase; therefore mice at an age of 5 months were used. We balanced the treatment and control groups by gender-, age and body weight matched mice. Afterwards, we analysed possible difference in plaque burden and behavioural deficits between genders within a group and found no significant differences. But the brain sizes of female mice are generally smaller than that of males. Therefore, we introduced analysis of IR percentage to rule out interference of brain size differences between genders or single individuals of the same gender. Mice were deeply anesthetized with ether and perfused intracardially with 4°C, 4% paraformaldehyde in PBS. Brains were quickly removed and post-fixed in 4% paraformaldehyde overnight at 4°C. Post-fixed brains were cut into two hemispheres; hemispheres were embedded in paraffin, serially sectioned (3 μm) and mounted on silane-covered slides. Hemispheres sections were stained with HE or IHC as described previously [20]. The following antibodies were used: anti-β-amyloid (1:100; Abcam, Cambridge, UK) for Aβ deposition, anti-Iba-1 (1:200; Wako, Neuss, Germany) for activated microglia, anti-GFAP (1:500; Chemicon (Millipore), Billerica, MA, US) for astrocytes and anti-transforming growth factor beta 1 (TGF-β1) (1:50, Santa Cruz, Dallas, Texas, US). The rabbit polyclonal anti-β-amyloid antibody (ab2539) was generated against the synthetic peptide DAEFRHDSGYEVHH conjugated to KLH, corresponding to amino acids 1–14 of Human β-amyloid. Positive and negative controls were routinely performed in each staining experiment to validate the immunohistochemical staining quality and results. Negative controls were performed by deletion of primary antibody and no unspecific staining was observed.

After immunostaining, sections were examined by light microscopy (Nikon, Düsseldorf, Germany). Aβ deposition, Iba-1 and GFAP immunostaining were evaluated at cross-sections of hemispheres, especially focused on cortex and hippocampus. All sections were randomly numbered and analysed by two observers independently, who were not aware of the treatment and time points. Aβ plaques, Iba-1$^+$ and GFAP$^+$ cells in cortex and hippocampus were counted under a microscope with a 50-fold magnification, by clear deposition for plaques and clear counter staining of cellular nuclei for cells. Further, images of hemisphere cross-sections were captured using Nikon Cool-scope (Nikon, Düsseldorf, Germany) with fixed parameters; cortex and hippocampus of the images were outlined and analyzed using the software MetaMorph Offline 7.1 (Molecular Devices, Toronto, Canada). Area percentages of specific immunoreactivity (IR) of interesting regions were selected by color threshold segmentation and calculated. All parameters were fixed for all images of a particular staining. Results were given as arithmetic means of plaque/cell counts or area percentages of IR to interest areas on cross-sections and standard errors of means (SEM).

Statistical analysis

Difference of plaque/cell counts, area percentages of staining and scores of nest construction between treatments and controls were analysed by unpaired t-tests (Graph Pad Prism 5.0 software). For all statistical analyses, significance levels were set at P<0.05.

Results

Effect of Icariin treatment on behaviour impairments

Effect of Icariin treatment was monitored using a harmless behavioural test, the nesting assay. As a born instinct, nesting behaviour is important for small rodents in heat conservation, reproduction and shelter. As a baseline of this assay, an impaired

nesting ability of these transgenic mice was confirmed in our previous study, compared to age- and gender matched naïve mice [21]. Nest construction with paper towel material was explored using a 3 point scaling system.

A significant difference between treatment and control groups of these age- and gender-matched APP/PS1 mice was already observed at Day 11, after 10 days of treatment (control = 1.5±0.1, Icariin = 2.1±0.2, p<0.05, n = 6) (Figure 2A). In the Icariin treatment group, relatively immediate chewing and tearing of the paper towels were observed; paper towels were torn into pieces and grouped into a corner of the cage. In contrast, transgenic mice from the control group investigated and slightly chew but did not really destruct the paper towels; paper towels were found all over in the cage, not grouped or not in the corners. As a control, no significant difference between treatment and control groups could be observed right at the beginning of treatment, namely at Day 1 (control = 1.4±0.1, Icariin = 1.5±0.2, p>0.5, n = 6) (Figure 2B).

After acquisition of these positive results, all mice were sacrificed and the effects of Icariin on neuropathological changes were further investigated.

Effects of Icariin on amyloid plaques

In brain of APP/PS1 transgenic mice from the control group, Aβ plaques were distributed throughout the whole cortex. Plaques were of different sizes, most small plaques had dense cores and larger plaques mainly consisted of a dense core surrounded by a large halo of diffuse amyloid (Figure 3A). In the hippocampus, plaque density was lower and most plaques were of smaller size (Figure 3E).

The transgenic mice received Icariin suspension or vehicle by gavage for 10 days. The treatment with Icariin attenuated neuropathological change, compared to the age- and gender-matched control mice. The Icariin treatment reduced the plaque counts significantly in cortex (control = 155.4±13.2, Icariin = 95.0±12.6, p<0.05, n = 6) and hippocampus (control = 21.6±3.4, Icariin = 10.4±2.1, p<0.05; n = 6) (Figure 4A and B). Further analysis of the micro photos showed highly significantly decreased Aβ IR area in both cortex and hippocampus from the Icariin treatment group (cortex: control = 0.72±0.06%, Icariin = 0.51±0.10%, p<0.05; hippocampus: control = 0.63±0.12%, Icariin = 0.33±0.09%, p<0.05; n = 6) (Figure 4C and D). Notably, Aβ plaques had a smaller size and fewer branches (Figure 3A and B).

Effects of Icariin on microglial activation

In both cortex and hippocampus of non-transgenic mice, Iba-1 staining could be barely observed. In transgenic mice, however, amoeboid Iba1$^+$ microglia were observed clustered around amyloid deposits in both control and treatment groups (Figure 3C and G, Figure 5A with higher magnification), according to comparison with Aβ staining of serial sections. Numbers of Iba-1$^+$ cells in Icariin treatment group were significantly less than those of the control group. Notably, in the cortex; decreased Iba-1$^+$ cells were also less clustered around plaques (Figure 3D and H, Figure 5B with higher magnification). Further analysis showed that Icariin treatment significantly reduced the IR area of Iba-1 in cortex (control = 0.35±0.06%, Icariin = 0.21±0.04%, p<0.05, n = 6), and hippocampus (control = 0.29±0.03%, Icariin = 0.16±0.05%, p<0.05, n = 6) (Figure 4E and F). These results indicate reduced microglial activation.

Effects if Icariin on astrocyte GFAP expression

Numerous GFAP$^+$ cells (GFAP IR) were widely distributed throughout the hippocampus and cortex. They all showed typical

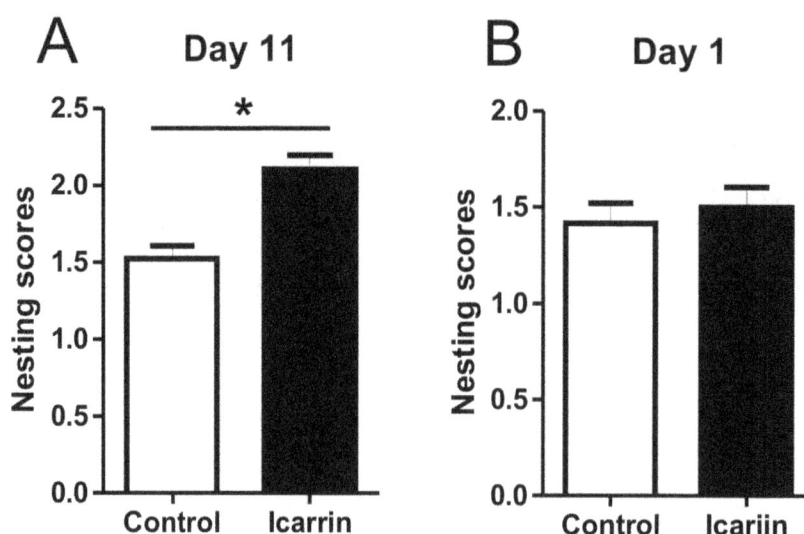

Figure 2. Effect of Icariin on impaired nesting ability. Therapeutic effects of Icariin were monitored using a nesting assay before and during the experimental treatment. Nest construction was explored with paper towel material using a 3 point scaling system in APP/PS1 mice. A: A significant difference between treatment and control group was observed after a 10-days treatment, namely at Day 11. B: No significant difference between the Icariin treatment and the control group could be observed right at the beginning of treatment, namely at Day 1.

morphology of astrocytes, including stellate shape and multiple branched processes; some of them had the typical morphology of perivascular astrocytes (Figure 5C and D). Double staining showed that all Aβ plaques were surrounded or covered by GFAP IR, but no massive aggregation of GFAP+ cells was related to Aβ plaques. Further analysis showed that no significant changes in GFAP IR were observed between Icariin treatment and controls; GFAP-IR area was only slightly decreased by Icariin treatment: in cortex (control = 0.51±0.05%, Icariin = 0.45±0.04%, p>0.05, n = 6), and in hippocampus (control = 0.38±0.04%, Icariin = 0.34± 0.05%, p>0.05, n = 6) (Figure 4G and H).

Expression pattern of TGF-β1 and effect of Icariin treatment on TGF-β1 IR

Until 3 months of age, TGF-β1 IR could not be seen in cortex or hippocampus of transgenic and naïve mice (data not shown). At age of 5 months, increased TGF-β1 IR was observed and mainly located on or around Aβ plaques, but could barely be seen on glial cells or neurons, according to the results of double staining and in comparison with serial sections of Aβ staining (Figure 6A and B). TGF-β1 IR therefore presented an Aβ plaque-like distribution pattern, but with much less intensity and smaller IR area compared to Aβ staining, as presented in the figure 6 (Figure 6A and B). TGF-β1 IR was more concentrated at the range of plaques and diffused branches, less in the center of plaques.

Following the 10 days-treatment with Icariin, TGF-β1 IR was obviously reduced in brains of transgenic mouse, especially in cortex (Figure 6C and D for control and Icariin group respectively). Numbers of TGF-β1 stained plaques in cortex were significantly reduced (control = 109.0±11.7, Icariin = 82.6±8.8, p<0.05, n = 6) and areas of TGF-β1 IR in cortex were also significantly decreased (cortex: control = 0.19±0.02%, Icariin = 0.13±0.01%, p<0.05) (Figure 6E and F).

Discussion

In this work we report therapeutically beneficial effect of Icariin, a flavonoid from the Chinese medicinal herb Epimedium

brevicornum (Herba epimedii), in a rodent APP/PS1 model of cerebral amyloidosis for AD. A 10-days oral treatment by Icariin significantly attenuated Aβ deposition, microglial activation and TGF-β1 IR in both cortex and hippocampus of APP/PS1 mice at an age of 5 months, and restored impaired nesting behavior as well.

As a tonic herbal in traditional Chinese medicine, Epimedium brevicornum has been used to treat various kinds of disorders, such as hypertension, coronary heart disease, osteoporosis, menopause syndrome, breast lump, rheumatism, arthritis or hypogonadism [1]. The most important active constituents of this herb are prenylated, isopentenyl flavonoids. Icariin is one of these isopentenyl flavonoids, a diglycoside with a glucose group at C-7, a rhamnose group at C-3 position, and a methoxyl group at C0-4, as shown in the (Figure 1) [22,23].

Recently, the structure-activity relationship (SAR) of flavonoids has been extensively investigated by theoretical calculations and experimental studies, including antitumor, antioxidant, vasorelax-ant, antibacterial and anti-inflammatory activities [1]. As a multifunctional flavonoid, Icariin shows antioxidant [24], antide-pressant [25], neuroprotective [7], anti-inflammatory activity/ effects [26,27]. Icariin also has functions in regulating bone remodeling, including enhancing osteoblastic differentiation and mineralization, inhibiting bone resorption, and inducing apoptosis of osteoclasts [28,29]; and in controlling sexual, especially erectile dysfunction [30].

Icariin has shown efficient anti-neuroinflammatory activity against learning and memory deficits in animal models [7], through attenuating microglial activation by inhibiting NF-kappaB and p38 MAPK pathways [31]. This suggests Icariin's promising potential in aging-related neurodegenerative disease, since mi-croglia are regarded as macrophage-like cells resident in the CNS and their activation has been implicated in many neurological disorders because of their inflammatory effects [32]. Therapeutic strategies controlling microglial activation and the excessive production of pro-inflammatory factors/molecules may be valu-able to control neurodegeneration in dementia [33]. An important role of neuroinflammation involved in AD pathology has been

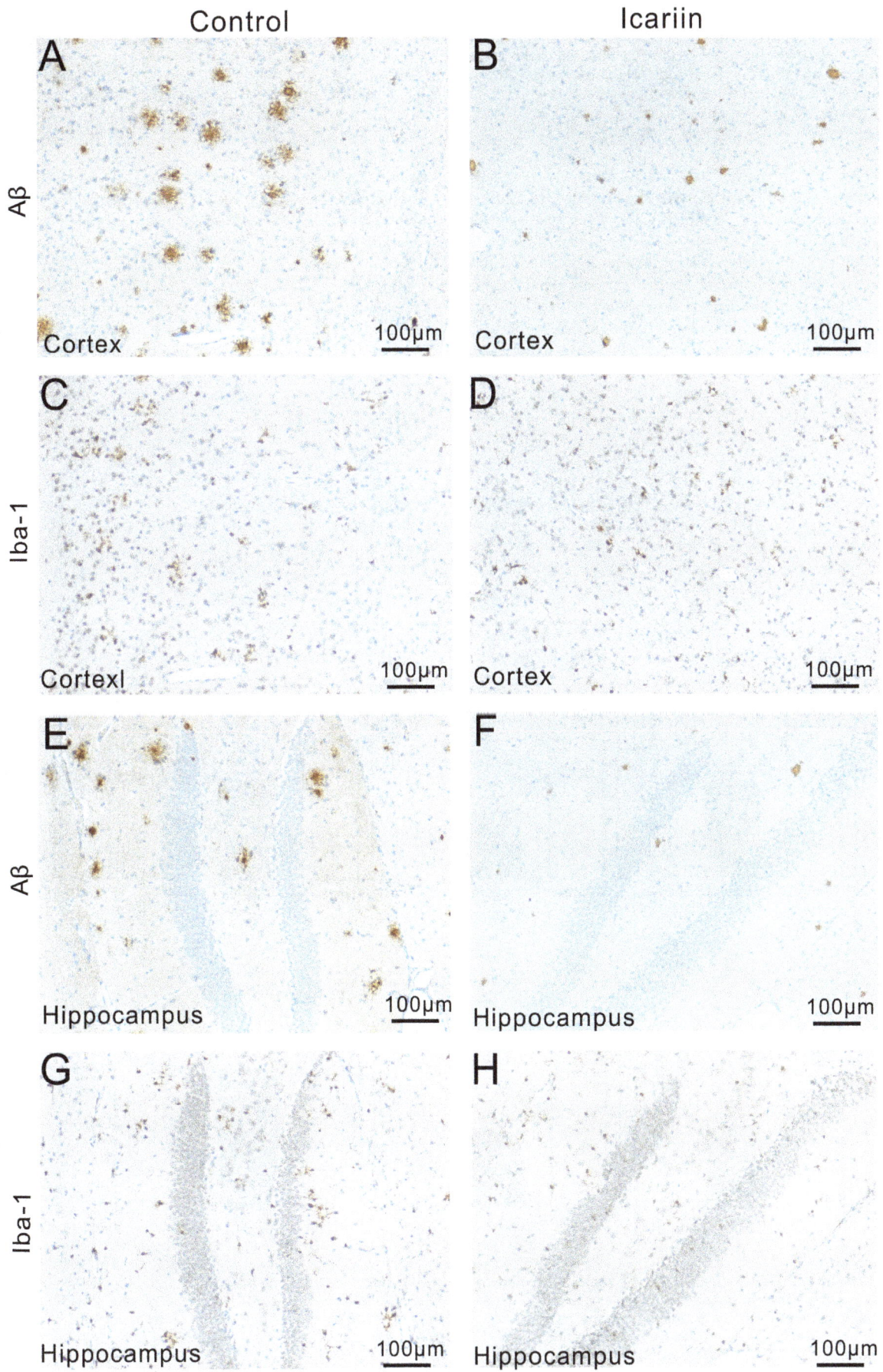

Figure 3. Therapeutic effect of Icariin on Aβ deposition and microglial activation. Representative microimages show the changes in Aβ deposition and microglial activation in cortex and hippocampus following Icariin treatment. A–B and E–F: In both cortex (A–B) and hippocampus (E–F) of APP/PS1 mice from the Icariin group (B and F), reduced numbers of Aβ plaques with relatively smaller size were observed, compared to the control group (A and E). C–D (cortex) and G–H (hippocampus): According to serial sections of Aβ staining, most Iba-1[+] microglia accumulated at or surrounding Aβ plaques. In the treatment group (D and H) fewer numbers of Iba-1[+] cells and smaller IR area of Iba-1 could be seen, compared to the control group (C and G).

reported from rodent models and humans [13], and attenuated neuroinflammation has been proven to contribute to reduced hallmark features of AD pathology, including Aβ-plaque accumulation [34]. Icariin significantly attenuated microglial activation in cortex of APP/PS1 transgenic mice, suggesting an inhibitory effect of Icariin on neuroinflammation, which may contribute to the ameliorated pathology and improved behaviors.

Aβ deposition in brains of transgenic mice is also significantly reduced by Icariin, following a relatively short term treatment of 10-days. It may be attributed to direct down-regulated amyloid loading [35] and attenuated neuroinflammation, because it has been reported that Icariin administrated by gavage negatively regulated beta-amyloid peptide segment 25–35 production in a rat model [35] and attenuated neuroinflammation also contributed to reduced Aβ deposition during progression of AD pathology [34].

Further, Amelioration of behavioral deficits was observed following Icariin treatment. Distinct behavioral and cognitive deficits are the most characteristic clinical feature of AD. Not only cognitive impairment but also deficits in non-cognitive/non-

mnemonic behaviors are found in most AD mouse models, they are therefore considered very valuable for modeling human AD [36]. Toxic Aβ peptides and amyloid precursor protein [37], inflammatory reaction and inflammatory cytokines/molecules are directly associated with these deficits in behaviors [38]. Nesting behavior is an affiliative, social behavior and deficit in this non-mnemonic behavior is a debilitating feature of neurodegenerative diseases, including AD. A previous study reported very early (3 months) occurrence of impaired nesting ability in transgenic APP and APPPS1 mice and suggested that social deficits precede other neuropsychiatric and cognitive AD-like symptoms and can be employed as early markers of AD pathology in transgenic mouse models [39]. Cognitive impairment, however, cannot be observed in the APP/PS1 mice until 8 months of age [17].

Amelioration in impaired nesting ability was observed following our Icariin treatment. The hippocampus and the prefrontal cortex damage in mice are demonstrated to lead to reduced nesting material consumption and nest quality, indicating that the impairment of nesting behavior in these transgenic mice might

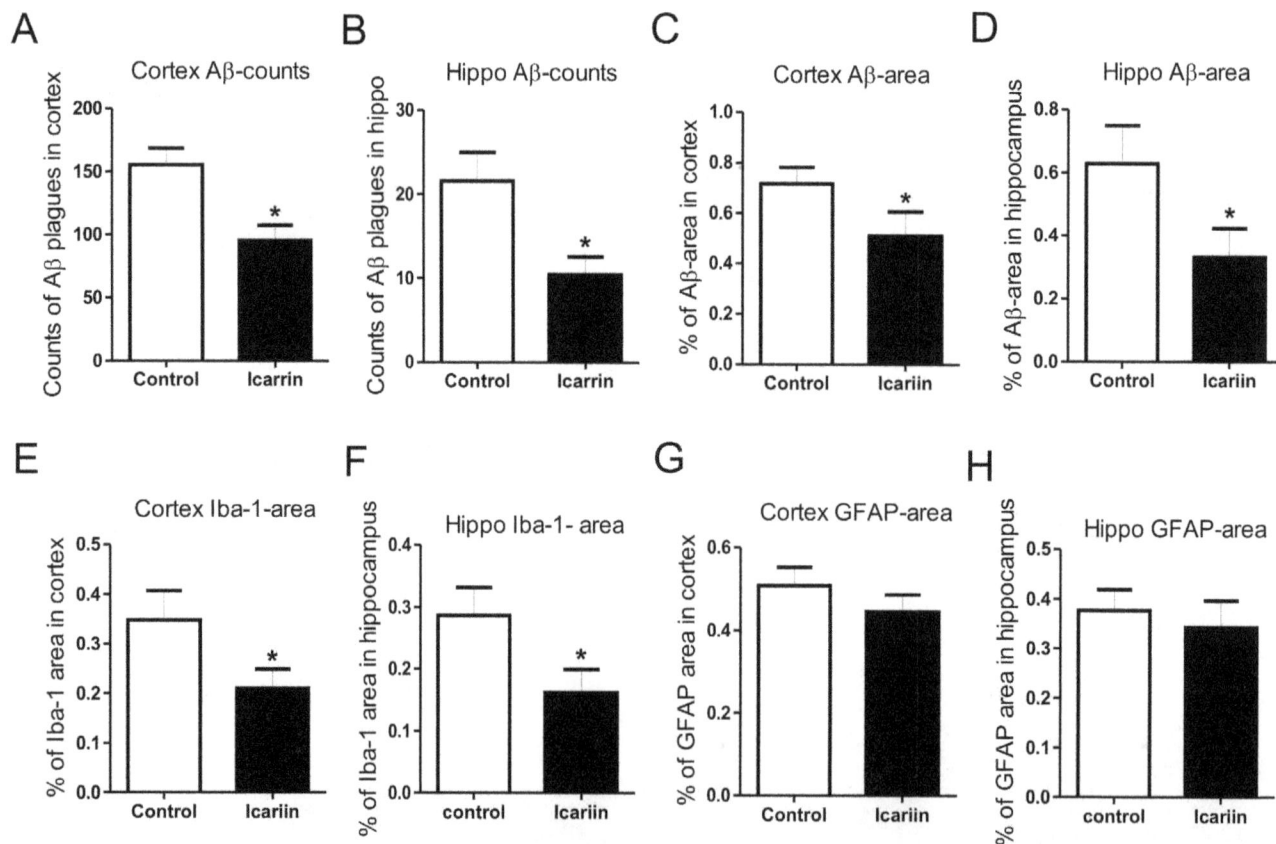

Figure 4. Icariin reduced β-amyloid (Aβ) deposition and microglial activation. Differences of Aβ plaque counts and Aβ[+]/Iba-1[+]/GFAP[+] area percentages between treatment and control were analysed by unpaired t-test and results are represented in the bar graphs. A and B: In cortex and hippocampus of transgenic mouse brains from the Icariin group, numbers of amyloid plaques were significantly reduced. C and D: IR area percentages of Aβ staining were highly significantly reduced. E and F: Following Icariin treatment, Iba-1 IR was significantly reduced in cortex and hippocampus. G and H: GFAP IR was not significantly changed by Icariin treatment.

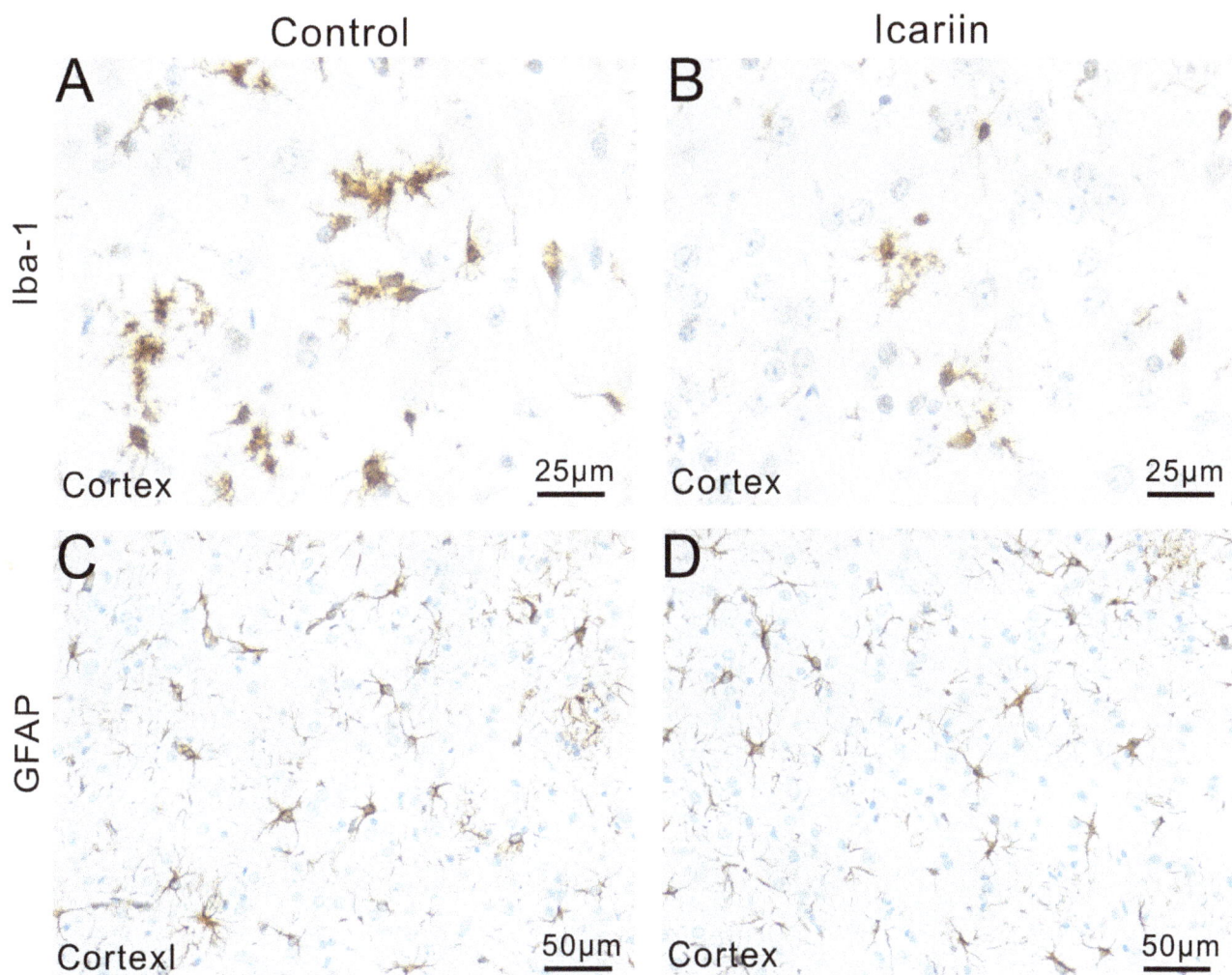

Figure 5. Therapeutic effect of Icariin on microglial and astrocytic activation. A and B: Representative microimages with higher magnification show the changes in microglial activation in cortex following Icariin treatment. C and D: Representative microimages show typical morphology of astrocytes, including stellate shape and multiple branched processes. No massive aggregation of GFAP⁺ cells and no obvious changes of GFAP IR were observed.

be caused by toxic injury of Aβ and accompanying neuro-inflammation in related brain areas [19]. In addition to reduced amyloid accumulation and decreased neuroinflammation, Icariin can also attenuate β-amyloid-induced neurotoxicity and neurite atrophy [31,40], all these may contribute to the improved affiliative/social behavior.

Moreover, Icariin also interferes with phosphodiesterases (PDEs), especially PDE4 and PDE5 [41,42]. This is of interest to AD research because inhibition of PDE4 is known to reverse Aβ-induced memory deficits [43]; PDE5 inhibition is known to ameliorate synaptic sprouting [44] and axonal remodelling [45] and PDE3 inhibition is known to enhance neurogenesis [46].

TGF-β1 is a potent immunomodulator and plays a central role in the response of the brain to inflammation and injury, but the specific role of TGF-β1 in AD pathogenesis still stays elusive. Some previous studies indicated a possibly protective effect of TGF-β1 in AD by its neuroproctive function [47] and by promoting Aβ clearance through activation of microglia cells [15]. But on the other hand, up-regulated expression of TGF-β1 has been reported from brains of AD patients and AD animal models [14,15]. Serveral studies further reported, that TGF-β1

potentiates/triggers increased production of APP and subsequent Aβ generation in murine and human astrocyte cultures [48,49] and transgenic mice overexpressing TGF-β1 in astrocytes elicit Aβ deposition [50]. Moreover, TGF-β1 can promote vascular abnormalities and Aβ deposition in cerebral blood vessels and cerebrovascular TGF-β may contribute to inflammation in AD brains [15,51,52]. We observed that TGF-β1 IR in cortex of transgenic mice was reduced following Icariin treatment. This may suggest a possible mechanism of Icariin's therapeutic effect through inhibition of local TGF-β deposition, which might be in part due to the reported inhibition of TGF-β expression [53]. In our study, TGF-β1 IR was observed in or around amyloid plaques and could barely be seen on glial cells or neurons, which is in accordance with previous studies of AD brains and animal models [15,51,54,55]. TGF-β1 is considered to be the most potent and ubiquitous profibrogenic cytokine and stimulates extracellular matrix (ECM) accumulation [56]. It has been reported, that Icariin protects tissue/organs from vascular pathological alter-ations by modulating expression of TGF-β1 [7,57]. ECMs such as collagen are important components of Aβ plaques [58], reduced TGF-β1 IR by Icariin may therefore, at least partialy, contribute

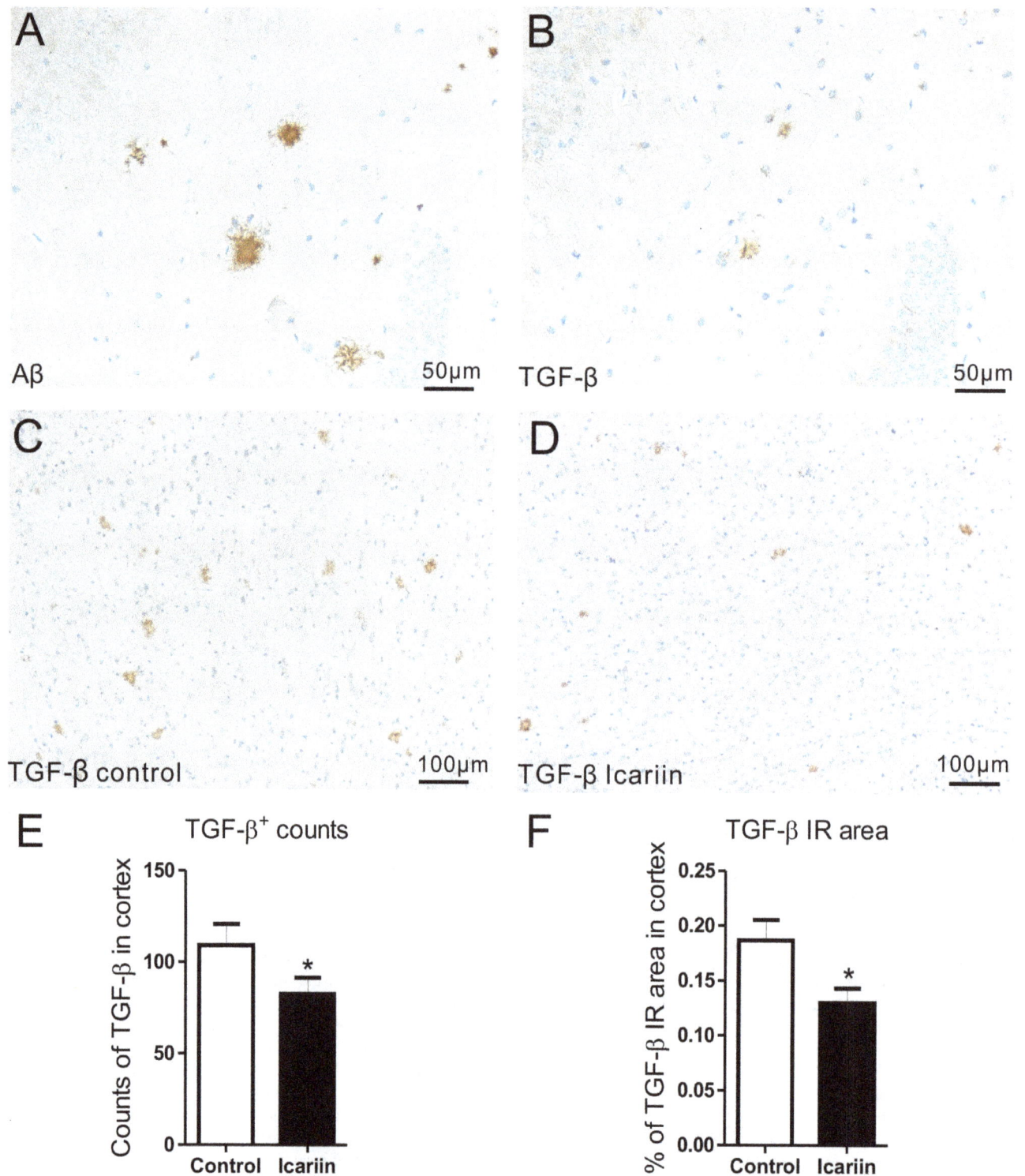

Figure 6. Effect of Icariin on TGF-β IR. At age of 5 months, increased TGF-β1 IR was observed in cortex of transgenic mice. A and B: In comparison of serial section of Aβ staining (A), TGF-β1 IR (B) had generally less intensity and smaller IR area, and was mainly located at or around Aβ plaques, but could be barely seen on glial cells or neurons. C and D: Much less intensity and smaller TGF-β1 IR area were seen in the Icariin treated group (D), compared to the control group (C). E and F: Statistical evaluation showed significant reductions in counts of TGF-β1[+] plaques and area of TGF-β1 IR.

to reduction of Aβ plaques by decreasing the formation of the plaque's matrix.

Taken together, treatments with Icariin by gavage effectively ameliorated neuroinflammatory reaction and cerebral amyloid-

osis, reduced IR of TGF-β1 in cortex and hippocampus of transgenic APP/PS1 mice, and restored impaired nesting ability. All these results suggest that Icariin may be considered a promising therapeutic option for human AD.

Acknowledgments

The authors would like to thank Prof. M. Jucker for providing male transgenic APP/PS1-21 mice.

Author Contributions

Conceived and designed the experiments: ZZ HS. Performed the experiments: ZZ CL CZ. Analyzed the data: ZZ CL CZ. Contributed reagents/materials/analysis tools: ZZ CZ. Wrote the paper: ZZ.

References

1. Zhang D, Zhang J, Fong C, Yao X, Yang M (2012) Herba epimedii flavonoids suppress osteoclastic differentiation and bone resorption by inducing G2/M arrest and apoptosis. Biochimie 94: 2514–2522.
2. Guo J, Li F, Wu Q, Gong Q, Lu Y, et al. (2010) Protective effects of icariin on brain dysfunction induced by lipopolysaccharide in rats. Phytomedicine 17: 950–955.
3. Zhang H, Liu B, Wu J, Xu C, Tao J, et al. (2012) Icariin inhibits corticosterone-induced apoptosis in hypothalamic neurons via the PI3-K/Akt signaling pathway. Mol Med Rep 6: 967–972.
4. Li L, Zhou QX, Shi JS (2005) Protective effects of icariin on neurons injured by cerebral ischemia/reperfusion. Chin Med J (Engl) 118: 1637–1643.
5. Kuroda M, Mimaki Y, Sashida Y, Umegaki E, Yamazaki M, et al. (2000) Flavonol glycosides from Epimedium sagittatum and their neurite outgrowth activity on PC12h cells. Planta Med 66: 575–577.
6. Pan Y, Kong L, Xia X, Zhang W, Xia Z, et al. (2005) Antidepressant-like effect of icariin and its possible mechanism in mice. Pharmacol Biochem Behav 82: 686–694.
7. Luo Y, Nie J, Gong QH, Lu YF, Wu Q, et al. (2007) Protective effects of icariin against learning and memory deficits induced by aluminium in rats. Clin Exp Pharmacol Physiol 34: 792–795.
8. Wang Z, Zhang X, Wang H, Qi L, Lou Y (2007) Neuroprotective effects of icaritin against beta amyloid-induced neurotoxicity in primary cultured rat neuronal cells via estrogen-dependent pathway. Neuroscience 145: 911–922.
9. Selkoe DJ (2002) Alzheimer's disease is a synaptic failure. Science 298: 789–791.
10. Herrmann N, Chau SA, Kircanski I, Lanctot KL (2011) Current and Emerging Drug Treatment Options for Alzheimer's Disease: A Systematic Review. Drugs.
11. Akiyama H, Barger S, Barnum S, Bradt B, Bauer J, et al. (2000) Inflammation and Alzheimer's disease. Neurobiol Aging 21: 383–421.
12. Wyss-Coray T, Mucke L (2002) Inflammation in neurodegenerative disease–a double-edged sword. Neuron 35: 419–432.
13. Martin-Moreno AM, Brera B, Spuch C, Carro E, Garcia-Garcia L, et al. (2012) Prolonged oral cannabinoid administration prevents neuroinflammation, lowers beta-amyloid levels and improves cognitive performance in Tg APP 2576 mice. J Neuroinflammation 9: 8.
14. Lippa CF, Flanders KC, Kim ES, Croul S (1998) TGF-beta receptors-I and -II immunoexpression in Alzheimer's disease: a comparison with aging and progressive supranuclear palsy. Neurobiol Aging 19: 527–533.
15. Wyss-Coray T, Lin C, Yan F, Yu GQ, Rohde M, et al. (2001) TGF-beta1 promotes microglial amyloid-beta clearance and reduces plaque burden in transgenic mice. Nat Med 7: 612–618.
16. Gengler S, Hamilton A, Holscher C (2010) Synaptic plasticity in the hippocampus of a APP/PS1 mouse model of Alzheimer's disease is impaired in old but not young mice. PLoS One 5: e9764.
17. Radde R, Bolmont T, Kaeser SA, Coomaraswamy J, Lindau D, et al. (2006) Abeta42-driven cerebral amyloidosis in transgenic mice reveals early and robust pathology. EMBO Rep 7: 940–946.
18. Moehlmann T, Winkler E, Xia X, Edbauer D, Murrell J, et al. (2002) Presenilin-1 mutations of leucine 166 equally affect the generation of the Notch and APP intracellular domains independent of their effect on Abeta 42 production. Proc Natl Acad Sci U S A 99: 8025–8030.
19. Wesson DW, Wilson DA (2011) Age and gene overexpression interact to abolish nesting behavior in Tg2576 amyloid precursor protein (APP) mice. Behav Brain Res 216: 408–413.
20. Zhang Z, Zhang ZY, Fauser U, Schluesener HJ (2008) FTY720 ameliorates experimental autoimmune neuritis by inhibition of lymphocyte and monocyte infiltration into peripheral nerves. Exp Neurol 210: 681–690.
21. Zhang ZY, Schluesener HJ (2013) Oral administration of histone deacetylase inhibitor MS-275 ameliorates neuroinflammation and cerebral amyloidosis and improves behavior in a mouse model. J Neuropathol Exp Neurol 72: 178–185.
22. Chen Y, Wang J, Jia X, Tan X, Hu M (2011) Role of intestinal hydrolase in the absorption of prenylated flavonoids present in Yinyanghuo. Molecules 16: 1336–1348.
23. Sun P, Liu Y, Deng X, Yu C, Dai N, et al. (2013) An inhibitor of cathepsin K, icariin suppresses cartilage and bone degradation in mice of collagen-induced arthritis. Phytomedicine 20: 975–979.
24. Xie J, Sun W, Duan K, Zhang Y (2007) Chemical constituents of roots of Epimedium wushanense and evaluation of their biological activities. Nat Prod Res 21: 600–605.
25. Pan Y, Kong LD, Li YC, Xia X, Kung HF, et al. (2007) Icariin from Epimedium brevicornum attenuates chronic mild stress-induced behavioral and neuroendocrinological alterations in male Wistar rats. Pharmacol Biochem Behav 87: 130–140.
26. Chen SR, Xu XZ, Wang YH, Chen JW, Xu SW, et al. (2010) Icariin derivative inhibits inflammation through suppression of p38 mitogen-activated protein kinase and nuclear factor-kappaB pathways. Biol Pharm Bull 33: 1307–1313.

27. Wu J, Zhou J, Chen X, Fortenbery N, Eksioglu EA, et al. (2012) Attenuation of LPS-induced inflammation by ICT, a derivate of icariin, via inhibition of the CD14/TLR4 signaling pathway in human monocytes. Int Immunopharmacol 12: 74–79.
28. Ming LG, Chen KM, Xian CJ (2013) Functions and action mechanisms of flavonoids genistein and icariin in regulating bone remodeling. J Cell Physiol 228: 513–521.
29. Huang J, Yuan L, Wang X, Zhang TL, Wang K (2007) Icaritin and its glycosides enhance osteoblastic, but suppress osteoclastic, differentiation and activity in vitro. Life Sci 81: 832–840.
30. Ho CC, Tan HM (2011) Rise of herbal and traditional medicine in erectile dysfunction management. Curr Urol Rep 12: 470–478.
31. Zeng KW, Fu H, Liu GX, Wang XM (2010) Icariin attenuates lipopolysaccharide-induced microglial activation and resultant death of neurons by inhibiting TAK1/IKK/NF-kappaB and JNK/p38 MAPK pathways. Int Immunopharmacol 10: 668–678.
32. Xu Y, Xue Y, Wang Y, Feng D, Lin S, et al. (2009) Multiple-modulation effects of Oridonin on the production of proinflammatory cytokines and neurotrophic factors in LPS-activated microglia. Int Immunopharmacol 9: 360–365.
33. Agostinho P, Cunha RA, Oliveira C (2010) Neuroinflammation, oxidative stress and the pathogenesis of Alzheimer's disease. Curr Pharm Des 16: 2766–2778.
34. Tweedie D, Ferguson RA, Fishman K, Frankola KA, Van Praag H, et al. (2012) Tumor necrosis factor-alpha synthesis inhibitor 3,6′-dithiothalidomide attenuates markers of inflammation, Alzheimer pathology and behavioral deficits in animal models of neuroinflammation and Alzheimer's disease. J Neuroinflammation 9: 106.
35. Nie J, Luo Y, Huang XN, Gong QH, Wu Q, et al. (2010) Icariin inhibits beta-amyloid peptide segment 25–35 induced expression of beta-secretase in rat hippocampus. Eur J Pharmacol 626: 213–218.
36. Alexander G, Hanna A, Serna V, Younkin L, Younkin S, et al. (2011) Increased aggression in males in transgenic Tg2576 mouse model of Alzheimer's disease. Behav Brain Res 216: 77–83.
37. Rangasamy T, Cho CY, Thimmulappa RK, Zhen L, Srisuma SS, et al. (2004) Genetic ablation of Nrf2 enhances susceptibility to cigarette smoke-induced emphysema in mice. J Clin Invest 114: 1248–1259.
38. Ownby RL (2010) Neuroinflammation and cognitive aging. Curr Psychiatry Rep 12: 39–45.
39. Pietropaolo S, Delage P, Lebreton F, Crusio WE, Cho YH (2012) Early development of social deficits in APP and APP-PS1 mice. Neurobiol Aging 33: 1002 e1017–1027.
40. Sha D, Li L, Ye L, Liu R, Xu Y (2009) Icariin inhibits neurotoxicity of beta-amyloid by upregulating cocaine-regulated and amphetamine-regulated transcripts. Neuroreport 20: 1564–1567.
41. Xin ZC, Kim EK, Lin CS, Liu WJ, Tian L, et al. (2003) Effects of icariin on cGMP-specific PDE5 and cAMP-specific PDE4 activities. Asian J Androl 5: 15–18.
42. Ning H, Xin ZC, Lin G, Banie L, Lue TF, et al. (2006) Effects of icariin on phosphodiesterase-5 activity in vitro and cyclic guanosine monophosphate level in cavernous smooth muscle cells. Urology 68: 1350–1354.
43. Cheng YF, Wang C, Lin HB, Li YF, Huang Y, et al. (2010) Inhibition of phosphodiesterase-4 reverses memory deficits produced by Abeta25-35 or Abeta1-40 peptide in rats. Psychopharmacology (Berl) 212: 181–191.
44. Zhang R, Wang Y, Zhang L, Zhang Z, Tsang W, et al. (2002) Sildenafil (Viagra) induces neurogenesis and promotes functional recovery after stroke in rats. Stroke 33: 2675–2680.
45. Zhang L, Zhang RL, Wang Y, Zhang C, Zhang ZG, et al. (2005) Functional recovery in aged and young rats after embolic stroke: treatment with a phosphodiesterase type 5 inhibitor. Stroke 36: 847–852.
46. Zhao J, Harada N, Kurihara H, Nakagata N, Okajima K (2010) Cilostazol improves cognitive function in mice by increasing the production of insulin-like growth factor-I in the hippocampus. Neuropharmacology 58: 774–783.
47. Wyss-Coray T (2006) Tgf-Beta pathway as a potential target in neurodegeneration and Alzheimer's. Curr Alzheimer Res 3: 191–195.
48. Gray CW, Patel AJ (1993) Regulation of beta-amyloid precursor protein isoform mRNAs by transforming growth factor-beta 1 and interleukin-1 beta in astrocytes. Brain Res Mol Brain Res 19: 251–256.
49. Lesne S, Docagne F, Gabriel C, Liot G, Lahiri DK, et al. (2003) Transforming growth factor-beta 1 potentiates amyloid-beta generation in astrocytes and in transgenic mice. J Biol Chem 278: 18408–18418.
50. Wyss-Coray T, Masliah E, Mallory M, McConlogue L, Johnson-Wood K, et al. (1997) Amyloidogenic role of cytokine TGF-beta1 in transgenic mice with Alzheimer's disease. Nature 389: 603–606.
51. Grammas P, Ovase R (2002) Cerebrovascular transforming growth factor-beta contributes to inflammation in the Alzheimer's disease brain. Am J Pathol 160: 1583–1587.

52. Wyss-Coray T, Lin C, Sanan DA, Mucke L, Masliah E (2000) Chronic overproduction of transforming growth factor-beta1 by astrocytes promotes Alzheimer's disease-like microvascular degeneration in transgenic mice. Am J Pathol 156: 139–150.

53. Li YC, Ding XS, Li HM, Zhang C (2013) Icariin attenuates high glucose-induced type IV collagen and fibronectin accumulation in glomerular mesangial cells by inhibiting transforming growth factor-beta production and signalling through G protein-coupled oestrogen receptor 1. Clin Exp Pharmacol Physiol 40: 635–643.

54. Peress NS, Perillo E (1995) Differential expression of TGF-beta 1, 2 and 3 isotypes in Alzheimer's disease: a comparative immunohistochemical study with cerebral infarction, aged human and mouse control brains. J Neuropathol Exp Neurol 54: 802–811.

55. van der Wal EA, Gomez-Pinilla F, Cotman CW (1993) Transforming growth factor-beta 1 is in plaques in Alzheimer and Down pathologies. Neuroreport 4: 69–72.

56. Liu RM, Gaston Pravia KA (2010) Oxidative stress and glutathione in TGF-beta-mediated fibrogenesis. Free Radic Biol Med 48: 1–15.

57. Qi MY, Kai C, Liu HR, Su YH, Yu SQ (2011) Protective effect of Icariin on the early stage of experimental diabetic nephropathy induced by streptozotocin via modulating transforming growth factor beta1 and type IV collagen expression in rats. J Ethnopharmacol 138: 731–736.

58. Brandan E, Inestrosa NC (1993) Extracellular matrix components and amyloid in neuritic plaques of Alzheimer's disease. Gen Pharmacol 24: 1063–1068.

11

Derivation of a Germline Competent Transgenic Fischer344 Embryonic Stem Cell Line

Hongsheng Men, Elizabeth C. Bryda*

Rat Resource and Research Center, Department of Veterinary Pathobiology, University of Missouri, Columbia, Missouri, United States of America

Abstract

Embryonic stem (ES) cell-based gene manipulation is an effective method for the generation of mutant animal models in mice and rats. Availability of germline-competent ES cell lines from inbred rat strains would allow for creation of new genetically modified models in the desired genetic background. Fischer344 (F344) males carrying an enhanced green fluorescence protein (EGFP) transgene were used as the founder animals for the derivation of ES cell lines. After establishment of ES cell lines, rigorous quality control testing that included assessment of pluripotency factor expression, karyotype analysis, and pathogen/sterility testing was conducted in selected ES cell lines. One male ES cell line, F344-Tg.EC4011, was further evaluated for germline competence by injection into Dark Agouti (DA) X Sprague Dawley (SD) blastocysts. Resulting chimeric animals were bred with wild-type SD mates and germline transmissibility of the ES cell line was confirmed by identification of pups carrying the ES cell line-derived EGFP transgene. This is the first report of a germline competent F344 ES cell line. The availability of a new germline competent ES cell line with a stable fluorescence reporter from an inbred transgenic rat strain provides an important new resource for genetic manipulations to create new rat models.

Editor: Lygia V. Pereira, Universidade de São Paulo, Brazil

Funding: The project was supported by grant funding to the Rat Resource and Research Center (RRRC) from the National Institutes of Health (8P40 OD011062-12) (URL http://dpcpsi.nih.gov/orip/cm/index.aspx). The funders had no role in study design, data collection and analysis, decision to publish, or preparation of the manuscript.

Competing Interests: The authors have declared that no competing interests exist.

* E-mail: brydae@missouri.edu

Introduction

Because rats share similarities in their anatomy and physiology with humans, they are often a model animal in biomedical research as well as drug discovery and development. Rats have been widely used in the areas of hypertension, aging, infectious diseases, cancer and neurological disorders [1]. Besides physiological similarities, the larger size of the rat increases ease of procedures, such as surgery, sampling, pharmacological development, stereotaxic neurological studies, neuroimaging and cardiovascular monitoring [1,2]. Inbred strains of rats are often preferred due to their identical and fixed genetic background among individuals.

Mouse models generated using embryonic stem (ES) cell-based gene engineering technologies have significantly contributed to advances in biomedical research. Derivation of germline competent rat ES cells will allow the production of rat models with targeted genetic alterations using the same methods that have been so successful in the mouse [3,4,5]. For example, ES cell-based genetic modification has been proven to be an effective method for the production of animal models with complicated designs, such as conditional or inducible knockouts [6,7].

Germline competent rat ES cell lines have been derived from Dark Agouti [3,4], Sprague Dawley [3,8], Wistar [9], and LEA [9]. Fischer344 rats are a popular strain for biomedical research in the areas including oncology, toxicology, carcinogenicity, aging and autoimmunity. However, proven germline competent ES cells

lines from the F344 strain have yet to be established despite efforts by multiple laboratories worldwide [3,4,10].

In these studies, we describe the isolation of a novel germline competent rat ES cell line derived from Fischer344 rats carrying an EGFP transgene. We describe the characterization of ES cell lines using various prescreening tests to select rat ES cell lines that have a higher probability for germline transmissibility and the use of hybrid recipient embryos to improve the efficiency of germline competency testing.

Materials and Methods

Ethics Statement

This study was carried out in strict accordance with the recommendations in the Guide for the Care and Use of Laboratory Animals of the National Institutes of Health. The protocol was approved by the Animal Care and Use Committee of the University of Missouri.

Derivation of ES Cell Lines from Transgenic Rats

Unless specifically indicated, all chemicals were obtained from Sigma-Aldrich (Sigma-Aldrich, St Louis, MO). Male F344-Tg (EGFP) F455/Rrrc (RRRC# 307) rats were obtained from the Rat Resource and Research Center (University of Missouri, www.rrrc.us) and were used as founder animals for the derivation of rat ES cell lines. This strain is homozygous for a single copy of an EGFP transgene under control of a human Ubiquitin C promoter with the woodchuck hepatitis virus post-transcriptional regulatory element (WRE) on a Fischer 344 (F344) genetic background [11]. The transgene insertion site is on Chromosome 5 (www.rrrc.us) [12]. Wild-type F344/Hsd females (Harlan, Indianapolis, IN) were mated to homozygous F344-Tg (EGFP) F455/Rrrc males. Blastocysts, all of which were hemizygous for the transgene and were positive for EGFP expression, were collected on Day 4.5 post mating in mRiECM+22 mM HEPES [13]. After collection, ES cells were isolated from blastocysts using a protocol described previously [3]. For ES cell derivation, blastocysts were treated briefly with acidic Tyrode's solution to remove zona pellucidae and then cultured in N2B27+3 μM CHIR99021 (Axon Medchem BV, Groeningen, The Netherlands) +0.5 μM PD0325901 (Sell-eckchem, Houston, TX) [14] on CF-1 mouse feeder cells (Millipore, Billerica, MA) in Nunc 4-well plates (Thermo Scientific, Roskilde, Denmark) at 37°C in an incubator with 5% CO_2 and maximal humidity. The CF-1 feeder cells were plated onto Nunc 4-well plates one day prior to the culture of embryos at a density of approximately 50,000 cells/cm2 as suggested by the manufacturer. On Day 5, outgrowths of the embryos were individually disassociated into single cell suspension using accutase and then cultured in 24-well plates. ES cells were passaged every 48–72 hours.

ES Cell Genotyping

Selected ES cell lines were genotyped by PCR to identify their sex chromosome composition. Primers used to detect the X chromosome were 5'-GTG AAG GAG GAA TTA GGT GG-3' and 5'-GAT GTG GTA ATT GTC ATC AC-3' [15]. Primers used to detect the Y chromosome were 5'-GTA GGT TGT TGT CCC ATT GC-3' and 5'-GAG AGA GGC ACA AGT TGG C-3' [16]. PCR was performed on 20 μl reactions containing ~10 ng genomic DNA, 1 unit of FastStart Taq DNA Polymerase (Roche), 750 nM of each primer, 200 μM each dNTP, and 1X Reaction buffer containing $MgCl_2$ (Roche). PCR conditions were 94°C for 5 minutes, then 35 cycles of 94°C for 1 min., 61°C for 1 min. and 72°C for 1 minute followed by 72°C for 7 min. Amplicons of 272 bp (Y chromosome) and 1100 bp (X chromosome) were detected by gel electrophoresis on 1% 1X TBE agarose gels. Male ES cell lines were selected for subsequent assays.

Expression of Pluripotency Factors

The expression of Oct4, Sox2, and Nanog in the established ES cell lines were examined by RT-PCR analysis using rat specific primers: Oct4, 5'-CCCAGCGCCGTGAAGTTG-GA-3' and, 5'-ACCTTTCCAAAGAGAACGCCCAGG-3'; Sox2, 5'-AT-TACCCGCAGCAAAATGAC-3' and, 5'-AT-CGCCCGGAGTCTAGTTCT-3'; Nanog, 5'-GACTAG-CAACGGCCTGACTCA-3' [3] and, 5'-CTGCAATG-GATGCTGGGATA-3'; GAPDH, ATCACTGCCACTCA-GAAG-3' and, AAGTCACAGGAGACAACC-3' [3]. Germline-competent rat ES cell line DAc8 [7] (RRRC# 464) obtained from the Rat Resource and Research Center served as a positive control. The negative controls were rat embryonic fibroblasts (made in house), mouse embryonic fibroblasts (feeder cells, Millipore) and a no template control. RNA was extracted from up to 5×10^5 cells using RNeasy Plus Micro Kit (QIAGEN, Valencia, CA). The High Capacity First Strand Synthesis Kit from Applied Biosystem (Carlsbad, California) was used to synthesize cDNA from 1 μg of RNA. RT-PCR was performed in 25 μl reactions containing 250 pg −250 ng cDNA, 1X PCR Buffer (Roche, Indianapolis, IN), 1.5 mM $MgCl_2$, 0.2 mM dNTPs, 0.2 μM of each primer and 2.5 U of Roche FastStart Taq polymerase. Thermal cycling conditions were 1 cycle at 95°C,

2 min; 35 cycles of 95°C, 30 sec., 61°C, 30 sec and 72°C, 30 sec; 1 cycle at 72°C, 5 min. The DNA samples were analyzed using the QIAxcel (QIAGEN) with the QIAxcel DNA Screening Kit, QX Alignment Marker 15 bp/3 kb, and QX DNA Size Marker 100 bp-3 kb. The method was AM320 with an injection of 10 s at 5 kV and a separation of 320 s at 6 kV.

ES Cell Karyotyping

Rat ES cells were treated with 0.1 μg/ml colcemid (Irvine Scientific, Santa Ana, CA) for 1 h at 37 °C when they reach 60–70% confluent. At the end of colcemid treatment, ES cell colonies were harvested and disassociated into single cell suspension with accutase and then pelleted by centrifugation at 200×g for 8 min in a 15 ml conical tube. After removing the supernatant, the cells were resuspended with 4–5 ml hypotonic solution (0.075 M KCl solution) and incubated at room temperature for 15 min. A few drops of freshly made fixative consisting of methanol: acetic acid (Fisher Scientific, Pittsburg, PA) in a ratio of 3:1 were then added to the hypotonically treated cell suspension and mixed by inversion. The cells were pelleted at 200×g for 8 min and were then resuspended in 4–5 ml fixative and re-pelleted at 200×g for 8 min. After one more repetition of the fixation step, the fixed ES cells were pelleted by centrifugation at 200×g and resuspended in 1 ml fixative. Preparation of chromosome spreads and karyotype analysis of the fixed cells were performed by Dr. Chin-Lin Hsieh (Arcadia, CA). ES cell lines were analyzed by Giemsa-Trypsin-Wrights (GTW) banding and at least 20 metaphase spreads were counted. A cell line with 70% or higher metaphase spreads exhibiting a normal number of chromosomes was considered to have a normal karyotype. The passage numbers at the time of karyotyping for each cell line were passage 6 and 13 for F344-Tg(EGFP).EC4011 and passage 7 for F344-Tg(EGFP).EC4013.

Pathogen Screening of Rat ES Cells

Both the culture media and the cell lines were subjected to pathogen screening. One milliliter of culture medium from each cell line was submitted to IDEXX-RADIL (Columbia, MO) for microbiological evaluation. The medium was placed on blood agar (BA) and Brain Heart Infusion broth (BHI) broth for 10 days to evaluate bacterial growth. Pathogen screening for ES cell lines was usually conducted after examination of the expression of pluripotent factors and karyotyping. One million cells from selected ES cell line were submitted to IDEXX-RADIL for a comprehensive pathogen testing. This included screening for the presence of H1 parvovirus, Kilham's rat virus, Mycoplasma spp., rat minute virus, and rat parvovirus in the cell lines. A portion of the cell sample was also grown on BA/BHI broth for 10 days to examine any potential bacterial contamination in the cell lines.

Chimeric Animal Production and Breeding

Male ES cell line, F344-Tg(EGFP).EC4011 was selected for the production of chimeric animals. Six days prior to blastocyst injection, cells frozen at passage 13 were thawed and cultured in N2B27+2i with CF-1 mouse feeder cells in 60 mm culture dishes and passaged every 48 h to ensure that the ES cells were fully recovered from any stress resulting from cryopreservation. On the day of injection, rat ES colonies were detached from the feeders by gently pipetting the media up and down followed by collection into a 15 ml centrifugation tube. After centrifugation at 200×g for 3 min and removing the supernatant, the pelleted ES cell colonies were disassociated with accutase into a single cell suspension followed by centrifugation at 200×g for 3 min. The cell pellet was resuspended in N2B27+20 mM HEPES and incubated on ice. Donor blastocysts were collected from Day 4.5 pregnant SD females that had been mated with DA males (Harlan). These females were synchronized using GnRH at 40 µg/rat 4 days before the mating. Donor blastocysts were cultured in mRiECM +10% fetal bovine serum (FBS) after collection. Blastocyts were injected in groups of 10, in 20 ul m-RECM-1-HEPES.ES cells were freshly added to each injection drop. Ten (10) to 12 rat ES cells were injected into single blastocysts using a beveled Transfertip (Eppendorf, Hauppauge, NY). After injection of each group, injected blastocysts were immediately transferred into mRiECM +10% FBS and cultured for about 1 hour. Approximately 20–30 blastocysts were transferred into the uterine horns of Day 3.5 pseudo-pregnant SD females (10–15 blastocysts per uterine horn). All surgical procedures were approved by the Animal Care and Use Committee of the University of Missouri-Columbia.

Chimeric animals were identified from the resulting pups by coat color chimerism (presence of albino hairs against an agouti coat color background). Upon sexual maturation, chimeric animals were bred with SD mates to verify germline transmissibility. The inheritance of ES cell genetics in the offspring was assessed by the presence of the EGFP transgene using an insertion site specific PCR genotyping assay developed by RRRC (www.rrrc.us). DNA was extracted from tail biopsies using the Extract-N-Amp Tissue PCR kit and PCR was performed using the manufacturer's protocol and reagents. Primers are LWS 455 5F: 5′-AAC CTC CCA GTG CTT TGA ACG CTA-3′; LWS 455 5R: 5′-GGT GCC AAG CCT CAA CTT CTT TGT-3′ and U3r-4 5′-ATC AGG GAA GTA GCC TTG TGT GTG-3′. Thermal cycling conditions were 1 cycle at 94°C, 3 min; 35 cycles of 94°C, 30 sec., 64°C, 30 sec and 72°C, 1 min; 1 cycle at 72°C, 10 min. The wild type product is 438 bp and the mutant product is 129 bp. Recovery of animals that inherited the transgene from their chimeric parent was evidence of germline competency of the ES cell line. Failure to produce any transgenic offspring in three consecutive litters was taken as lack of germline competency.

Results

Derivation of ES Cell Lines from F344-Tg (EGFP) F455/Rrrc Transgenic Rats

A total of 34 blastocysts were collected from 12 F344 females mated with F344-Tg (EGFP) F455/Rrrc males. After removal of zona pellucidae, these 34 blastocysts were successfully cultured and showed outgrowths. After several passages, a total of 27 ES cell lines were established from the 34 blastocysts.

Characterization of the Novel ES Cell Lines

All 27 cell lines could be maintained in an undifferentiated state in rat ES medium (N2B27+2i) and showed compact colonies with smooth boundaries and retained GFP fluorescence (Fig. 1A and 1B). Of these 27 cell lines, 14 were chosen randomly for further analysis. Genotyping results for the fourteen lines (F344-Tg.EC4001 to F344-Tg.EC4014) showed that F344-Tg.EC4011 and F344-Tg.EC4013 are male cell lines and the other 12 lines were female lines. The two male lines expressed pluripotency factors *Oct4*, *Sox2*, and *Nanog* by RT-PCR analysis and had normal karyotypes at Passage 6 and 7, respectively (Fig. 1C and 1D). F344-Tg.4011 was karyotyped again at Passage 13 with 15/20 cells examined exhibiting a normal karyotype. Pathogen screening indicated that the two male lines (F344-Tg.EC4011 and F344-Tg.EC4013) were free of H1 parvovirus, Kilham's rat virus, *Mycoplasma* spp., rat minute virus, and rat parvovirus. There was also no bacterial or fungal growth after 10 days of sterility testing for both the culture media as well as the cell lines.

Generation of Chimeras

Based on karyotyping results, ES cell line F344-Tg.EC4011 had a higher percentage of cells with a normal male karyotype than line F344-Tg.EC4013, and therefore this line was chosen for further analysis. Ten to twelve F344-Tg.EC4011 ES cells at passage 16 were injected into hybrid DA X SD blastocysts. A total of 199 blastocysts were injected and transferred. Seventy-five live pups were produced. A total of 13 animals (11 males and 2 females) showed coat color chimerism (Table 1 and Figure 2A).

Demonstration of Germline Competency

Eleven male chimeric animals derived from the F344-Tg.EC4011 cell line were bred to SD mates. Once a chimeric animal produced a GFP positive pup, indicating its ability to pass on the transgene derived from the ES cells, breeding was stopped. Breeding was also discontinued after a chimeric animal failed to produce a GFP positive pup within three litters. One animal (468RII) did not produce any offspring after the second litter. The results showed that 6 out of the 11 chimeric animals derived from cell line F344-Tg.EC4011 were able to transmit the ES cell-derived EGFP gene through the germline (Table 2 and Figure 2B).

Discussion

In the present study, we report derivation of a novel rat ES cell line with germline transmissibility from transgenic F344 rats carrying a ubiquitously expressed EGFP gene on Chromosome 5. We also demonstrate that hybrid recipient embryos with a SD × DA genetic background were able to support the germline competence testing of the F344-Tg.EC4011 embryonic stem cell line.

The establishment of germline competent ES cell lines from inbred rats will provide valuable resources for the creation of new genetically engineered rat models on inbred genetic backgrounds. It has been shown in mice that variations in genetic background can have a profound influence on the phenotypes of genetically altered animals [17]. Therefore, inbred animals are usually the preferred animals for the generation of mutant animal models [18]. F344 rats are a popular strain used in both biomedical and drug discovery. However, similar to mice, the derivation of germline competent ES cell lines in rats is also highly strain-dependent [3,4,19,20]. ES cell lines from F344 genetic background have been generated in at least three laboratories. None of these ES cell line has been demonstrated to be able to transmit through the germline. In fact, the existing putative F344 ES cell lines even failed to give rise to chimeric animals [3,4,20].

The genetic combination of ES cells and the host embryos is a key factor governing the ES cells' ability to colonize the gonads. There are several factors that have been demonstrated to affect the ES cell's ability to transmit their genetic material through the germline including the genetic background, stemness, normality of karyotype, pathogen status of the ES cell line as well as the genetic background of recipient embryos [7,21]. Among these factors, the combination of the genetic background of the ES cells and the recipient embryos has been demonstrated to be a critical factor affecting the germline transmissibility of the ES cells [7,21,22]. Ideally, the host background should allow the ES cells to have an optimal developmental advantage when injected into the blastocyst. This allows the ES cells to contribute to the germline of the chimeric animals which consequently transmit the ES cell-derived genetic material to their offspring [3,7,19]. Because a relatively few number of ES cell lines have been isolated to date and even fewer have been shown to be germline competent, relatively little is known about the optimal combinations of ES cell genetic

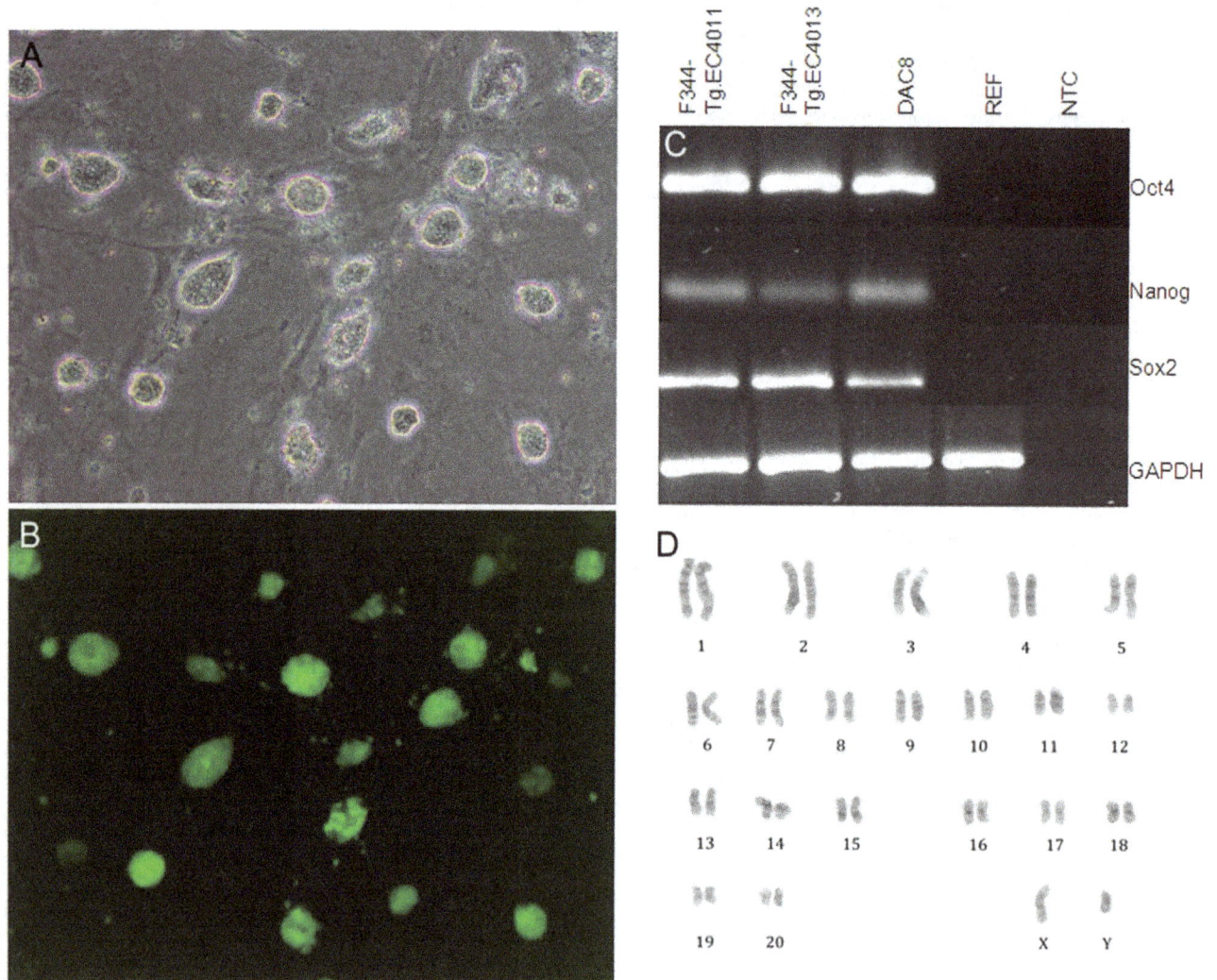

Figure 1. ES cell morphology and karyotype. The morphology and karyotype of F344-Tg.EC4011 is shown and is representative of the other ES cell lines. (A) Phase contrast image shows cultured ES cells forming compact colonies with smooth edges. (B) Fluorescence microscopy image of same field of view as (A). Cultured ES cells express the EGFP transgene. Scale bar represents 100 μm. (C) RT-PCR analysis of *Oct4*, *Nanog*, and *Sox2* gene expression using rat specific primers. DAc8, a proven germline competent rat ES cell line (Li et al., 2008) is included as a positive control; rat embryonic fibroblasts (REFs), mouse embryonic fibroblasts (MEFs) as well as a no template control (NTC) are also shown. (D) Cytogenetic analysis. ES cells have a normal male karyotype (42, XY).

background and recipient blastocyst genetic background in the rat. ES cell lines from F344 genetic background have been injected into host blastocysts from DA rats [3], F344 [20] or SD rats [20]. These ES cell lines failed to generate chimeric animals. Similarly, recipient embryos from SD rats have been used as recipient embryos for DA ES cell lines to generate chimeric animals, however, these chimeric animals failed to produce offspring with an ES cell genetic contribution [3]. Chimeric animals resulting

Table 1. Production of chimeric animals via blastocyst injection with rat ES cell line F344-Tg.EC4011.

Blastocyst injected	ES cells (passage #)	Embryos		Total pups	Chimeric animals (sex)
		injected	transferred		
1st injection	F344-Tg.EC4011(P16)	55	55	24	8 (M)
2nd injection	F344-Tg.EC4011(P16)	39	39	17	2 (1 M and 1 F)
3rd injection	F344-Tg.EC4011(P16)	61	61	17	2 (M)
4th injection	F344-Tg.EC4011(P16)	44	44	17	1 (F)

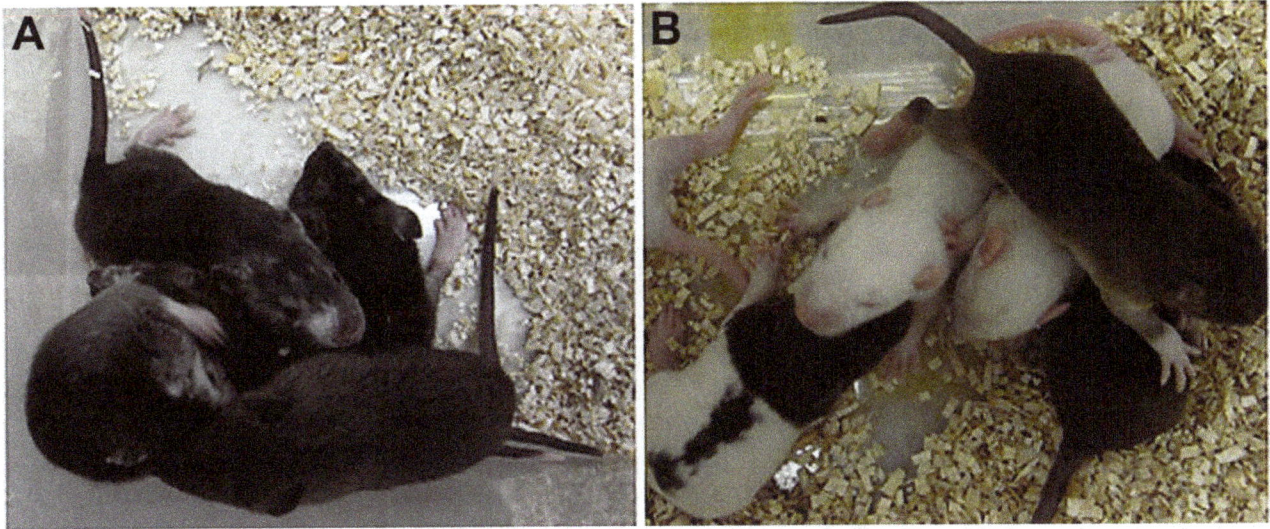

Figure 2. Coat color chimeras and their offspring. (A) Chimeric animals (albino patches on face) from F344-Tg.EC4011 ES cell injections into SD X DA blastocysts. (B) Offspring from chimeric animal breeding.

from SD blastocysts injected with ES cells from Brown Norway rats also failed to produce offspring with the Brown Norway genetic background [19].

In our study, we successfully used DA x SD hybrid blastocysts as recipient embryos. SD female rats were selected because of their high fecundity while male DA rats were selected for their pigmented coat color to aid in detection of chimeric animals. Chimeric animals were generated through blastocyst injection of ES cells from F344-Tg.EC4011 and the ES cells were able to colonize the gonads and produce sperm as evidenced by the transmission of the ES cell-derived EGFP gene from the chimeras to their offspring.

In conclusion, novel germline competent F344 rat ES cell line is now available and this particular ES cell line has the added advantage that it carries an EGFP transgene. The cell line has been deposited in the Rat Resource and Research Center

(RRRC), assigned stock number RRRC#654, and is available for distribution to the research community. This fluorescently-tagged F344 ES cell line provides an extremely useful tool for investigators who want to make genetically engineered rat models directly in a F344 genetic background.

Acknowledgments

We thank Beth Bauer for advice, Angela Goerndt for assistance with animal care, Miriam Hankins for technical assistance, and Howard Wilson for assistance with graphics.

Author Contributions

Conceived and designed the experiments: ECB HM. Performed the experiments: HM. Analyzed the data: ECB HM. Contributed reagents/materials/analysis tools: ECB. Wrote the paper: ECB HM.

Table 2. Breeding results of chimeric animals derived from rat ES cell line F344-Tg.EC4011.

Chimeric animals	1st litter		2nd litter		3rd litter		Germline competence
	Total	GFP+	Total	GFP+	Total	GFP+	
465RII	11	7					+
466RII	8	2					+
467RII	13	1					+
468RII	2	0	10	0			−
469RII	4	0	3	0	5	0	−
470RII	7	1					+
471RII	13	1					+
472RII	12	0	8	1			+
817RII	1	0	17	0	6	0	−
913RII	14	0	11	0	13	0	−
914RII	16	1					+

References

1. Gill TJ III, Smith GJ, Wissler RW, Kunz HW (1989) The rat as an experimental animal. Science 245: 269–276.
2. Abbott A (2004) Laboratory animals: the Renaissance rat. Nature 428: 464–466.
3. Li P, Tong C, Mehrian-Shai R, Jia L, Wu N, et al. (2008) Germline competent embryonic stem cells derived from rat blastocysts. Cell 135: 1299–1310.
4. Buehr M, Meek S, Blair K, Yang J, Ure J, et al. (2008) Capture of authentic embryonic stem cells from rat blastocysts. Cell 135: 1287–1298.
5. Tong C, Li P, Wu NL, Yan Y, Ying QL (2010) Production of p53 gene knockout rats by homologous recombination in embryonic stem cells. Nature 467: 211–213.
6. Capecchi MR (2005) Gene targeting in mice: functional analysis of the mammalian genome for the twenty-first century. Nat Rev Genet 6: 507–512.
7. Tong C, Huang G, Ashton C, Li P, Ying QL (2011) Generating gene knockout rats by homologous recombination in embryonic stem cells. Nat Protoc 6: 827–844.
8. Men H, Bauer BA, Bryda EC (2012) Germline transmission of a novel rat embryonic stem cell line derived from transgenic rats. Stem Cells Dev 21: 2606–2612.
9. Kawamata M, Ochiya T (2010) Generation of genetically modified rats from embryonic stem cells. Proc Natl Acad Sci U S A 107: 14223–14228.
10. Hong J, He H, Weiss ML (2012) Derivation and characterization of embryonic stem cells lines derived from transgenic Fischer 344 and Dark Agouti rats. Stem Cells Dev 21: 1571–1586.
11. Lois C, Hong EJ, Pease S, Brown EJ, Baltimore D (2002) Germline transmission and tissue-specific expression of transgenes delivered by lentiviral vectors. Science 295: 868–872.
12. Bryda EC, Pearson M, Agca Y, Bauer BA (2006) Method for detection and identification of multiple chromosomal integration sites in transgenic animals created with lentivirus. Biotechniques 41: 715–719.
13. Oh SH, Miyoshi K, Funahashi H (1998) Rat oocytes fertilized in modified rat 1-cell embryo culture medium containing a high sodium chloride concentration and bovine serum albumin maintain developmental ability to the blastocyst stage. Biol Reprod 59: 884–889.
14. Nichols J, Ying QL (2006) Derivation and propagation of embryonic stem cells in serum- and feeder-free culture. Methods Mol Biol 329: 91–98.
15. Xu J, Burgoyne PS, Arnold AP (2002) Sex differences in sex chromosome gene expression in mouse brain. Hum Mol Genet 11: 1409–1419.
16. Kakinoki R, Bishop AT, Tu YK, Matsui N (2002) Detection of the proliferated donor cells in bone grafts in rats, using a PCR for a Y-chromosome-specific gene. J Orthop Sci 7: 252–257.
17. Linder CC (2001) The influence of genetic background on spontaneous and genetically engineered mouse models of complex diseases. Lab Anim (NY) 30: 34–39.
18. Schoonjans L, Kreemers V, Danloy S, Moreadith RW, Laroche Y, et al. (2003) Improved generation of germline-competent embryonic stem cell lines from inbred mouse strains. Stem Cells 21: 90–97.
19. Zhao X, Lv Z, Liu L, Wang L, Tong M, et al. (2010) Derivation of embryonic stem cells from Brown Norway rats blastocysts. J Genet Genomics 37: 467–473.
20. Hong J, He H, Weiss ML (2012) Derivation and Characterization of Embryonic Stem Cells Lines Derived from Transgenic Fischer 344 and Dark Agouti Rats. Stem Cells Dev 21: 1571–1586.
21. Carstea AC, Pirity MK, Dinnyes A (2009) Germline competence of mouse ES and iPS cell lines: Chimera technologies and genetic background. World J Stem Cells 1: 22–29.
22. Schwartzberg PL, Goff SP, Robertson EJ (1989) Germ-line transmission of a c-abl mutation produced by targeted gene disruption in ES cells. Science 246: 799–803.

12

An α-Smooth Muscle Actin (acta2/αsma) Zebrafish Transgenic Line Marking Vascular Mural Cells and Visceral Smooth Muscle Cells

Thomas R. Whitesell[1,9], Regan M. Kennedy[1,9], Alyson D. Carter[1], Evvi-Lynn Rollins[1], Sonja Georgijevic[1], Massimo M. Santoro[2], Sarah J. Childs[1]*

1 Department of Biochemistry and Molecular Biology, and Smooth Muscle Research Group, University of Calgary, Calgary, Alberta, Canada, 2 VIB Vesalius Research Center, University of Leuven (KU Leuven), Leuven, Belgium

Abstract

Mural cells of the vascular system include vascular smooth muscle cells (SMCs) and pericytes whose role is to stabilize and/or provide contractility to blood vessels. One of the earliest markers of mural cell development in vertebrates is α smooth muscle actin (acta2; αsma), which is expressed by pericytes and SMCs. In vivo models of vascular mural cell development in zebrafish are currently lacking, therefore we developed two transgenic zebrafish lines driving expression of GFP or mCherry in acta2-expressing cells. These transgenic fish were used to trace the live development of mural cells in embryonic and larval transgenic zebrafish. acta2:EGFP transgenic animals show expression that largely mirrors native acta2 expression, with early pan-muscle expression starting at 24 hpf in the heart muscle, followed by skeletal and visceral muscle. At 3.5 dpf, expression in the bulbus arteriosus and ventral aorta marks the first expression in vascular smooth muscle. Over the next 10 days of development, the number of acta2:EGFP positive cells and the number of types of blood vessels associated with mural cells increases. Interestingly, the mural cells are not motile and remain in the same position once they express the acta2:EGFP transgene. Taken together, our data suggests that zebrafish mural cells develop relatively late, and have little mobility once they associate with vessels.

Editor: Ben Hogan, University of Queensland, Australia

Funding: This work was supported by grants to SJC from the National Science and Engineering Research Council (RGPIN/312496-2009) and the Heart and Stroke Foundation of Alberta, Nunavut and NWT (M06092). The funders had no role in study design, data collection and analysis, decision to publish, or preparation of the manuscript.

Competing Interests: The authors have declared that no competing interests exist.

* E-mail: schilds@ucalgary.ca

9 These authors contributed equally to this work.

Introduction

New blood vessels form during angiogenesis from angioblasts that migrate into position and differentiate into endothelial cells. These 'naked' endothelial tubes then undergo a maturation process. In the next stage of angiogenesis, endothelial cells attract perivascular mural cells including pericytes found on smaller vessels, and smooth muscle cells (SMCs) found on larger vessels. The role of the mural cells is to physically support vessels, secrete extracellular matrix, provide vascular tone and induce vessel quiescence [1].

Hemorrhage results from breakage of contacts between endothelial cells, and can be due to a variety of mechanisms, either poor junctional contacts, defective extracellular matrix contacts, or lack the association of mural cells with endothelial cells [1,2]. Reciprocal signalling events between endothelium and mural cells are critical for the maturation and stabilization of new vessels [3]. Endothelial cells express the chemoattractant PDGF-B, to attract mesenchymal cells expressing the PDGFRβ receptor to vessels [4]. In turn, these mesenchymal cells secrete Angiopoietin1 [5], which binds to Tie2 receptors expressed on endothelial cells and promote their differentiation [6]. The mutual attraction of the

mesenchymal and endothelial cells results in the two layers forming close contacts, followed by maturation of the mesenchymal cells into smooth muscle or pericyte cells. In addition, both pericytes and SMCs require Sonic hedgehog signalling (Shh) for normal vascular development [7],and for the induction of Angiopoietin1 expression [8,9]. The requirement for Shh extends throughout the lifetime for some SMCs as it is indispensable for their survival [10,11]. Finally, signalling through Notch3 and Sphingosine1 phosphate pathways promotes the investment of mural cells on endothelial tubes [12,13].

In the head, pericytes and vascular SMCs derive from the ectomesenchymal lineage of the cranial neural crest (CNC), at least in the chick and mouse [14,15].The ectomesenchymal lineage also produces cartilage and bone of the face, mesenchyme and some cells of the heart [16]. These neural crest cells migrate ventrally from the hindbrain rhombomeres to populate the region around the eye and around the pharyngeal arches, arriving by 24 hpf in the zebrafish [16]. FoxD3, TFAP2 and Sox10 are three genes that promote specification of neural and neural crest pigment derivatives, and repress ectomesenchymal fates [15,17,18]. However transcription factors that actively specify SMCs from ectomesenchymal cells are currently unknown.

Vascular SMCs (vSMCs) are found in large blood vessels where a continuous single or multilamellar SMC layer surrounds the endothelial cell lining and provides contractility to modulate blood flow and stability. SMCs are separated from the endothelium by a basement membrane. SMCs secrete a large amount of the blood vessel ECM, consisting mainly of Laminin, Collagen IV, Nidogen, Perlecan, and Fibulins. Secretion of ECM from vSMCs is vital, as loss of the collagen Col4a1 leads to perinatal hemorrhage [19], while loss of Fibulin4 leads to aneurysms [20,21].

Pericytes are mural cells found in microvessels (smaller arterioles and capillaries), particularly in the brain, eye and kidney [22]. Unlike SMCs, pericytes do not form a continuous layer and are present as isolated cells. Pericytes are embedded within the basement membrane. Recent findings show that they also provide contractility to blood vessels [23]. Interestingly once pericytes cover vessels, this halts vascular remodelling and prevents further proliferation of endothelial cells [24,25]. Pericytes thus control fundamental behaviours of endothelial cells.

Although there are suggestions that the lineage of pericytes and smooth muscle cells might be identical, and they express overlapping sets of molecular markers [3], the two cell types are defined as morphologically distinct. One of the main criteria used to distinguish pericytes and SMCs is whether the mural cell lies within or outside of the basement membrane [3]. In embryonic development, however, this cannot always be applied as the basement membrane is not always present during angiogenesis. For instance, in early zebrafish vessels, ultrastructure shows no evidence of a basement membrane, nor convincing expression of pericyte or smooth muscle specific molecular markers [26]. In this case, mural cells have been referred to as 'mesenchymal' or 'perivascular support cells' until they can be properly identified [8,26].

We and others have developed markers for the early vascular mural cell lineage in zebrafish, and have shown that these markers are expressed at much later equivalent developmental stages than in other organisms such as the mouse [27,28,29]. Another early smooth muscle marker, transgelin (tagln) is first visualized by antibody around 80 hpf in zebrafish [30]. An tagln/sm22α-b transgene can also be seen in the ventral head vessels on the late third and fourth day of development [28]. This timing suggests that mural cells are developing concomitantly with angiogenesis, however, these cells are difficult to visualize without molecular markers. Miano et al. used electron microscopy to conclude that undifferentiated mural cells were in place around the dorsal aorta at 7 dpf, but could not likely be identified by histology [27]. Thus the development of molecular markers that can identify vascular mural cells in vivo in zebrafish is urgent.

In contrast, early visceral smooth muscle development has been well characterized in zebrafish. RNA for *smooth muscle myosin heavy chain* and *non-muscle myosin heavy chain-b* begins to be expressed around 50 hpf [29,31]. We previously showed that mRNA for the early smooth muscle marker *tagln/sm22α-b* turns on at 56 hpf and *acta2* turns on at 60 hpf in the gut [29]. More mature smooth muscle markers such as *cpi17* and *smoothelin-b* turn on at 72 hpf. Transgenic zebrafish generated using the *tagln/sm22α-b* promoter highlight visceral smooth muscle development, but in these animals, vascular smooth muscle is difficult to visualize [28]. This useful animal model still lacks reagents for in vivo imaging which would highlight vascular mural cell developmental processes at the cellular level. Here we develop transgenic animals expressing GFP or mCherry under the mural cell promoter α-smooth muscle actin and trace the development of these vSMC in living embryos, showing that although vascular mural cells arise late in development, they form smooth muscle layers around blood vessels, and are associated with vascular stabilization.

Materials and Methods

Ethics statement

Zebrafish were maintained and staged as previously described [32]. All procedures in this study were specifically approved by the University of Calgary Animal Care Committee. Wild type Tupfel long fin (TL) zebrafish or Tg(6.5kdrl:mCherry)[ci5], Tg(fli1a:EGFP)[y1], Tg(fli1a:nEGFP)[y7] were used for all experiments [33,34,35].

Whole-mount in situ hybridization and immunostaining

Digoxigenin-labelled antisense RNA probes were used in whole-mount in situ hybridization as previously described, with the exception that embryos older than 7 days post- fertilization (dpf) were fixed in Dietrich's fixative [36]. The probe for *acta2* has been described [29]. For histological analysis, embryos were embedded in JB4 (Polysciences) and 7 μm sections were cut on a Leica microtome. For antibody staining, the Vectstain ABC Kit (Vector Labs) with mouse αGFP (JL-8, BD Clontech). For histology, sections were stained with hematoxylin and aqueous eosin. The transgelin rabbit polyclonal antibody has previously been described [30].

Identification of Acta2 promotor and enhancer sequences

We used the Santa Cruz genome browser to identify regions of cross-species conservation focused on the proximal promoter and first intron of the mouse *acta2* which drives smooth muscle-specific expression [37]. We used available software prediction programs (PATCH1.0) from TRANSFAC [38] to predict one potential CArG site. Other CArG sites were identified by manual inspection using validated CArG sites from mice.

Acta2 transgenic zebrafish

A 300 bp proximal promoter for *acta2* and 2165 bp fragment from the *acta2* intron 1 was cloned using the primers described [39]. The two genomic fragments were fused in a PCR reaction to make a 2465 bp enhancer/promoter construct (referred to as the *acta2* promoter from this point on) which was cloned into pDONRp4p1r and then using the three way Tol2 Gateway cloning system upstream of GFP or mCherry [40]. Both promoter-fluorophore constructs were then isolated from the Tol2 backbone by digestion with Xho I and Cla I before injection into early one stage embryos. All founders had similar expression patterns, and a single founder was chosen for further analysis (Tg(acta2:EGFP)[ca7] or Tg(acta2:mCherry)[ca8]. We note that Tg(acta2:mCherry)[ca8] embryos have substantially weaker fluorescence. Both transgenic lines have been deposited to the Zebrafish International Resource Center with the Catalog IDs ZL4966 and ZL4967.

Imaging

Sections were photographed using a Leica DMR microscope equipped with an Optronics Magnafire camera and Nomarski optics. Adult heart was photographed on a Zeiss Stemi SV11 microscope equipped with a Zeiss HR camera. For confocal microscopy, embryos were live imaged after mounting in low melt agarose on glass bottom dishes on a Zeiss LSM 510 Meta or Zeiss LSM 700 microscope. Slices were taken at intervals ranging from 1- 3 μm on a 10, 20 or 40× objective and subject to 2 times averaging. Image stacks were processed using a Kalman stack filter in ImageJ or in Zen Blue and are presented as maximal intensity projections. For timelapse imaging, embryos were mounted in low melt agarose in a heated chamber and imaged at 60 minute intervals.

Morpholino knockdown

FoxD3 MO1 (5′ tgctgctggagcaacccaaggtaag 3′) and Tfap2a 5.1 MO (5′ cctccattcttagatttggccctat 3′) published morpholinos [18] were obtained from Gene Tools LLC and dissolved in water. 2.5 ng of morpholino was injected into 1–4 cell stage embryos.

Results

Creation of an Acta2 promoter/enhancer construct for in vivo expression

In mouse, both the proximal promoter and first intron of *acta2* contribute to its expression [37]. In particular 'CArG box' (CC (A/T)$_6$ GG) motifs are critical for mural cell expression of *acta2*, including two CArG boxes in the *acta2* proximal promoter (CArG-A and CArG-B) and one in the first intron (intron CArG). We manually aligned the 300 bp proximal promoter and first intron of the zebrafish *acta2* gene with that of mouse. Although there is very poor sequence identity of these two genomic regions (42% overall identity), the relative position of the CArG elements is conserved (CArG-A is at -83 bp in fish and −70 in mouse; CArG-B is at −135 bp in fish and −121 in mouse; intron CArG is at +655 bp in fish and +1039 in mouse). Furthermore, the sequence of these elements is highly conserved from zebrafish to mouse, with only one conservative nucleotide difference in CArG-B, and absolute conservation in CArG-A and intron CArG (Figure 1A). In comparison with other fishes, an identical CArG-B box was also found in tilapia and medaka, as well as a completely conserved CArg-A box in tilapia (Figure 1B). These elements have not been functionally tested in zebrafish, although their conservation with those in mouse, tilapia and medaka, suggests they have been conserved through a long evolutionary period and are likely to be functionally important.

acta2 is strongly expressed in visceral smooth muscle precursors in the gut and swim bladder at 72 hpf and 100 hpf (Figure 1C, D). Non-smooth muscle expression of *acta2* is also observed in the ventral eye in the site of the optic fissure (Figure 1C) and in the floor plate (Figure 1D).

Acta2 is a marker of vascular mural cells and visceral smooth muscle

Hypothesizing that the first intron of *acta2* would act as a transcriptional enhancer in zebrafish as in mouse, we cloned the entire intron 1 upstream of 300 bp of proximal *acta2* promoter to make a compact promoter/enhancer construct that could easily be used for transgenesis (Figure 1A). The promoter/enhancer was adapted with Gateway cloning sites and inserted into the Tol2 transposon, driving either GFP or mCherry [40,41]. To make transgenic animals, the promoter/enhancer:EGFP or promoter/enhancer:mCherry cassette was digested away from the transposon vector and the DNA injected into single cell embryos in a traditional transgenesis approach.

Multiple founders were identified from injection of the *acta2* promoter/enhancer construct. The alleles Tg(acta2:EGFP)^{ca7} and Tg(acta2:mCherry)^{ca8} were maintained for further study. Tg(acta2:EGFP)^{ca7} and Tg(acta2:mCherry)^{ca8} show identical expression patterns, although expression in Tg(acta2:mCherry)^{ca8} is weaker in intensity. For this reason, most experiments were conducted using Tg(acta2:EGFP)^{ca7} animals. We note that we have previously used the Tg(acta2:mCherry)^{ca8} line to demonstrate expression in the gut

smooth muscle [39]. We also have previously used the Tg(acta2:GFP)^{ca7} line to show a decrease in vascular mural cells in the ventral head of βPix and integrin morphant animals [42] without characterizing the spatiotemporal expression of the transgene in the context of mural cell development.

To demonstrate the fidelity of the transgene, we compared transgene expression (as detected by an anti-GFP antibody) to that of native *acta2* transcript as detected by in situ hybridization. We find good concordance between both staining (Figure 1 E and F). We then compared acta2:EGFP expression in transgenic fish to native Tagln expression using antibody staining [30] (Figure 1 G–I), and find co-localization of Tagln and GFP.

To demonstrate that smooth muscle was being labelled in the acta2:EGFP line, we first examined the morphology of cells expressing EGFP in vascular and visceral beds in double transgenic acta2:EGFP; kdrl:mCherry fish. In the pharyngeal region of 4 dpf embryos, the ventral aorta shows numerous acta2:EGFP cells exterior to the endothelial lining of the blood vessel, marked by Tg(kdrl:mCherry)^{ci5} shown in red (Figure 2A, Figure S1). Thus, acta2:EGFP cells meet the classic definition of 'vascular mural cells' in being closely associated with the endothelial cell wall. As the coverage of the endothelium is not continuous at this stage of development, these cells are morphologically more similar to pericytes than smooth muscle cells. It is interesting that even though the ventral aorta is the vessel receiving the highest blood pressure in the zebrafish embryo, during embryonic development there is no clear evidence of a multilamellar vascular muscle wall as would be present in mammalian species at an equivalent developmental stage [28,43]. To determine whether these pericyte-like cells eventually mature into smooth muscle, we dissected whole adult ventral aorta and associated vessels of an adult transgenic animal. We find that there is extensive coverage of the ventral aorta and associated vessels in the adult acta2:EGFP transgenic fish, suggesting that these early pericyte-like acta2:EGFP positive cells mature into a conventional smooth muscle layer as the fish grows (Figure 2B).

Clear evidence of radial and circumferential smooth muscle cells is seen in both the gut and the swim bladder of 14 dpf acta2:EGFP in the trunk region (Figure 2C). In contrast to the well-developed visceral smooth muscle, very few acta2:EGFP cells are seen on the dorsal aorta of the trunk, even at this juvenile stage. Skeletal muscle fiber expression of the acta2:EGFP transgene is present but highly variable from embryo to embryo at early stages (Figure S2).

acta2 expression in the developing heart outflow tract

Since the region of the outflow tract has the highest blood pressure exposure, we hypothesized that it would be the first vessels to develop vascular smooth muscle. We examined native *acta2* transcript by in situ hybridization and compared it to acta2:EGFP transgene expression. At 56 hpf *acta2* mRNA is restricted to the bulbus arteriosus (Figure 3A). In contrast, the acta2:EGFP transgene is expressed in the myocardium of the atrium and ventricle, but not in the bulbus arteriosus at this time (Figure 3B,C). The lack of expression of the acta2:EGFP transgene in the bulbus arteriosus at 56 h hpf likely reflects a delay in GFP protein expression, as both *acta2* mRNA and acta2:EGFP transgene are expressed in the bulbus arteriosus at 78 hpf (Figure 3D–F). At 100 hpf, mRNA for *acta2* and the acta2:EGFP transgene are both visible in the bulbus arteriosus and ventral aorta (Figure 3G–I). We note persistent expression of the acta2:EGFP transgene in the atrium and ventricle at every time point examined including 78 and 100 hpf and beyond, times at which native *acta2* expression is not observed by in situ

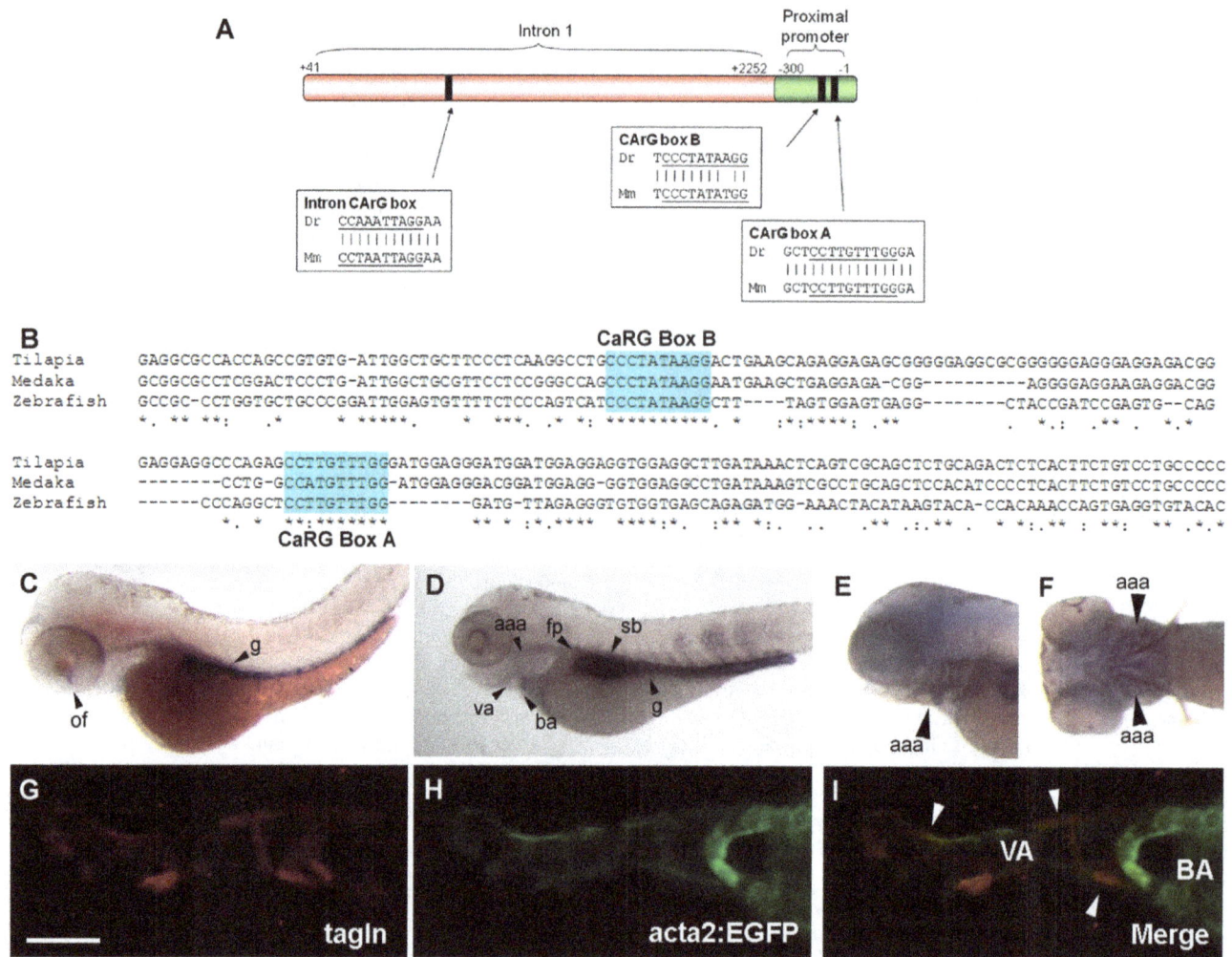

Figure 1. Acta2 promoter/enhancer construct design and expression in zebrafish. (A) A zebrafish (Dr) enhancer/promoter construct was constructed from the proximal promoter and first intron sequence of the zebrafish *acta2* gene, and contains three highly conserved CArG binding sites also found in the mouse (Mm) *acta2* proximal promoter and first intron. (B) Comparison of zebrafish CaRG boxes A and B in zebrafish, tilapia and medaka. (C,D) By wholemount in situ hybridization, *acta2* shows strong expression in the gut (g) at 72 hpf (B), and expressed in the gut, swim bladder (sb), ventral aorta (va), floor plate (fp), aortic arch arteries (aaa), and bulbus arteriosus (ba) at 100 hpf (C). (E,F) Co-localization of wholemount in situ hybridization *acta2* and anti-GFP staining of the acta2:GFP transgene shows strong expression in the aortic arch arteries (aaa) at 100 hpf. (G,H,I) 4 dpf acta2:EGFP transgenic fish (H) stained with Tagln rabbit polyclonal antibody (G). Merge (I) shows co-localization between acta2:GFP and Tagln. Arrowheads in G–I depict vascular mural cells. Scale bar in G represents 20 μm.

hybridization (Figure3 F, I). As myocardial expression continues into the adult, this suggests our *acta2* promoter-enhancer construct is lacking additional sequence to properly downregulate its expression in the heart. Additional non-muscle sites of *acta2* mRNA expression are seen in the tip of the notochord and floorplate (Figure 3 G–H). The anterior tip of the notochord is strongly labelled in acta2:EGFP transgenic fish (Figure 3F,I).

We next studied larval stages. At 22 dpf, multiple layers of smooth muscle are observed in the bulbus arteriosus by histology (Figure 3J), while the ventral aorta still has only a single layer (Figure 3M). This smooth muscle is positive for *acta2* expression by in situ hybridization (Figure 3K, N) and expresses the acta2:EGFP transgene (Figure 3L, O). Anatomical context for Figure 3 K and N is provided in Figure S3.

We then observed *acta2* expression in the adult heart. In situ hybridization of the wholemount heart shows strong staining of *acta2* in the bulbus arteriosus, but not atrium or ventricle (Figure 3P). Sections of the heart reveal that the *acta2* staining is localized to the muscle wall (Figure 3Q). In comparison, a whole heart isolated from an acta2:EGFP transgenic animal shows intense acta2:EGFP expression in the bulbus arteriosus, but maintains weak GFP expression in the ventricle (Figure 3R).

acta2:GFP vascular mural cells gradually increase in number

To examine the progression of mural cell association with endothelium, we imaged acta2:EGFP cells in the ventral head at 4, 7, 11 and 14 days post fertilization. Zebrafish grow at variable rates during larval periods [44], and thus we imaged a minimum of 3 fish for each time point and present a representative image. The location of acta2:EGFP cells is compared to the pattern of blood vessels as marked by the Tg(kdrl:mcherry)[ci5] transgene. At 4 dpf in ventral views (Figure 4), both chambers of the heart are strongly acta2:EGFP positive, as is the bulbus arteriosus. Along the ventral aorta, coverage by acta2:EGFP cells is dense, but does not completely cover the blood vessel. Scattered cells are seen on

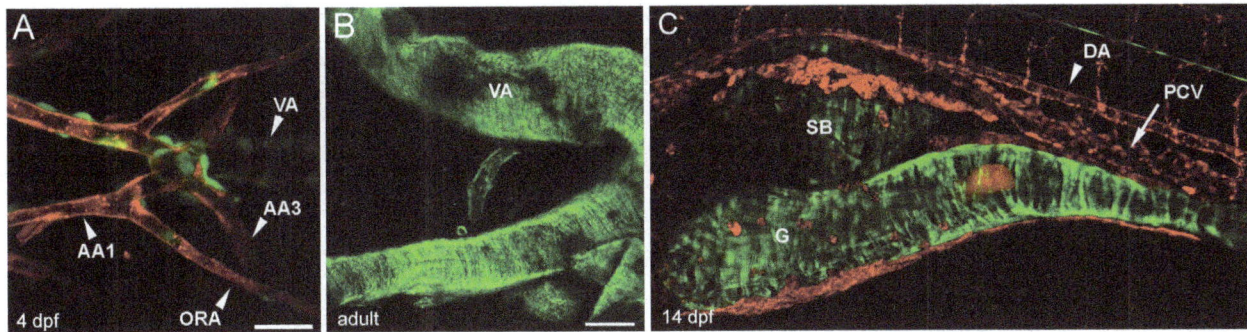

Figure 2. Morphology of vascular and visceral mural cells in acta2:EGFP transgenic fish. (A) Ventral pharyngeal region of a 4 dpf double transgenic Tg(acta2:EGFP)^{ca7}; Tg(kdrl:mCherry)^{ci5} (mural cells are green and endothelial cells are red) zebrafish shows extensive mural cell coverage of the ventral aorta (VA) and lesser coverage on the smaller aortic arches (AA) or opercular artery (ORA). (B) Wholemount adult ventral aorta and attached afferent branchial arteries shows extensive smooth muscle coverage. (C) Lateral view of the gut (g) and swim bladder (b) of a 14 dpf double transgenic Tg(acta2:EGFP)^{ca7}; Tg(kdrl:mCherry)^{ci5} zebrafish shows radial and circumferential smooth muscle on both gut and swim bladder, but sparse mural cells on the dorsal aorta (DA) and no visible cells on the posterior cardinal vein (PCV). Scale bar in A represents 25 μm. Scale bar in B and C represents 100 μm.

aortic arch arteries. At 7, 11 and 14 dpf the ventral aorta is still undergoing morphogenesis and increases in length. During this period, the coverage of acta2:EGFP cells increases modestly. Over this period we observe an increasing complexity in vascular pattern of the gill arches, although the majority of acta2:EGFP positive cells are associated with larger vessels and not newly formed small vessels.

In dorsal view at 4 dpf only a few acta2:EGFP cells are observed (Figure 5B,D,F) despite extensive vascularization of the brain (Figure 5A,C,E). At 7 dpf and 11 dpf large head vessels are associated with acta2:EGFP cells.

In lateral view, we observe a striking scarcity of acta2:EGFP cells in the brain at early stages from 4 dpf to 14 dpf (Figure 6A,C,E,G). However there is extensive association of acta2:EGFP cells with ventral aortic arch vessels with the coverage becoming more complete over time (Figure 6B,D,F,H).

Vascular trunk acta2:EGFP positive vascular mural cells are scarce

In the trunk, vascular acta2:EGFP cells are scarce and are associated only with the ventral surface of the dorsal aorta at 4 dpf (Figure 7A,B). At this stage, these mural cells do not encircle the vessel however they are located outside the endothelium (Figure S1). At 14 dpf the morphology of the acta2:EGFP cells on the ventral aorta is similarly sparse, although a few acta2:EGFP cells can be seen on the ventral aspect of some intersegmental arteries suggesting that these angiogenic vessels are beginning to develop associations with mural cells (Figure 7C,D). In contrast, at 80 hpf, visceral smooth muscle is well developed and strongly expresses acta2:EGFP (Figure 7E). Variable, scattered skeletal muscle fibres express the acta2:EGFP transgene, although this diminishes over time.

Embryonic origins of acta2:EGFP vascular mural cells

In mouse, chicken, and frog, vascular mural cells of the head originate from a migratory population of neural crest cells [14,45,46,47]. There is no information on the origins of zebrafish head mural cells, therefore we crossed our transgenic acta2:mCherry^{ca8} fish with transgenic fli1a:nEGFP^{y7} zebrafish. Fli1a:-nEGFP^{y7} labels endothelial cells and ectomesenchymal neural crest derivatives of the ventral head but not mesodermal or endodermal derivatives [34,48]. If mural cells derive from a neural

crest lineage, we might expect co-localization of fli1a:EGFP and acta2:mCherry, however, at 4, 7, and 10 dpf (Fig 8A, 8B, data not shown, respectively), we do not see co-localization of markers. We observe mCherry positive mural cells in proximity with GFP positive endothelial cells, but no obvious co-localization. This includes cells on the ventral aorta (Fig 8A' and B'), and the aortic arch arteries (Fig 8A'' and B''), which are some of the first vessels to be covered with mural cells. These experiments could suggest that vascular mural cells of the ventral head arise from a non-neural crest origin, however we cannot rule out that *fli1a* expression has been downregulated in this lineage, or that the cells are neural crest derived but the *fli1a* transgenic fish line do not have the correct promoter/enhancer elements to express GFP in these cells.

As an independent test of origin we also tried ablating neural crest specification by knockdown of the transcription factors FoxD3 and TFAP2 [18]. We observed severely decreased numbers of acta2:EGFP positive cells; however embryos with double knockdowns of these transcription factors also had severely disrupted ventral head and endothelial patterning and circulation was compromised (Figure S4). Single knockdown of either FoxD3 or TFAP2A resulted in atypical, but less severely affected vessel patterning, with reduced mural cell coverage. These experiments are therefore inconclusive as mural cell differentiation could have secondarily been affected by the lack of robust circulation.

acta2:EGFP expressing mural cells are stably associated with blood vessels

Data from in vitro models of endothelial-pericyte co-assembly suggests that pericytes are highly motile and migrate along nascent endothelial tubes in these culture systems [49]. We thus wanted to examine the behaviour of mural cells in vivo to determine their motility and proliferation when associated with vessels. We used a timelapse confocal microscopy strategy to follow acta2:GFP expressing cells on the ventral aorta and aortic arch vessels for several 12 hour windows from 3.5 through 5 days of development. In contrast to in vitro observations, we observe that cells expressing acta2:EGFP for the most part do not migrate, alter their cellular morphology or have observable cytokinesis during this window, and are therefore morphologically stable (Figure 9 A–E; Movie S1; n = 5 embryos). As the ventral aorta eventually

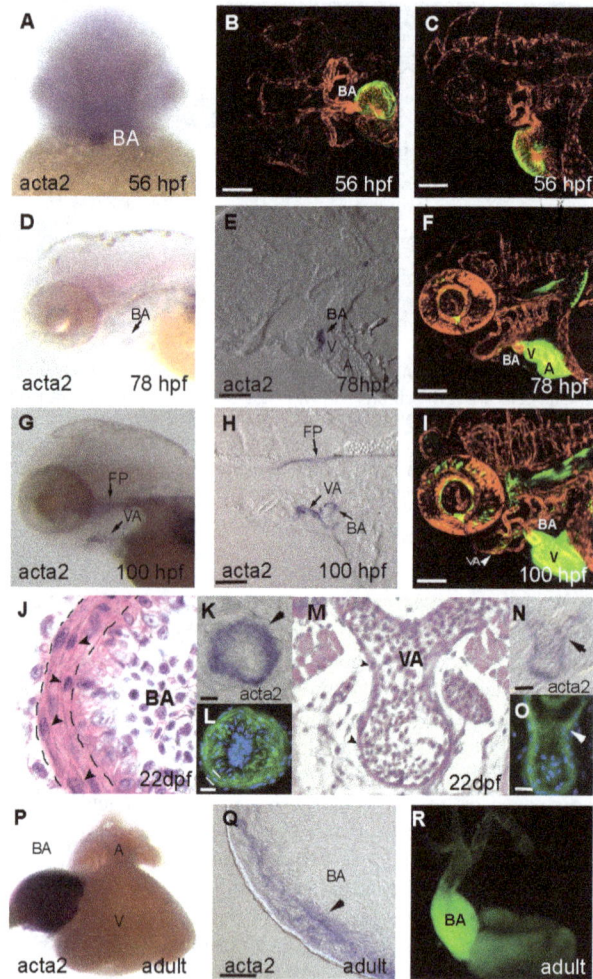

Figure 3. Smooth muscle markers are restricted to the developing cardiac outflow tract by 56 hpf. (A) At 56 hpf, *acta2* expression is restricted to the developing BA. (B,C) Double transgenic Tg(acta2:EGFP)[ca7]; Tg(kdrl:mCherry)[ci5] embryo shows expression of EGFP in both the atrium and ventricle of the heart at 56 hpf, but not in the BA. (D) *acta2* expression is evident at 78 hpf in the BA in both wholemount and cross section (E) and in transgenic animals (F). (G–I) Expression of *acta2* continues to be restricted to the BA and ventral aorta (VA) at 100 hpf by in situ hybridization and in transgenic fish. (J–O): Cross sections of the 22 dpf BA show a multilamellar arterial phenotype as visualized by hematoxylin and eosin staining (J), in situ hybridization of *acta2* (K) and transgenic GFP (nuclei stained blue with DAPI, L). The bulbus vascular wall consists of three layers: an inner intima, middle media, and outer adventitia (Ad, separated by black lines in J). The intima is endothelial (arrowheads point to nuclei of endothelial cells). The media consists of 3–4 cell-thick layers of vascular smooth muscle cells (M, arrows point to nuclei of SMCs). In comparison to the BA, the vascular wall of the VA at 22 dpf is thin (M) but expresses *acta2* by in situ hybridization (N) and GFP in transgenic animals (O). The endothelium of VA is covered by a thin layer of SMCs (arrowheads point to nuclei of SMCs). (P) In situ hybridization of the wholemount adult heart shows strong staining in the bulbus arteriosus, but not ventricle or atrium, which is localized to the myocardial wall in cross section (Q). (R) Wholemount dissected acta2:EGFP transgenic heart shows stronger expression of GFP in the bulbus arteriosus as compared to ventricle. Staining is also continuous with the ventral aorta. In B,C, F, I, and R, green expression is acta2:EGFP transgene. Scale bar in B, C, F, and I is 100 μm. Scale bar in E, H, and Q is 50 μm. Scale bar in K, L, N, and O is 20 μm.

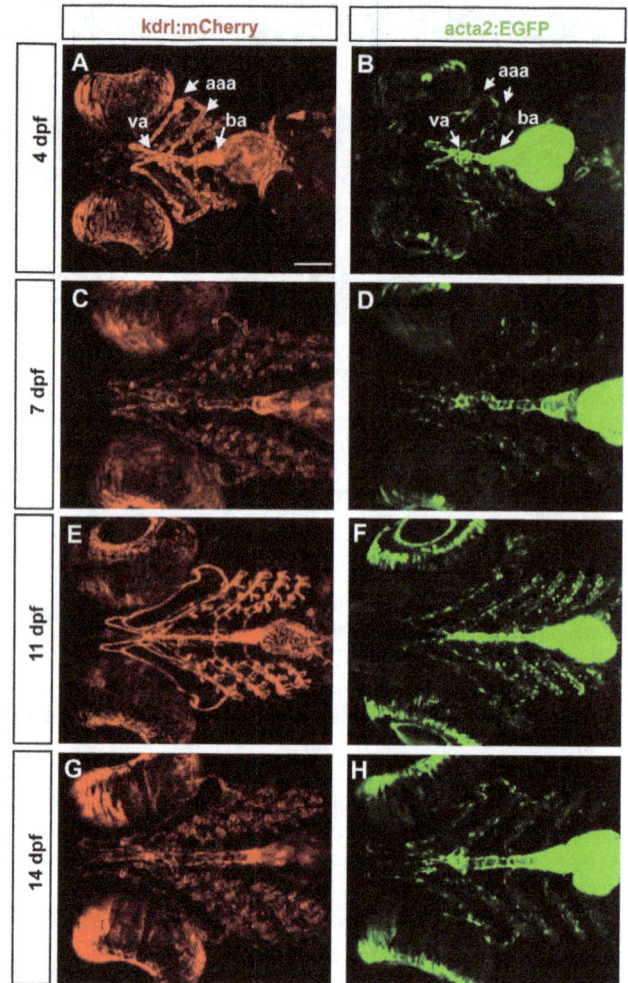

Figure 4. Mural cell and endothelial development in the ventral head of larval zebrafish. Confocal micrographs collected from a ventral point of view show a progressive increase in vessel complexity (red, A, C, E, G) and in density of mural cell coverage of aortic arch vessels (green, B, D, F, H) from 4 dpf (A, B), 7 dpf (C, D), 11 dpf (E, F) through 14 dpf (G, H). Heart expression of acta2:EGFP is maintained. aaa = aortic arch arteries; va = ventral aorta; ba = bulbus arteriosus. Scale bar in A represents 100 μm.

becomes covered in multilamellar smooth muscle, we suggest that mural cells proliferate on very long time scale.

Discussion

We show here that an acta2:EGFP transgenic line derived from an enhancer/promoter fusion is expressed in vascular mural cells and visceral smooth muscle cells in the zebrafish embryo, larva and adult. Visceral smooth muscle development has been well described by others using Tagln/SM22 transgenic zebrafish, and we will not discuss its formation further here as our observations in acta2:EGFP transgenic fish are similar [28]. Although *tagln* and *acta2* are both early smooth muscle markers with an essentially identical expression pattern in zebrafish [29], *acta2* turns on slightly later in development as detected by in situ hybridization. However, the temporal development of vascular smooth muscle as assayed by the Tagln/SM22 transgenic and our acta2:EGFP transgenic is similar. Seiler and Pack show expression of the Tagln/SM22 transgene at 4 dpf in the bulbus arteriosus and

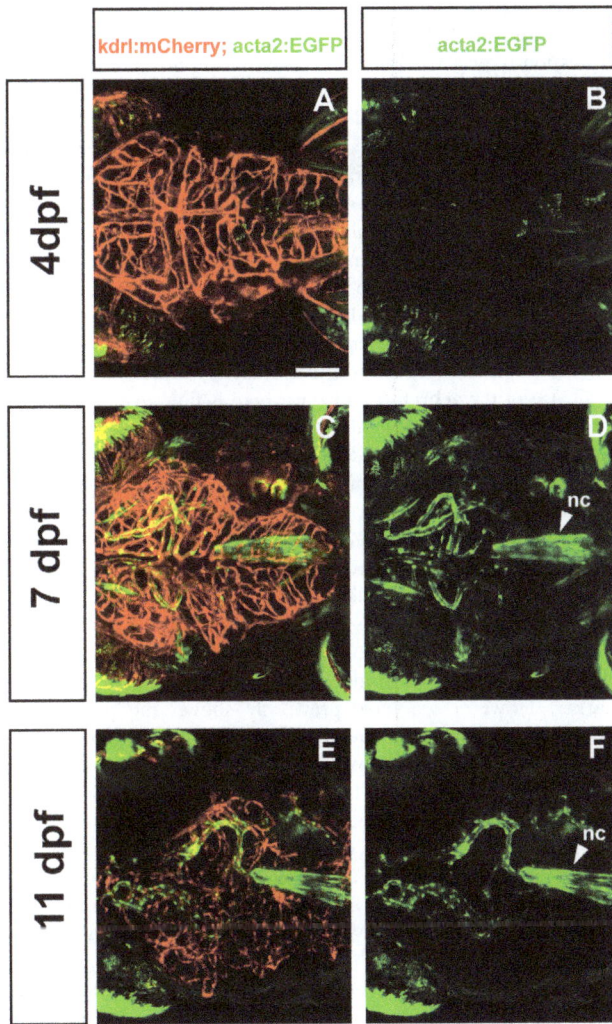

Figure 5. Development of mural cells and endothelial cells as seen in dorsal view. Confocal micrographs collected from a dorsal point of view show a progressive increase in vessel complexity (red, A, C, E) and in density of mural cell coverage of head vessels (green, B, D, F) at 4 dpf (A, B), 7 dpf (C, D), and 11 dpf (E, F). nc = notochord. Scale bar in A represents 100 μm.

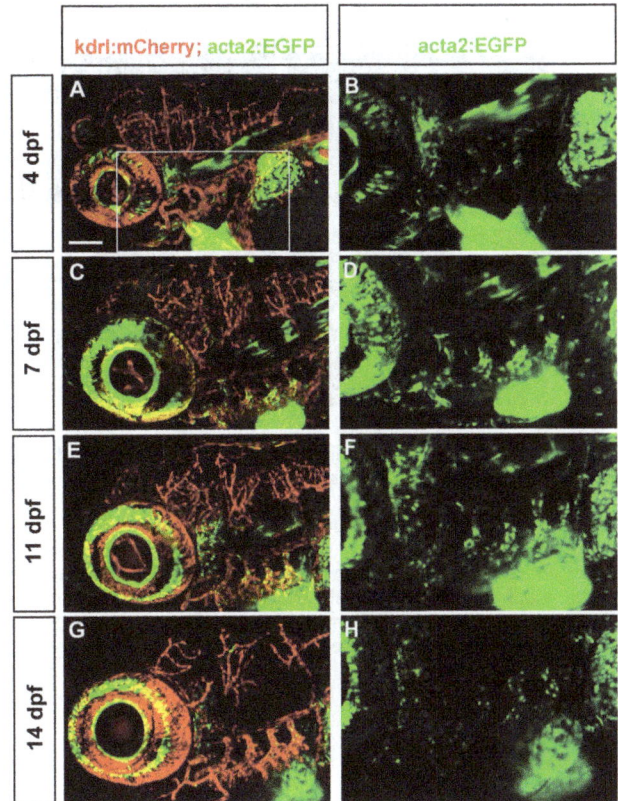

Figure 6. Development of mural cells and endothelial cells as seen in lateral view. Confocal micrographs collected from a lateral point of view show a progressive increase in vessel complexity (red, A, C, E, G) and in density of mural cell coverage of aortic arch vessels (green, all panels, inset is enlarged in B, D, F, H to show coverage of aortic arches) at 4 dpf (A, B), 7 dpf (C, D), and 11 dpf (E, F). Scale bar in A represents 100 μm.

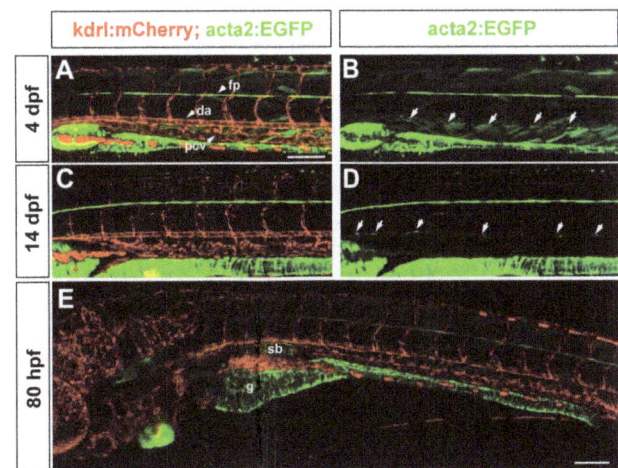

Figure 7. Vascular and visceral mural and smooth muscle cells in the trunk. (A–B) At 4 dpf, acta2:EGFP positive cells (arrows in B) are seen in the ventral portion of the dorsal aorta, but not in other vessels of the trunk. Floor plate (fp) expression of acta2:EGFP is observed in all images. (C–D) At 14 dpf, the distribution of vascular mural cells to the ventral portion of the dorsal aorta only, is still observed. (E) In contrast to the scarce vascular smooth muscle coverage, visceral smooth muscle cells strongly express the acta2:EGFP transgene at 80 hpf. Scale bars represent 100 μm. Green striations are skeletal muscle fibres.

ventral aorta, similar to the acta2:EGFP transgenic [28]. However, the strong expression of acta2:EGFP in ca7 transgenic fish has allowed us to trace vascular mural cell development in live animals from late embryogenesis through larval and adult stages for the first time in this important model organism.

We demonstrate that vascular mural cells turn on acta2:EGFP several days after the initiation of circulation and initially show a pericyte-like morphology. Furthermore, it is only in late larval stages that multilamellar smooth muscle is observed. We corroborate transgenic expression with morphology and in situ hybridization for native acta2 mRNA. The most developed vascular smooth muscle occurs in bulbus arteriosus, with the next greatest coverage occurring in the ventral aorta. In contrast, the dorsal aorta of the trunk has single acta2:EGFP cells on the ventral surface only, at an equivalent time point. Hu et al., have shown that the ventral aorta has a 40% greater blood pressure than the dorsal aorta in the head of adult zebrafish due to the resistance in the fine branchial (gill) arteries [50]. Our observation of much greater mural cell coverage in the vessels adjacent to the heart

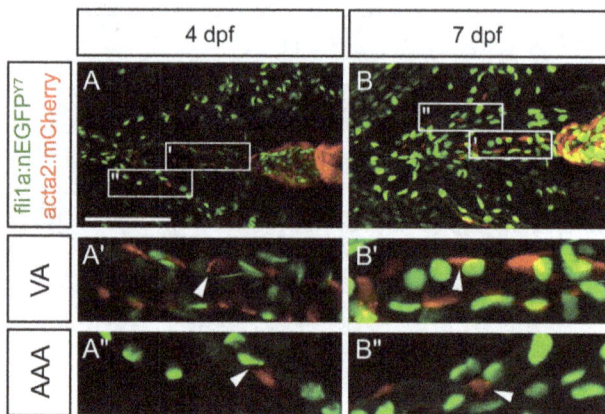

Figure 8. Lack of co-localization of mural cell and the ectomesenchymal neural crest marker Fli1a. Confocal images of 4 and 7 dpf embryos using a cytoplasmic mural cell marker (acta2:mCherry) and nuclear neural crest marker (fli1a:nEGFP[y7]) using ventrally staged embryos. (A) Mural cell and neural crest markers are expressed along the ventral aorta (A') and aortic arch artery region (A'') of the 4 dpf embryo. (B) Mural cell and neural crest markers are expressed along the ventral aorta (B') and aortic arch region (B'') at 7 dpf. There appears to be little to no co-localization of fluorescent markers at both 4 and 7 dpf. Scale bar in A represents 100 µm. Insets (A', A'', B', B'') are 100 µm in length. VA = Ventral Aorta, AAA= Aortic Arch Arteries. Arrowheads depict cells that no do not co-localize.

Expression of *acta2* between mice and zebrafish also differs. Mice expressing a BAC-derived acta2:mCherry transgene show expression of acta2 at E8.5 in the myocardium of the heart [43]. At E9.5 and 10.5, signal is observable in the aorta and somites, while visceral smooth muscle expression begins at E13.5 in these mice. Thus mouse acta2 is expressed in vascular mural cells during embryonic stages concomitantly with the onset of circulation, while zebrafish *acta2* is expressed in larval and juvenile stages, long after the initiation of circulation. While the cell types that are labelled are similar between mouse and zebrafish, the order of their appearance differs. For instance, zebrafish show expression first in the myocardium, then in visceral smooth muscle and skeletal muscle before vascular smooth muscle expression. The slow development of vascular mural cells may be a reflection of the small size of the zebrafish and thus there is little need to develop contractile smooth muscle at an early stage. Unlike our zebrafish transgenic where acta2:EGFP is expressed in the myocardium through life, expression of the mouse acta2 transgene was maintained in adult smooth muscle, but not cardiac muscle.

acta2:EGFP expression seems to simply turn on in cells with a pericyte-like morphology as opposed to a gradual increase in intensity. We therefore predict there is an immature mural cell in place that is associated with vessels, which switches on acta2:EGFP expression when mature. We cannot visualize immature mural cells with our current transgenic model, or with any current marker. However we have previously shown the presence of vascular mural cells on the dorsal aorta of the hindbrain by transmission electron microscopy at 48–52 hpf. These cells lack contacts with endothelial cells in models of hemorrhage such as *igu* and *bbh* genetic mutants [8,26]. Mural cells are therefore present and functional as early as 2 dpf, but do not express *acta2*, *tagln* or other early mural cell genes at this early stage. In the mouse, *acta2* and *tagln* are some of the earliest markers of the vascular mural cell lineage [53], but clearly, novel markers of even earlier mural cells are required in the zebrafish.

As head mural cells are thought to derive from neural crest, we wanted to test whether ectomesenchymal neural crest markers co-localized with the acta2 transgenic line. We show here that Tg(fli1a:nEGFP)[y7], which is an ectomesenchymal neural crest marker, does not co-localize with acta2:mCherry. However, there are caveats to our double transgenic experiment because by the time that the acta2:mCherry transgene turns on at 4 dpf, fli1a:nEGFP may either never be expressed in vascular mural cells, or may be already downregulated. However, experiments to rule out that the vascular mural cells in the ventral head originate

(such as the ventral aorta) and much lower coverage in distal vessels (such as the dorsal aorta of the trunk) mirrors this blood pressure difference.

Why is there less vascular mural cell coverage of vessels in fish embryos as compared to mouse embryos? The reported ventricular systolic pressure in a 5 dpf zebrafish is 0.47 mmHg, while that of an adult is 2.5 mmHg [50,51]. In comparison, the mouse blood pressure is around 2 mmHg at E9.5, but rises to 11.5 mmHg at E11.5 and 30 mmHg at P2[52]. Thus the blood pressure of a developing mouse is considerably higher than that of zebrafish. The need for vascular stabilization and control of blood vessel tone is clearly greater in a larger organism with much higher blood pressures [27]. Relatively low blood pressure may also explain why the first mural cells we observe have a morphology more similar to pericytes than smooth muscle as their function might be more important for vascular stabilization than in vascular tone.

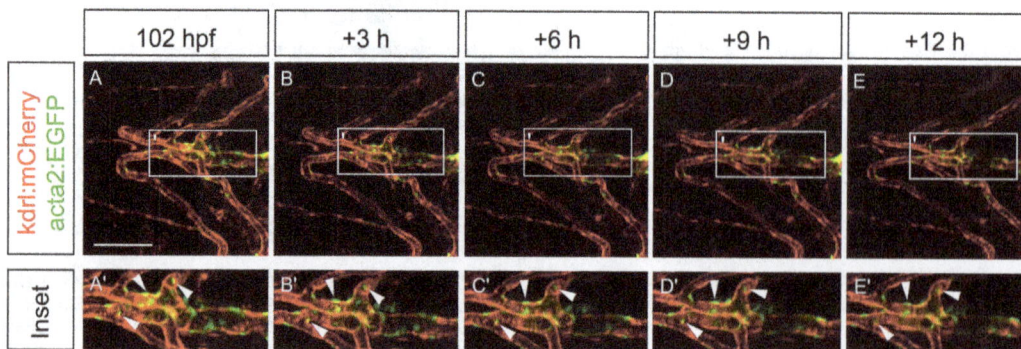

Figure 9. Vascular mural cells of the ventral head are very stable over time. (A–E) Single images taken from a confocal microscopy timelapse video. Images were collected at 102 hpf (A) and every three hours for 12 hours (B–E). Insets (A'–E') show a higher magnification of the ventral aorta, where mural cells that are present at the beginning of the timelapse are still present at the end of the timelapse with no cytokinesis. Arrowheads depict mural cells throughout the timelapse that appear to have little movement. Scale bar represents 100 µm.

from lateral mesoderm are still required. This interesting question awaits a thorough lineage analysis.

We show that early vascular mural cells of the ventral aorta have the appearance of pericytes as they appear as single isolated cells on this vessel. Mural cells with a pericyte-like morphology can be found on large vessels in adult zebrafish. Some of these vessels would normally be covered in smooth muscle in other organisms. For instance, pericytes, not smooth muscle, are observed on adult coronary vessels [54]. Even the bulbus arteriosus, which is argued to be an enlarged artery, has been reported to have smooth muscle at 4 weeks, but not before [27,55] although here we observe multilamellar smooth muscle at 22 dpf. As an aquatic organism with low blood pressure, more mural cell coverage may not be required.

The description of the acta2:EGFP transgenic line will now open the door to many unanswered questions in zebrafish vascular biology, allowing the simultaneous imaging of endothelial and mural cells during larval stages. This should allow us to address questions of origins, gene expression and behaviour of mural cells in this tractable model system.

Supporting Information

Figure S1   The acta2:GFP transgene is expressed surrounding endothelium in the ventral aorta. Single slices of confocal micrograph stacks of double transgenic Tg(kdrl:mCherry; acta2:EGFP) zebrafish embryos at 7 dpf show green acta2:EGFP cells surrounding red kdrl:mCherry expressing endothelial cells in two different regions of the ventral aorta, distal (A) and proximate (B) to the heart outflow tract. The dorsal aorta is depicted in 11 dpf embryos, with green acta2:EGFP cells surrounding red kdrl:mCherry expressing endothelial cells (C) and with individual fluorescent markers (C' - green acta2:EGFP cells; C'' red kdrl:mCherry endothelial cells). Scale bar in B represents 20 μm, scale bar in C represents 50 μm.

Figure S2   Wholemount image of 4 dpf acta2 transgenic zebrafish shows constant smooth muscle and heart expression and variable skeletal muscle expression. Wholemount images of two independent 4 dpf zebrafish embryos using brightfield and fluorescent microscopy. While embryo 1 shows strong visceral smooth muscle expression and heart expression of the transgene, embryo 2 also shows scattered skeletal muscle fiber expression. The expression in skeletal muscle is variable from embryo to embryo and decreases over developmental time.

Figure S3   In situ hybridization shows expression of acta2 in the Bulbus Arteriosus and Ventral Aorta. Cross sections of 22 dpf zebrafish showing strong acta2 expression in the bulbus arteriosus and ventral aorta. This provides context to Figure 3 K and N. Scale bars are 50 μm.

Figure S4   Single or double knockdown of FoxD3 or TFAP2a to block neural crest specification results in a reduction in acta2:GFP cells, but also severe ventral head and blood vessel patterning defects. Representative brightfield images of 2 dpf zebrafish embryos show that both double knockdown (dMO) of FoxD3 and TFAP2A (C) or single knockdown (sMO) of FoxD3 (E) or TFAP2A (G), results in hemorrhage which is not present in control (A). Hydrocephalus of the hindbrain ventricle is also observed in dMO and sMO FoxD3. At 4 dpf, confocal microscopy shows that the control has a well-defined heart outflow tract, with mural cell coverage (kdrl:mCherry – red vessels; acta2:EGFP – green mural cells) (B). In dMO there are severe vessel malformations and a reduction in mural cell coverage (D). In the single FoxD3 (F) and TFAP2A (H) morphants, there are also malformations and reduced mural cell coverage, although these are less severe than the double morphant. Scale bar for A, C, E, G represents 200 μm. Scale bar for B, D, F, H represents 100 μm.

Movie S1   Timelapse imaging of vascular mural cells reveals a stable phenotype over time. Timelapse confocal microscopy of 102 hpf embryos (kdrl:mCherry – red vessels; acta2:EGFP – green mural cells) over a 12 hour timeframe, allowing for visualization of zebrafish embryo development. During this time period, mural cells do not appear to move or proliferate. Movie is representative of n = 5

Acknowledgments

We would like to thank Dr. Jae-Ryeon Ryu, Corey Arnold, Peter Spice, Michela Goi, and Paniz Davari for helpful comments on the manuscript.

Author Contributions

Conceived and designed the experiments: RMK TRW ADC ELR SG SJC. Performed the experiments: RMK TRW ADC ELR SG SJC. Analyzed the data: RMK TRW SG SJC. Contributed reagents/materials/analysis tools: RMK TRW ADC ELR SG SJC MMS. Wrote the paper: TRW SJC.

References

1. Gaengel K, Genove G, Armulik A, Betsholtz C (2009) Endothelial-mural cell signaling in vascular development and angiogenesis. Arterioscler Thromb Vasc Biol 29: 630–638.
2. Giannotta M, Trani M, Dejana E (2013) VE-Cadherin and Endothelial Adherens Junctions: Active Guardians of Vascular Integrity. Dev Cell 26: 441–454.
3. Armulik A, Genove G, Betsholtz C (2011) Pericytes: developmental, physiological, and pathological perspectives, problems, and promises. Dev Cell 21: 193–215.
4. Lindahl P, Johansson BR, Leveen P, Betsholtz C (1997) Pericyte loss and microaneurysm formation in PDGF-B-deficient mice. Science 277: 242–245.
5. Patan S (1998) TIE1 and TIE2 receptor tyrosine kinases inversely regulate embryonic angiogenesis by the mechanism of intussusceptive microvascular growth. Microvasc Res 56: 1–21.
6. Davis S, Aldrich TH, Jones PF, Acheson A, Compton DL, et al. (1996) Isolation of angiopoietin-1, a ligand for the TIE2 receptor, by secretion-trap expression cloning. Cell 87:1161-9: 87:1161–1169.
7. Nielsen CM, Dymecki SM (2010) Sonic hedgehog is required for vascular outgrowth in the hindbrain choroid plexus. Dev Biol 340: 430–437.
8. Lamont RE, Vu W, Carter AD, Serluca FC, MacRae CA, et al. (2010) Hedgehog signaling via angiopoietin1 is required for developmental vascular stability. Mech Dev 127: 159–168.
9. Pola R, Ling LE, Silver M, Corbley MJ, Kearney M, et al. (2001) The morphogen Sonic hedgehog is an indirect angiogenic agent upregulating two families of angiogenic growth factors. Nat Med 7: 706–711.
10. Lavine KJ, Kovacs A, Ornitz DM (2008) Hedgehog signaling is critical for maintenance of the adult coronary vasculature in mice. J Clin Invest 118: 2404–2414.
11. Passman JN, Dong XR, Wu SP, Maguire CT, Hogan KA, et al. (2008) A sonic hedgehog signaling domain in the arterial adventitia supports resident Sca1+ smooth muscle progenitor cells. Proc Natl Acad Sci U S A 105: 9349–9354.
12. Domenga V, Fardoux P, Lacombe P, Monet M, Maciazek J, et al. (2004) Notch3 is required for arterial identity and maturation of vascular smooth muscle cells. Genes Dev 18: 2730–2735.
13. Paik JH, Skoura A, Chae SS, Cowan AE, Han DK, et al. (2004) Sphingosine 1-phosphate receptor regulation of N-cadherin mediates vascular stabilization. Genes Dev 18: 2392–2403.

14. Etchevers HC, Vincent C, Le Douarin NM, Couly GF (2001) The cephalic neural crest provides pericytes and smooth muscle cells to all blood vessels of the face and forebrain. Development 128: 1059–1068.

15. Mundell NA, Labosky PA (2011) Neural crest stem cell multipotency requires Foxd3 to maintain neural potential and repress mesenchymal fates. Development 138: 641–652.

16. Olesnicky Killian EC, Birkholz DA, Artinger KB (2009) A role for chemokine signaling in neural crest cell migration and craniofacial development. Dev Biol 333: 161–172.

17. Dutton KA, Pauliny A, Lopes SS, Elworthy S, Carney TJ, et al. (2001) Zebrafish colourless encodes sox10 and specifies non-ectomesenchymal neural crest fates. Development 128: 4113–4125.

18. Wang WD, Melville DB, Montero-Balaguer M, Hatzopoulos AK, Knapik EW (2011) Tfap2a and Foxd3 regulate early steps in the development of the neural crest progenitor population. Dev Biol 360: 173–185.

19. Gould DB, Phalan FC, Breedveld GJ, van Mil SE, Smith RS, et al. (2005) Mutations in Col4a1 cause perinatal cerebral hemorrhage and porencephaly. Science 308: 1167–1171.

20. Huang J, Davis EC, Chapman SL, Budatha M, Marmorstein LY, et al. (2010) Fibulin-4 deficiency results in ascending aortic aneurysms: a potential link between abnormal smooth muscle cell phenotype and aneurysm progression. Circ Res 106: 583–592.

21. Renard M, Holm T, Veith R, Callewaert BL, Ades LC, et al. (2010) Altered TGFbeta signaling and cardiovascular manifestations in patients with autosomal recessive cutis laxa type I caused by fibulin-4 deficiency. Eur J Hum Genet 18: 895–901.

22. von Tell D, Armulik A, Betsholtz C (2006) Pericytes and vascular stability. Exp Cell Res 312: 623–629.

23. Peppiatt CM, Howarth C, Mobbs P, Attwell D (2006) Bidirectional control of CNS capillary diameter by pericytes. Nature 443: 700–704.

24. Benjamin L, Hemo I, Keshet E (1998) A plasticity window for blood vessel remodelling is defined by pericyte coverage of the preformed endothelial network and is regulated by PDGF-B and VEGF. Development 125: 1591–1598.

25. Hellstrom M, Gerhardt H, Kalen M, Li X, Eriksson U, et al. (2001) Lack of pericytes leads to endothelial hyperplasia and abnormal vascular morphogenesis. J Cell Biol 153: 543–553.

26. Liu J, Fraser SD, Faloon PW, Rollins EL, Vom Berg J, et al. (2007) A bPix-Pak2a signaling pathway regulates cerebral vascular stability in zebrafish. Proc Natl Acad Sci U S A 104: 13990–13995.

27. Miano JM, Georger MA, Rich A, De Mesy Bentley KL (2006) Ultrastructure of zebrafish dorsal aortic cells. Zebrafish 3: 455–463.

28. Seiler C, Abrams J, Pack M (2010) Characterization of zebrafish intestinal smooth muscle development using a novel sm22alpha-b promoter. Dev Dyn 239: 2806–2812.

29. Georgijevic S, Subramanian Y, Rollins EL, Starovic-Subota O, Tang AC, et al. (2007) Spatiotemporal expression of smooth muscle markers in developing zebrafish gut. Dev Dyn 236: 1623–1632.

30. Santoro MM, Pesce G, Stainier DY (2009) Characterization of vascular mural cells during zebrafish development. Mech Dev 126: 638–649.

31. Wallace KN, Akhter S, Smith EM, Lorent K, Pack M (2005) Intestinal growth and differentiation in zebrafish. Mech Dev 122: 157–173.

32. Westerfield M (1995) The Zebrafish Book: A Guide for the Laboratory Use of Zebrafish (Danio Rerio). Eugene, OR: University of Oregon Press.

33. Proulx K, Lu A, Sumanas S (2010) Cranial vasculature in zebrafish forms by angioblast cluster-derived angiogenesis. Dev Biol 348: 34–46.

34. Lawson N, Weinstein B (2002) In vivo imaging of embryonic vascular development using transgenic zebrafish. Dev Biol 248: 307.

35. Roman BL, Pham VN, Lawson ND, Kulik M, Childs S, et al. (2002) Disruption of acvrl1 increases endothelial cell number in zebrafish cranial vessels. Development 129: 3009–3019.

36. Lauter G, Soll I, Hauptmann G (2011) Multicolor fluorescent in situ hybridization to define abutting and overlapping gene expression in the embryonic zebrafish brain. Neural Dev 6: 10.

37. Mack CP, Owens GK (1999) Regulation of smooth muscle alpha-actin expression in vivo is dependent on CArG elements within the 5′ and first intron promoter regions. Circ Res 84: 852–861.

38. Matys V, Kel-Margoulis OV, Fricke E, Liebich I, Land S, et al. (2006) TRANSFAC and its module TRANSCompel: transcriptional gene regulation in eukaryotes. Nucleic Acids Res 34: D108–110.

39. Zeng L, Carter AD, Childs SJ (2009) miR-145 directs intestinal maturation in zebrafish. Proc Natl Acad Sci U S A 106: 17793–17798.

40. Kwan KM, Fujimoto E, Grabher C, Mangum BD, Hardy ME, et al. (2007) The Tol2kit: a multisite gateway-based construction kit for Tol2 transposon transgenesis constructs. Dev Dyn 236: 3088–3099.

41. Kawakami K, Shima A, Kawakami N (2000) Identification of a functional transposase of the Tol2 element, an Ac-like element from the Japanese medaka fish, and its transposition in the zebrafish germ lineage. Proc Natl Acad Sci U S A 97: 11403–11408.

42. Liu J, Zeng L, Kennedy RM, Gruenig NM, Childs SJ (2012) betaPix plays a dual role in cerebral vascular stability and angiogenesis, and interacts with integrin alpha(v)beta(8). Developmental Biology 363: 95–105.

43. Armstrong JJ, Larina IV, Dickinson ME, Zimmer WE, Hirschi KK (2010) Characterization of bacterial artificial chromosome transgenic mice expressing mCherry fluorescent protein substituted for the murine smooth muscle alpha-actin gene. Genesis 48: 457–463.

44. Parichy DM, Elizondo MR, Mills MG, Gordon TN, Engeszer RE (2009) Normal table of postembryonic zebrafish development: staging by externally visible anatomy of the living fish. Dev Dyn 238: 2975–3015.

45. Korn J, Christ B, Kurz H (2002) Neuroectodermal origin of brain pericytes and vascular smooth muscle cells. J Comp Neurol 442: 78–88.

46. Wasteson P, Johansson BR, Jukkola T, Breuer S, Akyurek LM, et al. (2008) Developmental origin of smooth muscle cells in the descending aorta in mice. Development 135: 1823–1832.

47. Gittenberger-de Groot AC, DeRuiter MC, Bergwerff M, Poelmann RE (1999) Smooth muscle cell origin and its relation to heterogeneity in development and disease. Arterioscler Thromb Vasc Biol 19: 1589–1594.

48. Crump JG, Maves L, Lawson ND, Weinstein BM, Kimmel CB (2004) An essential role for Fgfs in endodermal pouch formation influences later craniofacial skeletal patterning. Development 131: 5703–5716.

49. Stratman AN, Malotte KM, Mahan RD, Davis MJ, Davis GE (2009) Pericyte recruitment during vasculogenic tube assembly stimulates endothelial basement membrane matrix formation. Blood 114: 5091–5101.

50. Hu N, Yost H, EB C (2001) Cardiac Morphology and Blood Pressure in the Adult Zebrafish. Anat Rec 264: 1–12.

51. Hu N, Sedmera D, Yost HJ, Clark EB (2000) Structure and function of the developing zebrafish heart. Anat Rec 260: 148–157.

52. Le VP, Kovacs A, Wagenseil JE (2012) Measuring left ventricular pressure in late embryonic and neonatal mice. J Vis Exp.

53. Li L, Miano JM, Cserjesi P, Olson EN (1996) SM22 alpha, a marker of adult smooth muscle, is expressed in multiple myogenic lineages during embryogenesis. Circ Res 78: 188–195.

54. Kim J, Wu Q, Zhang Y, Wiens KM, Huang Y, et al. (2010) PDGF signaling is required for epicardial function and blood vessel formation in regenerating zebrafish hearts. Proc Natl Acad Sci U S A 107: 17206–17210.

55. Grimes AC, Stadt HA, Shepherd IT, Kirby ML (2006) Solving an enigma: arterial pole development in the zebrafish heart. Dev Biol 290: 265–276.

13

Targeted Overexpression of Amelotin Disrupts the Microstructure of Dental Enamel

Rodrigo S. Lacruz[1], Yohei Nakayama[2], James Holcroft[2], Van Nguyen[2], Eszter Somogyi-Ganss[2], Malcolm L. Snead[1], Shane N. White[3], Michael L. Paine[1], Bernhard Ganss[2]*

1 School of Dentistry, Center for Craniofacial Molecular Biology, University of Southern California, Los Angeles, California, United States of America, **2** Matrix Dynamics Group, Faculty of Dentistry, University of Toronto, Toronto, Ontario, Canada, **3** School of Dentistry, University of California Los Angeles, Los Angeles, California, United States of America

Abstract

We have previously identified amelotin (AMTN) as a novel protein expressed predominantly during the late stages of dental enamel formation, but its role during amelogenesis remains to be determined. In this study we generated transgenic mice that produce AMTN under the amelogenin (*Amel*) gene promoter to study the effect of AMTN overexpression on enamel formation in vivo. The specific overexpression of AMTN in secretory stage ameloblasts was confirmed by Western blot and immunohistochemistry. The gross histological appearance of ameloblasts or supporting cellular structures as well as the expression of the enamel proteins amelogenin (AMEL) and ameloblastin (AMBN) was not altered by AMTN overexpression, suggesting that protein production, processing and secretion occurred normally in transgenic mice. The expression of Odontogenic, Ameloblast-Associated (ODAM) was slightly increased in secretory stage ameloblasts of transgenic animals. The enamel in AMTN-overexpressing mice was much thinner and displayed a highly irregular surface structure compared to wild type littermates. Teeth of transgenic animals underwent rapid attrition due to the brittleness of the enamel layer. The microstructure of enamel, normally a highly ordered arrangement of hydroxyapatite crystals, was completely disorganized. Tomes' process, the hallmark of secretory stage ameloblasts, did not form in transgenic mice. Collectively our data demonstrate that the overexpression of amelotin has a profound effect on enamel structure by disrupting the formation of Tomes' process and the orderly growth of enamel prisms.

Editor: Vincent Laudet, Ecole Normale Supérieure de Lyon, France

Funding: This work was supported by Grants DE006988 and DE013045 (MLS), DE013404 and DE019629 (MLP) and DE014189 (SNW) from the NIDCR, National Institutes of Health and by an Operating Grant MOP-79449 (BG) from the Canadian Institutes of Health Research (CIHR). The funders had no role in study design, data collection and analysis, decision to publish, or preparation of the manuscript.

Competing Interests: The authors have declared that no competing interests exist.

* E-mail: b.ganss@utoronto.ca

Introduction

Dental enamel is the hardest tissue in vertebrates. If properly formed and cared for it is, unlike synthetic restorative materials, designed to last a lifetime under immense mechanical stress, in spite of constant challenges within the oral cavity through changes in temperature, pH, and exposure to aggressive cariogenic microorganisms. The formation of dental enamel is a prototype of functional organ development through a biomineralization process. The three main structural proteins of the forming enamel, the most abundant amelogenin (AMEL), as well as ameloblastin (AMBN) and enamelin (ENAM), are collectively referred to here as enamel matrix proteins (EMPs). The EMPs are produced at their highest levels by ameloblasts during the secretory and transition stages of amelogenesis and collectively orchestrate the proper assembly and growth of crystals within the mineralized enamel. The proteins are almost completely degraded by specific proteases such as MMP-20 mainly during the secretory/transition stage and KLK4 mainly during the transition/maturation stage, respectively, resulting in a highly ordered and purposefully designed meshwork of carbonated hydroxyapatite crystals with astonishing mechanical properties [1].

Amelotin (AMTN) is a recently discovered enamel protein [2] with very limited sequence similarity to the EMPs, although there is evidence that AMTN and other ameloblast-expressed genes, as well as secreted calcium-binding phosphoproteins (SCPPs), have evolved from a common ancestral gene [3]. *Amtn* mRNA expression is transient in ameloblasts of rodent molars from postnatal day 2 to the time of tooth eruption approximately 14 days later, but persists in the continuously erupting incisors. A detailed expression profiling of AMTN protein [4] has shown that its expression is dramatically upregulated during the transition stage in ameloblasts and continues throughout the maturation stage in a fashion similar to another recently described ameloblast gene, *Odam/Apin* [5,6]. The expression profile of *Amtn* (and *Odam*) is thus clearly distinct from that of the three classical EMPs (*Amel*, *Ambn*, *Enam*), but parallels that of *Klk4*, which codes for a protein associated with EMP degradation and mineral maturation. The AMTN protein is secreted and has been localized to a basal lamina-like layer between ameloblasts and the enamel mineral surface in incisors of rats [7] and mice [4]. Thus it has been suggested to be involved in cell attachment, control of ion and peptide transport to and from maturing enamel, or the formation of the distinct aprismatic, specialized, superficial layer referred to

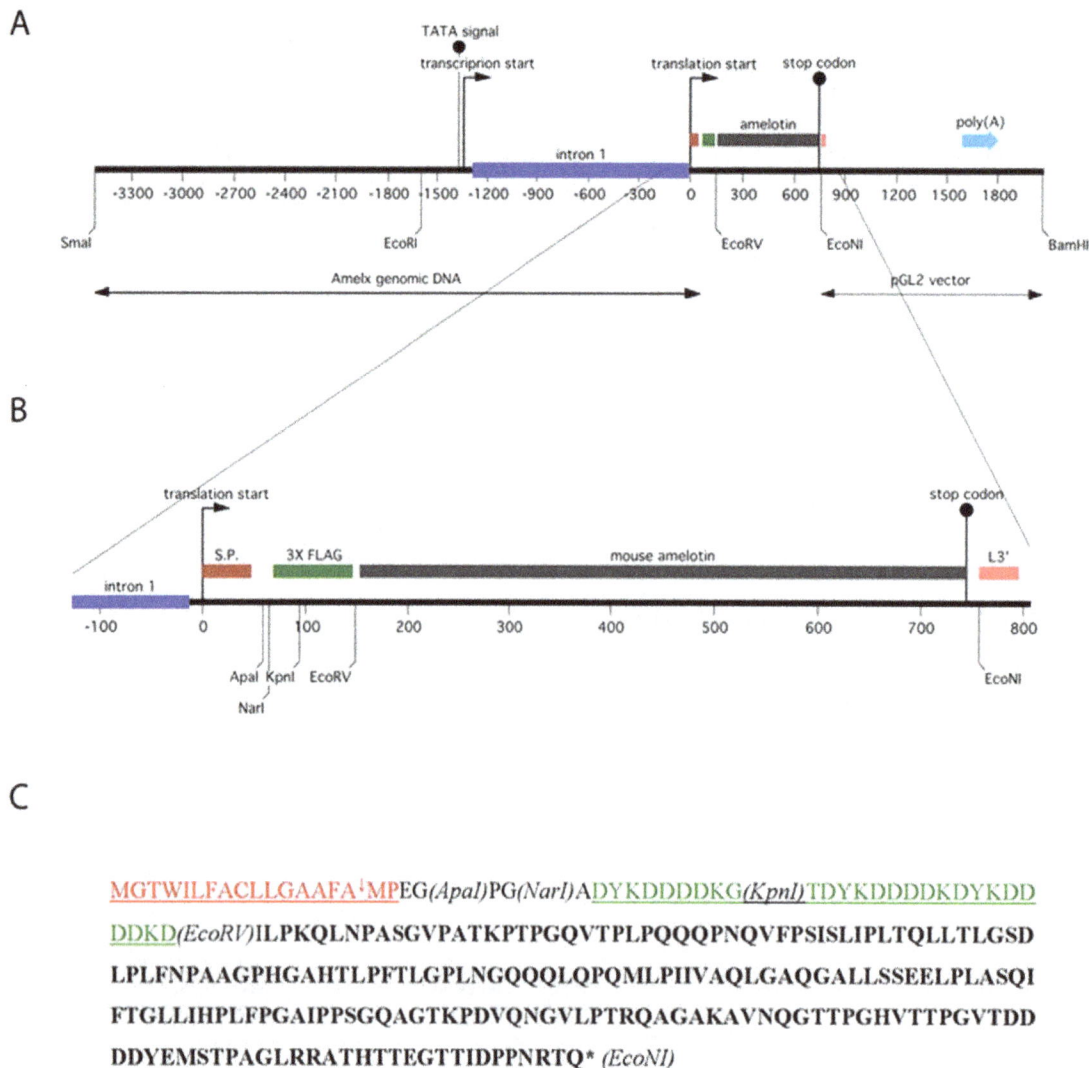

MGTWILFACLLGAAFA↓MPEG(Apal)PG(Narl)ADYKDDDDKG(KpnI)TDYKDDDDKDYKDD

DDKD(EcoRV)ILPKQLNPASGVPATKPTPGQVTPLPQQQPNQVFPSISLIPLTQLLTLGSD

LPLFNPAAGPHGAHTLPFTLGPLNGQQQLQPQMLPIIVAQLGAQGALLSSEELPLASQI

FTGLLIHPLFPGAIPPSGQAGTKPDVQNGVLPTRQAGAKAVNQGTTPGHVTTPGVTDD

DDYEMSTPAGLRRATHTTEGTTIDPPNRTQ* (EcoNI)

Figure 1. Construction and schematic representation of the amelotin transgene. (A) The 3.5 kb mouse amelogenin promoter region extending through intron 1 (blue) and the signal peptide region (red) was used to drive the triple FLAG epitope (green) containing mouse amelotin transgene construct (grey). (B) An enlarged region of the transgene's open reading frame. (C) Transgene protein sequence identifying the amelogenin signal peptide (red) and the cleavage site (arrow), the three repeats of the FLAG epitope at the amino terminus, and amelotin. The positions of relevant restriction enzyme sites are also included. Numbers refer to nucleotides with +1 being the first adenosine in the translated gene product; L3' refers to the 3' region of luciferase cDNA; poly(A) refers to the SV40 late poly(A) signal; S.P. refers to the amelogenin signal peptide; and *Amelx* genomic DNA refers to the mouse amelogenin gene promoter region and intron 1. The transgene was released from the vector backbone with restriction enzymes SmaI and BamHI prior to the generation of animals. The theoretical molecular weight of the unmodified transgene product is 24.4 kDa without, 26.1 kDa with signal peptide.

as "final" enamel [8,9]. The lack of cell culture models for maturation stage ameloblasts has hampered functional studies *in vitro*, but functional studies of specific proteins, including those controlling the biomineralization process of dental enamel, have been greatly facilitated by the development of genetically engineered mice as *in vivo* gain- and loss-of-function models. Due to the confined expression of enamel genes, unlike many other genes, gene "knock-outs" have been particularly useful in revealing the specific roles of individual enamel proteins without thwarting the interpretation of experimental phenotypes by early lethality or compensation mechanisms. Conversely, gain-of-function by the tissue-specific overexpression of enamel proteins, driven by a well-characterized 2.3 kb mouse amelogenin promoter, has significantly advanced our understanding of the contribution of individual proteins and protein fragments to amelogenesis

[10,11,12,13,14]. In this paper we have used the latter technology to create several lines of experimental mice that overexpress the amelotin protein in ameloblasts. The rationale for choosing this strategy was that the transgenic mice would not only produce higher levels of AMTN, but also at an earlier stage of amelogenesis, and would thus allow us to determine the effect of AMTN on enamel prism growth. We describe the resulting effects on enamel structure and mechanical properties, protein expression patterns and cellular morphology of ameloblasts.

Results

Production of transgenic animals

The vector for the site-specific overexpression of TgAMTN in transgenic mice is shown schematically in Figure 1. The hallmarks of the construct include the ~2.3 kb murine amelogenin promoter, followed by intron 1 to allow optimal processing of the mRNA transcript, and the coding sequences for the amelogenin signal peptide to achieve efficient protein secretion into the extracellular space. Three repeats of the FLAG epitope (DYKDDDDK) were engineered into the N-terminal region of the transgene which were followed in frame by the murine amelotin coding sequence. The *Amel* promoter region was recovered from a lambda phage cDNA library, while the FLAG epitope and amelotin cDNA regions were amplified by PCR. All DNA derived from PCR were sequenced in their entirety to ensure sequence integrity of the final plasmid construct. The production of transgenic animals resulted in two independent lines (57 and 457). The presence and relative abundance of the transgene were validated and assessed by Southern blot analysis (Figure 2), confirming lines 57 and 457 as harboring similar copy numbers of the transgene (Figure 2A). Further qualitative genotyping by PCR verified the transgene status, producing a single 429 bp product only in transgenic animals, regardless of which transgenic line was analyzed (Figure 2B). Corresponding protein levels were assessed by Western blot, confirming the presence of the FLAG-tagged transgenic protein in molar tooth extracts of lines 57 and 457, but not in wild type animals at 3–4 days of age (Figure 3A). Western blot analysis with the anti-AMTN antibody FL-rmAMTN, after enrichment of the expressed AMTN protein in cell lysates by immunoprecipitation, also confirmed the presence of TgAMTN in transgenic animals of the line 57 (Fig. 3B, Amtn) and 457 (not shown), but not in wild type animals. Competition by an approximately 200-fold molar excess of recombinant murine AMTN abolished the specific signal around 24 kDa in molar tooth extracts from transgenic mice and the recombinant murine AMTN protein as positive control, but did not affect the signal of the rabbit immunoglobulin heavy chain at about 50 kDa (Fig. 3B; Amtn+Comp). Western blots with the mAMTN-1 antibody did not produce any signal (not shown).

Immunohistochemistry

To verify the various levels and to determine the specific sites of TgAMTN overexpression in transgenic animals immunohisto-chemical analyses were conducted, which confirmed overexpression of TgAMTN in ameloblasts at the early secretory stage. Two time points, postnatal days 4 and 30 (P4 and P30), were chosen for detailed analyses in molars to reflect pre-eruptive and mature stages of tooth formation (Figure 4). The overexpression of TgAMTN in molars of transgenic line 57 was confirmed at P4, but the expression pattern and levels of other enamel proteins (AMEL, AMBN, ODAM) were not significantly altered by the early overexpression of TgAMTN. Both AMEL and AMBN showed a fairly uniform signal in ameloblasts, enamel matrix and, to a lesser extent, pulp, while ODAM was practically undetectable at this stage in all animals. At P30, where the Amel promoter driving the transgene expression is essentially inactive as seen by the lack of immunostaining for AMEL in wild type animals (Fig. 4, day 30, Amel, WT), AMBN, ODAM and AMTN produced signals in the interradicular bone crest region, possibly staining epithelial rests of Malassez. ODAM was predominantly found in the junctional epithelium, consistent with previous reports [6]. In transgenic animals (line 57) the signals for AMBN, ODAM and AMTN in the interradicular bone crest were abolished, but ODAM expression persisted in the junctional epithelium. Identical results were obtained when analyzing line 457 animals (results not shown). In the continuously erupting incisor at P30 (Fig. 5), where the entire developmental spectrum of enamel formation is present, AMTN was detected in ameloblasts from the late secretory stage to the

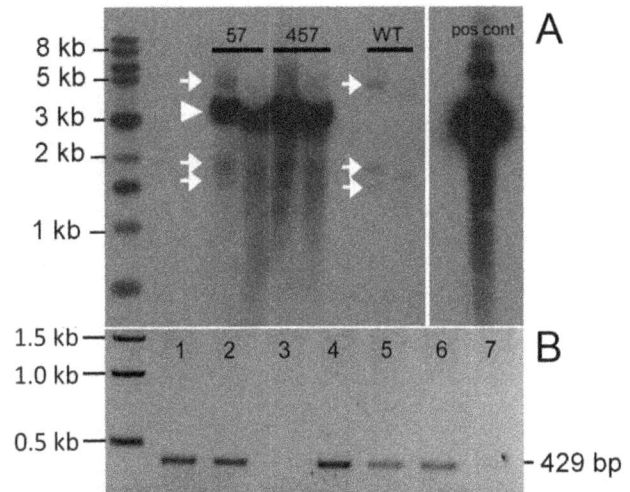

Figure 2. Southern Blot analysis of transgenic animals. Genomic DNA isolated from various transgenic founder lines (57, 457) and wild type (WT) littermates were digested with PstI, separated by agarose gel electrophoresis and (A) autoradiographed after hybridization to a radioactively labeled amelotin cDNA probe isolated from the EcoRV-PstI 528 bp region of the transgene cassette (see Figure 1). Three specific bands correlating to genomic DNA, each of the predicted size (4.5, 1.7 and 1.4 kb), are seen in WT samples and all transgenic animals (arrows) and serve as loading control between lanes; the arrowhead indicates the position of the transgene-specific signal at 3.115 kb correlating to a PstI-PstI transgene fragment, which is all-inclusive of the probe. (B) PCR genotyping using a forward primer contained within intron 1 and a reverse primer completely contained within the amelotin cDNA region verifies the generation of a 429 bp amplicon in transgenic mice (lanes 1, 2, 4, 5, 6), but not in non-transgenic littermates (lanes 3, 7). Samples for lanes 1 to 7 were randomly selected from litters from both transgenic lines (57 and 457).

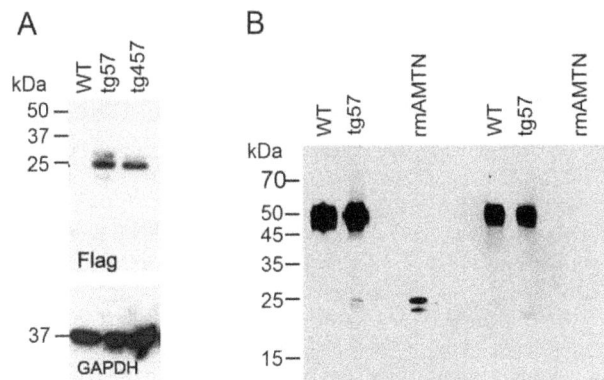

Figure 3. Western Blot analysis of wild type (WT) and transgenic (tg57 and tg457) 3–4-day old mice. Total protein extracts from mandibular first and second molars were probed with anti-Flag (A, upper panel) and anti-GAPDH (A, lower panel) antibodies, indicating overexpression of the transgenic protein at comparable levels in both transgenic lines. Probing extracts with an anti-amelotin antibody after enrichment via IgG immunoprecipitation (B) yielded, in addition to the IgG heavy chain band at ~50 kDa, two immunoreactive bands at ~24 kDa and ~22 kDa in transgenic animals only, which was similar in size to that of bacterially expressed recombinant mouse AMTN (rmAMTN) as a positive control (B, left panel). Competition with excess rmAMTN protein abolished the 24 and 22 kDa signals in tg57 and rmAMTN, confirming specificity of the signal (B, right panel).

maturation stage in wild type animals, but TgAMTN was found at higher levels and at earlier stages in the transgenic line 57 (Fig. 5A). An immunoreactive signal for the FLAG peptide was only detected in transgenic animals. The expression of AMEL, detected in wild type animals from the early secretory to the mid-maturation stage, was not altered in transgenic animals. AMBN labeling in wild type animals paralleled that of AMEL in early stage ameloblasts, but persisted farther into the maturation zone. AMBN expression was also unchanged in transgenic animals. ODAM signals in wild type animals were undetectable in apical ameloblasts, but intense in cells of the late secretory to maturation stage. In transgenic animals, the expression in late secretory to maturation stage ameloblasts was unaltered, but an ODAM signal that was slightly increased compared to wild type animals could be detected in secretory stage ameloblasts (Fig. 5A, arrow). Magnified views of the apical end of P30 incisors (Fig. 5B) illustrate the increased expression of TgAMTN in apical ameloblasts of transgenic mice at much earlier stages than in wild type animals. Notably, the pattern of TgAMTN protein expression in adjacent apical ameloblasts of line 57 animals often appeared patchy, with alternating high and low TgAMTN signal intensities in adjacent cells (Figure 5B, arrow).

Histological analysis

The histological analysis of hematoxylin and eosin (H&E) stained sagittal incisor sections of mice at P30 revealed significant differences in the structural arrangement of the organic extracellular enamel matrix in the apical portion of the tooth, corresponding to the early maturation stage at the anatomical level of the second molar (Figure 6 a, b). While the eosin-stained enamel matrix in wild type animals showed a striated pattern indicating the presence of organized, interwoven structures typical of prismatic enamel (Fig. 6a, insert), the matrix in line 57 animals was much thinner and completely unstructured (Fig. 6b, insert). The enamel matrix at this anatomical location also appeared to contain less organic material in transgenic animals. Dental pulp, dentin, stratum intermedium, stellate reticulum and alveolar bone structures were not significantly altered in transgenic mice. Ameloblast cell bodies and nuclei aligned similarly in both wild type and transgenic animals, but the normally straight demarcation between the apical membrane of ameloblasts and the enamel matrix was highly irregular in tg57 animals (Fig. 6b). Closer to the tip of the incisor, at an anatomical location between the mesial side of the first molar and the alveolar bone margin (Fig. 6c, d), only little eosinophilic matrix was left behind after decalcification in wild type animals, while in line 57 animals the enamel space contained much more such residual organic material. The thickness of the enamel space was clearly reduced in transgenic animals, but the morphology of ameloblasts appeared unaltered and the underlying papillary and reticular layers that form from stratum intermedium and stellate reticulum were indistinguishable between wild type and transgenic mice. The apical surface of ameloblasts, however, showed discontinuities in tg57 animals with occasionally disconnected adjacent cells (Fig. 6c and d, inserts).

Scanning Electron Microscopy

SEM analyses of incisors from 9 week-old mice, which were fractured at the level of the mesial surface of the first molar, revealed the typical enamel rod/interrod structure in wild type animals (Figure 7A) with normal decussation patters, producing a fairly even plane of fracture perpendicular to the long incisal axis. In contrast, the enamel in line 57 animals was much thinner (~30% of wild type) and showed a highly irregular microstructure with no visible decussation patters. The outer enamel surface in

wild type animals was generally smooth and hard, but the much thinner enamel in animals from lines 57 displayed an irregular and pitted outer surface, and was easily damaged during dissection (Fig. 7E). SEM imaging of teeth from 4 month-old wild type (Fig. 7B–D) and transgenic mice (Fig. 7F–H) revealed severe attrition in first and second molars with all cusps affected and in incisors, where attrition had progressed to the point of pulp exposure (Fig. 7H). More detailed analyses of the enamel microstructure in ground and etched transverse incisor sections at a distance of 6 and 2 mm from the incisal edge (Fig. 8) showed that the organized interdigitated arrangement of enamel prisms, which was observed in wild type animals at the 6 mm level, was severely disrupted in tg57 animals. Although the structure of the initial aprismatic enamel layer immediately adjacent to the DEJ was similar between WT and tg57 samples (Fig. 8m, n), the prismatic structure in bulk enamel (Fig. 8k) never forms in tg57 animals (Fig. 8l). Consistent with results from Fig. 6c and 6d, at an anatomical position similar to the 6 mm level shown in Fig. 8, the presence of residual organic matrix reduced the exposition of enamel prisms by phosphoric acid etching. At the erupted portion of the incisor (2 mm level), where enamel is fully matured, the regular pattern of interlocking enamel prisms was revealed in WT animals, but the arrangement of prisms in tg57 was completely disorganized (Fig. 8y, z). The initial enamel layer at the DEJ was again similar between WT and tg57 animals, but the transition to highly organized prismatic enamel, which is observed within ~8 μm from the DEJ in WT animals, never occured in tg57 enamel (Fig. 8aa, ab). The appearance of the enamel surface layer, which is known to be more acid resistant than outer or bulk enamel [15], was also significantly altered in tg57 enamel (Fig. 8w, x)

Transmission Electron Microscopy

Transmission electron microscopy studies were conducted to investigate the effect of TgAMTN overexpression on the cellular morphology of ameloblasts at various stages of enamel formation (Fig. 9). In wild type enamel, prominent Tomes' processes as hallmarks of secretory ameloblasts were found at the 10 mm level (Fig. 9a). During the early, mid and late maturation stages at 8, 6 and 4 mm levels, respectively, the apical surface of ameloblasts formed a linear interface with the enamel space (Fig. 9 b–d). In contrast, Tomes' process never formed in secretory stage ameloblasts of tg57 mice (Fig. 9e), and the compartmentalization of the enamel matrix into rod and interrod enamel [16] never occurred. During subsequent post-secretory stages of amelogenesis, the interface between the apical ameloblast surface and the enamel space maintained its irregular shape (Fig. 9f–h). No obvious differences were observed in the cellular structures of ameloblasts (secretory vesicles, pigment granules, Golgi, mitochondria, desmosomes, terminal webs) between wild type and transgenic animals.

Mechanical hardness testing

Knoop's hardness of the enamel in wild type animals was found to be 2.2 ± 0.4 GPa, consistent with previous data [17], but in tg57 animals the enamel was found to be too brittle to obtain any defined and reproducible indentation profiles. While it was thus impossible to derive values for enamel hardness, this result clearly indicates the severely compromised mechanical properties of enamel in the line 57 animals.

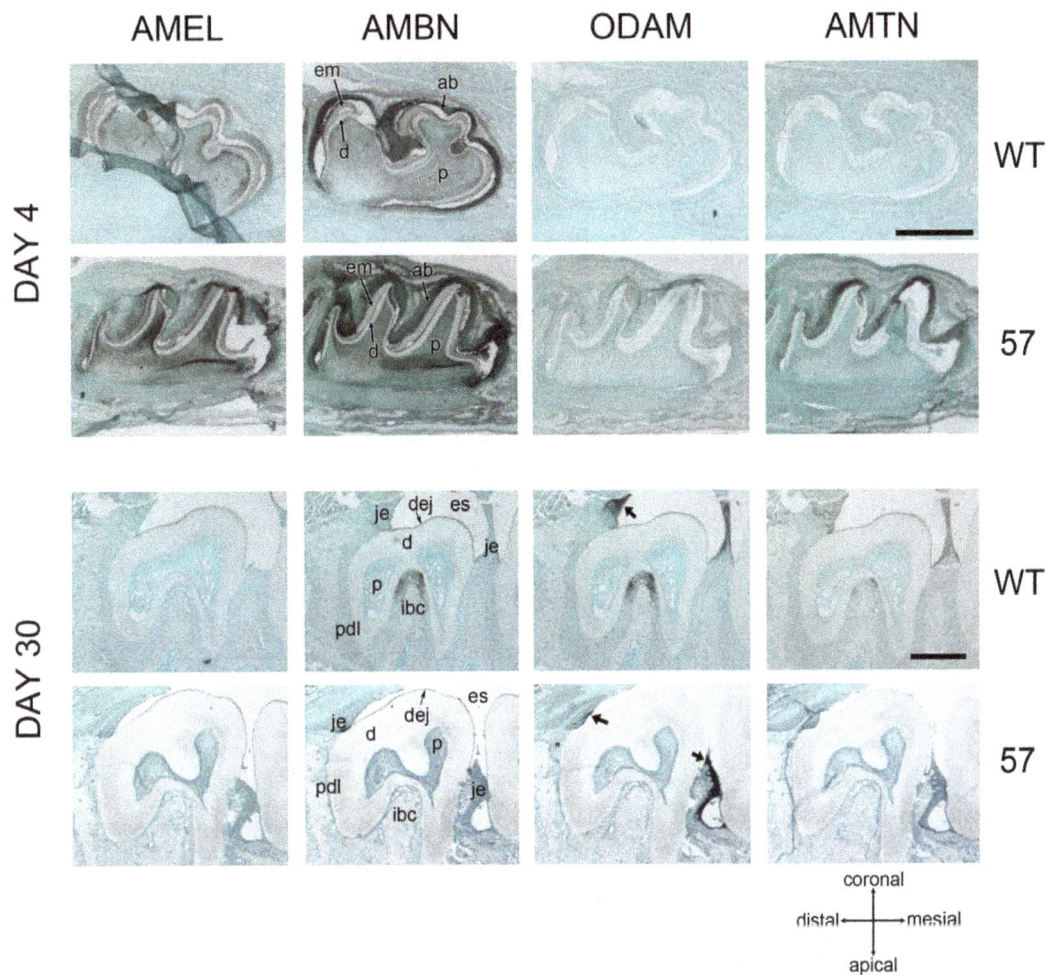

Figure 4. Immunohistochemical (IHC) analysis of molar teeth at pre- and post-eruptive stages. Sagittal sections of mandibular first molars at postnatal day 4 (upper panel) and second molars at day 30 (lower panel) were probed with antibodies against amelogenin (AMEL), ameloblastin (AMBN), ODAM and amelotin (AMTN) in wild type (WT) and transgenic (57) animals. Only AMTN is significantly overexpressed in transgenic animals at day 4. After tooth eruption at day 30 when the amelogenin promoter shows minimal activitythe expression levels of all proteins have returned to very low levels, except ODAM, which continues to show a signal in the junctional epithelia (arrows in lower panel). The orientation of all sections is indicated in the lower right portion. Tissue structures are indicated with the following abbreviations: ab-ameloblasts; d-dentin; dej-dentinoenamel junction; em-enamel matrix; es-enamel space; ibc-interradicular bone crest; je-junctional epithelium; p-pulp; pdl-periodontal ligament. Scale bars = 500 μm.

Discussion

The purpose of this study was to generate transgenic mice that overexpress AMTN under the amelogenin promoter to investigate the effects on EMP expression, and ameloblast and hierarchical enamel structure. The *Amel* promoter has been shown to drive expression of several target genes in secretory stage and transition zone of ameloblasts [10,11,12,13]. Considering that AMTN is predominantly expressed in maturation stage ameloblasts, such animals would not only overexpress, but also misexpress TgAMTN at an earlier stage than normal. Such overexpression allows for the analysis of AMTN effects during enamel growth. The rationale for incorporating the individual components of the transgenic construct has been detailed elsewhere [12]. In this report we have used three repeats of the FLAG epitope in frame with and located at the N-terminus of the AMTN sequence to detect transgene expression by Western blot and immunohisto-chemistry. We have created two different transgenic lines (57 and 457), which show comparable genomic copy numbers based on

Southern blot analysis, and protein levels based on Western blot analysis. For reasons of space constraint we provide data for one representative line (line 57) only; animals from line 457 produced identical results to those presented here for line 57. Previous reports using similar constructs for the overexpression of other enamel genes have produced, compared and analyzed several transgenic lines [10,11,12,13]. Most of them have not found any significant dose-dependent effects of transgene overexpression with the exception of transgenic mice overexpressing tuftelin, where an increased disruption to the rod/interrod architecture was observed in parallel with higher tuftelin expression levels [10] and mice overexpressing dentin phosphoproteins (Dpp), where increasing levels of transgene expression led to increasingly severe enamel abnormalities [11]. In our analysis, two lines with similar transgene copy numbers were obtained, although the absolute number of transgene copies has not been determined. The incorporation of the transgene at similar levels also resulted in correspondingly similar protein expression levels as demonstrated by Western blot. The AMTN signal in Western blots appeared as

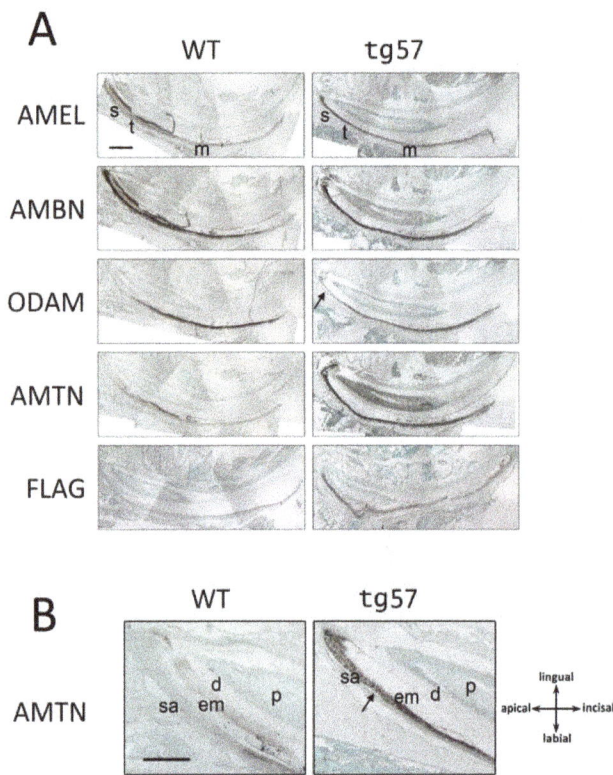

Figure 5. IHC analysis of mandibular incisors. (A) Sagittal sections at P30, probed with antibodies against amelogenin (AMEL), ameloblastin (AMBN), ODAM, amelotin (AMTN) and the FLAG epitope in wild type (WT) and transgenic (tg57) animals. Overexpression of AMTN in tg57 animals extends to the apical portion of the incisor. A slightly increased ODAM signal (arrow) in secretory stage ameloblasts is observed, while the signal patterns and intensities for AMEL and AMBN remained unchanged. (B) Magnified views of the apical portion of incisors, showing little expression of Amtn in wild type apical enamel tissues, but significantly increased amounts in transgenic lines. The orientation of all sections is indicated. Abbreviations: d, dentin; em, enamel matrix; m, maturation stage; p, pulp; sa, secretory stage ameloblasts; s, secretory stage; t, transition stage. Scale bars: 500 µm (A); 250 µm (B).

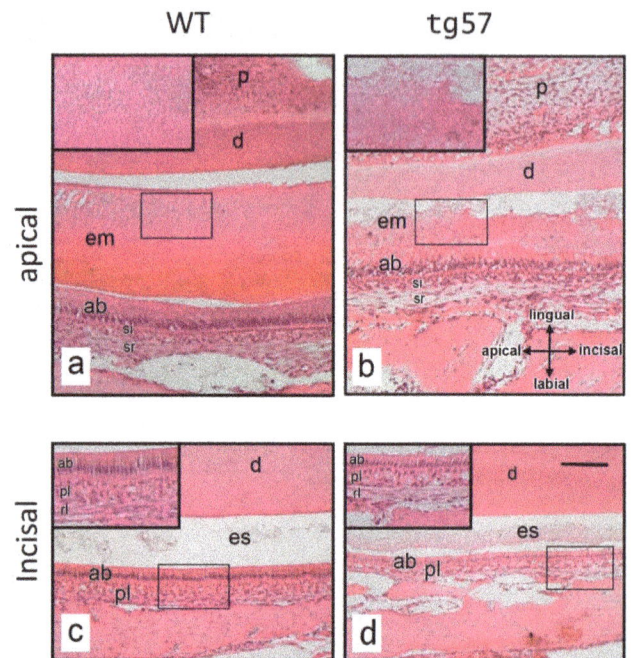

Figure 6. Histological analysis. Sagittal sections of mandibular incisors of wild type (WT) and transgenic (tg57) mice were stained with hematoxylin and eosin. Comparable anatomical regions from apical (late secretory stage; a, b) and incisal (maturation stage; c, d) portions along the apical/incisal axis are shown and magnified in inserts. At the late secretory stage, the eosinophilic enamel matrix appears completely disorganized and thinner in tg57 animals, compared to WT, where an organized, striated pattern is visible. At the maturation stage (c, d), tg57 animals display a thinner enamel space containing more residual organic matrix compared to WT animals. The orientation that applies to all sections is indicated in (b). Abbreviations: ab-ameloblast layer; d-dentin; em-enamel matrix; es-enamel space; p-pulp; pl-papillary layer; rl-reticular layer; si-stratum intermedium; sr-stellate reticulum. Scale bar in (d): 200 µm.

double bands when using both anti-Flag and anti-Amtn antibodies (Fig. 3). The lower molecular weight band is likely a product of proteolytic processing, as it appeared in the recombinant murine AMTN protein only after storage at 4°C for several weeks. The expected size of the transgenically produced protein is 22.4 kDa, and the observed signals with both Flag and Amtn antibodies correspond well with this size, indicating that the protein is not post-translationally modified to any significant degree. The appearance of only two main immunoreactive bands for the transgenically produced protein at the theoretically expected molecular weight also indicates that AMTN is not subjected to extensive proteolytic degradation by enamel proteases such as MMP20 and KLK4 [18], although the structural differences between both immunoreactive bands remain to be determined. Immunohistochemistry experiments clearly confirmed the overexpression of AMTN, although AMTN signal intensities were somewhat variable between different histological preparations. This is likely due to subtle differences in tissue manipulation and level of sectioning, inherent to immunohistochemical experimentation.

Significant differences in qualitative expression patterns at various stages of amelogenesis in molars and incisors (secretory, transition and maturation stages) between wild type and transgenic animals were obvious for AMTN only (Fig. 4 and 5). Prior to this study, the different expression profile of AMEL in secretory stage ameloblasts and AMTN mainly in maturation stage ameloblasts suggested the possibility that AMTN might act as a signal to terminate AMEL expression during amelogenesis. However, the expression patterns of EMPs (AMEL, AMBN) and ODAM were found to be essentially unaffected by TgAMTN overexpression, indicating that AMTN does not have a direct effect on the expression of these proteins. Rather, the observed effect on the structural architecture of the enamel matrix appears to be the result of a direct effect of AMTN, which, when increased in the model presented here, likely alters the structural organization of the organic matrix that provides instructive information for controlled mineralization. Whether this effect is the result of a direct binding of AMTN to mineral, or to other EMPs, remains to be determined. Since the primary sequence of AMTN contains only limited amounts of acidic amino acids, which have been shown to mediate hydroxyapatite binding in other mineral-associated proteins [19], it will be particularly important to determine whether AMTN binds to other extracellular matrix proteins that are expressed during, and affect, enamel formation. The fact that other proteins have been expressed at high levels from similar constructs in ameloblasts of transgenic mice, and as a

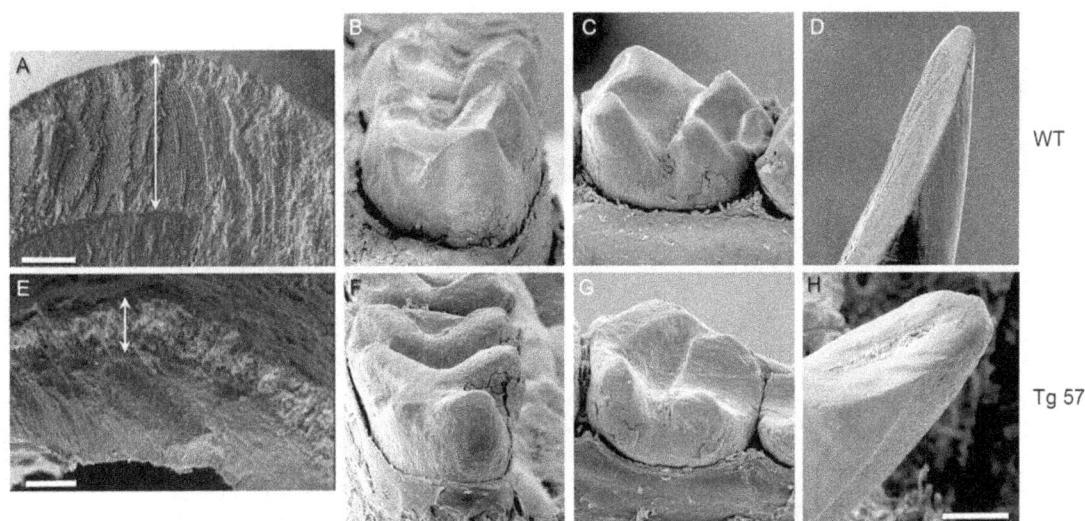

Figure 7. SEM analyses of teeth from wild type (A–D) and tg57 (E–H) animals. Images A and E show incisors fractured in cross section at the same anatomical region, approximately half way between the incisal edge and the gingival margin of the labial surface. Double-headed arrows indicate the substanital difference in enamel thickness. Enamel from wild type animals also showed the typical decussation pattern with a smooth enamel surface (A), while enamel in tg57 animals showed no decussation, and the surface was irregular and pitted (E). We also imaged whole teeth of WT and tg57 animals at 4 months of age, which revealed severe attrition of mandibular first molars (B, F; M1, mesial aspect), second molars (C, G; M2, lingual aspect) and incisors (D, H; buccal aspect) due to the loss of functional enamel. The positions of individual molar cusps, which are attrited in transgenic samples, are indicated by asterisks in wild type teeth. Scale bars: A and E: 50 μm; H (applicable to all other panels): 300 μm.

result show no obvious alterations of the enamel structure [13], suggests that the observed effects are specific to AMTN.

The molecular mechanisms that control the structural transition from the typical rod/interrod structure of bulk enamel to the more compact, aprismatic layer of last-formed surface ("final") enamel are not well understood. Since current studies have localized AMTN predominantly at the enamel/ameloblast interface in incisors of rats [7] and mice [4] it is conceivable that AMTN, possibly in co-operation with ODAM, is involved in the structural transition from bulk to surface enamel, promoting the formation of the compact enamel surface layer. If this is indeed the case then the overexpression of AMTN at a stage earlier than normal would be expected to result in structural alterations that resemble those observed at the outer and surface enamel layers. The formation of these final enamel layers coincides spatially and temporally with the retraction of Tomes' process. Indeed, the irregular structure of enamel that forms in the presence of TgAMTN supports a role for amelotin in terminating the regular arrangement of hydroxyapatite prisms. The clearly reduced enamel thickness in line 57 animals may also support a functional role for AMTN in disrupting the longitudinal enamel prism growth. In wild type rodent incisors, the formation of the aprismatic enamel surface coincides with the disappearance of Tomes' process in ameloblasts at the transition stage [8] and the expression of AMTN increases dramatically at the same location [4,7]. Although Tomes' process is a critical hallmark of secretory stage ameloblasts, little is known about the molecular mechanisms that regulate its formation, function and retraction. This lack of knowledge is largely due to the unavailability of in vitro model systems that display – and allow study of - this anatomical feature of ameloblasts. It is known, however, that during the formation of initial enamel at the dentin-enamel junction (DEJ) the distal portion of Tomes' process does not yet exist, and only appears when the typical bulk enamel structure with rod and interrod compartments forms [20]. We thus speculate that the presence of AMTN at the initial stage of enamel formation disrupts the development of such an organized rod/

interrod structure (as seen in Fig. 8), and leads to the failure of Tomes' process to develop (Fig. 9 a, e). The absence of Tomes' process is thus likely a consequence of disturbed crystal growth at interrod areas, which normally lead to the "picket fence" appearance of the ameloblast/enamel interface at this stage. However, whether there is a causal relationship between AMTN expression and the loss of Tomes' process at the transition stage under normal conditions, and the nature of such a relationship, is beyond the scope of this contribution but opens interesting areas for future research.

In our current study we have shown that AMTN overexpression does not notably affect the expression of two EMPs, AMEL and AMBN, which are normally expressed during the secretory stage. In contrast, AMTN overexpression resulted in a slight increase of ODAM expression in secretory stage ameloblasts. AMEL and AMBN on one hand, and AMTN and ODAM on the other hand, are expressed during the early and late stage of amelogenesis, respectively. It thus appears that distinct proteins activities are responsible for the formation and maturation of enamel and that AMTN and ODAM expression may not be independent of each other. The effects of overexpression or disruption of any one particular enamel gene on the expression level(s) of other enamel genes have not been elucidated in detail, and we expect the transgenic model presented here to be a useful tool for such studies. It also remains to be determined whether the overexpression of AMTN affects the expression level, pattern or activity of enamel-related proteolytic enzymes such as KLK4 or MMP20 to ultimately affect enamel matrix dynamics. Some clues to the activity of AMTN might also be related to the identification of novel proteases involved in enamel development such as chymotrypsin-c (encoded by CTRC) [21], since the expression profiles of AMTN and CTRC are essentially identical. Based on its localization at the enamel/ameloblast interface [4] it is possible that AMTN may also play a role in modulating transport phenomena to and from the mineralized enamel to regulate pH,

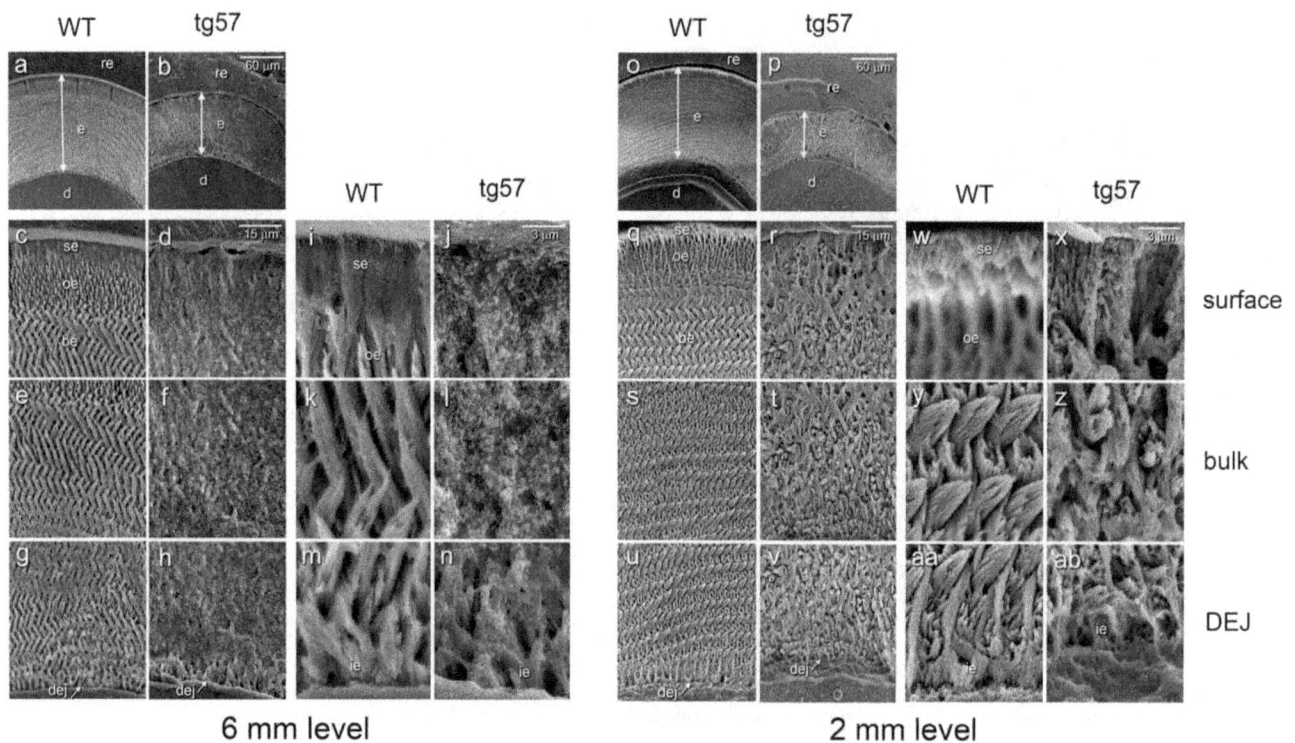

Figure 8. SEM analysis of transverse, etched enamel sections from wild type (WT) and transgenic (tg57) mice. Sections prepared at a distance of 6 mm (left) and 2 mm (right) from the incisal edge were imaged at various magnifications. At the 6 mm level, corresponding to the late maturation stage, enamel thickness was reduced by approximately 50% in transgenic animals (a, b). Higher magnification revealed the irregular enamel structure and residual organic matrix in transgenic mice at the enamel surface (c,d and i, j) and the bulk enamel (e, f and k, l). In contrast, the structure of the aprismatic initial enamel (ie) layer at the dentin-enamel junction (dej) was similar between wild type and tg57 animals (m, n). At the 2 mm level, where enamel is fully mature, its thickness in tg57 animals (p) was reduced to approximately 30% of that in wild type littermates (o), and higher magnification showed dramatic structural differences at the surface (q, r and w, x) and bulk (s, t and y, z) enamel layers. The enamel crystal structure at the initial enamel layer (ie in aa and ab) was again similar between WT and tg57, but the transition to the regular prismatic enamel structure was not observed in tg57 animals. At the enamel surface (w, x) the discrete aprismatic surface layer of about 3–5 µm (se in w), which was more acid resistant than the underlying outer enamel layer as seen by the abrupt slope of acid-etched material, and the outer enamel layer featuring parallel enamel prisms (w) in wild type animals were replaced by a densely mineralized, compact layer in tg57 mice (x). Abbreviations: be, bulk enamel; d, dentin; dej, dentin-enamel junction; e, enamel; ie, inner enamel; oe, outer enamel; re, embedding resin; se, surface enamel; ep, enamel prism). The size of scale bars (a, b and o, p: 60 µm; c–h and q–v: 15 µm; i–n and w–ab: 3 µm) is indicated.

ion transport, and/or the removal of organic matrix from the maturing enamel.

While we are still in the process of understanding the specific individual and collective contributions of enamel matrix proteins to enamel biomineralization, we have demonstrated that AMTN has a profound effect on the structure and mechanical properties of this functional bioceramic material. It is to be expected that more detailed future analyses of its role in establishing the enamel microstructure will provide a more comprehensive picture of the mechanisms involved in enamel biomineralization. A complete understanding of this process is required to devise strategies to re-mineralize, regenerate or engineer natural dental enamel, and to meet the clinical demands for natural restorative biomaterials.

Materials and Methods

Plasmid construction

The mouse amelogenin promoter [22,23] was used in the construction of a plasmid cassette to express the amelotin protein, and this plasmid has been used to produce transgenic mice. Mouse amelotin cDNA with appropriate restriction sites was prepared by PCR using the forward primer 5′- G GATATC TTA CCA AAG CAG CTT AAC CCT which included an EcoRV restriction site

(underlined) and a reverse primer 5′- GCCTCATAAAGG TGA AAC AGC TTA CTG AGT TCT that include an EcoNI restriction site and the stop codon (both underlined). This amplified DNA product was cut with EcoRV and EcoNI and then inserted into a previously generated vector containing the mouse amelogenin promoter, intron 1, the amelogenin signal peptide, and 3 repeats of the FLAG epitope. The final plasmid construct (Figure 1) was sequenced through the entire open reading frame to ensure that there were no PCR or cloning errors. This plasmid construct and its translated product will be referred to as pTgAMTN and TgAMTN, respectively.

Transgenic animal production and verification of transgene status

All animal manipulation was performed under approved protocols (University of Toronto AUP Nr. 20008384 and University of Southern California AUP Nr. 9666) and complied with institutional and federal guidelines. Transgenic mouse lines were prepared as described elsewhere [13,22,24,25]. Briefly, fertilized eggs for microinjection were harvested from super-ovulated six-week old female F1(C57Bl/6J×CBA/J) mice impregnated by adult male F1(C57Bl/6J×CBA/J) mice. For embryo transfer, pseudo-pregnant females were produced by mating CD1

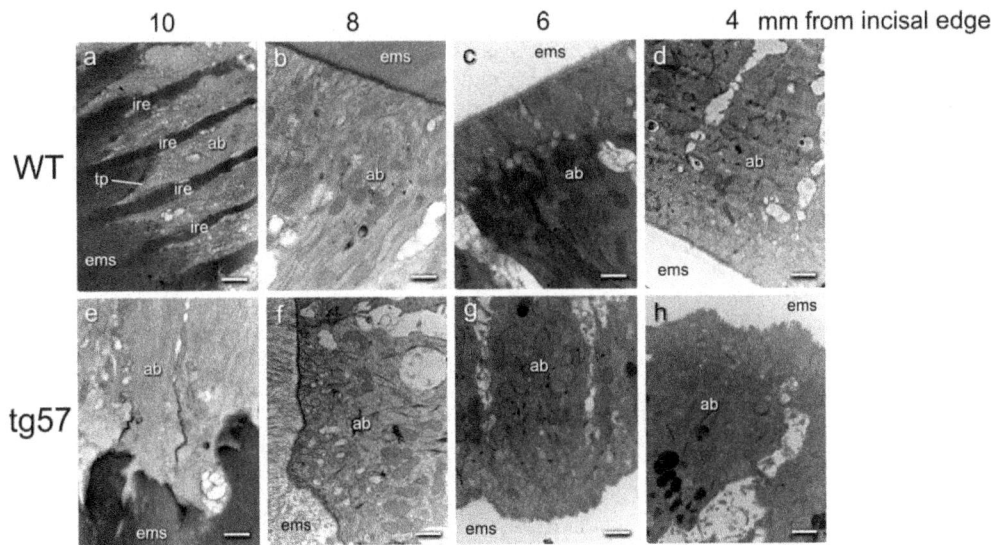

Figure 9. TEM analyses of the interface between the apical ameloblast surface and enamel matrix space (ems) at various positions along the apical-incisal axis in four week-old wild type (WT) and transgenic (tg57) animals. Images were taken from cross-sections made at a distance of 10, 8, 6 and 4 mm from the incisal edge of mandibular incisors. At the 10 mm level (a, e), corresponding to the secretory stage, the picket fence structure of orderly arranged Tomes' processes (tp), separated by interrod enamel (ire) is well developed in wild type animals, but absent in tg57 mice (e). The linear demarcation line between ameloblasts and the enamel space that forms after retraction of Tomes' process at the early (8 mm), mid (6 mm) and late (4 mm) maturation stages in wild type animals (b, c, d) appears highly irregular at all stages in transgenic animals (f, g, h). Abbreviations: ab, ameloblast; ems, enamel matrix space; tp, Tomes' process. Scale bars represent 500 nm.

adult females with a vasectomized CD1 adult male. Microinjection of DNA and oviduct transfer of injected zygotes was performed as described [24] using the ~5.6 kb Sma I/Bam HI fragment of the plasmid pTgAMTN (Figure 1). Two independent lines obtained from this microinjection experiment (numbers 57 and 457), both with a presumed unique site of DNA integration, were bred beyond three generations before conducting any analyses. Animals were assessed for transgene status by Southern blot hybridization of genomic DNA digested with PstI, and hybridized to random primed ^{32}P-labeled Eco RV/Pst I 528 bp amelotin single-stranded cDNA. Subsequent genotyping analysis of animals was performed by PCR using 2 µl of genomic DNA solution prepared from tail tissue (DNeasy Tissue Kit, Qiagen) with Titanium Taq polymerase (Titanium Taq Kit, Clontech Inc.) and the primers F4 (5′-C TTT TGG TCC TCT AAC TCG TTA-3′; a forward primer located in amelogenin intron 1 region) and R3 (5′-GA AGC AGG GTT AAG CTG CTT-3′; a reverse primer located in the amelotin cDNA region) to generate an amplicon of 429 bp. PCR was performed in a final volume of 25 µl under the following conditions: 94°C/5 minutes, followed by 40 cycles of (94°C/ 45 sec; 65°C/45 sec; 72°C/90 sec) and a final extension at 72°C for 10 minutes.

Western blotting

For Western blots using anti-FLAG and anti-GAPDH primary antibodies (Fig. 3A) the following procedure was applied: First and second mandibular molars were dissected from 3–4 day old pups using a standard dissecting stereoscopic microscope. Tissues were lysed in 2× SDS-PAGE loading buffer (50 µl per mouse) using a tissue homogenizer. The samples were boiled for five minutes, chilled on ice and centrifuged to remove residual debris. The protein lysate was resolved according to size on 12% SDS-PAGE (20 µl each) and a Western blot was performed using a monoclonal antibody recognizing the FLAG epitope DYKDDDDK (Sigma-Aldrich Anti-FLAG M2, peroxidase con-

jugated, catalogue # A8592) and a mouse anti-glyceraldehyde-3-phosphate dehydrogenase (anti-GAPDH; Millipore; catalogue #MAB374) with the Amersham ECL kit (GE Healthcare Bio-Sciences Corporation). Prior to Western blots using an anti-AMTN primary antibody (Fig. 3B) the amelotin protein was enriched by immunoprecipitation as follows: First and second mandibular molars from 3–4 day old pups were extracted with 0.5 M acetic acid and the acid extracts desalted over a Sephadex G-25 column (PD10, GE Healthcare, catalogue # 17-0851-01). Extracts were then incubated with protein G agarose (Invitrogen, Carlsbad,CA, USA, catalogue# 15920-010), centrifuged and the supernatant mixed with approximately 1 µg of an affinity-purified rabbit antibody against the full-length recombinant mouse amelotin protein (FL-rmAMTN). AMTN-immunoglobulin complexes were precipitated with protein G agarose, washed with PBS, eluted with SDS-PAGE loading buffer and subjected to SDS-PAGE as above. Western blotting was performed using the FL-rmAMTN primary antibody and a goat-anti-rabbit HRP-conjugated secondary antibody (Biorad, Mississauga, ON, Canada; catalogue # 170-6515) using the Immun-Star Western kit (Biorad, catalogue # 170-5070) according to the manufacturer's recommendations. Primary antibodies FL-rmAMTN, FLAG and GAPDH were used at a dilution of 1:4,000; 1:2,000; and 1:200, respectively. For protein competition, the FL-rmAMTN antibody was incubated with a 200-fold molar excess of the rmAMTN protein at 4°C over night and cleared by centrifugation prior to use.

Histology and Immunohistochemistry

Animals at 4 and 30 days of age from each genotype (wild type, line 57 and line 457) were sacrificed for this analysis to reflect pre-eruptive and mature stages of tooth development. Animals were euthanized by CO_2 inhalation, mandibular arches dissected, freed from excess skin and soft tissue and separated along the midline. After fixation in 4% paraformaldehyde (PFA) in phosphate buffered saline (PBS) at 4°C over night hemimandibles were decalcified in 12.5% EDTA (pH 7.2) for three days (tissues from four day-old animals) or ten days (30 day-old animals) with daily solution changes, then embedded in paraffin. Six micrometer thick sagittal sections were prepared as described previously [2,13,26] and used for histological and immunohistochemical analyses (Envision+ kit HRP, DAB, DAKO Canada, Burlington, ON). Sections were either stained with hematoxylin/eosin (H&E) for general histological evaluation, or probed with a rabbit anti-FLAG antibody (catalogue # F7425; Sigma-Aldrich,; dilution 1:3,000) to demonstrate tissue specific expression of the introduced transgene. In addition, primary rabbit antibodies for mouse amelotin [4], diluted 1:500, amelogenin (Abcam Cat. # 59705; 1:2000), amelin/ameloblastin [27]; (a gift from Dr. M. Wendel, Centre for Oral Biology, Huddinge, Sweden; 1:10,000) and Odam [6]; (gift from Dr. A. Nanci, Université de Montréal, QC; 1:10,000) were used. Following color development with diaminobenzidine (DAB), sections were counterstained with methyl green (DAKO Canada) and mounted in Biomeda™ Gel/Mount media (Electron Microscopy Sciences, Hatfield, PA, USA). Images were captured with a Pixelink (Ottawa, ON, Canada) digital camera (model PL-A623C) mounted on an Eclipse E400 microscope (Nikon Canada, Mississauga, ON).

Scanning electron microscopy

All scanning electron microscopy (SEM) data was obtained from heterozygous animals. Age-matched non-transgenic animals from each animal line served as controls. Six-week old transgenic and non-transgenic animals were sacrificed by CO_2 asphyxiation. For fractured surface studies (to reveal the untreated enamel structure) the lower incisors were extracted and air-dried. Each sample was mechanically fractured cross-sectionally at the same anatomical region of the crown, in the mature enamel. For studies on sectioned and etched samples, hemimandibles from 9 week-old mice were fixed in 4% paraformaldehyde in PBS buffer at 4°C over night and transverse slices were prepared at every 2 mm from the incisal tip using a low speed saw (Buehler IsoMET®, Whitby, ON, Canada) with a diamond wheel (Part # DWH4123, South Bay Technology, Inc., San Clemente, CA, USA). Segments were dehydrated with graded acetone for about 12 h each, 100% acetone O/N at RT, then embedded with epoxy resin (Low Viscosity Embedding Media Spurr's Kit®, Electron Microscopy Sciences, Hatfield, PA, USA). The embedded samples were polished and etched with 38% phosphoric acid etching gel (Etch Rite®, Pulpdent Corp., Watertown, MA, USA). Samples were mounted on SEM stubs and coated with platinum (SC515 SEM Coating System, Polaron; Quorum Technologies, Ashford, Kent, UK). SEM pictures were captured using the Quartz PCI-Image Management System (Quartz Imaging Corporation, Vancouver, BC, Canada) on a S-2500 SEM machine (Hitachi, Roslyn Heights, NY, USA) operating at 10 kV.

Transmission Electron Microscopy

Nine week-old animals were sacrificed by cardiac perfusion with 4%PFA, 1% glutaraldehyde in 0.1 M sodium cacodylate, hemimandibles dissected and fixed with fresh fixative at 4°C over night. Samples were then rinsed in PBS at 4°C overnight and demineralized with 0.8% glutaraldehyde, 12.5% EDTA in PBS for 5 days with three solution changes per day. After the secondary fixation with $1\%OsO_4$, $0.5\%K_2Cr_2O_7$, $0.5\%K_4[Fe(CN)_6]\cdot3H_2O$ at RT for 2 hours, we performed *en bloc* staining with freshly prepared 2% uranyl acetate at RT for 2 hours. Samples were dehydrated in graded concentrations of ethanol and propylene oxide, then infiltrated and embedded in Jembed 812® resin (Canemco & Marivac, Lakefield, QC, Canada). Sections were prepared with a diamond knife (DiATOME ultra45, Diatome AG, Biel, Switzerland) in an EM UC6-NT microtome (Leica Microsystems, Concord, ON, Canada). Selected sections of 800 nm thickness were stained with Epoxy Tissue Stain® (Electron Microscopy Sciences) for quality control. Sections of 100 nm thickness were then post-stained with 2% uranyl acetate and mounted onto TEM grids (SPI, West Chester, PA, USA) for TEM analysis. Images were captured on a Tecnai 20 Transmission Electron Microscope (Philips, Amsterdam, The Netherlands).

Mechanical hardness testing

Microindentation techniques as previously described [28] were used to measure enamel hardness in 6-week old (42-day post-natal) animals, five line 57 mice and six wild type controls. Briefly, freshly extracted intact murine lower left incisor teeth, kept moist at all times, were mounted in slow-set epoxy resin and sequentially ground in cross-section to a 0.1 μm alumina finish using a semiautomatic polisher (Buehler). Loads of 50 g were used with dwell times of 20 seconds using a customized manually-operated Vickers microhardness tester. Indentations were made within the erupted incisal thirds of the teeth, excepting the incisal-most 1 mm, approximately half way between the DEJ and the facial enamel surface. Indentations were examined by light microscopy, using polarization, interference, light/dark field, and measurements were made using a digital micrometer. The mean of 6 indentations was used to calculate means and standard errors.

Acknowledgments

We would like to thank Mikael Wendel and Antonio Nanci for sharing antibodies for this study, Daiana Stolf, Nawfal Al Hashimi and Desiree Yazdanshenas for their help with histological procedures and Brian Quan for his help with transmission electron microscopy.

Author Contributions

Conceived and designed the experiments: MLP BG. Performed the experiments: RL YN JH VN ESG SNW. Analyzed the data: RL YN. Wrote the paper: BG RL MLS MLP.

References

1. Hu JC, Chun YH, Al Hazzazzi T, Simmer JP (2007) Enamel formation and amelogenesis imperfecta. Cells Tissues Organs 186: 78–85.
2. Iwasaki K, Bajenova E, Somogyi-Ganss E, Miller M, Nguyen V, et al. (2005) Amelotin–a Novel Secreted, Ameloblast-specific Protein. J Dent Res 84: 1127–1132.
3. Sire JY, Davit-Beal T, Delgado S, Gu X (2007) The origin and evolution of enamel mineralization genes. Cells Tissues Organs 186: 25–48.
4. Somogyi-Ganss E, Nakayama Y, Iwasaki K, Nakano Y, Stolf D, et al. (2011) Comparative Temporospatial Expression Profiling of Murine Amelotin Protein during Amelogenesis. Cells Tissues Organs.
5. Kestler DP, Foster JS, Macy SD, Murphy CL, Weiss DT, et al. (2008) Expression of odontogenic ameloblast-associated protein (ODAM) in dental and other epithelial neoplasms. Mol Med 14: 318–326.

6. Moffatt P, Smith CE, St-Arnaud R, Nanci A (2008) Characterization of Apin, a secreted protein highly expressed in tooth-associated epithelia. J Cell Biochem 103: 941–956.
7. Moffatt P, Smith CE, St-Arnaud R, Simmons D, Wright JT, et al. (2006) Cloning of rat amelotin and localization of the protein to the basal lamina of maturation stage ameloblasts and junctional epithelium. Biochem J 399: 37–46.
8. Nanci A (2008) Enamel: Composition, Formation and Structure. In: Dolan JJ, ed. Ten Cate's Oral Histology: Development, Structure and Function. 7 ed. St. Louis: Mosby Elsevier. pp 141–190.
9. Weile V, Josephsen K, Fejerskov O (1993) Scanning electron microscopy of final enamel formation in rat mandibular incisors following single injections of 1-hydroxyethylidene-1,1-bisphosphonate. Calcif Tissue Int 52: 318–324.

10. Luo W, Wen X, Wang HJ, MacDougall M, Snead ML, et al. (2004) In vivo overexpression of tuftelin in the enamel organic matrix. Cells Tissues Organs 177: 212–220.

11. Paine ML, Luo W, Wang HJ, Bringas P, Jr., Ngan AY, et al. (2005) Dentin sialoprotein and dentin phosphoprotein overexpression during amelogenesis. J Biol Chem 280: 31991–31998.

12. Paine ML, Wang HJ, Luo W, Krebsbach PH, Snead ML (2003) A transgenic animal model resembling amelogenesis imperfecta related to ameloblastin overexpression. J Biol Chem 278: 19447–19452.

13. Paine ML, Zhu DH, Luo W, Snead ML (2004) Overexpression of TRAP in the enamel matrix does not alter the enamel structural hierarchy. Cells Tissues Organs 176: 7–16.

14. Wen X, Zou Y, Luo W, Goldberg M, Moats R, et al. (2008) Biglycan overexpression on tooth enamel formation in transgenic mice. Anat Rec (Hoboken) 291: 1246–1253.

15. Moinichen CB, Lyngstadaas SP, Risnes S (1996) Morphological characteristics of mouse incisor enamel. J Anat 189(Pt 2): 325–333.

16. Kallenbach E (1973) The fine structure of Tomes' process of rat incisor ameloblasts and its relationship to the elaboration of enamel. Tissue Cell 5: 501–524.

17. White SN, Paine ML, Ngan AY, Miklus VG, Luo W, et al. (2007) Ectopic expression of dentin sialoprotein during amelogenesis hardens bulk enamel. J Biol Chem 282: 5340–5345.

18. Lu Y, Papagerakis P, Yamakoshi Y, Hu JC, Bartlett JD, et al. (2008) Functions of KLK4 and MMP-20 in dental enamel formation. Biol Chem 389: 695–700.

19. Goldberg HA, Warner KJ, Li MC, Hunter GK (2001) Binding of bone sialoprotein, osteopontin and synthetic polypeptides to hydroxyapatite. Connect Tissue Res 42: 25–37.

20. Inage T, Fujita M, Kobayashi M, Wakao K, Saito N, et al. (1990) Ultrastructural differentiation in the distal ends of ameloblasts from the presecretory zone to the early secretory zone. J Nihon Univ Sch Dent 32: 259–269.

21. Lacruz RS, Smith CE, Smith SM, Hu P, Bringas P, Jr., et al. (2011) Chymotrypsin C (caldecrin) is associated with enamel development. J Dent Res 90: 1228–1233.

22. Snead ML, Paine ML, Chen LS, Luo BY, Zhou DH, et al. (1996) The murine amelogenin promoter: developmentally regulated expression in transgenic animals. Connect Tissue Res 35: 41–47.

23. Zhou YL, Snead ML (2000) Identification of CCAAT/enhancer-binding protein alpha as a transactivator of the mouse amelogenin gene. J Biol Chem 275: 12273–12280.

24. Nagy A, Gertsenstein M, Vintersten K, Behringer R (2003) Manipulating the mouse embryo: A laboratory manual Cuddihy J, ed. New York: Cold Spring Harbor Laboratory Press.

25. Ignelzi MA, Jr., Liu YH, Maxson RE, Jr., Snead ML (1995) Genetically engineered mice: tools to understand craniofacial development. Crit Rev Oral Biol Med 6: 181–201.

26. Somogyi E, Petersson U, Hultenby K, Wendel M (2003) Calreticulin-an endoplasmic reticulum protein with calcium-binding activity is also found in the extracellular matrix. Matrix Biol 22: 179–191.

27. Fong CD, Cerny R, Hammarstrom L, Slaby I (1998) Sequential expression of an amelin gene in mesenchymal and epithelial cells during odontogenesis in rats. Eur J Oral Sci 106 Suppl 1: 324–330.

28. White SN, Luo W, Paine ML, Fong H, Sarikaya M, et al. (2001) Biological organization of hydroxyapatite crystallites into a fibrous continuum toughens and controls anisotropy in human enamel. J Dent Res 80: 321–326.

14

Overexpression of an Acidic Endo-β-1,3-1,4-glucanase in Transgenic Maize Seed for Direct Utilization in Animal Feed

Yuhong Zhang[1◗], Xiaolu Xu[2◗], Xiaojin Zhou[1], Rumei Chen[1], Peilong Yang[2], Qingchang Meng[3], Kun Meng[2], Huiying Luo[2], Jianhua Yuan[3], Bin Yao[2]*, Wei Zhang[1]*

1 Biotechnology Research Institute, Chinese Academy of Agricultural Sciences, Beijing, P. R. China, 2 Key Laboratory for Feed Biotechnology of the Ministry of Agriculture, Feed Research Institute, Chinese Academy of Agricultural Sciences, Beijing, P. R. China, 3 Institute of Food Crops, Jiangsu Academy of Agricultural Sciences, Nanjing, P. R. China

Abstract

Background: Incorporation of exogenous glucanase into animal feed is common practice to remove glucan, one of the anti-nutritional factors, for efficient nutrition absorption. The acidic endo-β-1,3-1,4-glucanase (Bgl7A) from *Bispora* sp. MEY-1 has excellent properties and represents a potential enzyme supplement to animal feed.

Methodology/Principal Findings: Here we successfully developed a transgenic maize producing a high level of Bgl7AM (codon modified Bgl7A) by constructing a recombinant vector driven by the embryo-specific promoter ZM-leg1A. Southern and Western blot analysis indicated the stable integration and specific expression of the transgene in maize seeds over four generations. The β-glucanase activity of the transgenic maize seeds reached up to 779,800 U/kg, about 236-fold higher than that of non-transgenic maize. The β-glucanase derived from the transgenic maize seeds had an optimal pH of 4.0 and was stable at pH 1.0–8.0, which is in agreement with the normal environment of digestive tract.

Conclusion/Significance: Our study offers a transgenic maize line that could be directly used in animal feed without any glucanase production, purification and supplementation, consequently simplifying the feed enzyme processing procedure.

Editor: M. Lucrecia Alvarez, TGen, United States of America

Funding: This research was supported by the Key Program of Transgenic Plant Breeding 2013ZX08003-002, 2009ZX08003-020B (http://www.nmp.gov.cn/zxjs/zjy/201012/t20101208_2129.htm) and the China Modern Agriculture Research System CARS-42 (http://english.agri.gov.cn/hottopics/five/201304/t20130421_19483.htm). The funders had no role in study design, data collection and analysis, decision to publish, or preparation of the manuscript.

Competing Interests: The authors have declared that no competing interests exist.

* E-mail: binyao@caas.cn (BY); zhangwei02@caas.cn (ZW)

◗ These authors contributed equally to this work.

Introduction

β-1,3-1,4-D-Glucans (β-glucans) are the main component of cereal cell walls, particularly in the endosperm cell walls of barley and other grains [1]. It is composed of β-D-glycosyl residues linked through irregular β-1,3 and/or β-1,4 glycosidic bonds. Ruminants can utilize β-glucans through enzyme digestion of rumen microbes. However, monogastric animals such as pig, poultry, and fish do not have such enzymes to decompose the β-glucans. By combining with water, β-glucans increase the viscosity of chyme, block the intestinal surface partially, and prevent the mixing of intestinal endogenous digestive juice with the chyme [2]. Thus β-glucan represents one of the intense anti-nutritional factors in wheat- and barley-based diets [3].

To overcome these problems, the most common and effective practice is to add exogenous endoglucanases into animal feed [3]. Majority of endoglucanases are grouped into glycoside hydrolase (GH) families 3, 5, 7, 12 and 16, based on the amino acid sequence and catalytic domain structures (http://www.cazy.org/). According-ing to the degradation mode against glycosidic linkage, endoglu-

canases have been grouped into four main categories: β-1,3-glucanase (laminarinase, EC 3.2.1.39), β-1,4-glucanases (cellulase, EC 3.2.1.4), β-1,3-1,4-glucanases (lichenase, EC 3.2.1.73), and β-1,3(4)-glucanase (EC 3.2.1.6) [4]. Among them, β-1,3-1,4-glucanase has received significant attention in feed industrial applications because of their hydrolysis ability against grain-based glucan and multiple enzymatic functions. β-1,3-1,4-Glucanase is able to catalyze the hydrolysis of β-glucan into low molecular weight glucose polymers, thus reducing the hydrophilicity and viscosity of chyme and eliminating the anti-nutritional negative effect. Moreover, addition of β-1,3-1,4-glucanase can improve feed intake, enhance animal production, regulate cecal microbiota and increase feed conversion ratio [5–8]. Besides, the hydrolysis products from glucans—glucooligosaccharides may serve as fermentable dietary fiber-like substrates and positively affect gastrointestinal tract health [9].

To date, commercial feed additive β-1,3-1,4-glucanases are generally from microbial expression systems, commonly *Aspergillus japonicus* [10], *Pichia pastoris* [11] and *Clostridium thermocellum* [8]. This process is flexible and convenient, but has disadvantages like

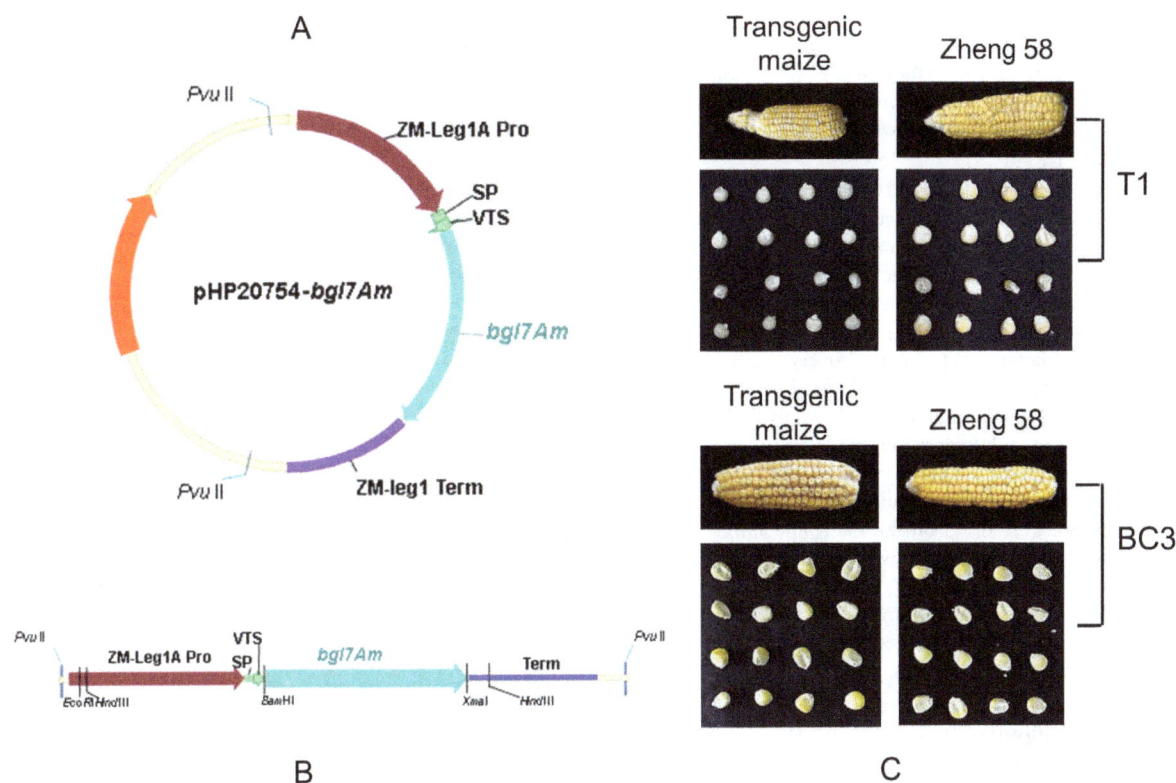

Figure 1. Construction of the recombinant vector and the transgenic maize seed. A The recombinant expression vector pHP20754-*bgl7Am*. **B** The chimeric gene cassettes for expression in maize. **C** Ears and seeds of transgenic maize (T1 and BC3) compared with that of wild-type Zheng58.

high energy consumption, high equipment cost and serious environmental pollution. Moreover, enzyme addition is a complex process involving enzyme isolation, purification and supplementation, which requires more energy and resources. Thus it's a good way to produce feed enzymes (e.g. β-1,3-1,4-glucanase) in transgenic feed grains directly without any industrial processing.

Transgenic plants are being developed for both commercial and environmental values. In 2011, the plantation area of transgenic plants reached about 160 million hectares worldwide and was distributed in 29 countries; transgenic maize accounted for nearly one third of the total genetically modified crops [12]. Maize (*Zea mays* L.) is the main ingredient of animal feed (nearly 50%), and represents an ideal bioreactor of feed enzymes because of its cultivation worldwide. A phytase gene *phyA2* from *Aspergillus niger* has been successfully overexpressed in maize seeds [13].

In this study, we developed a genetically stable maize line that had high β-glucanase activity in the seeds. The endo-β-1,3-1,4-glucanase, Bgl7A, from acidophilic *Bispora* sp. MEY-1 was selected due to its excellent properties as feed additive, such as acidic pH optimum, good thermostability and broad pH stability, highly resistance to proteases, and broad substrate specificity [11]. The gene codon was optimized for better expression in maize.

Materials and Methods

Plant materials

Maize Hi-II [14] was used for genetic transformation as host variety. The immature embryos, approximately 1.0–2.0 mm long, were preserved on N61-100-25 medium [14] containing 0.2% (w/v) phytagel (Sigma, St. Louis, MO) for callus induction. The

commercial maize inbred-line Zheng58 was used as recurrent parent to produce progenies.

Codon optimization of the β-1,3-1,4-glucanase gene *bgl7A*

To improve its expression level in transgenic maize, the DNA sequence of native endo-β-1,3-1,4-glucanase gene *bgl7A* from *Bispora* sp. MEY-1 (Genbank accession No. FJ695140) [11] was optimized according to the translationally optimal codon usage of maize [15]. Codon adaptation index (CAI), optimal codon usage, GC content and distribution, effective number of codons (Nc), negative CIS elements, negative repeat elements, and mRNA structure were used to evaluate the gene sequence (https://www.genscript.com/cgi-bin/tools/rare_codon_analysis). Low-usage codons (<15% frequency) were replaced by high-usage ones according to the known codon bias of maize [15]. The modified gene was named *bgl7Am* that encoded the same amino acid sequence as *bgl7A*. The optimized gene was synthesized by Genscript (Nanjing, China) and cloned into pUC57 vector to construct the recombinant plasmid pUC57-*bgl7Am*.

Plasmid construction

The transformation vector pHP20754 (Fig. 1a) consists of the corn *legumin1A* (leg1) promoter ZM-leg1A Pro, signal peptide (SP), vacuole targeting sequence (VTS) of corn Proaleurain and the corn leg1 terminator ZM-leg1 Term [16–18]. The β-glucanase gene was excised from pUC57-*bgl7Am* with *Bam*HI and *Xma*I and subcloned into pHP20754 to produce the expression construct pHP20754-*bgl7Am*, which was further digested with *Pvu*II to generate the chimeric gene expression cassette (Fig. 1b) for transformation. The plasmid pHP17042BAR carrying the maize histone H2B promoter, the maize Ubiquitin 5′-UTR intron-1, the bar gene and the potato protease II (PINII) terminator [13] was used as the selectable marker for screening of positive transgenic plants. The bar gene expression cassette was excised from pHP17042BAR by digest with HindIII, XhoI and SacI.

Maize transformation, selection and regeneration

The plasmid fragments containing the gene cassette of bgl7Am and bar, respectively, were mixed at the ratio of 1:1 and adjusted the concentration to 200 ng/μL. Maize transformation was carried out with high-velocity tungsten microprojectile (Bio-Rad, Hercules, CA) wrapped by the DNAs of bgl7Am and bar according to the method described before [19]. After recovery, embryonic calli were transferred onto the selective medium supplemented with bialaphos as the selectable marker. The positively transformed calli were cultivated in differentiation medium and rooting medium in succession. Seedlings (T0 plants) were transplanted into greenhouse and pollinated with the inbred-line Zheng58 to produce T1 seeds. Seeds were dried on the plant and harvested 35–45 days after pollination. Zheng58 was used as recurrent parent for backcrossing to produce filial generations (T1, BC1, BC2 and BC3). β-Glucanase activity determination in the kernels and PCR for the bgl7Am gene of seedlings were used in combination to screen the transgenic lines.

PCR detection of exogenous gene integration

The specific primers bgl7am-875F (5′-ACGGCAAGGT-CATCCAGAACGCGAAGG-3′) and 20754-398R (5′-TTCCTGGCAAATCACTCGGTGTATC-3′) were used for PCR detection of the positive plants harboring bgl7Am. The gene actin as control was amplified using primers AC326F (5′-ATGTTTCCTGGGGATTGCCGAT-3′) and AC326R (5′-GCATCACAAGCCAGTTTAACC-3′). Genomic DNA of the maize immature leaves was used as PCR templates. The recombinant plasmid pHP20754-bgl7Am and the genomic DNA of Zheng58 were used as the positive and negative controls, respectively.

Southern blot analysis

Five grams of maize leaves of generations T1 to BC3 were ground to powder in liquid nitrogen, and the genomic DNA was extracted with the CTAB method. Genomic DNA of Zheng58 was used as the negative control. About 50 μg of genomic DNA was digested by EcoRI and HindIII and then separated on a 0.8% (w/v) agarose gel. The agarose gel was transferred onto a hybond-N$^+$ nylon membrane (GE Healthcare, Uppsala, Sweden) with a Trans-Blot SD system followed by UV-crosslinking. A digoxin-labeled probe containing a 800-bp fragment of bgl7Am was used for southern-blot hybridization. Immunologic process was conducted following the instructions of DIG-high prime DNA labeling and detection starter kit II (Roche, Indianapolis, IN).

Western blot analysis

Five milligrams of lyophilized purified Bgl7A produced in Pichia pastoris GS115 [11] was used for the production of polyclonal antibody in rabbits. Recombinant proteins were extracted from seed meals. Kernels were ground with a high-throughput tissue homogenizer Geno/Grinder 2010 (SEPX CertiPrep, Metuchen, NJ).

To extract protein from seed meals, 30 mg of seed powder were placed into a 1.5-mL tube containing 300 μL extraction buffer (50 mM citric acid-Na$_2$HPO$_4$, pH 3.5). The tube was agitated on a shaker at room temperature for 1 h. After centrifugation at

5000× g for 10 min, the supernatant was incubated with 2-fold volume of pre-cooled acetone for 30 min, followed by centrifugation at 14,000× g for 10 min. The protein precipitate was dissolved in 30 μL of deionized water, and the protein sample was divided into two equal parts. One was deglycosylated with endo-β-N-acetylglucosaminidase (Endo H) according to the supplier's instructions (New England Biolabs, Ipswich, MA), the other remained intact. Protein extract of purified Bgl7A from P. pastoris and Zheng58 were used as the positive and negative controls, respectively. Proteins from the stem, root and leaf of a transgenic plant of generation BC1 were extracted in the same way and used for tissue specificity analysis.

Proteins were separated on SDS–PAGE (12% acrylamide, 0.4% acryl-bisacrylamide). and transferred onto PVDF membrane (Pall, Port Washington, NY). The polyclonal antibody raised in rabbits was added into the membrane confining liquid for prehybridization. The goat anti-rabbit IgG labeled with alkaline phosphatase was used as the secondary antibody. BCIP/NBT kit (Zomanbio, Beijing, China) was used for color development. To identify the proteins, bands were excised from the gel and analyzed using matrix assisted laser desorption/ionization time of flight mass spectrometry (MALDI-TOF-MS) at Tianjin Biochip Corporation (Tianjin, China).

β-Glucanase activity assay and enzyme characterization

Crude proteins of five randomly selected seeds were extracted with extraction buffer as described above, and the supernatant was subject to β-glucanase activity assay. β-Glucanase activity was determined by measuring the amount of reducing sugar released from lichenan with the method of 3,5-dinitrosalicylic acid (DNS) [11,20]. One unit of enzyme activity was defined as the amount of enzyme required to release 1 μmol of reducing sugar per minute from 1.0% lichenan in citric acid-Na$_2$HPO$_4$ (50 mM, pH 3.5) at 60°C for 10 min. β-Glucanase activities of generations T1, BC1, BC2 and BC3 of transgenic maize and Zheng58 were all evaluated. Each reaction and its control were run in triplicate. The enzyme properties of Bgl7AM derived from maize was determined using crude proteins from BC1 seeds as in Luo et al. (2010). The pH optimum of the protein was determined at 60°C and pH 1.0–6.0. The pH stability was determined by measuring the residual activity under standard conditions (pH 5.0, 60°C and 10 min) after pre-incubation at 37°C and pH 1.0–9.0 for 1 h. The optimal temperature was determined at 25–80°C at pH 5.0. Thermal stability of the enzyme was determined by assessing the residual enzyme activity under standard conditions after incubation of the enzyme at 70°C for various durations.

Evaluation of anti-inactivation stability over feed pelleting process

Feed pelleting was carried out with a twin-screw extruder (DSE-25 Extruder Lab-Station Brabender OHG, Duisburg, Germany). Part of the maize seeds were mixed and extruded at 70°C or 80°C, respectively. β-Glucanase activities and dry matter content (DM) values were determined before and after pelleting. Zheng58 seeds were treated as the non-transgenic control. Stability comparison was conducted with the β-glucanases derived from transgenic maize seeds and P. pastoris. Crude Bgl7A derived from P. pastoris with equal enzyme activity to transgenic maize was added into Zheng58 seeds, followed by pelleting treatment as described above. And the β-glucanase activity was detected after pelleting. One-way analysis of variance (ANOVA) was performed using the Duncan's multiple-range test to compare treatment means. Significance was defined at P<0.05.

Results

Construction and transformation of transgenic vector pHP20754-bgl7Am

The CAI value and GC content of *bgl7A* were 0.715 and 49.6%, respectively. After codon optimization and gene modification, the CAI value and GC content of *bgl7Am* was increased to 0.937 and 67.0%, respectively (Figure S1 and S2). These higher values are better for exogenous gene expression in maize. Furthermore, effective Nc, negative CIS elements, negative repeat elements, and mRNA structure of the target gene were also considered in gene modification (Figure S1, S2 and S3). As a result, native *bgl7A* and synthetic *bgl7Am* shared 82.2% nucleotide sequence identity but encoded identical amino acid sequences.

The 1221-bp *bgl7Am* was inserted into the expression vector pHP20754 between the embryo-specific ZM-leg1A promoter and ZM-leg1 terminator (Fig. 1b), which is a transcriptionally active spacer region that allows highly efficient transgene expression. The positive calli of maize Hi-II were regenerated on bialaphos medium and identified by PCR.

Plant regeneration and phenotypic evaluation

The regenerated young plants described above showed good growth in the greenhouse. A total of 27 independent transgenic lines were obtained. Based on β-glucanase activities of the seeds, 330 seeds of three independent transgenic events (40, 46 and 51-1) were selected to cultivate in fields and backcross with Zheng58 for progeny production. As shown in Fig. 1c, the ears and seeds of generation T1 showed significant phenotypic difference from Zheng58. This difference was generally subsided in the later generations because of the successive backcrossing with Zheng58. Up to transgenic generation BC3, the traits of transgenic maize were almost the same as that of non-transgenic Zheng58 through visual observation. The result suggests that the inserted exogenous gene has no negative impact on the maize seed.

Determination of exogenous gene integration

PCR assay with primers specific for *bgl7Am* was used to evaluate the inheritance of transgenic maize. Gene fragments of about 500 bp were detected in the transformation events 40, 46 and 51-1 (Fig. 2a). PCR results of *actin* gene (~300 bp) indicated the high quality of genomic DNA (Fig. 2b). To confirm the gene integration and the copy number of *bgl7Am* in transgenic plants, the genomic DNAs of three positive transgenic plants of event 40 were analyzed by southern blot after restriction digest with *Eco*RI and *Hind*III. The *bgl7Am* probe was prepared with a 800-bp fragment of the *bgl7Am* gene. There is only one *Eco*RI restriction site located between the promoter and the *bgl7Am* gene in the expression cassette of

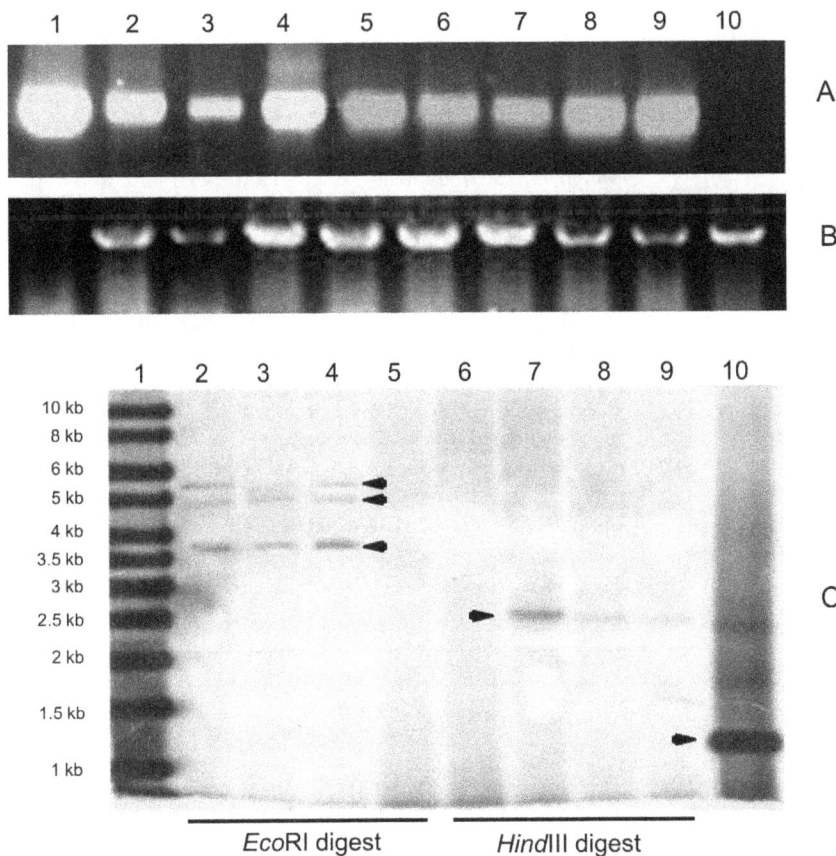

Figure 2. PCR analysis and southern blot analysis of the transgenic maize. A PCR detection of the gene *bgl7Am* in the genomic DNA of transgenic plant leaves of generation BC1. Lane 1, the plasmid PHP20754-*bgl7Am* as positive control; lane 2–9, the transgenic plants; lane 10, the non-transgenic Zheng58. **B** PCR detection of the gene *actin*. Lane 1, the plasmid PHP20754-*bgl7Am*; lane 2–9, the transgenic plants; lane 10, the non-transgenic Zheng58. **C** Southern blot analysis of *bgl7Am* in transgenic plants. The *Eco*RI and *Hind*III-digested genomic DNA was hybridized with the *bgl7Am* probe. Lane 1, the DIG-labeled molecular weight markers; lane 2–4, transgenic plants digested by *Eco*RI (arrowhead indicate the positive bands); lane 5, non-transgenic Zheng58 digested by *Eco*RI; lane 6, non-transgenic Zheng58 digested by *Hind*III; lane 7–9, transgenic plants digested by *Hind*III (arrowhead indicate the positive band); lane 10, PCR fragment of the *bgl7Am* as a positive control (arrowhead).

pHP20754-*bgl7Am*. A total of three bands of ~3.5, 5.0 and 6.0 kb, respectively, were detected in the positive lane via *Eco*RI digest, but not in non-transgenic Zheng58 (Fig. 2c). There is no *Hind*III site in the *bgl7Am*. While *Hind*III cut the gene expression cassette twice and released an internal fragment of 2.5 kb (Fig. 2c). These results indicate that there are three copies of *bgl7Am* in event 40.

Evaluation of site-specific expression

To determine the expression efficiency of exogenous Bgl7AM, proteins were extracted from two BC1 plants of event 40 that had high β-glucanase activities. In western blot analysis, no band was detected in the negative control of Zheng58 (Fig. 3a). The positive control, Bgl7A expressed by *P. pastoris*, showed a band of about 60 kDa, the same as that reported before [11]. One main band of ~60 kDa was identified on the PVDF membrane after hybridization with the antibody (Fig. 3a). This molecular weight (60 kDa) was much higher than the predicted molecular weight (45.3 kDa). After Endo H treatment, the band had no significant reduction in molecular weight (Fig. 3a). It suggested that other post-translation modifications rather than *N*-glycosylation, such as *O*-linked glycosylation, phosphorylation, acetylation or methylation, may occur in the transgenic maize. The band was verified to be Bgl7AM through MALDI-TOF-MS analysis (Figure S4). Except for the seeds, proteins extracted from the root, stem and leaf of the positive lines had no objective band (Fig. 3b), indicating the tissue specificity of Bgl7AM. Moreover, Bgl7AM present in seeds are more convenient for storage, transportation and direct utilization.

Evaluation of seed-derived β-glucanase activity

Positive transgenic plants of transgenic event 40 were selected for β-glucanase activity assay. Approximately 200–400 seeds of each generation were assessed using the DNS method (Table 1). Compared with the non-transgenic Zheng58 that had β-glucanase activity of 3300 U/kg of seeds, T1 seeds (207,800 U/kg) showed approximately 63-fold activities of Zheng58. Both the maximal and average activities of BC1, BC2 and BC3 seeds were increased slightly. The maximal β-glucanase activity of BC3 seeds was up to 779,800 U/kg, which was 236 folds of that of Zheng58. About 47% of the seeds showed over 200,000 U/kg of β-glucanase activity. The result further confirmed that *bgl7Am* is genetically stable over generations in maize.

Characterization of maize seed-derived Bgl7AM

The crude proteins of transgenic BC1 seeds were characterized (Fig. 4), and compared with Bgl7A of *P. pastoris* reported before

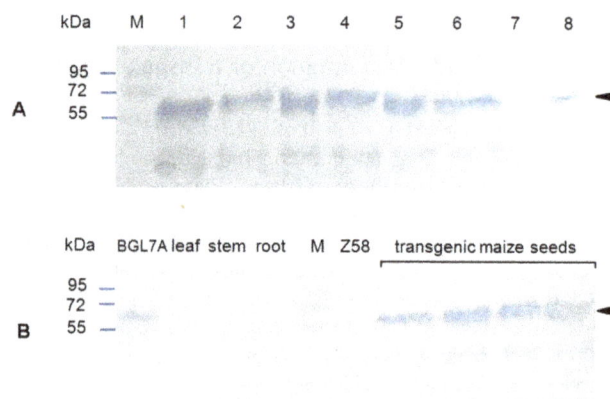

Figure 3. Western blot analysis of recombinant Bgl7AM from the transgenic maize. A Western blot analysis of the Bgl7AM from transgenic maize seeds. Lane M, the protein molecular markers; Lane 1, 3, 5, the proteins isolated from transgenic maize seeds; lane 2, 4, 6, the proteins isolated from transgenic maize seeds and treated with Endo H; lane 7, the non-transgenic Zheng58 as a negative control; lane 8, the purified *P. pastoris* Bgl7A as a positive control. **B** Western blot analysis of the Bgl7AM from different tissues of the transgenic maize (leaf, stem, root and seeds). Lane M, the protein molecular markers; purified *P. pastoris* Bgl7A as a positive control; Z58 refers to non-transgenic Zheng58 (negative control).

[11]. Bgl7AM had a pH optimum at 4.0, while Bgl7A exhibited high activity at pH 1.5, 3.5 and 5.0 (maxima). Both enzymes remained active at pH 1.0–8.0. The temperature optimum of Bgl7AM was 70°C, which was 10°C higher than that of Bgl7A. Moreover, thermostability of Bgl7AM was improved. After incubation at 70°C for 15 min, Bgl7AM retained 50% of the initial activity while Bgl7A remained less than 30%.

Evaluation of anti-inactivation stability in feed pelleting

The β-glucanase activities of Bgl7A (from *P. pastoris*) and Bgl7AM (from transgenic maize seeds) were determined after feed pelleting at 70°C and 80°C, respectively (Table 2). The initial β-glucanase activities in transgenic line or in Zheng 58 by supplementation of Bgl7A were set to 77,860 U/kg and 85,350 U/kg, respectively. After pelleting at each of the tested temperatures, Bgl7A lost more activities than Bgl7AM. In combination with the data of enzyme characterization, Bgl7AM

Table 1. β-Glucanase activities of the transgenic maize seeds within four generations.

Generation	Number of seeds with β-glucanase activity (U/kg)[*]					Maximal activity (U/kg)	Average activity (U/kg)[**]
	>200,000	100,000–200,000	50,000–100,000	10,000–50,000	<10,000		
T1	2	3	4	90	149	807,800	18,650±1,108[b]
BC1	167	77	11	48	297	734,100	112,700±10,33[c]
BC2	55	14	0	33	317	689,500	167,400±12,54[d]
BC3	103	7	1	11	98	779,800	239,300±8,646[e]
Zheng58	0	0	0	0	13	3,300	780±235[a]

[*]Five kernels from each ear were randomly selected, pooled, and glucanase activity assayed. One unit of enzyme activity was defined as the amount of enzyme required to release 1 μmol of reducing sugar per minute from 1.0% lichenan at 60°C for 10 min. U/kg, glucanase units per kilogram of seed.
[**]The values were means of three replicates±standard deviation.
[a,b,c,d,e]Means in the same column not sharing a common superscript are significantly different (P<0.05).

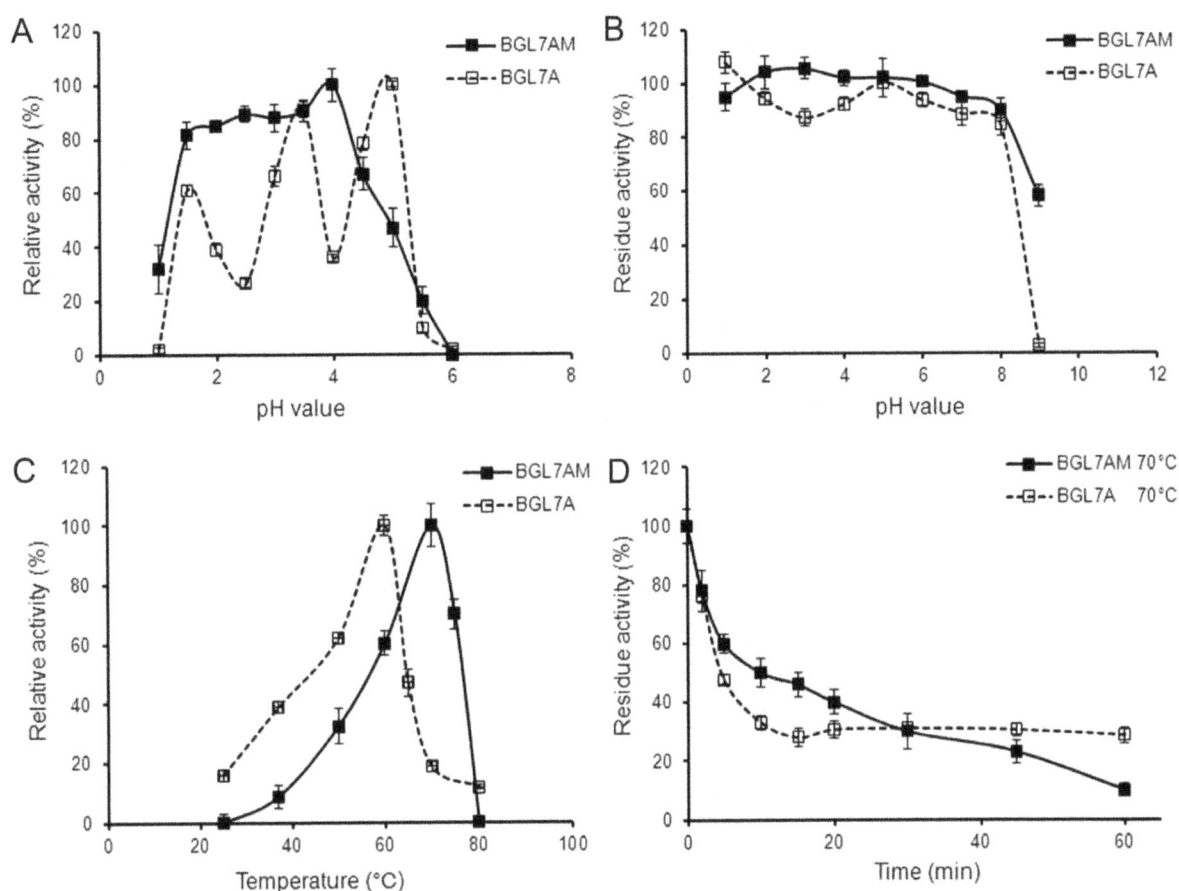

Figure 4. Property comparison of recombinant endo-β-1,3-1,4-glucanase expressed in maize (BGL7AM) and *P. pastoris* (BGL7A). A Effect of pH on β-glucanase activity of BGL7AM and BGL7A at 60°C. **B** pH stability of BGL7AM and BGL7A. After incubation at 37°C for 1 h in buffers ranging from pH 1.0 to 9.0, the β-glucanase activity was assayed in 100 mM citric acid-Na_2HPO_4 (pH 5.0) at 60°C. **C** Temperature-dependent activity profiles of BGL7AM and BGL7A in 100 mM citric acid-Na_2HPO_4 (pH 5.0). **D** Thermostability of BGL7AM and BGL7A pre-incubated at 70°C at pH 5.0. The aliquots were removed at different time points then measure residual β-glucanase activity at 60°C and pH 5.0. Error bars represent the standard deviation of triplicate measurements.

Table 2. Stability of of Bgl7A and Bgl7AM during feed pelleting[*].

	β-Glucanase activity (U/kg seeds)[**]		Activity loss (%)
	Before pelleting	After pelleting	
70°C			
Bgl7A	85.35±4.43	52.48±3.66	38.51±3.12[b]
Bgl7AM	77.86±6.30	56.82±2.31	27.02±4.83[a]
80°C			
Bgl7A	85.35±3.45	44.21±2.14	48.20±4.31[c]
Bgl7AM	77.86±5.36	49.61±3.22	36.28±3.32[b]

*Bgl7A was the recombinant protein expressed in *P. pastoris*; Bgl7AM was the recombinant protein expressed in transgenic maize seeds. The amino acid sequences of Bgl7A and Bgl7AM are totally identical.
**The values were means of three replicates±standard deviation.
a,b,cMeans in the same column not sharing a common superscript are significantly different (P<0.05).

was more excellent than Bgl7A even though they had complete identical amino acid sequences.

Discussion

So far several β-glucanase genes have been introduced into plants for different purposes. Endo-β-1,3-glucanase (laminarinase) can defend plants against fungal pathogens, introduction of its coding genes into crops is a plausible strategy to develop durable resistance against fungal pathogens [21]. Over the last two decades, transgenic plants harboring endo-β-1,4-glucanase (cellulase) genes have taken more attention for conversion of cellulosic biomass into fermentable sugars [22,23]. Production of recombinant endo-β-1,4-glucanases E1 in transgenic plants have been reported in *Arabidopsis* [24], leaf and root tissues of maize [25,26] and rice seeds [27]. Endo-β-1,3-1,4-glucanase (lichenase) is an important enzyme additive in monogastric animal feed to decompose β-glucan in cereals [3,6,28]. Up to now, β-1,3-1,4-glucanases have been expressed in transgenic barley [29,30] and potato [31] for feed purpose. However, maize seed has never been used for production of endo-β-1,3-1,4-glucanase. Here we developed a transgenic maize line that overexpressed an endo-β-1,3-1,4-glucanase from *Bispora* sp. MEY-1 in seeds. Compared with enzyme production by microbial fermentation and other transgenic crops, transgenic maize seed has several advantages, such as low cost production, cultivation worldwide and direct utilization in animal feed. On the other hand, the genetic manipulation of maize is more easily. For feed industry's interest, maize seed as the major feed ingredient represents an ideal bioreactor to produce feed enzymes.

About 65% of the maize seed produced in China is used as feed. If maize seeds express sufficient endo-β-1,3-1,4-glucanase, no supplementation of microbial glucanase will be required. To achieve high-level expression of *bgl7Am* in maize seed, several strategies have been utilized in combination, including a synthetic gene with preferred maize codons [15], a strong tissue-specific promoter, and an excellent transformation receptor with high competence and regeneration capacity that improves the transformation efficiency [13,14]. As a result, the average and maximum glucanase activities in maize seeds without purification and enrichment were up to 239,300 and 779,800 U/kg seeds, respectively (Table 1). Previous feeding trials have shown that the effectiveness of glucanase as a feed additive was maximized at approximately 30,000 units per kg of diets [5]. Typically maize grains constitute 50% of the animal diet, thus the transgenic maize seeds having an average glucanase activity of 239,300 U/kg is high enough to substitute the glucanase supplement. When the transgenic maize line developed in this study is propagated in field, it will enhance the nutritive values of glucan-abundant grains such as wheat and barley. The development of transgenic maize will not only reduce the loss of resources and simplify the production process, but also provide an environmental friendly approach to produce enzymes.

Moreover, Bgl7AM has good thermostability and excellent acidic stability, which are important factors for supplementation to animal feed. Thermostability is a key index of feed enzyme because of the high temperature during feed processing. Since most β-1,3-1,4-glucanases are not stable during coating of feed pellets (70–90°C), selection of a thermostable β-1,3-1,4-glucanase with high activity is of great interest to the animal feedstuff industries [8,32,33]. Bgl7AM retains most activities after pelleting at 80°C. This thermostability allows it to survive the heat generated from maize pressing into feed pellets and pasteurization. Similar results that plant-derived enzymes showed better stability have been reported. [34,35]. This phenomenon might be ascribed to the different folding patterns and disulphide bond formations in microbes and plants [35]. Protection from maize seed starch might be the other cause.

Furthermore, feed enzyme should be stable within the acid environment of monogastric animals' digestive tract, in which the pH value is lower than pH 3.0 in stomach (pH 1.3–3.5 for pigs and pH 2.8–4.8 for chickens) [36]. An acidic-tolerable β-glucanase

has been isolated from *Trichoderma koningii* ZJU-T, with optimal activity at pH 2.0 [33]. Molecular modification approaches have been employed to enhance the activity of a β-1,3-1,4-glucanase at acidic pH [37]. Compared with counterparts, Bgl7AM is highly active and stable within pH 1.0–4.0, and retains above 90% of its activity at pH 3.0, the average pH in the animal digestive tract.

This study provides an environment-friendly and low-cost approach to produce transgenic maize with social and ecological significance. It's the first report that produces a biologically active endo-β-1,3-1,4-glucanase in transgenic maize seeds. Approaches to increase the seed glucanase activities are preceding, including selection of more transgenic events and application of stronger promoters. In the future studies, we'll evaluate its direct application effectiveness in animal feed by comparison with traditional feed supplemented with glucanases.

Supporting Information

Figure S1   Codon related parameters of wild-type gene *bgl7A* and optimized *bgl7Am*. A Codon adaptation index (CAI), negative CIS elements, and negative repeat elements of the *bgl7A* and *bgl7Am*. **B** effective number of codons (Nc) of the *bgl7A* and *bgl7Am*.

Figure S2   Codon usage and GC content of wild-type gene *bgl7A* and optimized *bgl7Am*. A Relative codon frequency of *bgl7A*. **B** Relative codon frequency of *bgl7Am*. **C** GC content and distribution of *bgl7A*. **D** GC content and distribution of *bgl7Am*. **E** Percentage of high frequency used codons of maize in *bgl7A*. **F** Percentage of high frequency used codons of maize in *bgl7Am*.

Figure S3   mRNA structure prediction of wild-type gene *bgl7A* and optimized *bgl7Am*.

Figure S4   MALDI-TOF-MS analysis of the BGL7AM from transgenic maize seeds. A Peptide fragments produced by digestion with protease. **B** Analysis of the identified sequence by Mascot.

Author Contributions

Conceived and designed the experiments: PLY BY WZ. Performed the experiments: XLX YHZ QCM. Analyzed the data: RMC JHY. Contributed reagents/materials/analysis tools: XJZ KM HYL. Wrote the paper: YHZ.

References

1. Buliga GS, Brant DA, Fincher GB (1986) The sequence statistics and solution conformation of a barley (1,3-1,4)-β-D-glucan. Carbohyd Res 157: 139–156.

2. Almirall M, Francesch M, Perez-Vendrell AM, Brufau J, Esteve-Garcia E (1995) The differences in intestinal viscosity produced by barley and β-glucanase alter digesta enzyme activities and ileal nutrient digestibilities more in broiler chicks than in cocks. J Nutr 125: 947–955.

3. Choct M (2006) Enzymes for the feed industry: past, present and future. World Poultry Sci J 62: 5–16.

4. McCarthy T, Hanniffy O, Savage AV, Tuohy MG (2003) Catalytic properties and mode of action of three endo-β-glucanases from *Talaromyces emersonii* on soluble β-1,4- and β-1,3-1,4-linked glucans. Int J Biol Macromol 33: 141–148.

5. Mathlouthi N, Mallet S, Saulnier L, Quemener B, Larbier M (2002) Effects of xylanase and β-glucanase addition on performance, nutrient digestibility, and physico-chemical conditions in the small intestine contents and caecal microflora of broiler chickens fed a wheat and barley-based diet. Anim Res 51: 395–406.

6. Kiarie E, Owusu-Asiedu A, Peron A, Simmins PH, Nyachoti CM (2012) Efficacy of xylanase and β-glucanase blend in mixed grains and grain co-products-based diets for fattening pigs. Livest Sci 148: 129–133.

7. Jozefiak D, Rutkowski A, Kaczmarek S, Jensen BB, Engberg RM, et al. (2010) Effect of β-glucanase and xylanase supplementation of barley- and rye-based diets on caecal microbiota of broiler chickens. Brit Poultry Sci 51: 546–557.

8. Ribeiro T, Lordelo MMS, Prates JAM, Falcao L, Freire JPB, et al. (2012) The thermostable β-1,3-1,4-glucanase from *Clostridium thermocellum* improves the nutritive value of highly viscous barley-based diets for broilers. Brit Poultry Sci 53: 224–234.

9. Flickinger EA, Wolf BW, Garleb KA, Chow J, Leyer GJ, et al. (2000) Glucose-based oligosaccharides exhibit different in vitro fermentation patterns and affect in vivo apparent nutrient digestibility and microbial populations in dogs. J Nutr 130: 1267–1273.

10. Grishutin S, Gusakov A, Dzedzyulya E, Sinitsyn A (2006) A lichenase-like family 12 endo-(1, 4)-β-glucanase from *Aspergillus japonicus*: study of the substrate

specificity and mode of action on β-glucans in comparison with other glycoside hydrolases. Carbohyd Res 341: 218–229.

11. Luo HY, Yang J, Yang PL, Li J, Huang HQ, et al. (2010) Gene cloning and expression of a new acidic family 7 endo-β-1,3-1,4-glucanase from the acidophilic fungus *Bispora* sp. MEY-1. Appl Microbiol Biot 85: 1015–1023.

12. James C (2011) Global status of commercialized biotech/GM crops: 2011. ISAAA Brief No. 43. Ithaca, NY.

13. Chen RM, Xue GX, Chen P, Yao B, Yang WZ, et al. (2008) Transgenic maize plants expressing a fungal phytase gene. Transgenic Res 17: 633–643.

14. Armstrong C, Green C, Phillips R (1991) Development and availability of germplasm with high type II culture formation response. Maize Genet Coop Newslett 65: 92–93.

15. Liu HM, He R, Zhang HY, Huang YB, Tian ML, et al. (2010) Analysis of synonymous codon usage in *Zea mays*. Mol Biol Rep 37: 677–684.

16. Holwerda BC, Padgett HS, Rogers JC (1992) Proaleurain vacuolar targeting is mediated by short contiguous peptide interactions. Plant Cell 4: 307–318.

17. Woo YM, Hu DWN, Larkins BA, Jung R (2001) Genomics analysis of genes expressed in maize endosperm identifies novel seed proteins and clarifies patterns of zein gene expression. Plant Cell 13: 2297–2317.

18. Yamagata T, Kato H, Kuroda S, Abe S, Davies E (2003) Uncleaved legumin in developing maize endosperm: identification, accumulation and putative subcellular localization. J Exp Bot 54: 913–922.

19. Tomes D (1995) Direct DNA transfer into plant cell via microprojectile bombardment. In: Gamborg O, Philipps G, editors. Plant cell tissue and organ culture: fundamental methods. Berlin: Springer-Verlag. pp. 197–213.

20. Miller GL (1959) Use of dinitrosalicylic acid reagent for determination of reducing sugar. Anal Chem 31: 426–428.

21. Sridevi G, Parameswari C, Sabapathi N, Raghupathy V, Veluthambi K (2008) Combined expression of chitinase and β-1,3-glucanase genes in indica rice (*Oryza sativa* L.) enhances resistance against *Rhizoctonia solani*. Plant Sci 175: 283–290.

22. Taylor LE, Dai Z, Decker SR, Brunecky R, Adney WS, et al. (2008) Heterologous expression of glycosyl hydrolases in planta: a new departure for biofuels. Trends Biotechnol 26: 413–424.

23. Venkatesh B, Dale B, Ahmad R, Ransom C, Oehmke J, et al. (2007) Enhanced conversion of plant biomass into glucose using transgenic rice-produced endoglucanase for cellulosic ethanol. Transgenic Res 16: 739–749.

24. Ziegler MT, Thomas SR, Danna KJ (2000) Accumulation of a thermostable endo-1,4-β-D-glucanase in the apoplast of *Arabidopsis thaliana* leaves. Mol Breeding 6: 37–46.

25. Biswas GCG, Ransom C, Sticklen M (2006) Expression of biologically active *Acidothermus cellulolyticus* endoglucanase in transgenic maize plants. Plant Sci 171: 617–623.

26. Ransom C, Balan V, Biswas G, Dale B, Crockett E, et al. (2007) Heterologous *Acidothermus cellulolyticus* 1,4-β-endoglucanase E1 produced within the corn biomass converts corn stover into glucose. Appl Biochem Biotech 136-140: 207–219.

27. Zhang Q, Zhang W, Lin CY, Xu XL, Shen ZC (2012) Expression of an *Acidothermus cellulolyticus* endoglucanase in transgenic rice seeds. Protein Expres Purif 82: 279–283.

28. von Wettstein D, Mikhaylenko G, Froseth JA, Kannangara CG (2000) Improved barley broiler feed with transgenic malt containing heat-stable (1,3-1,4)-β-glucanase. Proc Natl Acad Sci 97: 13512–13517.

29. Horvath H, Huang J, Wong O, Kohl E, Okita T, et al. (2000) The production of recombinant proteins in transgenic barley grains. Proc Natl Acad Sci 97: 1914–1919.

30. Jensen LG, Olsen O, Kops O, Wolf N, Thomsen KK, et al. (1996) Transgenic barley expressing a protein-engineered, thermostable (1,3-1,4)-β-glucanase during germination. Proc Natl Acad Sci 93: 3487–3491.

31. Armstrong JD, Inglis GD, Kawchuk LM, McAllister TA, Leggett F, et al. (2002) Expression of a *Fibrobacter succinogenes* 1,3-1,4-β-glucanase in potato (*Solanum tuberosum*). Am J Potato Res 79: 39–48.

32. Hua CW, Yan QJ, Jiang ZQ, Li YN, Katrolia P (2010) High-level expression of a specific β-1,3-1,4-glucanase from the thermophilic fungus *Paecilomyces thermophila* in *Pichia pastoris*. Appl Microbiol Biot 88: 509–518.

33. Wang JL, Ruan H, Zhang HF, Zhang Q, Zhang HB, et al. (2007) Characterization of a thermostable and acidic-tolerable β-glucanase from aerobic fungi *Trichoderma koningii* ZJU-T. J Food Sci 72: C452–C456.

34. Agrawal P, Verma D, Daniell H (2011) Expression of *Trichoderma reesei* β-mannanase in tobacco chloroplasts and its utilization in lignocellulosic woody biomass hydrolysis. PloS one 6: e29302.

35. Verma D, Kanagaraj A, Jin S, Singh ND, Kolattukudy PE, et al. (2010) Chloroplast-derived enzyme cocktails hydrolyse lignocellulosic biomass and release fermentable sugars. Plant Biotechnol J 8: 332–350.

36. Deng F, Wang J, Pu D, Liu K, Zhou L, et al. (2009) Advance in assessment of direct-fed microorganism. Chin Agric Sci Bull 25: 7–12.

37. Jia HY, Li YN, Liu YC, Yan QJ, Yang SQ, et al. (2012) Engineering a thermostable β-1,3-1,4-glucanase from *Paecilomyces thermophila* to improve catalytic efficiency at acidic pH. J Biotechnol 159: 50–55.

15

Galantamine Slows Down Plaque Formation and Behavioral Decline in the 5XFAD Mouse Model of Alzheimer's Disease

Soumee Bhattacharya[1], Christin Haertel[1], Alfred Maelicke[2], Dirk Montag[1]*

1 Neurogenetics Special Laboratory, Leibniz Institute for Neurobiology, Magdeburg, Germany, 2 Galantos Pharma GmbH, Nieder-Olm, Germany

Abstract

The plant alkaloid galantamine is an established symptomatic drug treatment for Alzheimer's disease (AD), providing temporary cognitive and global relief in human patients. In this study, the 5X Familial Alzheimer's Disease (5XFAD) mouse model was used to investigate the effect of chronic galantamine treatment on behavior and amyloid β (Aβ) plaque deposition in the mouse brain. Quantification of plaques in untreated 5XFAD mice showed a gender specific phenotype; the plaque density increased steadily reaching saturation in males after 10 months of age, whereas in females the density further increased until after 14 months of age. Moreover, females consistently displayed a higher plaque density in comparison to males of the same age. Chronic oral treatment with galantamine resulted in improved performance in behavioral tests, such as open field and light-dark avoidance, already at mildly affected stages compared to untreated controls. Treated animals of both sexes showed significantly lower plaque density in the brain, i.e., the entorhinal cortex and hippocampus, gliosis being always positively correlated to plaque load. A high dose treatment with a daily uptake of 26 mg/kg body weight was tolerated well and produced significantly larger positive effects than a lower dose treatment (14 mg/kg body weight) in terms of plaque density and behavior. These results strongly support that galantamine, in addition to improving cognitive and behavioral symptoms in AD, may have disease-modifying and neuroprotective properties, as is indicated by delayed Aβ plaque formation and reduced gliosis.

Editor: Thomas Arendt, University of Leipzig, Germany

Funding: This study was supported by the German Ministry for Education and Research (BMBF) special network program KMU-Innovativ-2(http://www.bmbf.de/en/986.php). The funders had no role in study design, data collection and analysis, decision to publish, or preparation of the manuscript.

Competing Interests: Dr. A. Maelicke is CEO of Galantos Pharma GmbH and Managing Director Europe of Neurodyn Inc. The authors SB, CH, and DM have declared that no competing interests exist.

* E-mail: montag@LIN-magdeburg.de

Introduction

Alzheimer's disease (AD) is a progressive neurodegenerative disorder and the most common cause of old-age dementia. Neuritic plaques containing amyloid β (Aβ) and neurofibrillary tangles composed of hyperphosphorylated Tau protein constitute major neuropathological hallmarks of AD. The amyloid cascade theory provides a rationale for many features of the disease including the pathological markers, the phenotypes caused by autosomal dominant disease genes, and the risk conferred by the APOE gene status [1]. Increased production of certain Aβ species, their aggregation, and deposition as insoluble plaques is regarded as an early and key pathology in the development of AD [2]. Aβ plaques may serve as reservoirs of soluble Aβ oligomers injuring surrounding neurites and synapses [3,4]. At a systemic level, therapeutic strategies to reverse or prevent Aβ deposits could lead to partial functional restoration of neural circuits [5]. Therefore, most AD treatment approaches aim at prevention or reversal of Aβ plaque deposition [6,7].

The acetylcholinesterase inhibitors donepezil, galantamine, and rivastigmine serve as first-line symptomatic drug treatment in mild to moderate Alzheimer's dementia [8]. Whereas donepezil and rivastigmine are designed acetylcholinesterase inhibitors, the plant alkaloid galantamine additionally acts as an allosterically poten-

tiating ligand of nicotinic receptors, increasing their sensitivity to the neurotransmitter acetylcholine [9]. Chronic low-level stimulation of nicotinic receptors might up-regulate their expression [10], slow down neurodegeneration [11], and confer protection against β-amyloid toxicity [12]. Furthermore, galantamine exerts in cell systems neuroprotective effects by anti-apoptotic action [13], by modulating amyloid precursor processing [14], and by inhibiting β-amyloid aggregation and cytotoxicity [15]. Galantamine activates microglia resulting in enhanced Aβ clearance [16]. Long-term galantamine treatment of AD patients slows down cognitive and global decline [17] and reduces behavioral symptoms, most strongly in patients with moderate or advanced forms of the disease [18]. Similar long-term positive effects are also reflected in PET measurements [19].

The 5X Familial Alzheimer's Disease (5XFAD) mouse line co-overexpresses APP with three FAD mutations (K670N/M671L, I716V, and V717I) and PS1 with two FAD mutations (M146L and L286V) under the control of the neuron-specific *thy1* promoter [20]. This model recapitulates a variety of AD features, including working memory impairment, reduced anxiety, extensive extracellular plaque formation beginning at 2 months of age, and selective neuron loss, making it a suitable research model for early-onset AD [20–24].

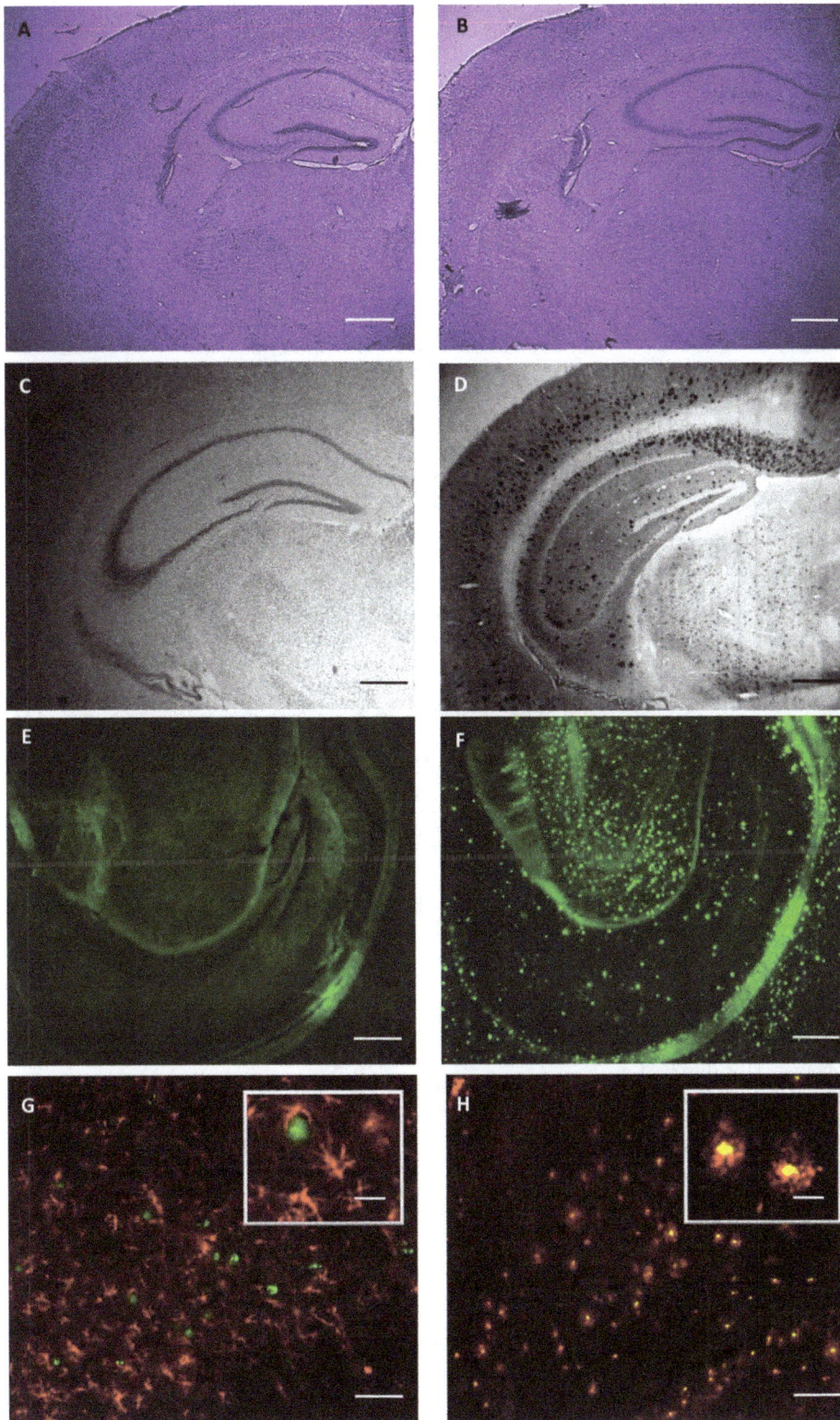

Figure 1. Brain morphology, amyloid plaque formation, gliosis, and microglial activation in 5XFAD transgenic mice. Coronal brain sections from 22-week-old wild-type control littermates (A, C, E) and 5XFAD transgenic mice (B, D, F, G, H) were subjected to Nissl staining (A, B) revealing similar brain morphology. Amyloid β immunohistochemistry (C, D) or thioflavin-S staining (E, F) detect numerous amyloid plaques in 5XFAD mice (D, F), whereas wild-type control brains (C, E) are completely devoid of plaques. GFAP immunohistochemistry (activated astrocytes, red in G) or GSA-lectin (activated microglia, red in H) in combination with thioflavin S staining (green in G and H) revealed plaques surrounded by reactive astrocytes and associated with activated microglia in 5XFAD mice. Scale bars: 500 μm (A–F), 250 μm (G, H), and 100 μm (inset in G and H).

In this study, we used the 5XFAD model to investigate the effects of chronic galantamine treatment on behavior and cognition, formation of β amyloid plaques, and gliosis. Our data show that galantamine slows down plaque deposition and improves behavioral performance.

Materials and Methods

Ethics Statement

Animal experiments were in line with the guidelines for the welfare of experimental animals and approved by the local authorities of Sachsen-Anhalt/Germany (numbers 42502/2-382 and -945) and carried out in accordance with the European Communities Council Directive of 24th November 1986 (86/609/EEC).

Mice

5XFAD   (B6SJL-Tg(APPSwFlLon,P-SEN1*M146L*L286V)6799 Vas/J mice were described by Oakley et al. [20] and were obtained from The Jackson Laboratory (Bar Harbor, stock number 006554). These "5XFAD" transgenic mice overexpress both mutant human APP(695) with the Swedish (K670N, M671L), Florida (I716V), and London (V717I) Familial Alzheimer's Disease (FAD) mutations and human PS1 harboring two FAD mutations, M146L and L286V. Expression of both transgenes is regulated by neural-specific elements of the mouse Thy1 promoter to drive overexpression in the brain. 5XFAD transgenic male mice were crossed with B6SJLF1/J female mice (Jackson Laboratory, stock number 100012). The resulting F2-offspring were used in all experiments. Transgenic mice were identified by PCR according to the supplier's protocol.

Galantamine

Galantamine hydrobromide was obtained from Macfarlan Smith (Edinburgh, UK). The naturally occurring alkaloid was extracted from daffodil bulbs (Narcissus pseudonarcissus) and was isolated and purified as the hydrobromide salt, as described in the related drug master file. Purity was >99%. The molecular formula is $C17H22NO3Br$ and the molecular weight is 368.28.

Galantamine Treatment

10 to 12-week-old mice received galantamineHBr dissolved in drinking water at a concentration of either 12 mg/l during four weeks followed by three weeks with 60 mg/l (low dose), or 36 mg/l during four weeks followed by three weeks with 120 mg/l (high dose). Thereafter, the mice were water deprived overnight and received in the morning drinking water containing 120 mg galantamineHBr/l until the behavioral experiments were terminated and animals were sacrificed for histological examination. Behavioral experiments were conducted after eight weeks of treatment and water deprivation was terminated 30 to 60 min before the behavioral test to ensure a high galantamine concentration during the experiment. Water consumption and body weight were monitored during the application period.

Behavior

For the behavioral analysis, sex- and age-matched littermate wild-type mice were used as controls. During the light phase (12h/12h light-dark cycle), mice were subjected to a series of behavioral tests [25,26] by an experimenter not aware of the genotype. First, general parameters indicative of the health and neurological state were addressed following the neurobehavioral examination described by Whishaw and colleagues [27] and the tests of the primary screen of the SHIRPA protocol except startle response

[28]. Then, the following behavioral paradigms were conducted in sequential order: *Grip strength*. Strength was measured with a high-precision force sensor to evaluate neuromuscular functioning (TSE Systems GmbH, Bad Homburg, Germany). *Rota-rod performance*. Animals received two training sessions (3 h interval) on a rota-rod apparatus (TSE) with increasing speed from 4 to 40 rpm for 5 min. After 4 days, mice were tested at 16, 24, 32, and 40 rpm constant speed. The latency to fall off the rod was measured. *Open field*. Exploration was assessed by placing mice in the middle of a 50×50 cm arena for 15 min. Using the VideoMot 2 system (TSE), tracks were analyzed for path length, visits, walking speed, and relative time spent in the central area (infield), in the area close to the walls (<10 cm, outfield), and in the corners. *O-Maze*. Mice were placed in the center of an open area of an O-maze (San Diego Instruments). Their behavior during 5 min was recorded on videotape. Number of entries into the closed or open areas was counted and the time spent in these compartments was determined using the VideoMot 2 system (TSE). *Light-dark Avoidance*. Anxiety-related behavior was tested by placing mice in a brightly lit compartment (250 lux, 25×25 cm) adjacent to a dark compartment (12.5×25 cm). The number of transitions between the compartments and the time spent within each were analyzed during 10 min. As a test for long-term memory [29], animals were placed at the last day of testing again in the light-dark avoidance box. The latency to enter the dark compartment was measured and compared to the latency at the first time in the box. *Acoustic startle response and prepulse inhibition (PPI)*. A startle stimulus (50 ms, 120 dB) was delivered to the mice in a startle-box system (TSE) with or without preceding prepulse stimulus (30 ms, 100 ms before the startle stimulus) at eight different intensities (73–94 dB, 3 dB increments) on a 70 dB white noise background. After habituation to the box (3 min), 2 startle trials were followed in pseudo-random order by 10 startle trials and 5 trials at each of the prepulse intensities with stochastically varied intertrial intervals (5–30 s). The maximal startle amplitude was measured by a sensor platform.

For conditioned fear testing of 5XFAD mice, the experimental protocol used for the study of Tg2576 mice by Comery *et al.* [30], Jacobsen *et al.* [31], and Schilling et al. [6] was followed closely. Mice were trained and tested on 2 consecutive days. Training consisted of placing the subject in an operant chamber (San Diego Instruments) and allowing exploration for 2 min. Afterwards, an auditory cue was presented for 15 sec followed by a footshock for 2 sec (1.5 mA un-pulsed). This procedure was repeated, and mice were returned to the home cage 30 sec later. 24 hours after training, mice were returned to the same chamber in which training had occurred (context), and freezing behavior (immobility) was recorded. At the end of the 5 min context test, mice were returned to their home cage. One hour later, mice were placed in a novel environment and freezing behavior (immobility) was recorded for 3 min. The auditory cue (CS) was then presented for 3 min and freezing behavior (immobility) was recorded. Freezing scores are expressed as percentage for each portion of the test.

Histology

Animals were anesthetized with CO_2 and perfused intracardially with PBS (10ml/min, pH7.4, 10 min) followed by freshly prepared 4% PFA in PBS (10ml/min, 10 minutes). Brains were post fixed in the same fixative at 4°C overnight, serially infiltrated with 0.5M and 1M saccharose for 24–48 hrs each and frozen in methylbutane at around −70°C. 60 consecutive coronal sections (40 μm wide) from approximately bregma 0 to bregma −3.5mm were obtained per animal. In order to normalize regional bias in

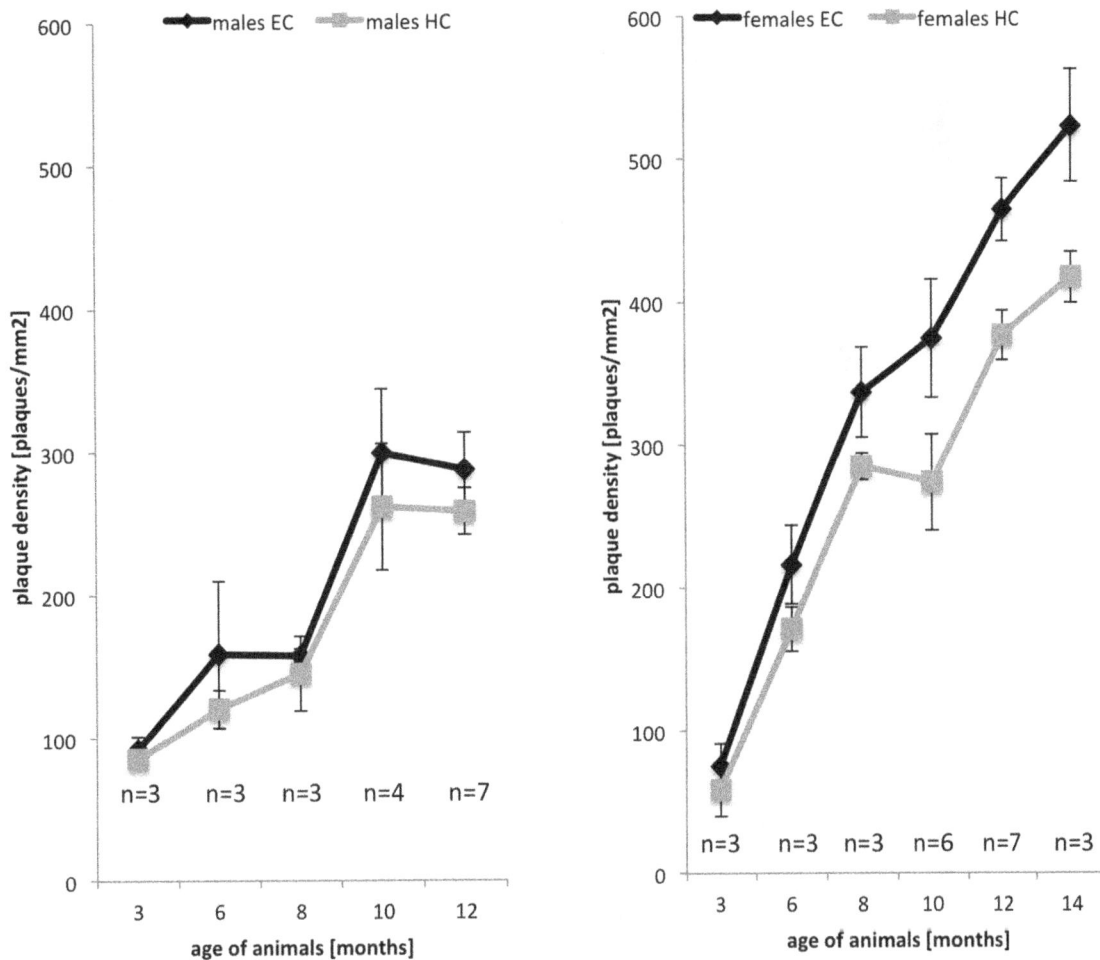

Figure 2. Progression of plaque formation with age in 5XFAD transgenic mice. The plaque density in the hippocampus (HC) and entorhinal cortex (EC) of male and female 5XFAD mice at various ages was determined using thioflavin S staining. In males, plaque density increased in both brain areas and reached saturation at 10 months of age. In female mice, plaque accumulation was faster, reached higher levels, and continued to increase after 14 months of age.

plaque load, 12 sections 200 μm apart from each other were selected from all animals in every experiment.

Nissl staining. Cryosections were subjected to the standard Nissl staining protocol using cresyl violet acetate (Sigma) and viewed using an Axioplan 2 microscope (Zeiss). Images were captured with Photometrix coolSNAP EZ (Visitron systems GmbH) and analyzed with ImageJ software (NIH).

Immunofluorescence. Slide-mounted cryostat sections were blocked with BSA (5% in phosphate-buffered saline) for 1 h and then incubated with primary antibody singly or in combination overnight at 4°C in a humidified chamber. Different combinations of primary antibodies used were; 4G8 (Covance, 1/10000) alone for detection of β amyloid, 4G8 together with GFAP (Sigma Aldrich, 1/10000) for detection of reactive astrocytes, and 4G8 with biotinylated Isolectin GSA (Sigma Aldrich, 10 μg/ml) for detection of microglia. Post incubation, the sections were washed with PBS and probed with the secondary antibody/reagents as required, in a sequential manner. Fluorochromated secondary reagents used were Streptavidin AlexaFluor 488 (Molecular Probes, 1/200), Cy3Goat Anti Rabbit (Abcam, 1/200) and Streptavidin Cy3 (Jackson Laboratories, 1 μg/ml). Sections were examined by Axioplan2 (Carl Zeiss) and images captured by Spot RT-KE (Diagnostic Instruments).

Immunoperoxidase. Free floating sections were collected, treated with a 1:1 solution of PBS and methanol with 1% H_2O_2, blocked with BSA (5% in phosphate-buffered saline) and incubated overnight at 4°C with 4G8 antibody (Covance, 1/10000) in a humidified chamber. The sections were then treated with Vectashield ABC kit (Vector laboratories) followed by the chromogenic substrate DAB (Vector laboratories). Sections were viewed using Axioplan 2 microscope (Carl Zeiss). Images were captured with Photometrix coolSNAP EZ (Visitron systems GmbH) and analyzed with ImageJ software (NIH).

Thioflavin-S staining. An improved thioflavin-S staining protocol [32] was used to ensure reduced photobleaching and tissue damage. Briefly, sections were treated with 0.25% potassium permanganate solution (quenching) at room temperature for 4 minutes followed by 1% sodium borohydride solution for 2–3 minutes. This was followed by incubation with 0.05% thioflavin-S (Sigma Aldrich T1892) solution in 50% ethanol at room temperature for 8 minutes in the dark. Sections were then subjected to 2 washes of 10 seconds each with 80% ethanol, 3 washes of 30 seconds each with water, and post treatment with 5X phosphate buffered saline (pH 7.4) at 4°C in the dark. Sections were mounted in Entellan (Merck) after a brief wash with water and viewed under the FITC filter set of Axioplan2 (Carl Zeiss).

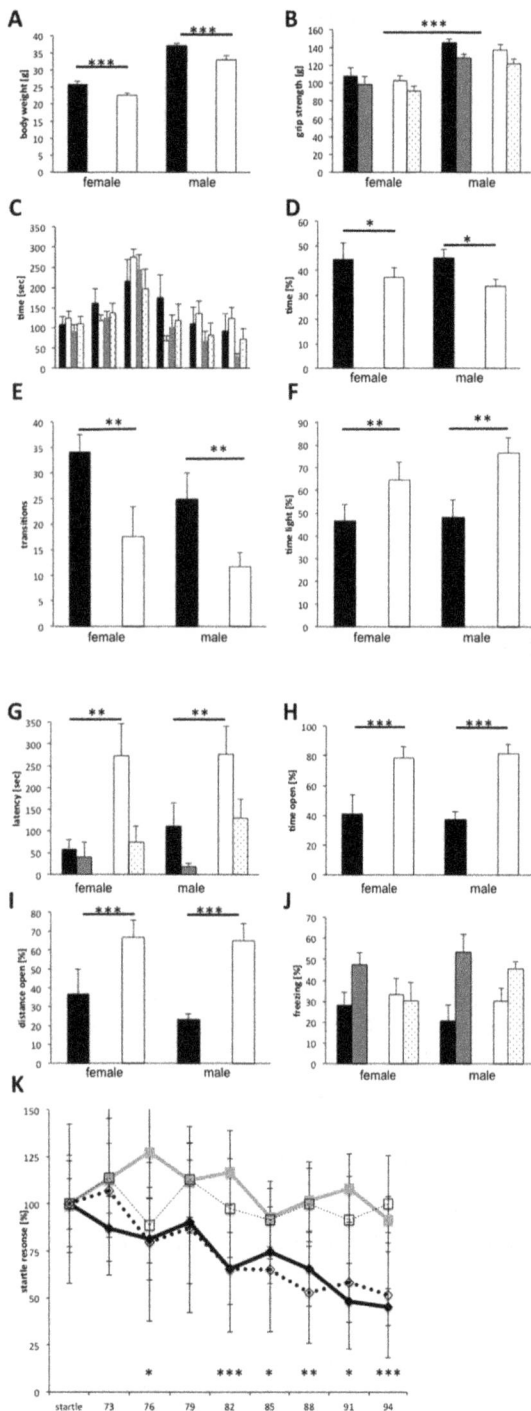

Figure 3. Behavioral analysis of 7-month-old 5XFAD transgenic mice. 5XFAD mice (10 female, 8 male) were analyzed in a series of behavioral tests in comparison to non-transgenic littermates (6 female, 8 male). A significant reduction in body weight (A) of transgenic (white columns) compared to non-transgenic (black columns) mice was observed for both sexes. The grip strength (B) differed between sexes but not between transgenic and non-transgenic mice. Maximum grip strength (black, white columns) and average grip strength (grey, stippled columns) were similar for non-transgenic (black, grey columns) and transgenic mice (white, stippled columns). The rota-rod (C; black control females, grey control males, white transgenic females, dotted transgenic males) did not reveal differences between transgenic and control littermate mice. In the open field (D) transgenic mice (white columns) of both sexes stayed less time in the corners compared to

their wild-type littermates (black columns). In the light-dark avoidance paradigm (E, F, G) transgenic mice (white columns) showed less transitions (E), stayed longer time in the light (F), and had a much greater latency to enter the dark compartment (G) both at the first encounter or tested for memory 3 weeks later (grey columns non-transgenic, stippled columns 5XFAD in G). In the O-Maze (H, I) 5XFAD mice (white columns) spend more time (H) and traveled longer distances (I) in the open areas compared to littermate controls (black columns). Fear conditioning (J) did not differ significantly between 5XFAD (white, stippled columns) and control littermate mice (black, grey columns) with respect to freezing in the same context (black, white columns) or when exposed to the tone in a novel environment (grey, stippled columns). Prepulse inhibition of the startle response (K) was not obtained in 5XFAD mice (squares) in contrast to wild-type littermates (diamonds) in both sexes (males open, females filled symbols).

Quantitative analysis of plaques and astrocytes. For quantification of plaque load, images were captured by a Spot RT-KE camera (Diagnostic Instruments) at a magnification of 2.5X so as to include the entire hippocampus/entorhinal cortex in a single frame. Plaque load was determined by counting thioflavin-S-positive plaques using ImageJ software (NIH) and Adobe Photoshop (CS3 version). Reactive astrocytes in the hippocampus and the entorhinal cortex were identified using GFAP staining and images were captured by a Spot RT-KE camera at a magnification of 10X. Five to seven sections from each animal were analyzed using ImageJ software (NIH) and GFAP positive astrocytes were counted.

Statistical Analysis

Behavioral and imaging data were analyzed using analysis of variance (ANOVA with genotype, sex, and treatment as factors) and *post hoc* analysis using Scheffe's test (STATVIEW Program, SAS Institute Inc., Cary, NC) to determine statistical significance. For the rota-rod, open field, and startle/PPI experiments, statistical analysis was additionally performed using repeated measures ANOVA (with between-subject factor genotype and within-subject factor session). A *P*-value smaller than 0.05 ($p<0.05$) was considered significant. Regression analysis (STATVIEW Program) was used for the correlation of plaque count and number of reactive astrocytes.

Results

Quantitative Analysis of Plaque Deposition in 5XFAD Mice

Early deposition of amyloid plaques is a characteristic feature of the 5XFAD mouse model [20]. To exploit this model for the study of drugs for the treatment of AD, we analyzed quantitatively the progression of plaque deposition with age and characterized the behavior at ages with moderate disease progression. In 22-week-old mice, the general brain morphology of 5XFAD transgenic mice is not affected by plaque deposition as exemplified in the hippocampus and cortex by Nissl staining (Figure 1A, B). However, Aβ deposits are easily identified by immunohistochemistry (Figure 1D) or by thioflavin-S staining (Figure 1F). Furthermore, β amyloid deposits are usually associated with reactive astrocytes identified by immunohistochemical staining for GFAP (Figure 1G) and microglia identified by isolectin GSA staining (Figure 1H). Using thioflavin-S as a marker, we investigated the progression of amyloid β deposition with age in two representative brain areas, the hippocampal formation and the entorhinal cortex (Figure 2). Plaques were detectable already in 2-month-old animals. The quantification revealed a strong increase in plaque

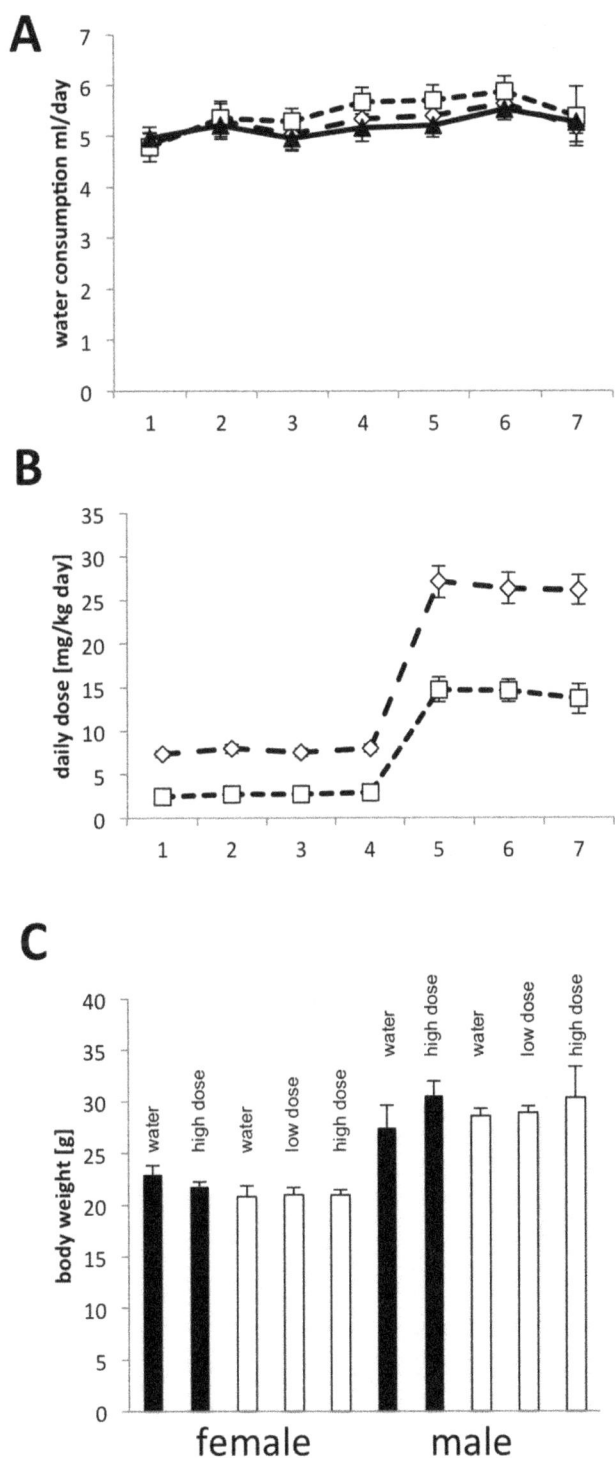

Figure 4. Water consumption, galantamine uptake, and body weight of 5XFAD-transgenic mice treated with galantamine. Water consumption (A) of 5XFAD transgenic was similar irrespective of added galantamine (black triangles: water (n = 16); diamonds: high dose (n = 16), squares: low dose (n = 8). Galantamine uptake (B) was calculated from the amount of drinking water consumed for the low dose (squares, 12 then 60 mg/l) and the high dose (diamonds, 36 then 120 mg/l). The body weight (C) of non-transgenic control (black) or 5XFAD (white) mice was not significantly influenced by the treatment and the behavioral tests.

number and density with age. Maximal density was not reached even at 14 months of age in females, whereas in males no increase in plaque density was observed after 10 months of age. Noteworthy, as also observed for other transgenic Alzheimer's disease models [33–35], we found a strikingly higher number of thioflavin-stained plaques in female compared to male mice. For both sexes, the plaque density was always higher in the entorhinal cortex than in the hippocampus. In control non-transgenic littermates, plaques were never observed at any age. In conclusion, the age before saturation at 10 months appeared to be best suited for the analysis of plaque deposition after a chronic treatment.

Behavioral Analysis of Untreated Medium Age 5XFAD Mice

Therefore, we analyzed whether behavioral deficits are detectable in mice with a moderate plaque load. 7-month-old transgenic (10 female, 8 male) and non-transgenic (6 female, 8 male) littermate mice were subjected to a series of behavioral tests. During the neurological examination, transgenic mice did not display obvious abnormalities with respect to body posture, reflexes (uprighting, eye-blink), or general sensory perception (vision, hearing, touch, pain). In contrast, the body weight of transgenic mice was approximately 10% less than that of their control littermates, both, before and after the behavioral tests (Figure 3A) (2-way ANOVA and post hoc analysis, weight before tests: factor genotype $F_{(1,28)} = 15.912$, p = 0.0004, factor sex $F_{(1,28)} = 136,797$, p<0.0001; Fisher's PLSD <0.0001 for each factor; weight after tests factor genotype $F_{(1,28)} = 24.426$, p< 0.0001, factor sex $F_{(1,28)} = 147.182$, p<0.0001; Fisher's PLSD < 0.0001 for each factor). As expected, the maximum and average grip strength were significantly higher for males compared to females ($F_{(1,28)} = 32.038$, p<0.0001; $F_{(1,28)} = 25.969$, p<0.0001, respectively), but did not differ between transgenic and non-transgenic mice (Figure 3B). Furthermore, motoric abilities examined on the Rota-Rod were similar for control and 5XFAD mice (Figure 3C), indicating that motor coordination and motoric capabilities are not generally impaired in 5XFAD transgenic mice of this age.

In the open field, transgenic mice spend significantly less time in the corners ($F_{(1,28)} = 5.245$, p = 0.0297, post hoc Fisher's PLSD p = 0.0311) compared to control littermates (Figure 3D). In the light-dark avoidance paradigm, 5XFAD transgenic mice made fewer transitions between compartments (Figure 3E) ($F_{(1,28)} = 9.077$, p = 0.0054; post hoc Fisher's PLSD p = 0.0084) and spend more time in the illuminated part (Figure 3F) ($F_{(1,28)} = 9.026$, p = 0.0056; post hoc Fisher's PLSD p = 0.0067). In addition, transgenic mice entered the dark compartment after longer latency (Figure 3G, $F_{(1,28)} = 8.635$, p = 0.0065; post hoc Fisher's PLSD p = 0.0071). The latency with which transgenic mice entered the dark compartment when tested for memory was still longer as of their non-transgenic littermates ($F_{(1,28)} = 4.345$, p = 0.0464; post hoc Fisher's PLSD p = 0.0491). The reduced latency in comparison to the first exposure to the box, however, indicates formation of long-term memories by 5XFAD transgenic mice. In the O-Maze, 5XFAD transgenic mice spent longer time (Figure 3H) and moved longer distances (Figure 3I) in the open areas compared to their control littermates (time $F_{(1,24)} = 23,679$, p<0.0001; distance $F_{(1,24)} = 15.319$, p = 0.0007). In summary, 5XFAD transgenic mice spend in these mazes more time in the open illuminated areas.

When analyzed for the startle response and its prepulse inhibition, 5XFAD transgenic mice displayed a significantly reduced startle response at 120 dB ($F_{(1,28)} = 4.578$, p = 0.0412; post hoc Fisher's PLSD p = 0.0265), which was not inhibited by

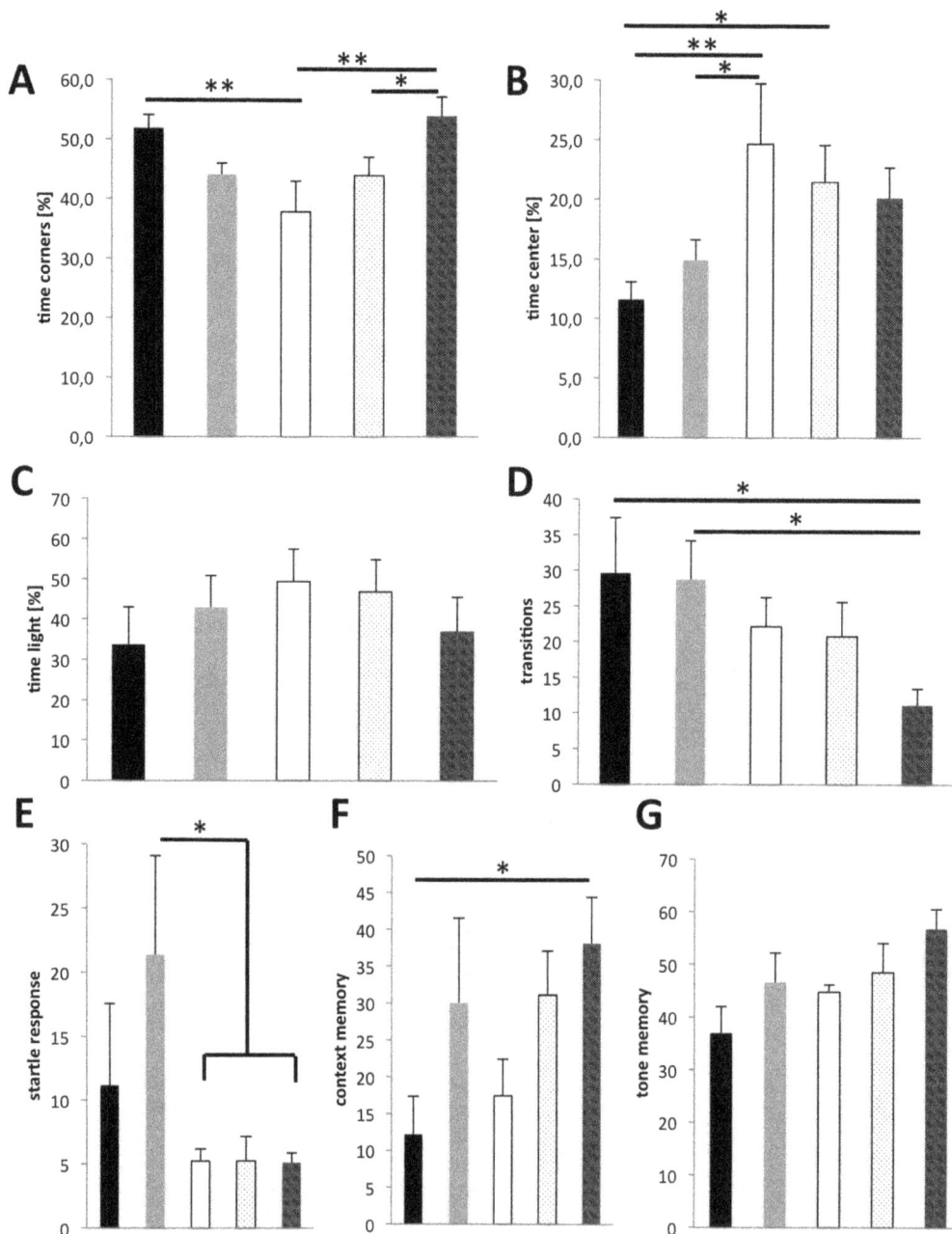

Figure 5. Behavioral analysis of 5XFAD-transgenic mice after treatment with galantamine. In the Open Field (A, B) galantamine treatment restored the preference for the corners (A) and the avoidance of the center (B). Non-transgenic mice receiving water (black columns) showed a significantly higher corner preference (**, p = 0.0045) and avoidance of the center (**, p = 0.0048) compared to untreated 5XFAD transgenic mice (white columns). Treatment of 5XFAD transgenic mice with low dose (stippled columns) or high dose (hatched columns) galantamine increased their corner preference and center avoidance in a dose dependent manner. High dose treated transgenic mice spend significantly (**, p = 0.002) more time in the corners compared to untreated transgenic mice or low dose treated mice (*, p = 0.0451). Non-transgenic mice receiving high dose galantamine (grey columns). In the light-dark avoidance paradigm (C, D), high dose galantamine treatment of 5XFAD transgenic mice (hatched columns) reduced the time in the light (C) to untreated non-transgenic (black columns) control levels. In contrast, the number of transitions (D) is even further reduced by galantamine treatment of 5XFAD transgenic mice (*, p < 0.005). The magnitude of the startle response (E) was not affected by galantamine treatment of 5XFAD transgenic mice. During fear conditioning (F, G), galantamine increased the context memory (F) and the tone memory (G) of non-transgenic and 5XFAD transgenic mice similarly.

prepulses of intensities between 73 and 94 dB (Figure 3K) (repeated measures ANOVA $F_{(1,196)} = 11.488$, p = 0.0021). Fear conditioning was analyzed using a different cohort of mice. 5XFAD transgenic mice (8 males, 10 females) displayed significantly more freezing in the context ($F_{(1,32)} = 4.542$, p = 0.0409) but

also in the neutral surrounding (not significant, p = 0.1641), thus the context memory (% freezing context - % freezing neutral, Figure 3J) was not significantly different from non-transgenic control littermates (9 males, 9 females). Likewise, tone memory (%

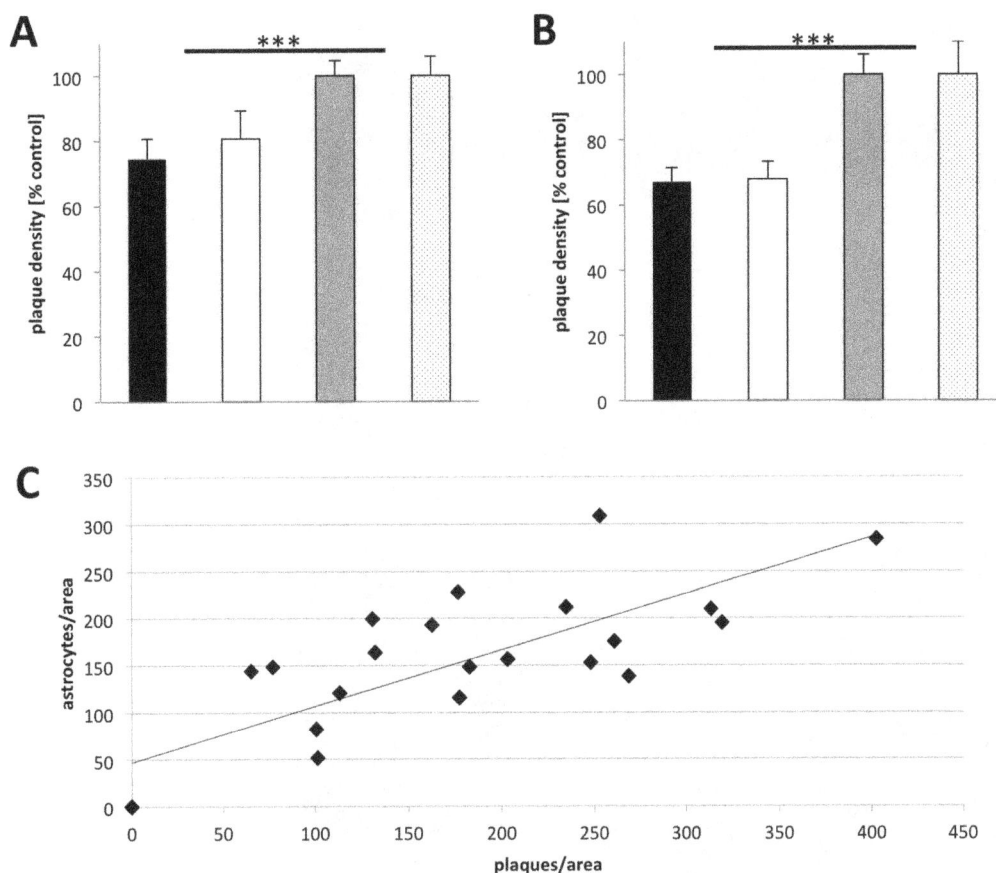

Figure 6. Quantification of plaque density after galantamine treatment. The plaque density in the hippocampus (A) of high dose galantamine treated female (black column, n = 22 sections) and male (white column, n = 47) 5XFAD transgenic mice is significantly reduced in comparison to littermate untreated 5XFAD transgenic mice (female grey column, n = 28; male stippled, n = 50) (***, p<0.0001). The reduction after treatment is even stronger for the entorhinal cortex (B) (treated female black column, n = 27, and male white column, n = 59) (untreated female grey column, n = 45, and male stippled column, n = 59). Astrocyte density and plaque density are strongly correlated in 5XFAD transgenic mice (C) ($R^2 = 0.62$; ANOVA $F_{(1,21)} = 34.232$; p<0.0001).

freezing with tone in neutral - % freezing neutral without tone) was slightly but not significantly (p = 0.09) less in transgenic mice.

In conclusion, we identified several significant AD related behavioral differences between 5XFAD and control mice already at the age of seven months. Therefore, the 5XFAD transgenic model appears to be a suitable system to study the effects of pharmacologically active substances both on behavior and plaque load.

Chronic Galantamine Treatment of 5XFAD Mice

In order to further explore this model for assessing symptomatic and disease-modifying properties of AD drugs, or drug candidates in development, we investigated the effects of chronic treatment with galantamine on behavior and plaque load. Because this model is characterized by early-onset plaque deposition, we chose to treat 10–12-week-old 5XFAD transgenic mice and non-transgenic littermates with galantamine for 2 months and during the following behavioral tests, which were characterized above for the untreated animals. To reduce any potential adverse side effects, we administered during the first 4 weeks a lower dose of galantamine that thereafter was followed by a much higher dose. During the first 4 weeks, one group of animals received 36 mg/l (high dose) and a second group 12 mg/l (low dose) galantamine in the drinking water. Thereafter, the dose in the first group of

animals was increased to 120 mg/l (high dose) and to 60 mg/l (low dose) in the second group. 7 weeks after onset of treatment, the animals were water deprived during the night to ensure a high uptake of drinking water in the morning. During behavioral experiments, animals received 30–60 min prior to testing the drinking solution, so as to ensure a high drug dosage during the experiment. Control animals received water without drug. During the treatment period the general condition, water uptake (Figure 4A), and body weight (Figure 4C) were monitored and did not reveal any difference between treated and control or transgenic and non-transgenic mice indicating that the treatment was well tolerated. This application scheme resulted in the daily uptake of 14 mg/kg body weight (low dose) and 26 mg/kg body weight during the last phase of the treatment (Figure 4B). Consumption of similar amounts of water indicates that galantamine at the concentration used did not induce any preference or avoidance of the drug. Behavioral tests were conducted with 4–5-month-old animals (130–150 days). After the behavioral tests, brains were sectioned for the analysis of plaque density using thioflavin S staining [32,36].

In the open field, mock-treated transgenic and non-transgenic mice showed similar differences as described above for the 7-month-old animals. Transgenic mice spend significantly less time in the corners of the maze, but treatment with galantamine

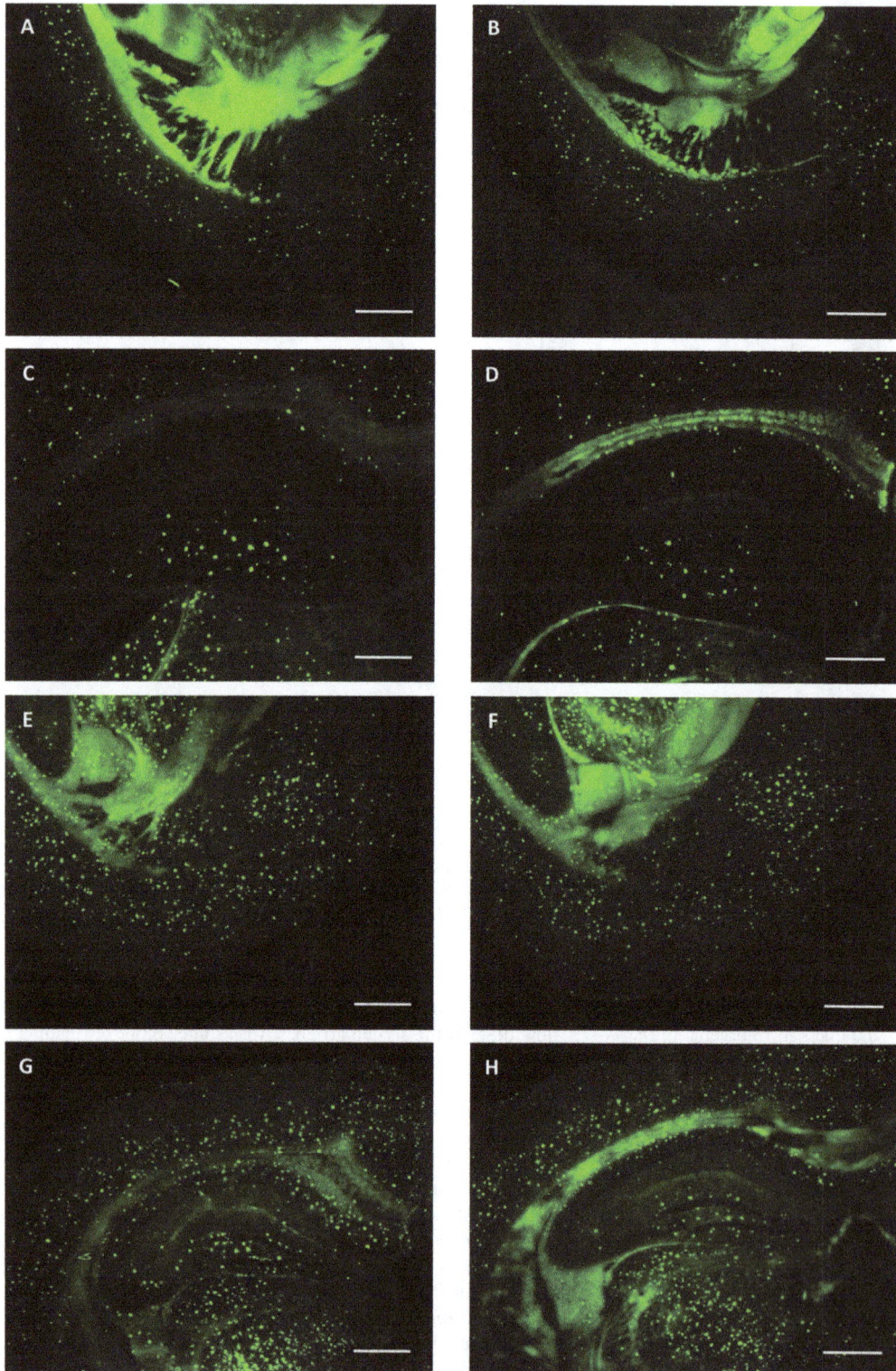

Figure 7. Plaque density after galantamine treatment. Thioflavin S staining of representative coronal brain sections from untreated (A, C, E, G) in comparison to chronically high dose galantamine treated (B, D, F, H) 22-week-old 5XFAD transgenic littermate mice. Fewer plaques are detected in treated animals. Males (A–D) show fewer plaques in comparison to females (E–H), both, in the entorhinal cortex (A, B, E, F) and in the hippocampus (C, D, G, H). Scale bars 500 μm.

elevated the preference for the corners to normal levels in a dose dependent manner (Figure 5A). Similarly, reduced avoidance of the center by transgenic mice was restored by galantamine

treatment in a dose dependent fashion (Figure 5B). Similar to the results with the older mice described above for the Light-Dark-Avoidance paradigm, mock-treated transgenic mice showed less transitions and longer presence in the illuminated compartment. Treatment with galantamine reduced the time spent in the light to normal levels (Figure 5C), but reduced the number of transitions even more (Figure 5D). The latency of transgenic mice compared to non-transgenic mice was higher on the first exposure and not significantly altered by galantamine treatment (data not shown). The latency at the second encounter with the test box was reduced irrespective of genotype or treatment indicating long-term memory formation. The lower magnitude of the startle response was reproduced with these younger transgenic mice (Figure 5E), but the treatment with galantamine had no effect on the magnitude of the startle response, or its inhibition by prepulses (data not shown). During fear conditioning, freezing in the shock context (Figure 5F) or after the tone in a neutral environment (Figure 5G) by mock-treated transgenic mice was slightly but not significantly higher compared to controls and treatment with galantamine resulted in increased freezing irrespective of the genotype.

In summary, the behavioral differences between transgenic and non-transgenic mice at this early age of 4–5 months confirmed the findings for the 7-month-old mice. The treatment with galantamine improved the behavior in the open field significantly and partially in the light-dark avoidance paradigm but was not able to normalize the startle response. In the fear conditioning paradigm, galantamine generally increased freezing potentially indicating a side effect on anxiety.

Following the behavioral tests, we quantified in the same animals the plaque load in the hippocampus and entorhinal cortex by thioflavin staining. Transgenic animals treated with the high dose of galantamine showed a highly significant lower number of plaques in both areas compared to untreated control mice (p≤ 0.0001 according to ANOVA) (Figure 6, 7). In the hippocampus of high dose treated transgenic males approximately 19% and in females approximately 25% less plaques were counted compared to untreated controls. In the entorhinal cortex, 32% less plaques for males and 33% less for females were observed. These data indicated that galantamine treatment might reduce the formation of plaques in this model system.

GFAP-positive reactive astrocytes are usually found in association with Aβ plaques (Figure 1). Therefore, we determined the number of GFAP-positive reactive astrocytes in 5XFAD transgenic mice and found a strong correlation between plaque density and the number of reactive astrocytes (Figure 6C).

Discussion

In this study, the effect of chronic oral administration of galantamine, an acetylcholinesterase inhibitor and allosteric nicotinic receptor modulator, in 5XFAD transgenic mice [20] was investigated.

Time Course of Plaque Deposition in 5XFAD Mice

The 5XFAD model, due to its early plaque development phenotype, has been investigated extensively to study the various aspects of early onset AD [20–24,37]. However, the pattern of progressive plaque deposition and development has not yet been described in a systematic manner. Here, we monitored for the first time quantitatively the increase in plaque load over time between the age of 3 to 14 months in the hippocampus and the entorhinal cortex, because these plaques are known to closely resemble the amyloid plaques in human AD patients [38]. Moreover, in AD loss of pyramidal cells in lamina 2 of the entorhinal cortex and in the CA1 region of the hippocampus was described [39], indicating that these brain structures are severely affected by the disease. In

agreement with previous studies [20,21,40], we observed very early onset of plaque deposition in 5XFAD mice beginning around 2 months of age. Furthermore, we show here that plaque deposition occurs differently in the two sexes with lower plaque density reaching plateau levels at 10 months of age in males, while still increasing in females at least until an age of 14 months. A higher plaque density in female transgenic mice has also been noted in other AD mouse models [33–35] and is possibly a consequence of decreased estrogen levels [41], modified BACE activity, or altered metal ion levels [42].

Behavior of Untreated 7-months-old 5XFAD Mice

As therapeutic intervention may be most promising at early stages, we investigated relatively young 5XFAD transgenic mice for behavioral abnormalities. Although, 5XFAD transgenic mice displayed already at the age of seven months a significantly lower body weight paralleling the weight loss always closely associated with AD [43], neuromuscular functioning and motor-coordination appeared normal, confirming previous studies reporting that 5XFAD mice do not exhibit sensory-motoric impairments in string hanging and beam walking before 9 months [24] or abnormal rota-rod coordination before 12 months [22] of age. Likewise, several other AD mouse models display normal motor coordination and grip strength [44]. In contrast, we detected behavioral abnormalities in the anxiety addressing paradigms, namely, open field, light-dark avoidance, and O-maze, similar to the tendency to spend longer times in the center of the open field or the open arms of the elevated plus maze reported previously for 5XFAD [22,23] and several other AD mouse models [45,46]. In addition, in the course of the light-dark avoidance test, 5XFAD mice showed a longer latency to enter the dark compartment for the first time. Such behavior could be a consequence of reduced anxiety levels, but in addition, it could also reflect the characteristic aversion to darkness as noticed in AD patients.

Reduced hippocampus dependent trace fear conditioning, but normal hippocampus independent delay fear conditioning, which is comparable to the tone memory assayed here, has been reported for 5XFAD mice [47]. During hippocampus dependent contextual fear conditioning we observed significantly more freezing of 5XFAD mice in the context but also in the neutral surrounding. Therefore, the calculated context memory and the hippocampus independent tone memory were not significantly different. It has also been shown previously, that contextual fear conditioning after 1 footshock is normal in 5XFAD mice younger than 4 months, but impaired in the 6 month old animals, which could be overcome using 3–5 footshocks [48]. Our paradigm used 2 footshocks, which may have been sufficient to overcome a slight deficit.

Suppression of sensorimotor gating has been repeatedly reported in AD patients [49,50]. In a similar APP/PS1 transgenic mouse model, sensorimotor gating deficits were reported for aged mice using prepulse inhibition of the acoustic startle response [51]. In the 5XFAD mice, we measured a very weak acoustic startle response, indicating possible defects in processing or responding to the stimulus. Furthermore, prepulses of various intensities did not inhibit this residual startle response indicating sensorimotor gating deficits.

In conclusion, 5XFAD mice show at the age of seven months several behavioral deficits, a phenotype with mild cognitive impairment, and progressive amyloid plaque deposition permitting the detection of therapeutic effects.

Galantamine Treatment Affects Behavior and Plaque Deposition

To investigate the potential effects on the retardation of disease progression, we started with chronic galantamine treatment at 3 months of age, when plaque deposition is considerable, assayed the behavior 2 months later, and then determined the plaque load before reaching saturation levels. Galantamine is known to have side effects with respect to cholinergically mediated gastrointestinal symptoms like other cholinesterase inhibitors [52]. To mimic the treatment regimen in humans, where the dosage is increased with time to minimize negative side effects [53], we increased the dosage gradually during the chronic treatment, and chose 2 different concentrations that were tolerated well by the mice. Also, galantamine did not alter the amount of water consumed, indicating the absence of an aversive response.

The chronic treatment with galantamine had positive effects on certain behavioral tasks but could not rescue all the abnormalities. It did not modify the poor startle response or the lack of its prepulse inhibition in 5XFAD mice. At the moment, we can only speculate that the low startle response results from a very early event that cannot be cured by later treatment or that the dosage or duration were not sufficient to protect these neuronal circuits. Furthermore, the treatment increased the overall freezing behavior in the fear conditioning paradigm irrespective of the genotype, which could be a consequence of the potential "nicotinic effect" of the drug. However, in the open field paradigm galantamine treatment restored the behavior of treated 5XFAD mice to normal. Likewise, the abnormal light-dark avoidance behavior of 5XFAD mice was partially normalized. According to the cholinergic hypothesis, a decreased production of acetylcholine or an amplified acetylcholinesterase activity results in impairment of the cholinergic transmission, in turn leading to the loss of intellectual abilities [54]. Hence, the positive effect of galantamine in restoring certain normal behavioral traits could result from its acetylcholinesterase inhibitor activity.

On analyzing the plaque load in the hippocampal formation and the entorhinal cortex, we found that treated mice show a significantly lower plaque density in both the structures, irrespective of the sex, when compared to the untreated controls. Potential mechanisms of action of the drug may be a reduced deposition of $A\beta$ into plaques as opposed to clearance of already existing deposits. Importantly, acetylcholinesterase has been shown to promote the aggregation of β-amyloid peptide fragments by forming a complex with the growing fibrils [54]. Also, galantamine has been shown to enhance microglial $A\beta$ clearance [16]. Another possible explanation could be that galantamine binds to amyloid oligomers leading to a significant conformational change at the turn region (Asp23-Gly29) disrupting interactions between individual β strands and promoting a nontoxic conformation of $A\beta$ (1–40) to prevent the formation of neurotoxic oligomers [55]. A previous study by Unger et al. [56] reported that galantamine at a concentration of 2mg/kg injected subcutaneously for 10 days had no effect on the levels of soluble or insoluble forms of $A\beta$ in 10 months old Tg2576 mice. In our study, we used a much higher dose, chronic oral application for more than 2 months, and assayed the number of plaque deposits before saturation. Hence, as an extension of this study, quantification of the total $A\beta$ and its different isoforms in chronically treated mice and their relation to plaque density may possibly be helpful to address the mechanisms of galantamine action. However, despite the debate about the most toxic form of $A\beta$, it is important to note that a number of current strategies e.g. glutaminyl cyclase inhibition and immunotherapy, involved in the treatment and/ or prevention of AD progression are indeed directed against the deposition of plaques [6,57]. Fewer plaques could have a multifold benefit through the reduced levels of all possible harmful forms of $A\beta$ (monomer, oligomer, and fibril). A reduced plaque load has been positively correlated with better performances in behavioral tests [6,57]. Additionally, regardless of the levels of $A\beta$, reduced plaque formation could have a role in preventing further cortical atrophy in AD brain, as plaque deposition in multiple studies has been correlated with accelerated cortical atrophy [58,59]. Furthermore, a reduced plaque burden has been shown to be accompanied by a reduced glial activation [6,60]. In agreement, we found a positive correlation between the number of plaques and the extent of gliosis, quantified by the number of GFAP positive astrocytes. This implies that chronic galantamine administration reduced the level of astrogliosis in the AD mice brain. Therapeutic targeting of neuroinflammatory pathways in which astrocytes have a prominent position may be a promising strategy to cure AD [61]. The potential of galantamine to interfere with the extent of gliosis is, therefore, another positive outcome of the treatment.

In summary, we report a significantly delayed progression of amyloid plaque deposition and improvement of certain behavioral symptoms associated with AD in the 5XFAD Alzheimer's disease model after chronic treatment with galantamine.

In contrast to other cholinesterase inhibitors e.g. donepezil and rivastigmine, the relatively hydrophilic galantamine poorly penetrates the blood brain barrier. However, the much more hydrophobic Gln-1062 (Memogain), a pro-drug of galantamine, possesses a more than 15-fold higher bioavailability in the brain [62]. Therefore, it will be interesting to investigate Memogain in comparison to galantamine in this AD model in the future.

Acknowledgments

The authors gratefully acknowledge expert technical assistance by Angelika Reichel, Karla Sowa, and Daniela Hill. This study was supported by the German Ministry for Education and Research (BMBF) special network program KMU-Innovativ-2. We thank all members of the Neurocure network for collaboration and fruitful discussions and Galantos Pharma GmbH for providing galantamine hydrobromide.

Author Contributions

Conceived and designed the experiments: DM AM. Performed the experiments: DM SB CH. Analyzed the data: DM SB CH. Contributed reagents/materials/analysis tools: DM AM. Wrote the paper: DM SB AM.

References

1. Karran E, Mercken M, De Strooper B (2011) The amyloid cascade hypothesis for Alzheimer's disease: an appraisal for the development of therapeutics. Nat Rev Drug Discov 10: 698–712.

2. Citron M (2004) Strategies for disease modification in Alzheimer's disease. Nat Rev Neurosci 5: 677–685.

3. Koffie RM, Meyer-Luehmann M, Hashimoto T, Adams KW, Mielke ML, et al. (2009) Oligomeric amyloid β associates with postsynaptic densities and correlates with excitatory synapse loss near senile plaques. Proc Natl Acad Sci USA 106: 4012–4017.

4. Spires TL, Meyer-Luehmann M, Stern EA, McLean PJ, Skoch J, et al. (2005) Dendritic spine abnormalities in APP transgenic mice demonstrated by gene transfer and intravital multiphoton microscopy. J Neurosci 25: 7278–7287.

5. Knowles RB, Wyart C, Buldyrev SV, Cruz L, Urbanc B, et al. (1999) Plaque-induced neurite abnormalities: Implications for disruption of neural networks in Alzheimer's disease. Proc Natl Acad Sci USA 96: 5274–5279.

6. Schilling S, Zeitschel U, Hoffmann T, Heiser U, Francke M, et al. (2008) Glutaminyl cyclase inhibition attenuates pyroglutamate $A\beta$ and Alzheimer's disease-like pathology in vivo. Nat Med 14: 1106–1111.

7. Citron M (2010) Alzheimer's disease: strategies for disease modification. Nat Rev Drug Discov 9: 387–398.

8. Patel L, Grossberg GT (2011) Combination therapy for Alzheimer's disease. Drugs Aging 28: 539–546.

9. Maelicke A, Samochock M, Jostock R, Fehrenbacher A, Ludwig J, et al. (2001) Allosteric Sensitization of Nicotinic Receptors by Galantamine, a New Treatment Strategy for Alzheimer's Disease. Biol Psychiatry 49: 279–288.

10. Peng X, Gerzanich V, Anand R, Wang F, Lindstrom J (1997) Chronic nicotine treatment up-regulates alpha3 and alpha7 acetylcholine receptor subtypes expressed by the human neuroblastoma cell line SH-SY5Y. Mol Pharmacol 51: 776–784.

11. Court JA, Perry EK (1994) CNS nicotine receptors. Possible therapeutic targets in neurodegenerative disorders. CNS Drugs 2: 216–233.

12. Kihara T, Shimohama S, Urushitani M, Sawada H, Kimura J, et al. (1998) Stimulation of alpha4beta2 nicotinic acetylcholine receptors inhibits beta-amyloid toxicity. Brain Res 792: 331–334.

13. Villarroya M, García AG, Marco-Contelles J, López MG (2007) An update on the pharmacology of galantamine. Expert Opin Investig Drugs 16: 1987–1998.

14. Lenzken SC, Lanni C, Govoni S, Lucchelli A, Schettini G, et al. (2007) Nicotinic component of galantamine in the regulation of amyloid precursor protein processing. Chem Biol Interact 165: 138–145.

15. Matharu B, Gibson G, Parsons R, Huckerby TN, Moore SA, et al. (2009) Galantamine inhibits beta-amyloid aggregation and cytotoxicity. J Neurol Sci 280: 49–58.

16. Takata K, Kitamura Y, Saeki M, Terada M, Kagitani S, et al. (2010) Galantamine-induced Amyloid-β Clearance Mediated via Stimulation of Microglial Nicotinic Acetylcholine Receptors. J Biol Chem 285: 40180–40191.

17. Wallin ÅK, Wattmo C, Minthon L (2011) Galantamine treatment in Alzheimer's disease: response and long-term outcome in a routine clinical setting. Neuropsychiatr Dis Treat 7: 565–576.

18. Kavanagh S, Gaudig M, Van Baelen B, Adami M, Delgado A, et al. (2011) Galantamine and behavior in Alzheimer disease: analysis of four trials. Acta Neurol Scand 124: 302–308.

19. Keller C, Kadir A, Forsberg A, Porras O, Nordberg A (2011) Long-term effects of galantamine treatment on brain functional activities as measured by PET in Alzheimer's disease patients. J Alz Dis 24: 109–123.

20. Oakley H, Cole SL, Logan S, Maus E, Shao P, et al. (2006) Intraneuronal β-Amyloid Aggregates, Neurodegeneration, and Neuron Loss in Transgenic Mice with Five Familial Alzheimer's Disease Mutations: Potential Factors in Amyloid Plaque Formation. J Neurosci 26: 10129–10140.

21. Ohno M, Cole SL, Yasvoina M, Zhao J, Citron M, et al. (2007) BACE1 gene deletion prevents neuron loss and memory deficits in 5XFAD APP/PS1 transgenic mice. Neurobiol Dis 26: 134–145.

22. Shukla V, Zheng YL, Mishra SK, Amin ND, Steiner J, et al. (2013) A truncated peptide from p35, a Cdk5 activator, prevents Alzheimer's disease phenotypes in model mice. FASEB J 27: 174–186.

23. Wirths O, Erck C, Martens H, Harmeier A, Geumann C, et al. (2010) Identification of low molecular weight pyroglutamate Aβ oligomers in Alzheimer's disease: a novel tool for therapy and diagnosis. J Biol Chem 285: 41517–41524.

24. Jawhar S, Trawicka A, Jenneckens C, Bayer TA, Wirths O (2012) Motor deficits, neuron loss, and reduced anxiety coinciding with axonal degeneration and intraneuronal Aβ aggregation in the 5XFAD mouse model of Alzheimer's disease. Neurobiol Aging 33: 196. e29–40.

25. Montag-Sallaz M, Schachner M, Montag D (2002) Misguided axonal projections, NCAM180 mRNA upregulation, and altered behavior in mice deficient for the Close Homolog of L1 (CHL1). Mol Cell Biol 22: 7967–7981.

26. Montag-Sallaz M, Montag D (2003) Severe cognitive and motor coordination deficits in Tenascin-R-deficient mice. Genes Brain Behav 2: 20–31.

27. Wishaw IQ, Kolb B (1999) Analysis of behavior in laboratory rodents. In : Windhorst, U. & Johansson, H., eds. Modern Techniques in Neurosci Berlin Springer 1243–1275.

28. Rogers DC, Fisher EM, Brown SD, Peters J, Hunter AJ, et al. (1997) Behavioral and functional analysis of mouse phenotype: SHIRPA, a proposed protocol for comprehensive phenotype assessment. Mamm Genome 8: 711–713.

29. Montag-Sallaz M, Montag D (2003) Learning-induced arg 3.1 expression in the mouse brain. Learn Mem 10: 99–107.

30. Comery TA, Martone RL, Aschmies S, Atchison KP, Diamantidis G, et al. (2005) Acute gamma-secretase inhibition improves contextual fear conditioning in the Tg2576 mouse model of Alzheimer's disease. J. Neurosci 25: 8898–8902.

31. Jacobsen JS, Wu CC, Redwine JM, Comery TA, Arias R, et al. (2006) Early-onset behavioral and synaptic deficits in a mouse model of Alzheimer's disease. Proc Natl Acad Sci USA 103: 5161–5166.

32. Sun A, Nguyen XV, Bing G (2002) Comparative analysis of an improved thioflavin-s stain, Gallyas silver stain, and immunohistochemistry for neurofibrillary tangle demonstration on the same sections. J Histochem Cytochem 50: 463–472.

33. Sturchler-Pierrat C, Staufenbiel M (2000) Pathogenic mechanisms of Alzheimer's disease analyzed in the APP23 transgenic mouse model. Ann N Y Acad Sci 920: 134–139.

34. Lewis J, Dickson DW, Lin WL, Chisholm L, Corral A, et al. (2001) Enhanced Neurofibrillary Degeneration in Transgenic Mice Expressing Mutant Tau and APP. Science 293: 1487–1491.

35. Callahan MJ, Lipinski WJ, Bian F, Durham RA, Pack A, et al. (2001) Augmented Senile Plaque Load in Aged Female b-Amyloid Precursor Protein-Transgenic Mice. Am J Pathol 158: 1173–1177.

36. Guntern R, Bouras C, Hof PR, Vallet PG (1992) An improved thioflavine S method for staining neurofibrillary tangles and senile plaques in Alzheimer's disease. Experientia 48: 8–10.

37. Annunziata I, Patterson A, Helton D, Hu H, Moshiach S, et al. (2013) Lysosomal NEU1 deficiency affects amyloid precursor protein levels and amyloid-β secretion via deregulated lysosomal exocytosis. Nat Commun 4: 2734.

38. Wengenack TM, Reyes DA, Curran GL, Borowski BJ, Lin J, et al. (2011) Regional differences in MRI detection of amyloid plaques in AD transgenic mouse brain. Neuroimage 54: 113–122.

39. Gomez-Isla T, Price JL, McKeel DW Jr, Morris JC, Growdon JH, et al. (1996) Profound Loss of Layer II Entorhinal Cortex Neurons Occurs in Very Mild Alzheimer's Disease. J Neurosci 16: 4491–4500.

40. Vassar RJ, Oakley H, Krishnamurthy S, Maus E, Shao P, et al. (2005) 5XFAD Tg mice that express five FAD mutations have high-cerebral aβ42 levels, rapid amyloid deposition, and intraneuronal aβ. Soc Neurosci Abstr 587: 2.

41. Grimm A, Lim YA, Mensah-Nyagan AG, Götz J, Eckert A (2012) Alzheimer's disease, Oestrogen and mitochondria: an ambiguous relationship. Mol Neurobiol 46: 151–160.

42. Schaeffer S, Wirths O, Multhaup G, Bayer TA (2007) Gender dependent APP processing in a transgenic mouse model of Alzheimer's disease. J Neural Transm 114: 387–394.

43. Sergi G, Rui MD, Coin A, Inelmen EM, Manzato E (2013) Weight loss and Alzheimer's disease: temporal and aetiologic connections. Proc Nutri Soc 72: 160–165.

44. Webster SJ, Bachstetter AD, Van Eldik LJ (2013) Comprehensive behavioral characterization of an APP/PS-1 double knock-in mouse model of Alzheimer's disease. Alzheimers Res Ther 5: 28.

45. Tong Y, Xu Y, Scearce-Levie K, Ptácek LJ, Fu YH (2010) COL25A1 triggers and promotes Alzheimer's disease-like pathology in vivo. Neurogenet 11: 41–52.

46. Lalonde R, Fukuchi K, Strazielle C (2012) APP transgenic mice for modelling behavioral and psychological symptoms of dementia (BPSD). Neurosci Biobehav Rev 36: 1357–1375.

47. Ohno M, Chang L, Tseng W, Oakley H, Citron M, et al. (2006) Temporal memory deficits in Alzheimer's mouse models: rescue by genetic deletion of BACE1. Eur Jour Neurosci 23: 251–260.

48. Kimura R, Ohno M (2009) Impairments in remote memory stabilization precede hippocampal synaptic and cognitive failures in 5XFAD Alzheimer mouse model. Neurobiol Dis 33: 229–235.

49. Jessen F, Kucharsk C, Fries T, Papassotiropoulos A, Hoenig K, et al. (2001) Sensory Gating Deficit Expressed by a Disturbed Suppression of the P50 Event-Related Potential in Patients With Alzheimer's Disease. Am J Psychiatry 158: 1319–1321.

50. Cancelli I, Cadore IP, Merlino G, Valentinis L, Moratti U, et al. (2006) Sensory gating deficit assessed by P50/Pb middle latency event related potential in Alzheimer's disease. J Clin Neurophys 23: 421–425.

51. Wang H, He J, Zhang R, Zhu S, Wang J, et al. (2012) Sensorimotor gating and memory deficits in an APP/PS1 double transgenic mouse model of Alzheimer's disease. Beh Brain Res 233: 237–243.

52. Prvulovic D, Hampel H, Pantel J (2010) Galantamine for Alzheimer's disease. Expert Opin. Drug Metab Toxicol 6: 345–354.

53. Seltzer B (2010) Galantamine-ER for the treatment of mild-to-moderate Alzheimer's disease. Clin Interv Aging 5: 1–6.

54. Singh M, Kaur M, Kukreja H, Chugh R, Silakari O, et al. (2013) Acetylcholinesterase inhibitors as Alzheimer therapy: From nerve toxins to neuroprotection. Eur J Med Chem 70: 165–188.

55. Rao PPN, Mohamed T, Osman W (2013) Investigating the binding interactions of galantamine with b-amyloid peptide. Bioorg Med Chem Lett 23: 239–243.

56. Unger C, Svedberg MM, Yu WF, Hedberg MM, Nordberg A (2006) Effect of subchronic treatment of memantine, galantamine, and nicotine in the brain of Tg2576 (APPswe) transgenic mice. J Pharmacol Exp Ther 317: 30–36.

57. Demattos RB, Lu J, Tang Y, Racke MM, Delong CA, et al. (2012) A plaque-specific antibody clears existing β-amyloid plaques in Alzheimer's disease mice. Neuron 76: 908–920.

58. Becker JA, Hedden T, Carmasin J, Maye J, Rentz DM, et al. (2011) Amyloid-β associated cortical thinning in clinically normal elderly. Ann Neurol 69: 1032–1042.

59. Chételat G, Villemagne VL, Villain N, Jones G, Ellis KA, et al. (2012) Accelerated cortical atrophy in cognitively normal elderly with high β-amyloid deposition. Neurol 78: 477–484.

60. Kalinin S, Polak PE, Lin SX, Sakharkar AJ, Pandey SC, et al. (2012) The noradrenaline precursor L-DOPS reduces pathology in a mouse model of Alzheimer's disease. Neurobiol Aging 33: 1651–1663.

61. Li C, Zhao R, Gao K, Wei Z, Yin MY, et al. (2011) Astrocytes: implications for neuroinflammatory pathogenesis of Alzheimer's disease. Curr Alzheimer Res 8: 67–80.

62. Maelicke A, Hoeffle-Maas A, Ludwig J, Maus A, Samochocki M, et al. (2010) Memogain is a galantamine pro-drug having dramatically reduced adverse effects and enhanced efficacy. J Mol Neurosci 40: 135–137.

16

Highly Efficient Generation of Transgenic Sheep by Lentivirus Accompanying the Alteration of Methylation Status

Chenxi Liu[1,2,4]**, Liqin Wang**[1,2,3]**, Wenrong Li**[1,2,3]**, Xuemei Zhang**[1,2,3]**, Yongzhi Tian**[1,2,3]**, Ning Zhang**[1,2,3]**, Sangang He**[1,2,3]**, Tong Chen**[1,2,3]**, Juncheng Huang**[1,2,3]**, Mingjun Liu**[1,2,3]*

1 Xinjiang Laboratory of Animal Biotechnology, Urumqi, Xinjiang, China, **2** Key Laboratory of Genetics, Breeding and Reproduction of Grass Feeding Livestock, Ministry of Agriculture, Urumqi, Xinjiang, China, **3** Animal Biotechnology Research Center, Xinjiang Academy of Animal Science, Urumqi, Xinjiang, China, **4** College of Life Science and Technology, Xinjiang University, Urumqi, Xinjiang, China

Abstract

Background: Low efficiency of gene transfer and silence of transgene expression are the critical factors hampering the development of transgenic livestock. Recently, transfer of recombinant lentivirus has been demonstrated to be an efficient transgene delivery method in various animals. However, the lentiviral transgenesis and the methylation status of transgene in sheep have not been well addressed.

Methodology/Principle Findings: EGFP transgenic sheep were generated by injecting recombinant lentivirus into zygotes. Of the 13 lambs born, 8 carried the EGFP transgene, and its chromosomal integration was identified in all tested tissues. Western blotting showed that GFP was expressed in all transgenic founders and their various tissues. Analysis of CpG methylation status of CMV promoter by bisulfate sequencing unraveled remarkable variation of methylation levels in transgenic sheep. The average methylation levels ranged from 37.6% to 79.1% in the transgenic individuals and 34.7% to 83% in the tested tissues. Correlative analysis of methylation status with GFP expression revealed that the GFP expression level was inversely correlated with methylation density. The similar phenomenon was also observed in tested tissues. Transgene integration determined by Southern blotting presented multiple integrants ranging from 2 to 6 copies in the genome of transgenic sheep.

Conclusions/Significance: Injection of lentiviral transgene into zygotes could be a promising efficient gene delivery system to generate transgenic sheep and achieved widespread transgene expression. The promoter of integrants transferred by lentiviral vector was subjected to dramatic alteration of methylation status and the transgene expression level was inversely correlative with promoter methylation density. Our work illustrated for the first time that generation of transgenic sheep by injecting recombinant lentivirus into zygote could be an efficient tool to improve sheep performance by genetic modification.

Editor: Tim Thomas, The Walter and Eliza Hall of Medical Research, Australia

Funding: The studies were supported by the grant of "National Key Program for Transgenic Sheep Breeding". Grant numbers: 2008ZX08006-003. http://www.nmp.gov.cn/. The funders had no role in study design, data collection and analysis, decision to publish, or preparation of the manuscript.

Competing Interests: The authors have declared that no competing interests exist.

* E-mail: mingjun_1@sina.com

Introduction

The generation of transgenic livestock holds considerable promise for the development of biomedical and agricultural systems [1,2]. The first transgenic livestock was produced via microinjection of foreign DNA into pronuclei of zygote in 1985 [3]. In 1986, cloning sheep was generated by nuclear transfer using embryonic stem cells as donors [4], and then cloning sheep Dolly was born in 1997 by somatic cell cloning (SCC) [5]. Concomitant with the success of SCC, the first cloning transgenic sheep was produced by nuclear transfer with stably transgenic somatic cells. In spite of the success in generation of transgenic livestock by pronuclear microinjection or SCC, concurrent techniques shows several significant shortcomings, such as low efficiency, high cost, random integration, and frequent incidence of mosaicism. Efficient generation of transgenic livestock with low cost remains to be developed in transgenic animal field. Recent development of lentiviral vector for gene transfer shows the great potentials to overcome limitations mentioned above [6,7], and accordingly is becoming a new efficient tool to produce transgenic livestock.

To date, various transgenic species including mice, fish, chicken, pig, non-human primate, cattle and sheep have been generated by lentiviral transgenesis [8,9]. Compared to traditional pronuclear DNA microinjection or somatic cell cloning (SCC), lentiviral transgenesis results in a four to eight fold higher generation rate of transgenic animals per embryo treated [10], and more than 90% transgenic founders can be observed transgene expression [11,12]. Furthermore, the transgene delivered by lentiviral vector also stably expressed in their offsprings with considerably low methylation level in transgene promoter under certain circumstances. This differed from retrovirus-induced globally *de novo* methylation, which resulted in widespread silence of transgene expression [13].

Transgenic swine was the first livestock produced by injecting lentivirus into zygote with generation rates of 19–33% [14], which was significantly higher than 1% such rate obtained by conventional pronuclear microinjection [3]. However, the same investigators who successfully introduced lentiviral transgene into swine failed to produce transgenic cattle by the same procedure although the transgenic embryos were gained [15]. In 2004, the first transgenic cattle was produced by lentivirus infection of oocyte instead of microjection with the generation rate of 8.3% per oocyte treated [15]. In 2012, the transgenic cattle generated by injection of lentiviral vector into zygotes was reported with the generation rate of 7.5% per embryo transferred [16]. These studies indicated that the infection and integration capability of recombinant lentivirus were quite disparate within different livestock species.

Previous studies on lentiviral transgenesis demonstrated that the transgene expression was associated with transgene epigenetic modification, integrant numbers and locus [17,18]. So far, the overall regulatory mechanism of lentiviral transgene expression has been poorly understood. DNA hypermethylation was considered as a critical factor resulting in silence of transgene expression. Concurrent report also showed that about one-third of integrated lentiviral transgenes in pigs were subjected to methylation and exhibited lower expression [19].

As for transgenic sheep, since the first one was produced in 1985 by Hammer with pronuclear microinjection [3], some new approaches have been used for production of transgenic sheep, for instance, somatic cell cloning (SCC) [5] and sperm-mediated gene transfer (SMGT) [20]. In the mass, the transgenic efficiency remains extremely low. Ritchie *et al* succeeded in producing transgenic sheep by transferring blastocysts derived from oocyte injection of lentivirus with 20% transgenic efficiency [21]. However there are few comprehensive studies on the diversity of transgene integration, expression and alteration of methylation in transgenic sheep. Especially, transgene efficiency, expression pattern and epigenetic state of transgenic sheep produced by lentiviral injection have not been well understood. Hereby, we demonstrated for the first time that the transgenesis by injection of EGFP-lentivirus into perivitelline space of sheep zygote is a high efficient tool for generation of transgenic sheep and transgene can be expressed in almost all transgenic founders and their various tissues. Furthermore, methylation status of transgene and its effect on transgene expression, as well as relationship between integrant numbers and its expression were firstly investigated in lentivirus-mediated transgenic sheep.

Materials and Methods

Animals

All animals used for this study are Xinjiang Merino Fine Wool Sheep raised in the farm of Sheep Breeding and Reproduction Center. All studies carried out in sheep were approved by the Committee of Animal Research Security and Ethics (CARSE), Xinjiang Academy of Animal Science.

Construction of Plasmids and Preparation of Lentiviral Particles

EGFP gene was digested from pEGFP-N1 plasmid (Clontech) with *BamH I* and *Hind III* (TAKARA) and cloned into lentiviral vector (pLEX-MCS, Openbiosystem), named as pLEX-EGFP.

The vector carries self-inactivating long terminal repeat (SIN LTR), internal ribosome entry site (IRES), mammalian selectable marker (puromycin) and woodchuck hepatitis virus posttranscriptional regulatory element (WPRE). EGFP gene is located at downstream of the CMV promoter.

To generate pLEX-EGFP lentiviral particles, HEK293T cells were seeded in a 100-mm dish at a density of 60,000 cells/cm^2 and co-transfected with pLEX-EGFP (12 µg) along with packaging plasmids (3.5 µg pMD2.G and 6 µg pPAX2) using Lipofectamine 2000 (Invitrogen) at a DNA/Lipofectamine ratio of 1 to 3. After 48 h transfection, the supernatant containing lentivirus particles was filtered through 0.45-µm syringe filter and concentrated by ultracentrifugation (Beckman) at 50,000 g for 2 hour at 4°C. Precipitation was resuspended in phosphate buffered saline, aliquoted and stored at −80°C. Lentivirus titre was determined by infecting HEK293T cells with serial dilutions of concentrated lentivirus, and thereafter quantitated by counting the GFP fluorescent cells with flow cytometry (Becton, Dickinson and Company) post of 48 h infection as previous described [22]. The titre of pLEX-EGFP was approximately 3×10^9 infectious units (IU) per ml in average.

Lentivirus Injection and Embryo Transfer

Transgenic sheep were generated via injection of lentivirus into perivitelline space of the zygote. In brief, embryos were obtained from Xinjiang Merino Sheep which were approximately 2 years old and weighed at least 50 kg. Superovulation was carried out within sheep breeding season from September to November and started on 3 days before oestrus induced by intramuscular injection of follicle stimulating hormone (FSH, Sigma-Aldrich). FSH was injected once per 12 hours lasting for 3 days. Briefly, twice injection of 40 IU FSH was performed on the first day, 30 IU on the second day and 20 IU on the last day. After 12 hour of oestrus, the donors were mated with rams and repeated mating another 12 hours later. At 60 hours, zygotes from mated donors were collected by flushing the umbrella of oviducts with warm phosphate buffered saline containing 2% FBS. Then they were removed from the PBS and cultured in SOF medium with 3 mg/mL BSA at 38°C in 5% CO_2.

For lentivirus injection, around 50–100 pl of concentrated lentivirus with 3×10^9 IU/ml titer were injected into perivitelline space of zygotes using a micromanipulator (ECLipse TE2000-U, Nikon). For embryo transfer, recipients were synchronized by the same treatment as donor ewes. Embryos injected at one or two cell stage were transferred to recipient ewes with mid-line laparotomy under general anaesthesia. During surgery, the reproductive tract was exposed and embryos were transferred into the oviduct of recipients using a displacement micropipette. To assess the expression of GFP in vitro, part of injected zygotes were cultured to blastula in SOF medium supplemented with 3 mg/ml BSA at 38°C in 5% CO_2 and observed under UV-microscope.

PCR Detection

Transgene integration was detected by PCR screening. Genomic DNA was obtained from tail tips using the DNeasy@ Blood and Tissue Kit (QIAGEN) according to the instruction manual. PCR analysis was carried out with 500 ng genomic DNA as template and PCR Master mix (Promega). Primers used to amplify the 638 bp transgene fragment spaning CMV prompter and EGFP gene were: forward 5'-CACCAAAATCAACGGGACTT-3' and reverse 5'-GATGTTGCC GTCCTCCTTGAAGT-3'. The PCR conditions were 94°C denaturation for 5 min followed by 40 cycles of 94°C for 30 sec, 60°C for 45 sec, and 72°C for 55 sec and a final extension at 72°C for 7 min.

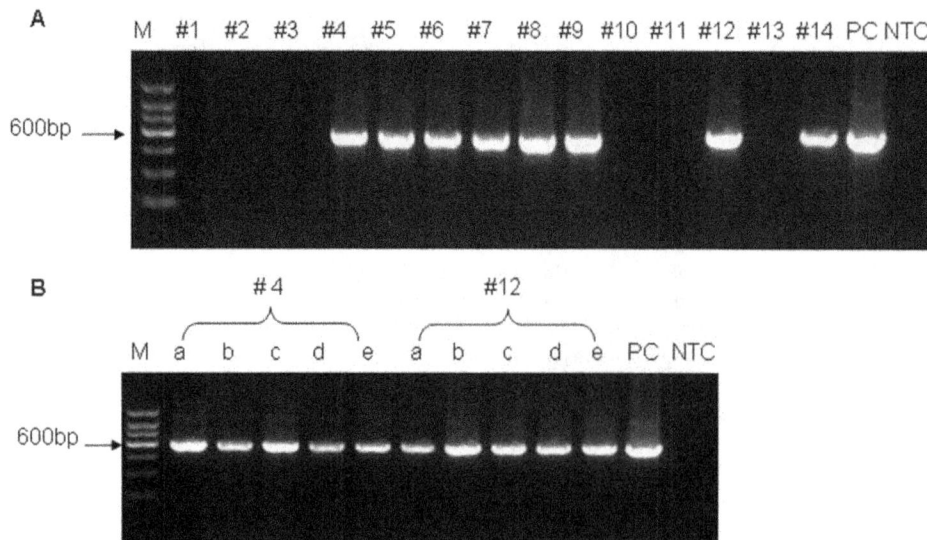

Figure 1. Analysis of EGFP-lentivirus transgene integration in transgenic sheep. (A) Amplification of EGFP transgene from genomic DNA extracted from tail tips of newborn lambs. #1–14: transgenic newborn lambs. (B) Amplification of EGFP transgene from tissues of #4 and #12 anatomized lambs. a-e: heart, liver, spleen, lung and kidney, respectively. Amplicons are 604 bp fragments spanning CMV promoter and EGFP sequences. M, DNA marker; PC, pLEX-EGFP vector as positive control; NTC, non-transgenic sheep DNA control.

Southern Blotting

Integration numbers of transgene were determined by Southern blotting analysis. Genomic DNA from tail tips was extracted by means of standard phenol-chloroform extraction and digested with

EcoRI (TAKARA) or double-digested with SfiI and HpaI (TAKARA). After precipitation with alcohol, 10 μg digested DNA was separated on 0.7% agarose gel with 25 volt electrophoresis overnight. Blotting was carried on by vacuum transfer to

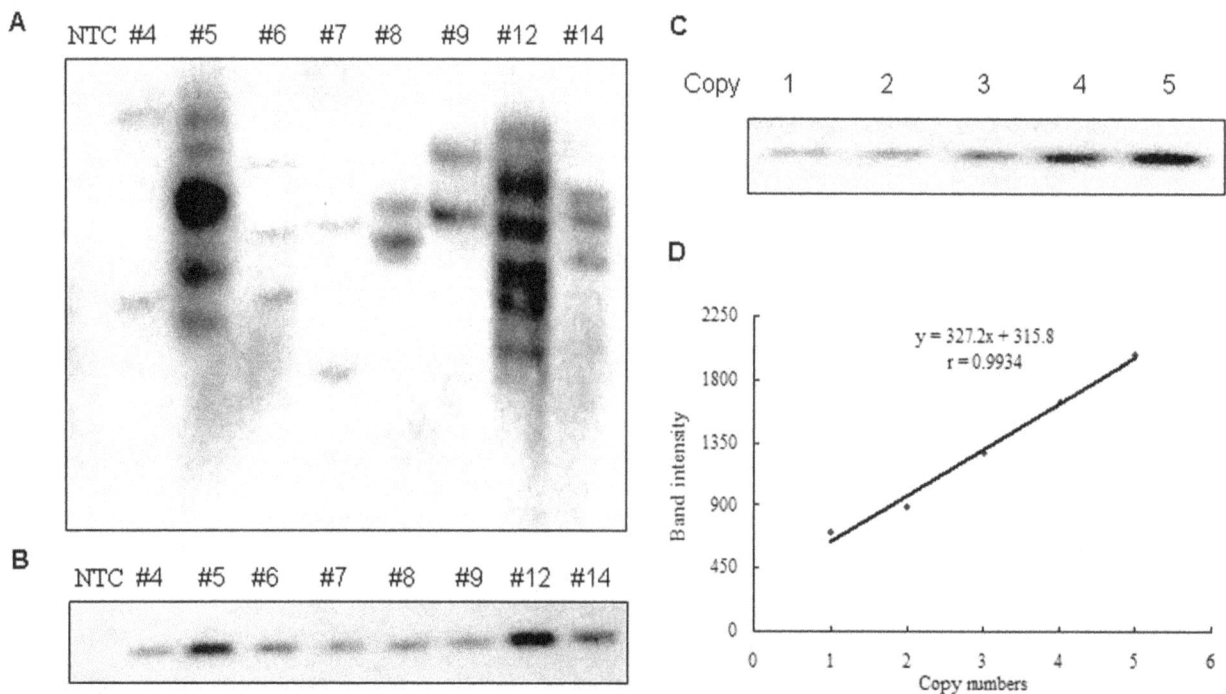

Figure 2. Southern blotting analysis of transgene integrants in genomic DNA of transgenic sheep. (A) Genomic DNA extracted from tail tips of transgenic sheep was digested with EcoRI and hybridized with ^{32}P labeled probe amplified from CMV promoter. (B) Genomic DNA extracted from tail tips of transgenic sheep was double-digested with SfiI/HpaI and hybridized with ^{32}P labeled probe. NTC, non-transgenic sheep control; # 4–14, transgenic lambs identified by PCR corresponding to Fig. 1A. (C) pLEX-EGFP plasmid was double-digested with SfiI/HpaI and diluted in serial concentrations matched to corresponding copies. Diluted plasmids with copies from 1 to 5 were hybridized with probe double-digested genomic DNA of transgenic lamb in parallel. (D) Standard curve of copy numbers in panel C was generated with diluted plasmid based on the quantification of the blots by densitometric measurement as described in the Materials and Method.

Table 1. Southern blot analysis of transgene copy numbers determined by standard curve with a double-digested genomic DNA sample.

Transgenic Sheep	#4	#5	#6	#7	#8	#9	#12	#14
Intensity	931	1949	1362	952	982	1013	2222	1442
Copy Numbers	1.88	4.99	3.2	1.94	2.03	2.13	5.83	3.44

nylon membrane (Amershan) in 10×SSC for 90 min. The 430 bp fragment of the CMV promoter was amplified as probe from pLEX-EGFP plasmid using primers: forward 5′-CGAGGGC-GATGCCACCTAC-3′ and reverse 5′-CTCCAGCAGGAC-CATGTGATC-3′. The probe was prepared by ^{32}P-dCTP labeling with random primer extension kit (Promega) and hybridized with blotting membrane by incubating overnight at 65°C in hybridization oven (Hoefer Scientific Instrument). The concentration of probe used for hybridization was 25 ng/μL. Membranes were washed three times at 65°C in 0.5×SSC buffer containing 1% SDS after hybridization and exposed against film in dark cassette at −80°C for 24 hours. Then the film was developed as general protocol.

To verify the integrant numbers observed in one-cut genomic DNA, the southern blot with double-digested genomic DNA was performed along with the standard curve which was generated by serial dilution of double-digested transgenic plasmid in parallel. For short, the plasmid was serially diluted from 120 pg (5 copies) to 24 pg (one copy). Each concentration of standard plasmid was converted into copy numbers per volume using the following equation: $N = \frac{C \times 10^{-9}}{M \times 660} \times 6.02 \times 10^{23}$, where N stands for copy number (copies/μL), C for concentration (ng/μL) and M for base pairs of the plasmid. Further, the integrants identified by counting the bands in single-digested genomic DNA southern blot was matched to the copy numbers determined in double-cut genomic DNA southern blot by quantification with standard curve.

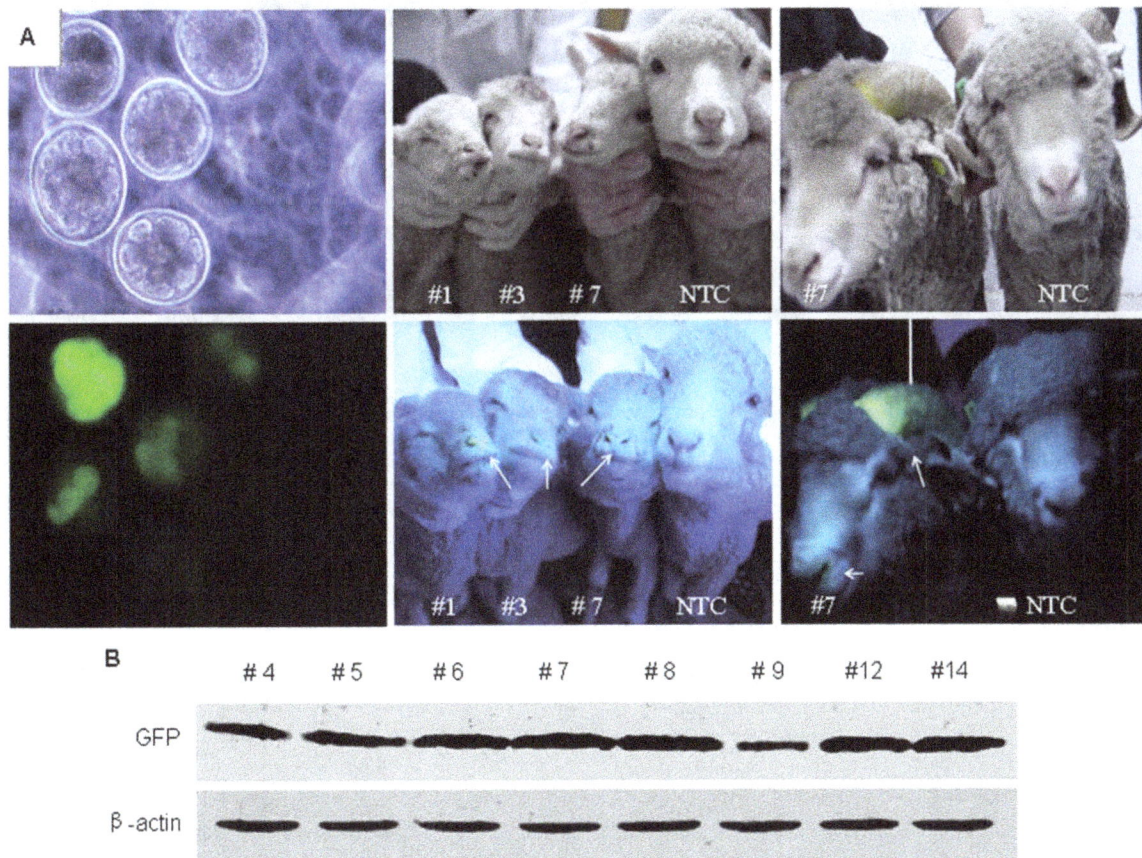

Figure 3. Analysis of the expression of GFP in transgenic lambs. (A) Embryos injected with lentivirus were cultured and developed to blastula and visualized by white and UV light (left panels) under microscope with magnification of 200×. Visualization of GFP expression in transgenic lambs (#1,#3,#7) and non-transgenic lamb control (NTC) were pictured under white light and UV light (middle panels). Visualization of GFP expression of horn in 1.5 year old transgenic lamb #7 and non-transgenic lamb pictured under white light and UV light (right panels). Arrows indicated the green fluorescence in transgenic sheep; (B) Proteins extracted from tail tips of eight transgenic lambs were subjected to immunoblotting with GFP antibody as described in Materials and Methods. β-actin levels were determined with an anti-β-actin antibody and used as loading control.

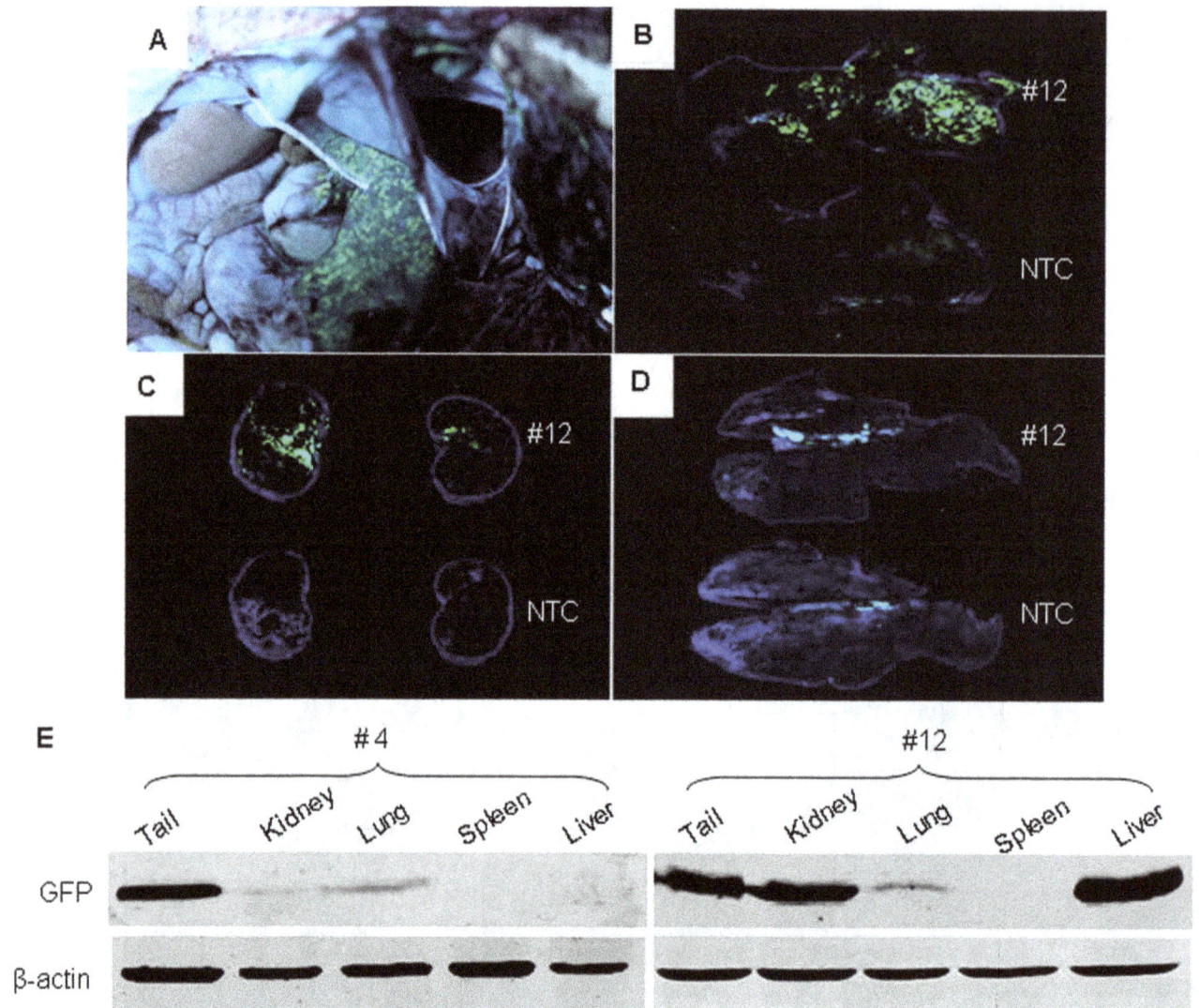

Figure 4. Expression of GFP in tissues of transgenic lambs observed by fluorescence imaging and assayed by western blotting. (A) Fluorescence imaging of the whole inner organs of transgenic sheep under white light. (B-D) Fluorescence imaging of liver, kidney and lung of transgenic or NTC sheep. The upper are organs of the transgenic sheep and the lower are organs of the NTC sheep. (E) Expression of GFP in tissues assayed by western bloting. Proteins extracted from tissues of tail tip, kidney, lung, spleen and liver of #4 and #12 lambs were subjected to immunoblotting with GFP antibody as described in Materials and Methods. β-actin levels were determined with an anti-β-actin antibody as loading control.

Fluorescence Imaging

Photomicrographs of embryo were taken under fluorescent microscope (ECLipse TE2000-U, Nikon) using Nis-Elements software. For transgenic sheep, GFP images were performed with a Wd-9403e UV portable device (61 Biological Instrument, Peking) fitted with UV filter and captured using a 5D-Mark 2 digital camera (Canon, 50 mm lens).

Western Blotting

Total proteins were extracted from tail tips or other tissues. Frozen samples were ground to powder by pestle and mortar grinding and solubilized in a solution of 62.5 mM Tris pH6.8, 10% glycerol, 2.5% sodium dodecyl sulfate (SDS), and Halt^{TM}-Protease Inhibitor Cocktail (Thermo Scientific). Quantification of total protein was carried out by Bicinchoninic acid assay with BSA (Sigma-Aldrich). The proteins (100 μg) were subjected to 12%

SDS-polyacrylamide gel electrophoresis. Separated proteins were transferred to nitrocellulose (NC) membrane (Bio-Rad) and immune-blotted with anti-GFP or anti-β-actin antibodies (Abcam). Immuno-reactive proteins were visualized using the Odyssey Infrared Imaging System and relatively quantified by densitometric analysis (Li-Cor, Lincoln, NE), as described by the manufacturer.

Bisulfite Sequencing

The genomic DNA was extracted from tail tips or other tissues by DNeasy@ Blood&Tissue Kit (QIAGEN) according to the instruction manual. Bisulfite modification was performed with 0.6 μg of DNA for each sample using the EpiTect@Bisulfite Kit (QIAGEN) according to the instruction manual. PCR primers used to amplify the CMV promoter were designed by MethPrimer software online (http://www.urogene.org/methprimer/), which

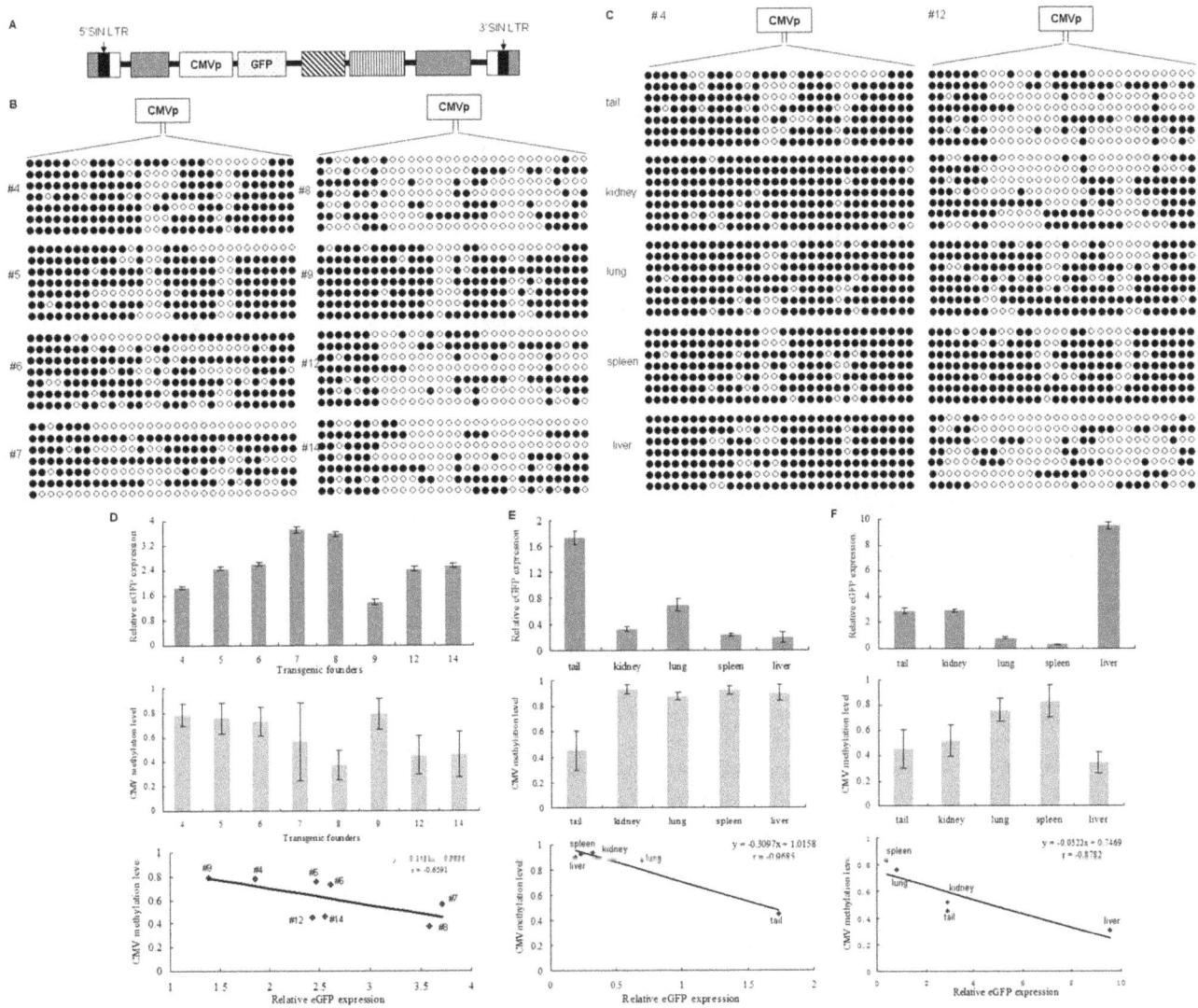

Figure 5. Correlation of CMV promoter methylation status with GFP expression level of transgenic sheep. (A) Schematic of pLEX-EGFP vector. (B) Status of the CMV promoter methylation in 8 transgenic sheep. The 487bp sequences of CMV promoter containing one CpG islands were targeted for methylation analysis. Genomic DNAs extracted from 8 transgenic lambs (#4–12) were treated with bisulfite and sequenced at least 7 clones for each sample. (C) Status of the CMV promoter methylation in tested tissues of two anatomized lambs (#4 and #12). Genomic DNAs extracted from tail tips, liver, lung, kidney and spleen were treated with bisulfite and sequenced at least 7 clones for each sample. The black cycles represented the methylated CpG and the white cycles represented the non-methylated CpG. (D) Correlation of GFP expression with methylation level of lentiviral CMV promoter of transgenic lambs. Densitometric quantification of the relative GFP expression was assayed by Western blotting (Fig. 3B) in tail tips of transgenic lambs #4–14 (up panel). Methylation levels were measured by the average ratios of methylated CpGs to total CpGs of the target CMV promoter sequence (middle panel). Correlation of the methylation levels of CMV promoter with GFP expression of 8 transgenic sheep was analyzed (low panel). (E and F) Correlation of GFP expression with methylation levels of CMV promoter in tested tissues of anatomized lambs (#4 and #12). Densitometric quantification of the relative GFP expression was assayed by Western blotting (Fig. 4B) in tissues of #4 (E, up panel) and #12 (F, up panel) lamb. Methylation levels of CMV promoter in tested tissues of #4 (E, middle panel) and #12 (F, middle panel) lambs were based on Fig. 5C. The average rate of methylated CpGs in the 487 bp region of CMV promoter was defined as the indicator of methylation status. Correlation of methylation levels of CMV promoter with GFP expression levels was analyzed in tested tissues of #4 (E, low panel) and #12 (F, low panel) lambs.

was also used to predict CpG site and CpG islands. The following PCR primers were used to amplify a 487-bp fragment containing one CpG islands with 30 CpGs: forward 5′-GGGTTATTAGTT-TATAG TTTATATATGG-3′ and reverse 5′-GATTCAC-TAAACCAACTCTACTTA-3′. The PCR of bisulfite-modified DNA was performed using PCR Master MIX (Promega). Amplicons were gel purified and cloned into pGEM-T vector (Promega), followed by sequencing at least 7 clones of each sample. For each DNA sample, the number of cytosine residues that remained as "C" was counted, and converted to a percentage

of the total 30 CpGs presented in 487 bp region of the CMV promoter.

Statistical Analysis

Statistical analysis were performed with the SPSS 13.0 software. Values were shown as Mean±SD and subjected to correlation analysis of Pearson. P value less than 0.05 was considered as statistically significant.

Results

Generation of EGFP Transgenic Sheep

Total of 46 zygotes were collected from FSH stimulated donors after artificial insemination. One or two cell stage embryos were injected with lentivirus (3.7×10^9 IU/mL) into perivitelline space. The injected embryos were then transferred to 22 hormonally synchronized recipients. In order to increase the productivity, all recipients were transferred with two embryos and resulted in the birth of 13 lambs from 9 pregnant ewes. Of the 13 newborn lambs, eight transgenic sheep were identified by PCR (Fig. 1A) and southern blotting (Fig. 2A and 2B). The rate of transgenic sheep to total of new born lambs and to embryos were 61.5% (8/13) and 17.4 (8/46) respectively. Except two lambs (#4 and #12) died after birth, the other 6 lambs survive normally. There was obvious variation of congenital malformation in dorsal keel of #4 lamb with death at birth. The other died lamb #12 displayed the anorexia and diarrhea before death, no other developmental abnormality was observed. Transgenic sheep mortality is 25% (2/8), which is the same as that of normal lamb of 25% (9/36). For the two died transgenic lambs, the genomic DNAs extracted from heart, liver, spleen, lung and kidney were subjected to PCR screening. The integration of transgene was observed in all tested tissues (Fig. 1B). The results inferred that the integration of lentiviral transgenesis may exist in all the tissues.

Analysis of the Transgene Integration

In order to analyze transgene integration and copy numbers, southern blot assay was carried out with genomic DNA digested with *EcoR*I or double-digested with *Sfi*I and *Hpa*I. In *EcoR*I digested genomic DNA samples, the number of integrants were visualized ranging from 2 to 6 copies and for most individuals with 2 to 3 copies (Fig. 2A). To exactly quantify the copy number of each transgenic sheep, we performed the southern blot with double-digested genomic DNA and quantified the copy number by standard curve. The standard curve was generated with pLEX-EGFP plasmid by concentration gradient southern blot, which was performed in parallel with double-digested genomic DNA derived from transgenic lambs (Fig. 2B). The copy number of each double-digested plasmid with serial dilution was linearly matched to the plasmid concentration (Fig. 2C and 2D). The copy number for each blot of transgenic lamb was calculated based on standard curve (Table.1).

The highest copy number was identified in #12 lamb with 6 copies, followed by #5 lamb with 5 copies. The copy numbers of other transgenic sheep were around 2 to 3. Copy number derived from these two approaches was consistent (Fig. 2A).

Analysis of EGFP Expression in Transgenic Lambs

The expression of EGFP transgene was analyzed by direct fluorescence observation and Western blotting. At first, we observed embryos injected with EGFP lentivirus in blastula stage under fluorescent microscope (Fig. 3A, left panels). Approximately 80% embryos subjected to injection of lentiviral transgene were presented green fluorescence. Further, we observed green fluorescence in hoof, lip and horn of newborn transgenic lambs (Fig. 3A, middle panels) and continuously to maturity (Fig. 3A, right panels), which suggested that the GFP could be expressed persistently in transgenic sheep. Additionally, we anatomized the died lamb (#4 and #12) to investigate the distribution of GFP expression in inner organs (Fig. 4A). Notably, the most intense GFP fluorescence was observed in liver (Fig. 4B) and then in kidney (Fig. 4C), weak GFP fluorescence was observed in lung of #12 lamb (Fig. 4D). To further analyze the GFP expression, we extracted the proteins

from tail tips of all transgenic sheep and inner organs from two died lambs to perform Western blotting. Expression of GFP was detected in tail tips of eight transgenic lambs (Fig. 3B), which indicated that GFP transgene expressed in all transgenic founders. The relative quantification of western blot showed that the levels of GFP expression of #7 and #8 were much higher than that of other founders. Consistent with green fluorescent intensity in inner organs of #12 lamb, the level of GFP protein measured by western blotting was highest in liver and lowest in lung, no GFP was detected in spleen (Fig. 4E, right panel). The expression of GFP in lamb #4 indicated that the expression of GFP was highest in tail and lower in lung and kidney, and no expression was detected in spleen and liver (Fig. 4E, left panel). These data indicated the disparity of transgene expression in different individuals and tissues.

Status of Promoter Methylation and Influence on Transgene Expression

Previous studies documented that transgene could be methylated in transgenic animals and resulted in repression of expression [23,24]. To investigate the methylation status and its influence on transgene expression in lentiviral-mediated transgenic sheep, we examined the methylation density of 487-bp region of the CMV promoter containing one CpG island with 30 CpGs in individuals and tissues. Firstly, CMV promoter methylation status in all transgenic founders was measured (Fig. 5B). The average methylation levels ranged from 37.6% to 79.1% in transgenic individuals (Fig. 5D, middle panel). Then the promoter methylation status in different tissues was measured (Fig. 5C) and the methylated CpG rate ranged from 34.7% to 93.3% (Fig. 5E and 5F, middle panels). Analysis of the correlation of methylation level with GFP expression in individuals (Fig. 5D, low panel) and tissues (Fig. 5E and 5F, low panels) revealed that the expression of GFP expression was inversely correlated with methylation status (r = -0.6591 for individules, $p < 0.05$; r = -0.9685 for #4 tissues, $p < 0.05$; r = -0.8782 for #12 tissues, $p < 0.05$). The lowest GFP expression (Fig. 5D, up panel) was observed in the transgenic sheep with the highest methylation level (79.1%, transgenic sheep #9). On the contrary, the lowest promoter methylation level was corresponding to the highest GFP expression level (transgenic sheep #8). In tissues, the highest methylation level was found in spleen of #4 lamb with 93.3% (Fig. 5E, middle panel), at which little GFP expression was detected (Fig. 5E, up panel), whereas the highest expression of GFP was found in liver of #12 lamb (Fig. 5F, up panel), in concomitant with the lowest methylation density (34.7%) (Fig. 5F, middle panel).

Discussion

Concurrent studies documented that lentiviral vectors had been successfully used to generate transgenic mice, rat, pig, cattle, chicken and nonhuman primate [8,14,25,26,27,28]. Different transgenic species generated by lentiviral vectors exhibited variability in gene transfer efficiency, transgene expression and epigenetic status. In this study, we generated 8 transgenic sheep by injection of lentiviral vector containing EGFP reporter into perivitelline space of ovine embryos with 17.4% transgenic efficiency, which was substantially higher than that of cattle produced using same method with rate of 7.5% (3/40) [16]. Previous reports on transgenic mice indicated that lentiviral injection should be performed at one-cell stage of zygotes [22,29]. As the variegation of response on the effect of superovulation treatment among donors, it is difficult to maintain the collected sheep embryos in the same stage. In our studies, approximate 60%

of zygotes gained were on one-cell stage, and the other stayed on two-cell stage. Based on our in vitro study by injection of GFP into IVF embryos at different stages, there is no significant difference of transgenic efficacy between one-cell and two-cell stage (76.9% versus 75.4%, data not shown). For the two lambs died postnatal, one (#4) was found with over-bend dorsal keel. The other lamb (#12) displayed the anorexia and diarrhea, which were the major causal that the non-transgenic sheep died from. The ratio of mortality was 25% in transgenic lambs, whereas the mortality of wild type investigated in the same reproductive term was 25% (9/36). There is no difference in mortality between transgenic sheep and non-transgenic sheep, which indicated lentiviral transgenesis has no obvious disturbance on development of transgenic sheep.

Multiple copies of integration are substantially observed in transgenic animals produced by lentiviral transgenesis [27,30]. Based on our analysis of lentiviral integration, we found that lentiviral transgene was occurred in various tissues of transgenic sheep. Moreover, the southern blotting illustrated that most of the transgenic lambs possessed more than one copy of integrant. The average integrant numbers were 3.1 (25/8), ranging from 2 to 7. Depending on literature investigation, the average integrant numbers in transgenic pigs generated by injection of recombinant lentivirus were 4.6, ranged from 1 to 20 copies [14]. The variability of integrant numbers was presumably associated with animal species and lentiviral titer injected.

Previous investigators had addressed unprecedented high rate of transgene expression [11,12]. To investigate the transgene expression, observation on whole lambs showed green fluorescence in hoof, lip and horn from birth continuously to maturity. Tissues from freshly dead transgenic lambs also presented green fluorescence in liver, kidney and lung. This was consistent with the results reported in pigs and cattles [15,31]. To further verify the expression of transgene, the proteins extracted from different tissues were subjected to western blotting analysis. The GFP protein expression varied among individuals and tissues of transgenic sheep, which was in consistence with the fluorescent intensity observed in vivo fluorescence imaging. In general, the overall expression of transgenic sheep derived from lentiviral transgenesis indicated the wide range of transgene expression in different individuals. However, dramatic disparity of transgene expression was identified in different tissues. Our results were similar to the report in transgenic pigs and birds [19,32]. Variegation of transgene expression might be explained by differences in the basic biology or lentiviral vector activity. On the other hand,these results were obtained from F0 founders. Based on previous report in pig [20], one-third of lentivirus-mediated transgenic pigs of F1 generation exhibited low expression levels and hypermethylation. Further studies are worthwhile to carry out to investigate the transgene expression in F1 generation of transgenic sheep in the future.

Previous reports showed that DNA methylation has been verified as a critical factor in regulating activity of transgenic vector [23,24,33]. Meanwhile, analysis of methylation status of

transgenic pigs found that the high degree of methylation in the promoter and coding region of lentiviral transgene was accompanied by low levels of transgene expression [20]. To explore the regulation mechanism of lentiviral transgene, we measured CMV promoter methylation levels and analysed the association of promoter methylation with transgene expression in all transgenic founders and part of tissues. Our results showed that the methylation levels ranged from 37.6% to 79.1% in transgenic individuals and 34.7% to 83% in tested tissues. The association of methylation level with GFP expression suggested that GFP expression was inversely correlated with methylation status, both in individuals and in tissues. This result was similar to the outcomes reported in transgenic mice [34] and pigs of somatic cell cloning [33]. The number of lentiviral transgene integrant was reported as an important factor involved in transgene expression besides the methylation [35]. Studies on lentiviral transgenic pigs found that the increase of transgene expression was almost linearly increasing with lentiviral integrant numbers [14]. In this study, there is no any inclination between integrant numbers and GFP expression levels (r = 0.128, p>0.05, datas not shown). We postulated that the lentiviral transgene expression was presumably influenced by integration loci and its context rather than integrant numbers. Further studies have been carried out to survey the integrant loci and study the association with transgene expression. Our results also inferred that the level of promoter methylation played much more important role in controlling transgene expression than that of integrant number in lentivirus-mediated transgenic sheep.

Since the first publication on generation of transgenic sheep by injection of lentivirus into oocytes in 2009 [21], no further studies have been reported so far. Hereby, we are the first time to comprehensively investigate the issues of transgenic integrant, expression and methylation in lentivirus-mediated transgenic sheep. Taken together, we demonstrated that lentiviral transgenesis by injection of recombinant lentivirus into perivitelline space of sheep zygote could achieve high transgenic efficiency and high rate of transgene expression. Furthermore, the lentiviral transgene was subjected to alteration of methylation status and the transgene expression was inversely correlative with promoter methylation, whereas has no association with integrant numbers in lentivirus-mediatied transgenic sheep.

Acknowledgments

We thank Dr. Zhanjun Hou for carefully inspection of the manuscript. We also thank the team of management of sheep breeding farm, Biao Li, Bing Han and Fan Yang.

Author Contributions

Conceived and designed the experiments: ML CL JH WL. Performed the experiments: CL LW TC YT. Analyzed the data: CL NZ SH. Contributed reagents/materials/analysis tools: XZ. Wrote the paper: ML CL WL.

References

1. Rudolph NS (1999) Biopharmaceutical production in transgenic livestock. Trends Biotechnol 17: 367–374.
2. Saeki K, Matsumoto K, Kinoshita M, Suzuki I, Tasaka Y, et al. (2004) Functional expression of a Delta12 fatty acid desaturase gene from spinach in transgenic pigs. Proc Natl Acad Sci U S A 101: 6361–6366.
3. Hammer RE, Pursel VG, Rexroad CE Jr, Wall RJ, Bolt DJ, et al. (1985) Production of transgenic rabbits, sheep and pigs by microinjection. Nature 315: 680–683.
4. Willadsen SM (1986) Nuclear transplantation in sheep embryos. Nature 320: 63–65.
5. Wilmut I, Schnieke AE, McWhir J, Kind AJ, Campbell KH (1997) Viable offspring derived from fetal and adult mammalian cells. Nature 385: 810–813.
6. Niemann H (2004) Transgenic pigs expressing plant genes. Proc Natl Acad Sci U S A 101: 7211–7212.
7. Kues WA, Niemann H (2004) The contribution of farm animals to human health. Trends Biotechnol 22: 286–294.
8. Wolfgang MJ, Eisele SG, Browne MA, Schotzko ML, Garthwaite MA, et al. (2001) Rhesus monkey placental transgene expression after lentiviral gene transfer into preimplantation embryos. Proc Natl Acad Sci U S A 98: 10728–10732.

9. Norgaard Glud A, Hedegaard C, Nielsen MS, Sorensen JC, Bendixen C, et al. (2010) Direct gene transfer in the Gottingen minipig CNS using stereotaxic lentiviral microinjections. Acta Neurobiol Exp (Wars) 70: 308–315.

10. Pfeifer A (2004) Lentiviral transgenesis. Transgenic Res 13: 513–522.

11. Lois C, Hong EJ, Pease S, Brown EJ, Baltimore D (2002) Germline transmission and tissue-specific expression of transgenes delivered by lentiviral vectors. Science 295: 868–872.

12. Pfeifer A, Ikawa M, Dayn Y, Verma IM (2002) Transgenesis by lentiviral vectors: lack of gene silencing in mammalian embryonic stem cells and preimplantation embryos. Proc Natl Acad Sci U S A 99: 2140–2145.

13. Jahner D, Stuhlmann H, Stewart CL, Harbers K, Lohler J, et al. (1982) De novo methylation and expression of retroviral genomes during mouse embryogenesis. Nature 298: 623–628.

14. Hofmann A, Kessler B, Ewerling S, Weppert M, Vogg B, et al. (2003) Efficient transgenesis in farm animals by lentiviral vectors. EMBO Rep 4: 1054–1060.

15. Hofmann A, Zakhartchenko V, Weppert M, Sebald H, Wenigerkind H, et al. (2004) Generation of transgenic cattle by lentiviral gene transfer into oocytes. Biol Reprod 71: 405–409.

16. Tessanne K, Golding MC, Long CR, Peoples MD, Hannon G, et al. (2012) Production of transgenic calves expressing an shRNA targeting myostatin. Mol Reprod Dev 79: 176–185.

17. Hino S, Fan J, Taguwa S, Akasaka K, Matsuoka M (2004) Sea urchin insulator protects lentiviral vector from silencing by maintaining active chromatin structure. Gene Ther 11: 819–828.

18. Mohamedali A, Moreau-Gaudry F, Richard E, Xia P, Nolta J, et al. (2004) Self-inactivating lentiviral vectors resist proviral methylation but do not confer position-independent expression in hematopoietic stem cells. Mol Ther 10: 249–259.

19. Hofmann A, Kessler B, Ewerling S, Kabermann A, Brem G, et al. (2006) Epigenetic regulation of lentiviral transgene vectors in a large animal model. Mol Ther 13: 59–66.

20. Lavitrano M, Bacci ML, Forni M, Lazzereschi D, Di Stefano C, et al. (2002) Efficient production by sperm-mediated gene transfer of human decay accelerating factor (hDAF) transgenic pigs for xenotransplantation. Proc Natl Acad Sci U S A 99: 14230–14235.

21. Ritchie WA, King T, Neil C, Carlisle AJ, Lillico S, et al. (2009) Transgenic sheep designed for transplantation studies. Mol Reprod Dev 76: 61–64.

22. Rubinson DA, Dillon CP, Kwiatkowski AV, Sievers C, Yang L, et al. (2003) A lentivirus-based system to functionally silence genes in primary mammalian cells, stem cells and transgenic mice by RNA interference. Nat Genet 33: 401–406.

23. Schweizer J, Valenza-Schaerly P, Goret F, Pourcel C (1998) Control of expression and methylation of a hepatitis B virus transgene by strain-specific modifiers. DNA Cell Biol 17: 427–435.

24. Pikaart MJ, Recillas-Targa F, Felsenfeld G (1998) Loss of transcriptional activity of a transgene is accompanied by DNA methylation and histone deacetylation and is prevented by insulators. Genes Dev 12: 2852–2862.

25. Yang SH, Cheng PH, Sullivan RT, Thomas JW, Chan AW (2008) Lentiviral integration preferences in transgenic mice. Genesis 46: 711–718.

26. Lo Bianco C, Schneider BL, Bauer M, Sajadi A, Brice A, et al. (2004) Lentiviral vector delivery of parkin prevents dopaminergic degeneration in an alpha-synuclein rat model of Parkinson's disease. Proc Natl Acad Sci U S A 101: 17510–17515.

27. Reichenbach M, Lim T, Reichenbach HD, Guengoer T, Habermann FA, et al. (2010) Germ-line transmission of lentiviral PGK-EGFP integrants in transgenic cattle: new perspectives for experimental embryology. Transgenic Res 19: 549–556.

28. Scott BB, Lois C (2005) Generation of tissue-specific transgenic birds with lentiviral vectors. Proc Natl Acad Sci U S A 102: 16443–16447.

29. Singer O, Tiscornia G, Ikawa M, Verma IM (2006) Rapid generation of knockdown transgenic mice by silencing lentiviral vectors. Nat Protoc 1: 286–292.

30. Ikawa M, Tanaka N, Kao WW, Verma IM (2003) Generation of transgenic mice using lentiviral vectors: a novel preclinical assessment of lentiviral vectors for gene therapy. Mol Ther 8: 666–673.

31. Whitelaw CB, Radcliffe PA, Ritchie WA, Carlisle A, Ellard FM, et al. (2004) Efficient generation of transgenic pigs using equine infectious anaemia virus (EIAV) derived vector. FEBS Lett 571: 233–236.

32. McGrew MJ, Sherman A, Ellard FM, Lillico SG, Gilhooley HJ, et al. (2004) Efficient production of germline transgenic chickens using lentiviral vectors. EMBO Rep 5: 728–733.

33. Kong Q, Wu M, Huan Y, Zhang L, Liu H, et al. (2009) Transgene expression is associated with copy number and cytomegalovirus promoter methylation in transgenic pigs. PLoS One 4: e6679.

34. Wang Y, Song YT, Liu Q, Liu C, Wang LL, et al. (2010) Quantitative analysis of lentiviral transgene expression in mice over seven generations. Transgenic Res 19: 775–784.

35. Zielske SP, Lingas KT, Li Y, Gerson SL (2004) Limited lentiviral transgene expression with increasing copy number in an MGMT selection model: lack of copy number selection by drug treatment. Mol Ther 9: 923–931.

17

Dre - Cre Sequential Recombination Provides New Tools for Retinal Ganglion Cell Labeling and Manipulation in Mice

Szilard Sajgo[1,2,☉], Miruna Georgiana Ghinia[1,2,☉], Melody Shi[1], Pinghu Liu[1], Lijin Dong[1], Nadia Parmhans[1], Octavian Popescu[2,3], Tudor Constantin Badea[1]*

1 National Eye Institute, NIH, Bethesda, Maryland, United States of America, 2 Biology Department, Babes-Bolyai University, Cluj-Napoca, Cluj, Romania, 3 Institute of Biology, Romanian Academy, Bucharest, Romania

Abstract

Background: Genetic targeting methods have greatly advanced our understanding of many of the 20 Retinal Ganglion Cell (RGC) types conveying visual information from the eyes to the brain. However, the complexity and partial overlap of gene expression patterns in RGCs call for genetic intersectional or sparse labeling strategies. Loci carrying the Cre recombinase in conjunction with conditional knock-out, reporter or other genetic tools can be used for targeted cell type ablation and functional manipulation of specific cell populations. The three members of the Pou4f family of transcription factors, Brn3a, Brn3b and Brn3c, expressed early during RGC development and in combinatorial pattern amongst RGC types are excellent candidates for such gene manipulations.

Methods and Findings: We generated conditional Cre knock-in alleles at the Brn3a and Brn3b loci, $Brn3a^{CKOCre}$ and $Brn3b^{CKOCre}$. When crossed to mice expressing the Dre recombinase, the endogenous Brn3 gene expressed by $Brn3a^{CKOCre}$ or $Brn3b^{CKOCre}$ is removed and replaced with a Cre recombinase, generating $Brn3a^{Cre}$ and $Brn3b^{Cre}$ knock-in alleles. Surprisingly both $Brn3a^{Cre}$ and $Brn3b^{Cre}$ knock-in alleles induce early ubiquitous recombination, consistent with germline expression. However in later stages of development, their expression is limited to the expected endogenous pattern of the $Brn3a$ and $Brn3b$ genes. We use the $Brn3a^{Cre}$ and $Brn3b^{Cre}$ alleles to target a Cre dependent Adeno Associated Virus (AAV) reporter to RGCs and demonstrate its use in morphological characterization, early postnatal gene delivery and tracing the expression of Brn3 genes in RGCs.

Conclusions: Dre recombinase effectively recombines the $Brn3a^{CKOCre}$ and $Brn3b^{CKOCre}$ alleles containing its roxP target sites. Sequential Dre to Cre recombination reveals Brn3a and Brn3b expression in early mouse development. The generated $Brn3a^{Cre}$ and $Brn3b^{Cre}$ alleles are useful tools that can target exogenously delivered Cre dependent reagents to RGCs in early postnatal development, opening up a large range of potential applications.

Editor: Alexandre Hiroaki Kihara, Universidade Federal do ABC, Brazil

Funding: This study was supported by the National Institutes of Health. Partial funding was provided to M.G.G. by the European Union doctoral fellowship system through POSDRU Program 107/1.5/S/76841. The funders had no role in study design, data collection and analysis, decision to publish, or preparation of the manuscript.

Competing Interests: The authors have declared that no competing interests exist.

* E-mail: badeatc@mail.nih.gov

☉ These authors contributed equally to this work.

Introduction

Understanding the development and functioning of neuronal circuits is greatly enhanced by mouse genetic tools, which enable the labeling, ablation or functional manipulation of individual neuronal cell types [1–3]. Genetically labeled mouse lines help in the definition of neuronal cell types, by allowing an integrated description of morphological, physiological and molecular features of the marked neuronal populations [4–7]. In addition, ablation, activation or silencing of specific neuronal populations, most easily achieved by genetic means, greatly helps in understanding their function within the circuit [8–11]. Our specific focus, RGCs, the neurons that carry the visual information from the eye to the brain, are a heterogeneous group comprised of about 20 different

individual cell types, which have unique combinations of anatomic and physiologic features, and carry out distinct visual functions [12–15]. Genetic marking methodologies have greatly advanced our understanding of the functioning of several RGC types. However, a majority of RGC types still cannot be approached by genetic manipulations. To expand the genetic tools enabling us access to RGC populations and RGC types, we are focusing on the three members of the POU4 family of transcription factors, Brn3a, Brn3b and Brn3c, which are expressed in a combinatorial pattern amongst partially overlapping but not identical populations of adult RGC types [16]. Brn3s are expressed early in RGC development, and play diverse roles in RGC type specification [17–21]. Thus, Brn3b plays a major role in the specification of a large fraction of RGCs (70%), and seems to be involved in a major fashion in axon formation and a more subtle one in dendritic arbor elaboration. On the other hand, Brn3a although broadly expressed in RGCs, especially in the adult, seems to play a role in specification of a particular subset of RGCs, with small and multistratified dendritic arbors. Brn3c, which is expressed in only three RGC types in the adult, currently has an ill-defined function in RGC development. To understand the exact RGC type distribution of Brn3 transcription factors, we have previously generated conditional reporter knock-in alleles ($Brn3^{CKOAP}$), in which the specific endogenous Brn3 gene is selectively ablated and replaced with the histochemical reporter, Alkaline Phosphatase (AP), in a Cre recombinase dependent manner [8]. These alleles enabled us to identify dendritic arbor morphologies and axonal targets within the brain, thus defining RGC types and eventually allowing us to correlate visual functions with particular types of RGCs. They also allowed us to study the fate of RGCs as a group or individually after cell specific ablation of the individual Brn3 gene ($Brn3^{AP/-}$), by tracking the axons and dendrites of targeted and hence AP labeled neurons.

Despite these insights, many questions remain unanswered. For instance, although Brn3s are expressed in partially overlapping RGC populations in the adult, it is not clear which fraction of RGC types express each Brn3 gene throughout development, and what role each gene plays in their specification. Also, although the AP reporter is extremely powerful for following morphology during development and through adulthood, as well as diagnosing anatomical defects resulting from genetic manipulations, there are a variety of other questions that will be better addressed by targeting a Cre recombinase to the expression domain of each Brn3 gene. These include lineage tracing, using fluorescent reporters of neuronal activity, performing RGC activation or inhibition using light inducible activation or inhibitory channels, etc. Besides having a very well characterized expression domain and history compared to all other genes in RGCs, Brn3s have the crucial advantage of being the first known RGC markers, expressed as soon as RGCs become postmitotic, at embryonic day E11– E12. Thus, Cre lines generated at the Brn3 loci ought to be extremely useful for the study of RGC development and function.

The Brn3 genes and RGCs however also demonstrate one of the major challenges we have when using genetics to manipulate neuronal cell types. In only very rare instances are gene expression patterns perfectly overlapping with individual RGC types. A majority of RGC associated genes are expressed in several types of RGCs. Thus, in the adult retina, Brn3b and Brn3a seem to be expressed each in 10 distinct RGC types, while Brn3c is expressed only in 3 RGC types. These subpopulations are partially overlapping, with different RGC types expressing either none, one, two or three of the Brn3 genes [16]. It is then a meaningful approach to seek intersectional genetic strategies, based on the Brn3 genes, in order to uniquely label or target a particular RGC type.

The spectrum of gene manipulation tools is continuously expanding, ranging from recombinases to transcriptional regulators [22–25]. Strategies based on recombinases are extremely powerful because of their binary switch capability, in which the target gene will be in one state before recombination and a different state afterwards [26]. The most broadly used recombinase is the Cre protein derived from the P1 phage [27]. It is extremely active and specific for its loxP targets sites, and a series of mutant loxP sites are available which react with each other but not with the wild type loxP, resulting in a multitude of gene ablation, reporter induction or gene manipulation strategies [26,28–30].

Two other recombinase – target systems, the FLP recombinase – FRT [31–33] and the PhiC31-att [34,35] system have been developed over the last two decades and are being used in a variety of applications and contexts. They show great specificity for the target sites, but have somewhat reduced efficiencies compared to Cre, although new improved variants have been developed [36]. Several other recombinases from different origins have been recently added to the toolbox [37–39]. Amongst these, we chose to use the Dre recombinase, a Cre homolog from the D6 phage [40]. Dre is a site specific recombinase that targets roxP sites of 32 bp containing inverted repeat arms and a central spacer of 4 bases, and has high levels of activity in eukaryotic cells, and transgenic mice [40,41]. Dre and Cre do not cross react with respect to their roxP and loxP target sites, and based on the high level of homology, their mechanism of action is likely to be very similar [40]. A codon optimized version has been recently developed, and a transgenic mouse line carrying the Dre under the control of the CAG promoter is available from mouse repositories [41]. Thus, it appears that Dre is an excellent candidate for generating combinatorial strategies in conjunction with the more established Cre, Flp and PhiC31 recombinases.

We therefore developed Dre dependent conditional Cre knock-in alleles for the Brn3a and Brn3b loci, trying to achieve several goals. First, we would like to be able to target RGC specific genes as early as feasible during development, taking advantage of the early expression of the Brn3 genes in RGCs. Second, we would like to perform lineage experiments in which the dynamic expression pattern of the Brn3 genes in RGCs is captured. Third, we would like to generate genetic intersections in which only RGCs expressing two of the Brn3 genes are labeled. Fourth, we would like to be able to do this in the context of wild type or mutant Brn3 alleles, and hope to develop Dre expressing drivers which will enable us to do this in a sparse or eye specific manner. Finally, we would like to direct genetic tools for neuronal ablation, activation or inhibition to specific RGC populations.

Unfortunately, due to technical limitations in the targeting constructs, as well as a newly documented expression of Brn3 genes in the germline and/or early embryo, not all these goals will be achievable with the generated alleles. However, besides demonstrating the efficiency and power of the Dre recombinase in mouse genetics, our work brings new insights and useful tools. Brn3a and Brn3b are broadly expressed in early stages of development, but are restricted to correct target populations thereafter, and thus conventional Cre alleles will result in whole embryo recombination when used in conjunction with Cre dependent conditional alleles and reporter lines. However, Cre dependent vectors such as AAV FLEX cassettes can be delivered during relevant stages of RGC development, resulting in accurate and efficient targeting in a Brn3 specific manner.

Materials and Methods

Mouse Lines and Targeting

Previously reported mouse lines used in this study are: a) CAG:Dre transgenic line, that drives early ubiquitous expression of the Dre recombinase [41], b) $ROSA26^{iAP}$, conditional knock-in line that expresses AP in a Cre recombination dependent manner from the early, ubiquitous ROSA26 locus [42,43], c) $ROSA26^{CreERt}$, expressing a 4-Hydroxytamoxyphen inducible Cre recombinase in an early ubiquitous manner [44], d) $ROSA26^{rtTACreERt}$, which expresses Cre recombinase in an early ubiquitous manner under dual pharmacological control of Doxycycline and 4-Hydroxyta-moxyphen [42], and e) $Brn3a^{CKOAP}$ and $Brn3b^{CKOAP}$, which allow for the conditional replacement of the *Brn3a* or *Brn3b* loci with AP in a Cre dependent manner [8].

The *Brn3a*^{CKOCre} and *Brn3b*^{CKOCre} conditional alleles were generated by homologous recombination in mouse embryonic stem cells using analogous strategies to the ones used to generate the *Brn3a*^{CKOAP} and *Brn3b*^{CKOAP} alleles previously reported [8]. The following changes were made from the original gene structure: a roxP site was inserted in the 5′ UTR 42 bp 50 (for Brn3a) or 98 bp 5′ (for Brn3b) before the initiator codon ATG; 3 repeats of the SV40 early region transcription terminator were added to the 3′ UTR 48 bp (for Brn3a) or 340 bp (for Brn3b) 3′ of the Brn3 translation termination codon, followed by a second roxP site and the coding region of the Cre recombinase. A positive selection cassette (PGK-Neo), flanked by FRT sites, followed the Cre cDNA and was subsequently removed by crossing to mice expressing FLP recombinase in the germline. All mice used were of mixed C57Bl6/SV129 background. All mouse handling procedures used in this study were approved by the National Eye Institute Animal Care and Use Committee (ACUC) under protocols NEI 640, NEI 651 and NEI 652.

Doxycycline, 4-hydroxytamoxifen (4HT) and AAV Treatments

Induction of recombination in the *ROSA26*^{rtTACreERt}; *Brn3b*^{CKOAP} mice was achieved by administering doxycycline food (200 mg/gr chow) to pregnant females on alternating days between embryonic days E2– E10 and one 12.5 µg 4HT IP injection at E10. For viral intraocular injection, postnatal day 0 (P0) pups were anesthetized by hypothermia for 30 seconds, a slit was cut in the eye lid, and roughly 0.2 µl of a mix of two viral vectors including AAV1.CAG.FLEX.tdTomato.WPRE.bGH (at 5.7 * e8, Gene Therapy Program at the University of Pennsylvania, Philadelphia, PA) expressing red fluorescent tdTomato in a Cre recombination dependent manner, and a AAV1.CAG.GFP, expressing GFP in a Cre independent manner (1*e8 GFP virus, gift of Dr. Peter Colosi and Zujhian Wu), were injected, using a femtojet (Eppendorf) fitted with a pulled glass capillary. Pups were then returned to their mothers and collected at P3, P7 and adult.

Histology and Immunohistochemistry

Alkaline Phosphatase histochemistry of retina flat mounts and brain vibratome sections were performed as previously described [8,16,44,45]. Wholemount embryos of genotypes indicated in the text were collected from timed pregnancies at E9.5 and E12.5, immersion fixed overnight in 4% Paraformaldehyde at 4 C, heat inactivated for one hour at 65 C in PBS, and then AP histochemistry was developed for 2–4 hours. Eyes from virus-infected pups or adult mice were enucleated, prefixed in 2% PFA for 15 minutes at room temperature, the cornea was cut out, the lens removed, and the eye cups were further fixed for 30 minutes. The retinas were then separated from the rest of the eye tissues, and flat mounted in Aquamount. For immunofluorescence experiments, eyes were fixed for 30 minutes in 2% PFA, equilibrated in sucrose and embedded in OCT. Endogenous fluorescence was detected in the red channel for tdTomato, while Brn3a, Brn3b (own production, previously reported [46] and NFL (Chemicon – Millipore, AB9568) were immunodetected with rabbit antibodies and Alexa 647 Donkey anti-rabbit secondary (Molecular Probes – life technologies). The Alexa 647 signals, detected with an infrared filter were pseudocolored in green for double immunofluorescence experiments. Imaging for all retina preparations were performed with a Zeiss AxioImager.2, with a 5x objective for survey of the entire retina, or 20x, 40x or 63x magnification and the Apotome for immunofluorescence detec-

tion, and collected with either color or black-and-white Zeiss Axiocams. Brain sections were imaged with a color Zeiss Axiocam adapted onto a Zeiss Discovery. V8 stereomicroscope. Images were imported with custom written ImageJ plugins, LUTs corrected and, where necessary z-stack projections were generated under ImageJ.

Statistical Analysis

For each immunostaining – genotype combination, 3 to 5 pictures at 20x magnification were taken from retina sections of multiple animals. For each image, the numbers of marker, tdTomato or double positive cells were expressed as percentages of DAPI positive cells in the section, and represented as Box-Whisker plots. Explanation of Box Whisker plots: the tops and bottoms of each "box" are the 25th and 75th percentiles of the samples, respectively. The distances between the tops and bottoms are the interquartile ranges. The line in the middle of each box is the sample median. Whiskers are drawn from the ends of the interquartile ranges to the furthest observations within the whisker length (the adjacent values). Student T tests under assumption of normal distribution and Kolmogorov-Smirnov (KS2) tests for comparison of two data sets of unknown distributions were performed, with comparable results for the relevant comparisons (tdT only versus tdT & Marker), using Matlab (The Mathworks, Inc.). All n, averages and standard deviations are reported in Tables (S1 and S2).

Results

Generation of *Brn3a*^{CKOCre} and *Brn3b*^{CKOCre} Conditional Cre Knock-in Alleles

Based on the previously described *Brn3*^{CKOAP} alleles, we have generated two new targeting constructs in which the loxP sites were replaced with roxP sites and the AP open reading frame with a Cre recombinase (Fig. 1). Thus, the coding region of each Brn3 gene is comprised between roxP sites, and followed by the cDNA for the Cre recombinase. The loci are designed such that, after Dre mediated recombination between the two roxP sites, the Brn3a or Brn3b gene is removed and replaced with the Cre recombinase, which will now be expressed under the control of the respective endogenous Brn3 locus. It carries most of the 3′ UTR of the respective Brn3 mRNA, and is expected to faithfully reproduce the expression profile of the endogenous Brn3 gene. The targeting constructs also contain a PGK-Neo selection cassette flanked by FRT sites.

For each Brn3a and Brn3b the diagrams in Figure 1 show the wild type loci (Fig. 1 D, J) and the targeted alleles before (Fig. 1 A, G) and after (Fig. 1 B, H) removal of the FRT- PGK-Neo-FRT selection cassette via Flp mediated recombination. Figure 1, C and I show schematics of the Brn3a and Brn3b loci after Dre mediated recombination, resulting in loss of the Brn3 coding exons, and their replacement with the Cre recombinase. Correct targeting and removal of the PGK-Neo selection cassette was tested for both Brn3a (Fig. 1 E and F) and Brn3b (Fig. 1 K and L) by southern blotting of genomic DNA with 5′ and 3′ probes, and the expected fragments confirmed. Both *Brn3a*^{CKOCre} and *Brn3b*^{CKOCre} lines can be maintained as homozygotes and are viable and fertile, suggesting, at least for the Brn3a locus, that the endogenous Brn3a gene expression has not been majorly disrupted by the conditional targeting event.

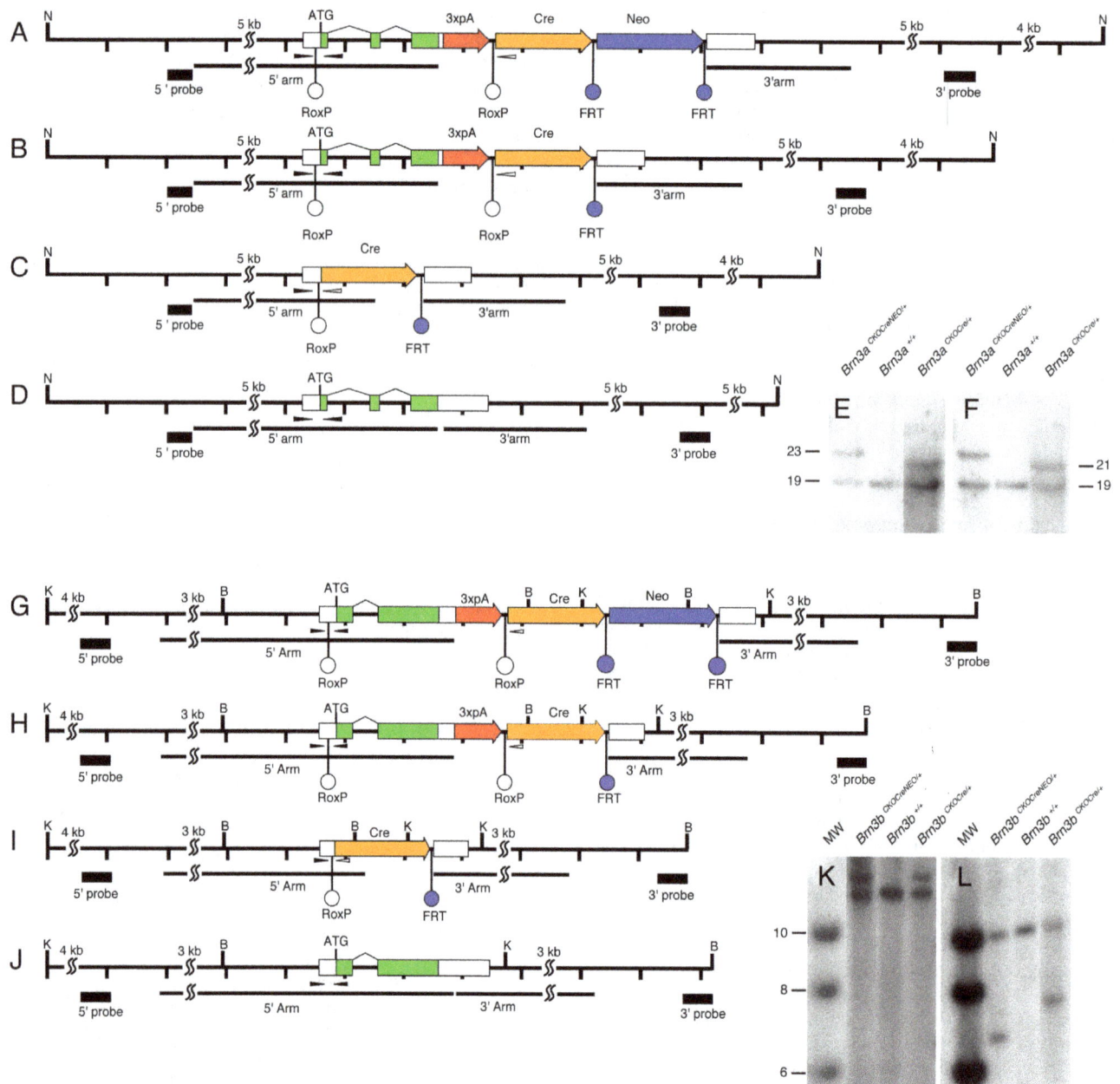

Figure 1. Conditional Cre knock-in constructs at the Brn3a and Brn3b loci. Diagrams represent the targeting strategy and recombination steps for the Brn3a (A–D) and Brn3b (G–J) locus. D, J, wild type configuration, A,G, locus after gene targeting, B,H, locus after removal of the Neo selection cassette by cross with germline Flp recombinase expressing mice, C,I, locus after removal of the endogenous gene by Dre mediated recombination, resulting in expression of the Cre recombinase gene from the original transcription start site of the Brn3 gene. Tick marks represent 1 kb distances, except for contracted longer stretches marked with SS marks. Green and white boxes represent coding regions and UTRs respectively, and splice junctions are indicated by angled black lines. Homologous recombination targeting arms are indicated by black lines, and southern blot probes as black boxes under the locus. Diagnostic restriction enzymes are represented as capital letters: Nde I (N), Kpn I (K), and Bam HI (B). A triple repeat of the SV40 polyA is inserted in the 3'UTR of the targeted locus, and indicated by a red thick arrow. RoxP sites (white circles), flank the endogenous gene while FRT sites (blue circles), flank the PGK-Neo Cassette. Genotyping primers (thin black arrowheads) are placed in the 5' UTR of the loci, and encompass the ATG in the wild type and ATG and 5' roxP site in the targeted locus. After Dre recombination (C, I) the reverse strand black primer is lost, and correct localization of the Cre can be tested by the black and white pair of primers indicated (see also figure 2). Southern blot fragment lengths are as follows: for Brn3a, the Nde I fragment is recognized by both 5' and 3' probes, and is 19116 bp for the wild type, 23542 for the targeted locus, and 21637 bp for the targeted locus after Flp mediated Neo cassette removal; for Brn3b, Kpn I digests are tested with the 5' probe, recognizing fragments of 12624 bp in wild type, and 14388 bp in the targeted construct, before and after Neo cassette removal; Bam HI digests, tested with the 3' probe yielding characteristic fragments of 10140 bp in the wild type, 6772 bp in the targeted locus, and 7682 bp after Neo removal. E, F, Genotyping by southern blot for the 5' (E) and 3' (F) ends of the Brn3a targeting event. For each E and F, targeted locus (A), wild type control (D), and targeted locus after Neo removal (B) are shown. K, L Genotyping by southern blot for the 5' (K) and 3' (L) ends of the Brn3b targeting event. For each K and L, molecular weight (MW) markers are followed by targeted locus (G), wild type control (J), and targeted locus after Neo removal (H).

Testing Sequential Dre to Cre Recombination in CAG:Dre; Brn3^{CKOCre}; ROSA26^{iAP} Triple Transgenic Mice

To date, only one publication reports successful Dre recombination in transgenic mice [41]. To test whether our conditional alleles are indeed susceptible to Dre mediated recombination, we crossed them with the ubiquitous Dre driver line, CAG:Dre, described by Annastasiadis and coworkers [41], and our previously published Cre reporter line, ROSA26^{iAP}, generating triple transgenic animals as well as appropriate double transgenic controls (Fig. 2). The expected succession of recombination events (Fig. 2A) involves early/germline expression of the Dre recombinase followed by conditional ablation of the Brn3 gene and induction of Cre expression from the Brn3 locus, which will be reported by expression of the Cre dependent AP, transcribed from the ubiquitous ROSA26 locus. Detection of the recombination event was done by two diagnostic genotyping primer pairs (Fig. 1, C, D, I, J and Fig. 2 A, D, and E). The first set of primers, shown as black arrow heads in Fig. 1, D and J and numbered as primer pairs Pr4 for Brn3a and Pr6 for Brn3b in Figure 2Ab and E, consists of a forward primer placed in the 5′UTR of the respective Brn3 gene, upstream of the insertion site of the roxP sequence, and a reverse primer, inserted in the open reading frame of the first exon, so downstream of the insertion site of the roxP. These primers will detect the wild type configuration of the loci, and a relative upward shift of 32 bp for the roxP insertion in Brn3a$^{CKOCre/+}$; ROSA26$^{iAP/+}$ and Brn3b$^{CKOCre/+}$; ROSA26$^{iAP/+}$ mice (Fig. 2E, iii and v). However, in the triple transgenic CAG:Dre; Brn3a$^{CKOCre/+}$; ROSA26$^{iAP/+}$ and CAG:Dre; Brn3b$^{CKOCre/+}$; ROSA26$^{iAP/+}$ mice, Dre recombination results in removal of the sequence between the two roxP sites, and hence of the reverse primer of primer pairs Pr4 and Pr6, and thus only the shorter product, generated from the wild type Brn3 allele, can be detected (Fig. 2E, ii and iv). In contrast, bringing the Cre recombinase cDNA in close apposition to the Brn3 5′UTR generates a novel PCR product between the forward primer and a reverse primer, placed in the Cre recombinase (black and white arrowheads in Figure 1, C and I and primer pairs Pr5 and Pr7 in Figure 2Ab and E, ii and iv). The specific removal of the diagnostic roxP products (Pr4 and Pr6) and generation of the Brn3-Cre products (Pr5 and Pr7), in CAG:Dre; Brn3$^{CKOCre/+}$; ROSA26$^{iAP/+}$ mice but not in Brn3$^{CKOCre/+}$; ROSA26$^{iAP/+}$ controls, (Fig. 2E) demonstrates that Dre recombination was successful and complete in these animals.

Sequential Dre to Cre Recombination Reveals Early/ubiquitous Expression of Cre from the Endogenous Brn3a and Brn3b Loci

We tested the expression pattern of the AP reporter in our triple transgenic CAG:Dre; Brn3a$^{CKOCre/+}$; ROSA26$^{iAP/+}$ and CAG:Dre; Brn3b$^{CKOCre/+}$; ROSA26$^{iAP/+}$ mice by histochemical staining of flat mount retinas and coronal brain sections (Figure 3). To our surprise, both retinas and brains of adult triple transgenic animals were completely and homogeneously stained, suggesting ubiquitous recombination of the ROSA26$^{iAP/+}$ Cre reporter locus (Fig. 3C, F, I, L). In contrast, tissues from CAG:Dre; ROSA26$^{iAP/+}$ littermates, stained and processed under identical condition, showed no staining, confirming previous results that the Dre recombinase by itself cannot mediate recombination of loxP sites. However, the retinas of Brn3a$^{CKOCre/+}$; ROSA26$^{iAP/+}$ and Brn3b$^{CKOCre/+}$; ROSA26$^{iAP/+}$ littermates showed a number of RGCs labeled with AP (Fig. 3 B, H), suggesting Dre independent production of Cre from the Brn3a$^{CKOCre/+}$ and Brn3b$^{CKOCre/+}$ loci. Consistent with this, brain sections from both Brn3$^{CKOCre/+}$; ROSA26$^{iAP/+}$ mice show labeling of the RGC and other Brn3-

expressing axonal tracts and nuclei (Fig. 3 E, K). Although the AP staining observed in the double transgenic Brn3$^{CKOCre/+}$; ROSA26$^{iAP/+}$ retinas and brains marks only a few percent of previously described Brn3 target neurons, these numbers make the two alleles less useful for sparse recombination approaches. We were also intrigued by the apparently complete neuronal recombination pattern seen in the triple transgenic animal. This is unlikely explained by a defect in our targeting design or execution, as the previously reported Brn3a$^{CKOAP/+}$ and Brn3b$^{CKOAP/+}$ alleles, identical from a gene regulation perspective, have been shown to correctly reflect the endogenous expression pattern of the Brn3a and Brn3b genes in several neuronal targets [8,16,45]. In addition, both Southern blot and PCR evidence suggest that the homologous recombination and Neo selection removal happened correctly (Fig. 1 and 2), and similar ubiquitous recombination patterns were observed with two independent Brn3$^{CKOCre/+}$ lines. Moreover, we bred out the recombined Brn3b$^{Cre/+}$ allele (Fig. 1H) from CAG:Dre; Brn3b$^{CKOCre/+}$ offspring, and crossed it to ROSA26$^{iAP/iAP}$ animals to generate Brn3b$^{Cre/+}$; ROSA26$^{iAP/+}$ mice, which had full recombination patterns identical to the triple transgenics (data not shown), suggesting that this effect is not dependent on the Dre to Cre interaction. A previous report had suggested expression of Brn3a and Brn3b in both the male and female germline (ref [47], see Fig. 1 for evidence of Brn3a and Brn3b expression in both testis and ovary). We therefore performed the crosses resulting in generation of triple transgenic animals, CAG:Dre; Brn3^{CKOCre}; ROSA26$^{iAP/+}$, providing the Brn3^{CKOCre} alleles from either the male or female parent, with similar results being obtained (Fig. 2 B and C). To test how early the Dre to Cre sequential recombination occurred, we stained E9–E10 CAG:Dre; Brn3$^{CKOCre/+}$; ROSA26$^{iAP/+}$, triple transgenic embryos in whole mount and found that indeed the entire embryo was AP positive (Fig. 4C, F), whereas the CAG:Dre; ROSA26$^{iAP/+}$ and Brn3$^{CKOCre/+}$; ROSA26$^{iAP/+}$ control embryos showed no or sparse recombination respectively (Fig. 4B, D, E). Finally, we stained E12.5 ROSA26$^{CreERt/+}$; Brn3a$^{CKOAP/+}$ (not shown) and ROSA26$^{rtTACreERt/+}$; Brn3b$^{CKOAP/+}$ (Fig. 4A) embryos, in which sparse recombination had been induced as previously [16,42], and noticed abundant AP expression in the germ ridge (see ref [48] for a comparison). Thus we can conclude that Brn3a$^{CKOCre/+}$ and Brn3b$^{CKOCre/+}$ alleles are expressed in ubiquitous fashion throughout the embryo at a time point before E9, and most likely in the germline of both males and females.

Brn3a^{Cre} and Brn3b^{Cre} are Driving RGC Specific Recombination at Postnatal Ages

Given the surprising early ubiquitous expression of Cre from the newly generated Brn3$^{CKOCre/+}$ alleles and the Brn3b$^{Cre/+}$ derivative, we wanted to know whether Cre expression would reproduce Brn3 expression at later stages of development. Since transgenic reporters would likely be converted by the early recombination effect, we used a Cre dependent reporter delivered by an AAV vector (Fig. 5). The viral vector was AAV1.CAG.FLEX.tdTomato.WPRE.bGH (henceforth AAVFLEXtdT), using an AAV1 capsid known to infect many retinal neuronal cell types, including RGCs [49], containing a ubiquitously expressing chicken beta actin enhancer CMV promoter fusion (CAG), and a red fluorescent tdTomato cDNA [50] in reverse orientation, flanked by the arms of the FLEX cassette [51], and followed by a WPRE (a posttranscriptional regulatory element derived from a woodchuck hepatitis virus [52]) and bovine Growth Hormone polyadenilation signal. This virus has been previously used to report Cre expression in transgenic models [53,54], and we tested its Cre dependent red fluorescence expression by infecting Cre-

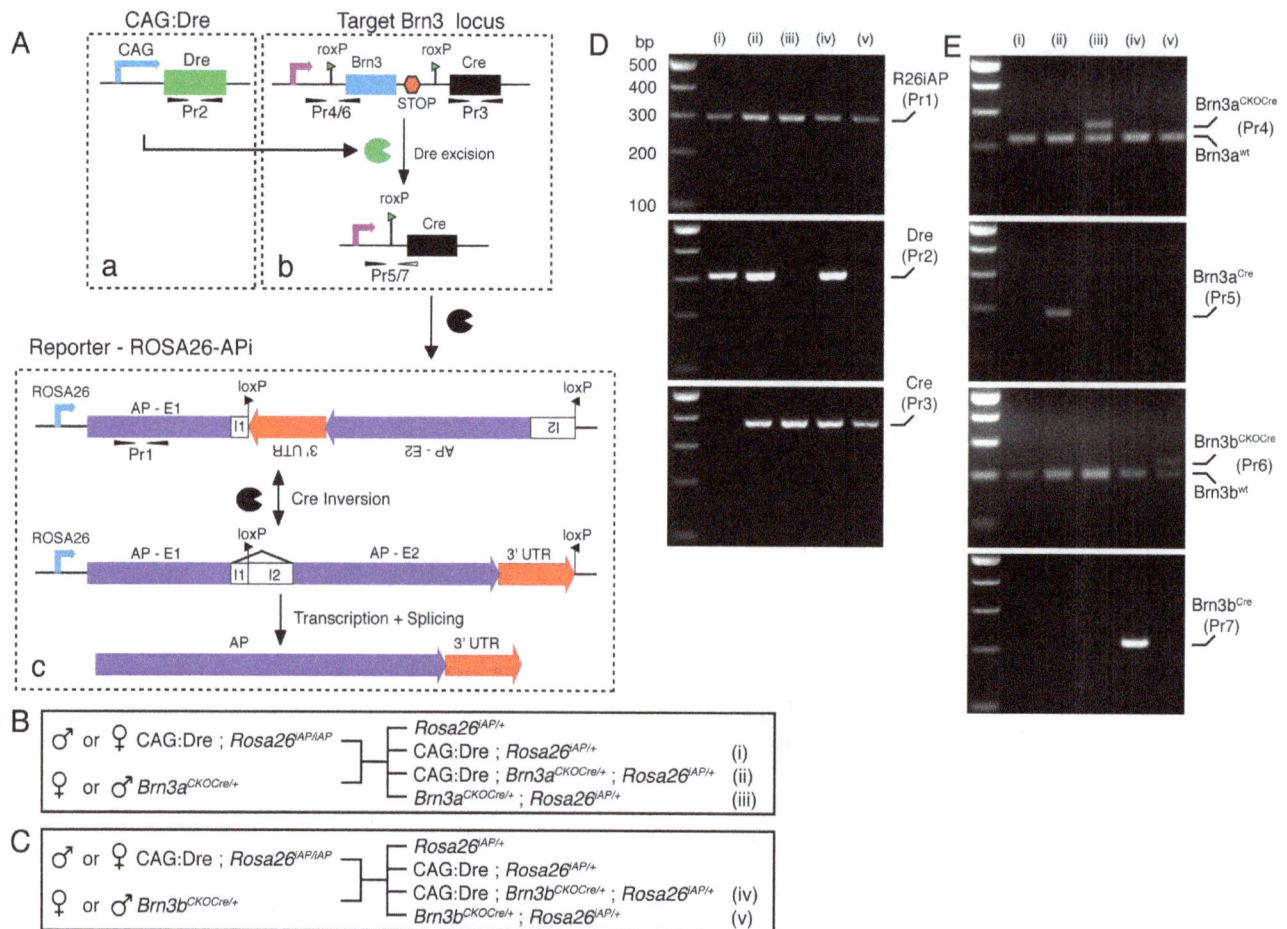

Figure 2. Recombination strategy used to test sequential Dre and Cre recombination. A, Genetic loci and recombination events. The CAG:Dre transgene (a), expresses ubiquitously Dre recombinase that targets the roxP sites of the $Brn3^{CKOCre}$ locus (b), removing the endogenous exons of the Brn3 gene and replacing them with the Cre open reading frame. After Dre recombination, primer pairs Pr4 (Brn3a) or Pr6 (Brn3b) are rendered nonproductive by the loss of the reverse strand primer and novel primer pairs Pr5 (Brn3a) and Pr7 (Brn3b) are generated, between the forward strand primer remaining 5' of the roxP site in the 5'UTR and a reverse strand primer placed in the Cre gene. The Brn3 locus starts expressing the Cre gene, which targets the inverted loxP sites of the $ROSA26^{iAP}$ locus (c), reversing the inverted exon, resulting in productive transcription of the AP protein. B, C, Crosses generating the triple transgenic mice and the appropriate controls, for $Brn3a^{CKOCre}$ and $Brn3b^{CKOCre}$ alleles. Note that, to control for the germline expression of the $Brn3^{CKOCre}$ loci, the crosses were performed in both male to female combinations. D, PCR genotyping demonstrating the presence of the AP (primer pair Pr1), Dre (primer pair Pr2) and Cre (primer pair Pr3) genes, in the various genetic knock-in combinations. E, PCR reactions demonstrating the insertion of the roxP site in the 5' UTR of the $Brn3a^{CKOCre}$ (iii, Pr4) and $Brn3b^{CKOCre}$ (v, Pr6) loci. Note that all samples have one wild type chromosome, showing a band of 220 bp for Brn3a (Pr4) and 200 bp for Brn3b (Pr6), but only the Dre – negative $Brn3a^{CKOCre}$ (iii) or $Brn3b^{CKOCre}$ (v) genetic combinations show a roxP insertion shifted by ~ 30 bp up (top band). In contrast, in the triple transgenic combinations (ii and iv) in which Dre recombination has occurred, the roxP insertion band is removed (top band, Pr4 and Pr6), and the Brn3-Cre reactions (Pr5 and Pr7) become positive.

expressing HEK293 cells and appropriate controls (data not shown). We co-injected the AAVFLEXtdT together with a constitutively GFP expressing AAV1 virus (AAV1.CAG.GFP, henceforth AAVGFP) intra-vitreously in the eyes of either $Brn3b^{Cre/+}$, $Brn3b^{Cre/Cre}$, $Brn3a^{Cre/+}$, or wild type (WT) P0.5 pups, and then tested their retinas for red and green fluorescence at postnatal day 3.5 (P3.5), 7.5 (P7.5) and in weaned adults (P21 and above) (Fig. 5A, B). Adult retinas of AAV1 injected $Brn3b^{Cre/+}$ and $Brn3a^{Cre/+}$ mice showed large numbers of tdTomato positive cells in the Ganglion Cell Layer (GCL) (Fig. 5 C, D, G, H). In $Brn3b^{Cre/Cre}$ retinas, in which the number of RGCs is greatly reduced as a result of loss of both copies of the Brn3b gene, tdTomato positive cells were still present, but in lower numbers compared to the $Brn3b^{Cre/+}$ mice, and we noticed a large number of looping axons at the periphery, consistent with previously reported RGC losses in these

mice (Fig. 5 E, F). A few isolated tdTomato positive cells were observed in WT retinas (Fig. 5 I, J). The overall success of P0 eye injections is documented by the presence of GFP positive cells, which are infected by the Cre independent AAVGFP virus. Thus it can be easily appreciated from the green channel fluorescence of the retinal flat mounts that plenty of cells were infected in all retinas (Fig. 5C–J). Taken together, these results suggest that the $Brn3a^{Cre}$ and $Brn3b^{Cre}$ alleles can indeed specifically induce the Cre dependent AAVFLEXtdT reporter. Since essentially all tdTomato positive cells are localized to the GCL (Fig. 6C), and have characteristic axonal and dendrite arbor morphologies, (Fig. 6B, D and E), we propose that a large fraction are RGCs. Moreover, virtually all tdTomato positive cells observed in adult $Brn3a^{Cre}$ or $Brn3b^{Cre}$ retinas were also positive for the RGC marker NFL (Fig. 7A, A' and D, D', compare tdT only to tdT & NFL columns, Tables S1

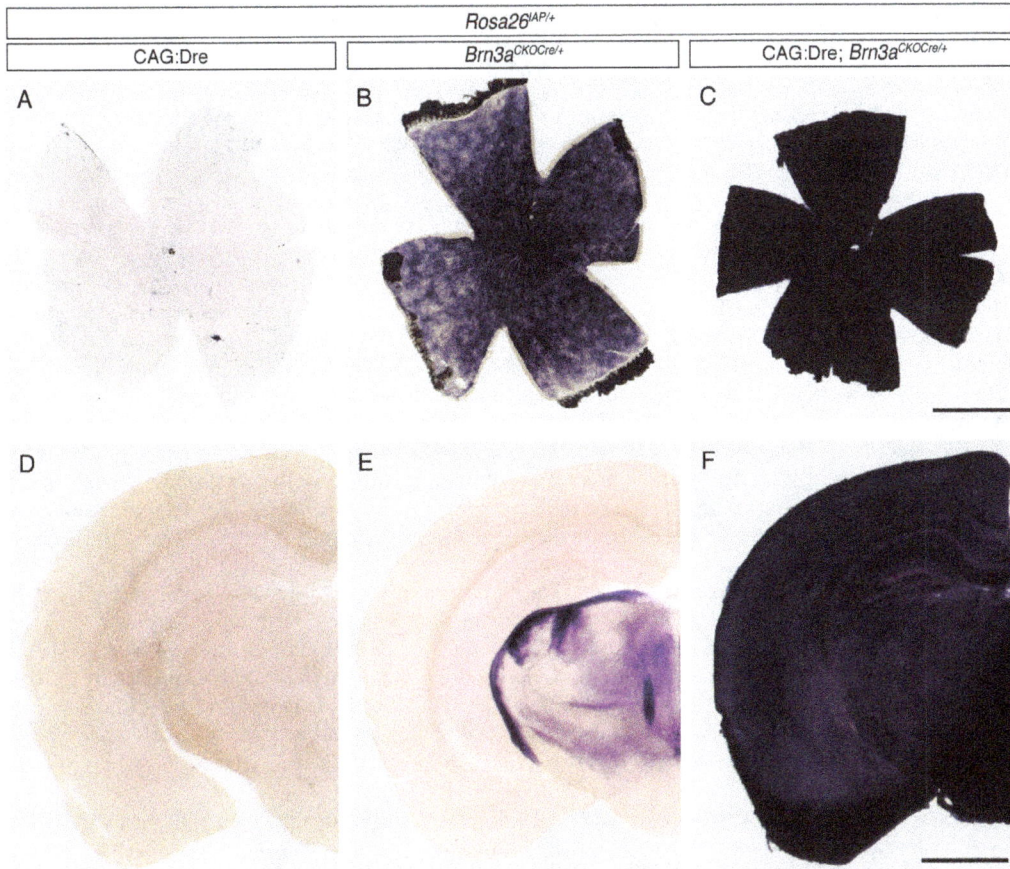

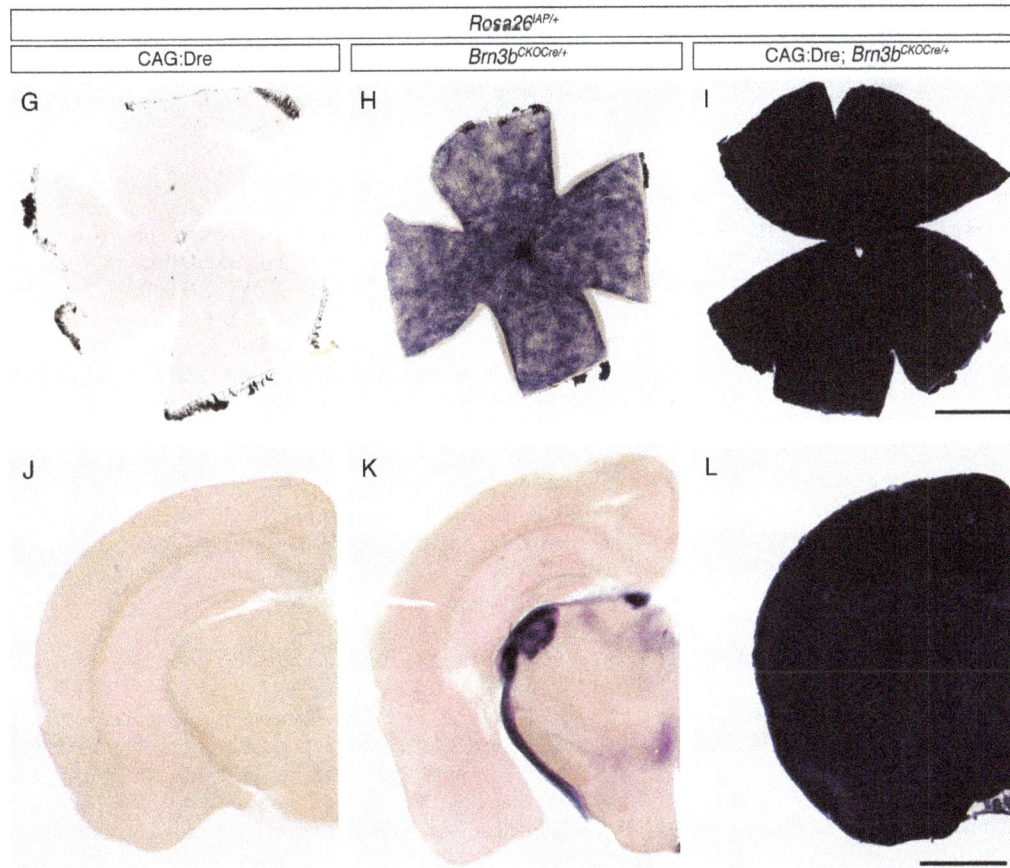

Figure 3. Sequential Dre to Cre recombination suggests ubiquitous Cre expression from the Brn3a and Brn3b loci. A–C, G–I, Adult retina flat mounts and D–F, J–L, hemispheres from coronal brain sections of mice with indicated genotypes. Note that, in CAG:Dre; ROSA26^{iAP} mice, germline Dre expression does not result in induction of AP positivity from the Cre dependent ROSA26^{iAP} locus (A, D - n = 4; G, J - n = 6). However, Brn3^{CKOCre}; ROSA26^{iAP} tissues show a reduced level of mosaic recombination in RGCs, and corresponding projection areas in the brain (B, E - n = 9; H, K – n = 13). In contrast, tissues from CAG:Dre; Brn3^{CKOCre}; ROSA26^{iAP} mice (C, F - n = 4; I, L – n = 13) show complete conversion to AP positivity, suggesting that the sequential Dre to Cre to AP recombination happened in the totality (or a vast majority) of the tissue. Scale bars in C, F, I and L are 1 mm.

and S2 for number of quantitated sections, counted cells and statistical significances), suggesting that the expression of the two Cre alleles is restricted to RGCs, at least in the early postnatal period. However, only about 50% (for Brn3a) and 66% (for Brn3b) of NFL positive cells were tdTomato positive. This could be due to incomplete infection, selective tropism of the AAV1 capsid, or restricted expression of the Brn3^{Cre} alleles in subpopulations of RGCs [16,55]. We then asked whether tdTomato cells labeled in either Brn3a^{Cre} or Brn3b^{Cre} retinas were indeed expressing the Brn3a and Brn3b proteins (Fig. 7 and Tables S1 and S2). Amongst tdTomato positive cells in Brn3b^{Cre} retinas, about 90% were Brn3b positive (Fig. 7C, C'), whereas only about 80% of tdTomato positive cells in Brn3a^{Cre} retinas were Brn3a positive (Fig. 7B, B'). Although a large majority of tdTomato positive cells generated by the Brn3a or Brn3b Cre drivers are indeed positive for the respective transcription factor in the adult, a substantive amount of cells are not. Since these cells became tdTomato positive presumably through Cre dependent recombination, the simplest explanation is that the expression domain of the Brn3a^{Cre} and Brn3b^{Cre} alleles, although restricted to RGCs, may be broader at early postnatal ages compared to the adult [56,57]. Interestingly, the degree of overlap between tdTomato and Brn3a positive cells in Brn3b^{Cre} retinas and tdTomato and Brn3b positive cells in Brn3a^{Cre} retinas were also in the range of 70–80% of all tdTomato positive cells, in keeps with the known large overlap of Brn3a and Brn3b expression in RGCs during development and in the adult [16,56,57].

Discussion

Sequential Dre to Cre Recombination

To our knowledge, this is the first report using a combinatorial genetic strategy in which a Dre and a Cre recombinase are both required to activate a reporter gene. Previous analogous strategies have been used by crossing a Flp and Cre dependent reporter to Flp and Cre expressing mouse drivers with distinct patterns of expression [58,59]. Our results show fully penetrant Dre recombination with essentially no cross-reactivity between Dre, Cre and their target sites. These results are very important, as they demonstrate the utility of Dre in combinatorial approaches. Placing the Dre recombinase at one genetic locus and intersecting it with a Dre dependent Cre recombinase expressed from another locus which will then target a reporter expressed from a third locus could theoretically allow us to target a specific cell population by the intersection of three specific gene expression patterns. Here we explore only the Dre and Cre recombination resulting from generally expressed Dre drivers and Cre reporters, but we hope in the future, we, and others will be able to develop more tissue or cell type specific alleles expressing either Dre or Dre dependent Cre expressing lines.

Dre Independent Cre Expression from the Conditional Brn3a^{CKOCre} and Brn3b^{CKOCre} Alleles

The results in figures 3 and 4 show relatively high level of AP induction from the Brn3$^{CKOCre/+}$; ROSA26$^{iAP/+}$, suggesting a leaky expression of the Cre reporter in the absence of the Dre inducible endogenous Brn3 ablation. The targeted alleles for both Brn3a^{CKOCre}

and Brn3b^{CKOCre} contain triple SV40 polyA transcription stop signals designed to prevent read through transcription from the endogenous Brn3 gene into the following Cre open reading frame. This strategy was used successfully in the previously published Brn3a^{CKOAP}, Brn3b^{CKOAP} and Brn3c^{CKOAP} alleles [8,16,45], and the loci are identical except for the substitution of the loxP sites and the AP cDNA with roxP sites and a Cre cDNA. In the case of the Brn3^{CKOAP} alleles, a moderate amount of read through can be detected even if no Cre is present, after prolonged AP staining. It may be that analogous amounts of Cre recombinase generated in the Brn3^{CKOCre} alleles are enough to recombine the reporter, thereby amplifying the effect of read through transcription. Whatever the cause, this Dre independent Cre expression makes these alleles less useful for analysis by sparse recombination, so it will have to be addressed in future designs, perhaps by creating FLEX type cassettes for roxP sites. It is worth noting that the pattern of expression revealed by the read through is consistent with the endogenous expression profiles of the Brn3a and Brn3b genes and very rarely reveals other types of neurons or cellular populations.

Early Developmental Expression of Brn3a and Brn3b

Previously only one report had documented expression of Brn3a and Brn3b in the germline [47], and both Brn3a and Brn3b KO mice develop normally throughout the intrauterine stages and Brn3b KO mice are viable and fertile, while Brn3a KO mice die at birth. However, microarray data sets from developing male and female gonad tissues, collected in the Genito Urinary Molecular Anatomy project (GUDMAP, http://www.gudmap.org/) or reported by Jameson et al, show expression of both Brn3a (Pou4f1) and Brn3b (Pou4f2) at several stages of male and female gonad development [60–62], so it is conceivable that enough Cre protein is deposited in the egg and the sperm to operate successful recombination upon fertilization. This expression profile might explain the whole animal recombination pattern we discovered in our triple transgenic animals, and which we confirmed at least for Brn3b using the derived Brn3b^{Cre} knock-in line. It is worth noting that embryos in which the Brn3a^{CKOAP} and Brn3b^{CKOAP} alleles were induced using the ubiquitous ROSA26$^{CreER/+}$ driver typically show AP expression only in the later, expected expression domain of the Brn3 genes. The early embryonic or germ line expression of the Brn3s might hence be transient, however immortalized by virtue of the Cre dependent conversion of the ubiquitous reporter. This unexpected expression profile justifies the combinatorial approach we took, and hopefully be circumvented in the future by the generation of tissue specific or sparse recombination Dre drivers.

Brn3 Loci as Drivers for RGC Specific Cre Lines

We demonstrate here that the Brn3a^{Cre} and Brn3b^{Cre} allele can successfully activate in a RGC dependent manner a fluorescent reporter delivered by an AAV virus. Since the Cre dependent AAV1-FLEX-tTomato virus is expressed only in Brn3a^{Cre} and Brn3b^{Cre} retinas but not WT controls, the labeled cell bodies are overwhelmingly restricted to the Ganglion Cell Layer, and sparsely or densely labeled cells have dendritic arbors with characteristic RGC morphologies, and axons tracking to the optic

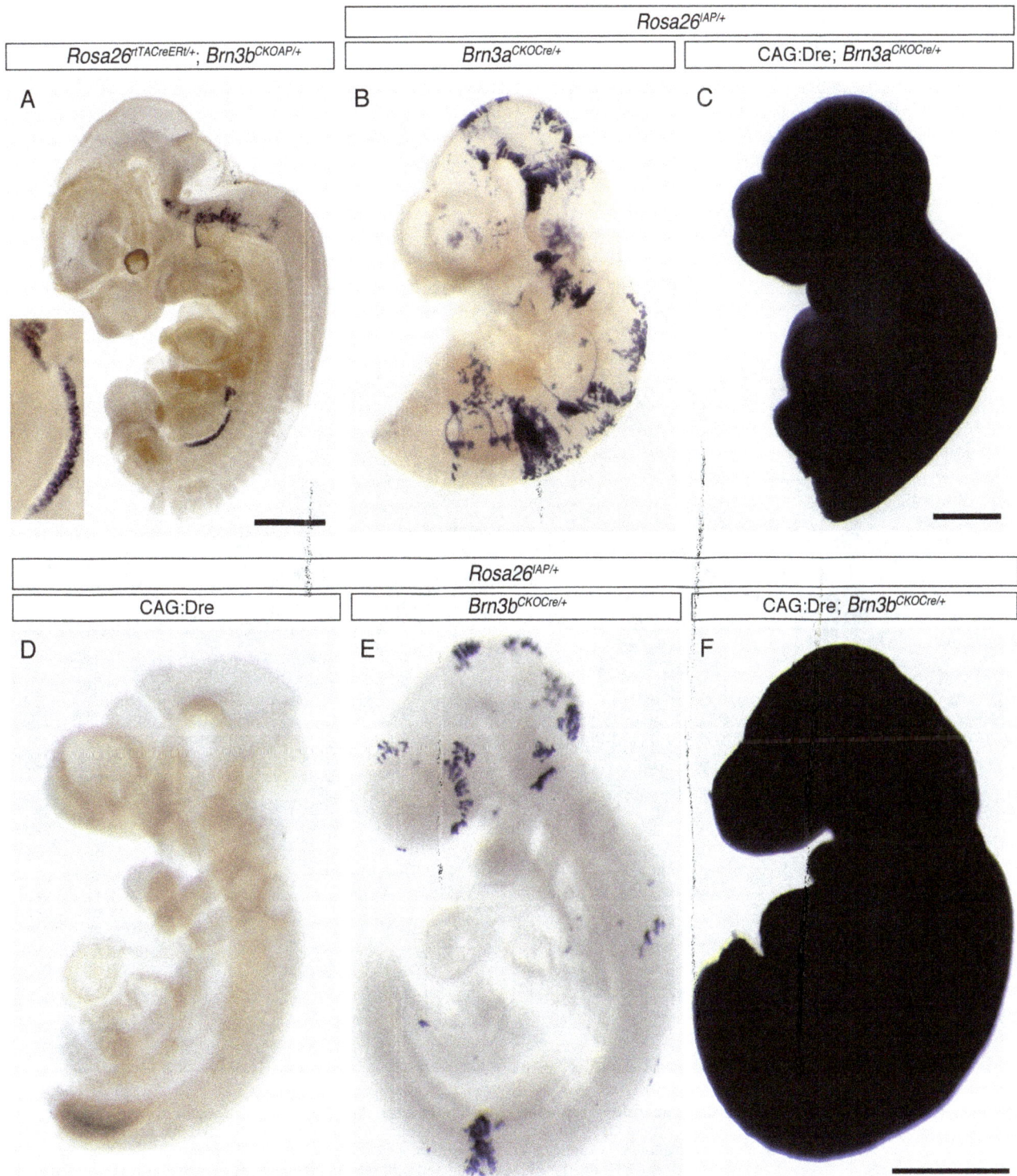

Figure 4. Whole tissue sequential Dre to Cre recombination in CAG:Dre; $Brn3^{CKOCre}$**;** $ROSA26^{iAP}$ **mice happens before E9.5.** A, wholemount staining of E12.5 $ROSA26^{rtTACreERt}$; $Brn3b^{CKOAP}$ embryo demonstrating $Brn3b^{AP}$ labeling of the germinal ridge. Inset shows 8x magnification of the gonad. B, C, wholemount E9.5 embryos showing complete AP recombination in CAG:Dre; $Brn3a^{CKOCre}$; $ROSA26^{iAP}$ mice (C, n = 6) and sparse mosaic recombination in $Brn3a^{CKOCre}$; $ROSA26^{iAP}$ (B, n = 9) controls. Whereas no AP positive cells are visible in E9 CAG:Dre; $ROSA26^{iAP}$ embryos (D, n = 4), sparse recombination can be seen in $Brn3b^{CKOCre}$; $ROSA26^{iAP}$ (E, n = 6) and full recombination in CAG:Dre; $Brn3b^{CKOCre}$; $ROSA26^{iAP}$ (F, n = 7) littermates. Scale bars in A, C, F are 0.5 mm.

disc, and exhibit positivity for the RGC markers NFL, Brn3a or Brn3b, we are fairly confident that our alleles are exclusively

expressed in RGCs in the early postnatal period. The fact that these are knock-in alleles, means that the mice in which they will

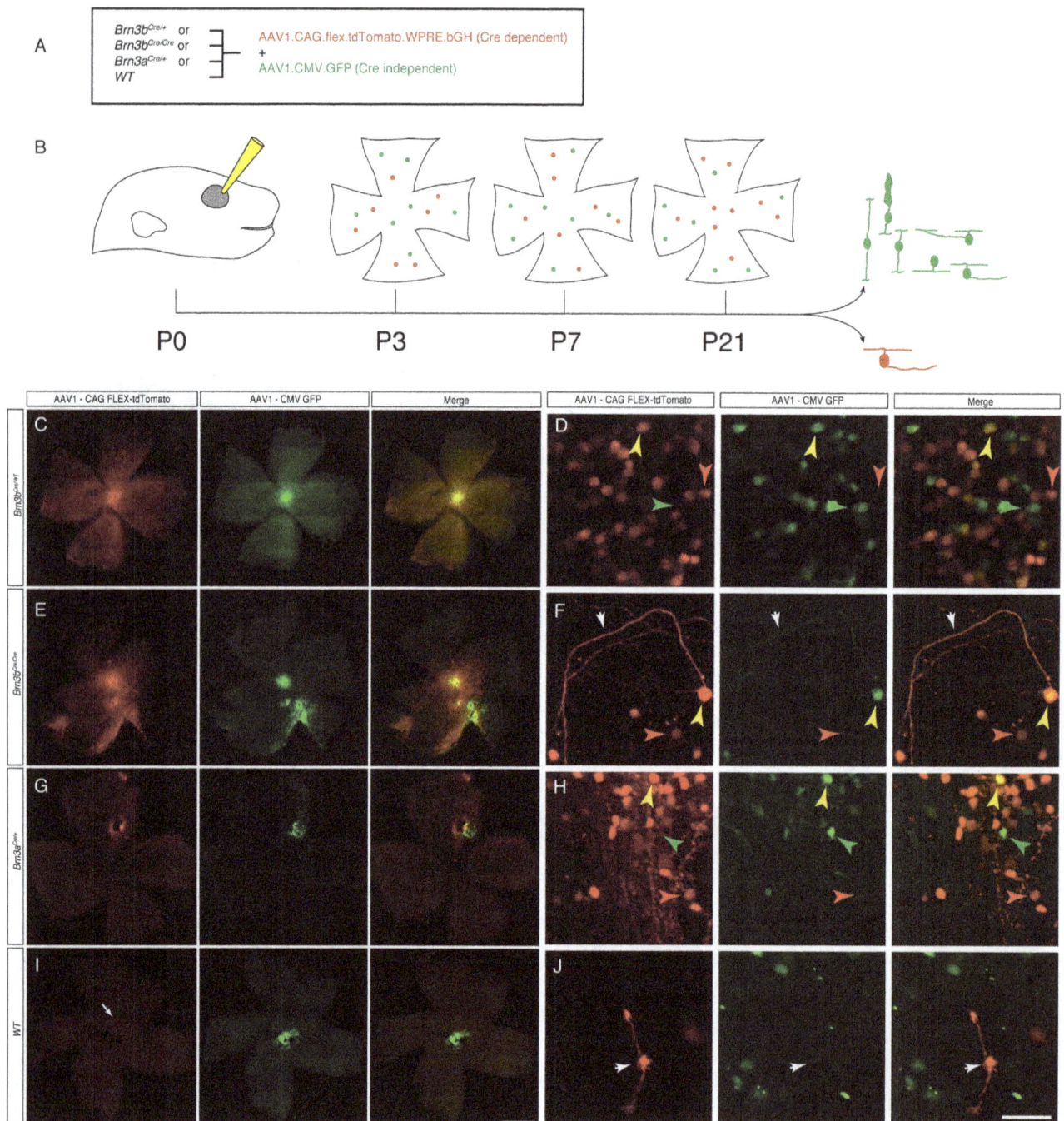

Figure 5. Retinal expression of viral Cre dependent reporters after postnatal day 0.5 (P0) intraocular infection. A, Genetic backgrounds and viruses used. $Brn3b^{Cre/+}$, $Brn3b^{Cre/Cre}$, $Brn3a^{Cre/+}$, or $Brn3b^{+/+}$ littermate pups were infected with a combination of Cre dependent AAV1.CAG.flex.tdTomato.WPRE.bGH virus and a constitutively expressed, Cre independent AAV1.CAG.GFP virus. B, Experimental time lines. Eyes of P0 pups were injected, and retinas were analyzed at either P3, P7 or after P21. Cells infected with the Cre independent virus should appear green, while Cre positive cells infected with the Cre dependent virus should appear red. Results from this experiment are shown in figures 5, 6 and 7. C, E, G, and I, flat mount preparations of adult retinas of indicated genotypes, demonstrating extensive expression of (Cre dependent) tdTomato red staining in $Brn3b^{Cre/+}$ (n = 6), $Brn3b^{Cre/Cre}$ (n = 4) or $Brn3a^{Cre/+}$ (n = 12) retinas, and only very isolated expression in Cre negative, WT retinas (n = 3 for the Brnb litters and n = 4 for the Brn3a litters). Note that in some cases, red and green fluorescence are evenly distributed over the entire retina, while in others there is an apparent segregation of red and green fluorescence, most likely by subretinal distribution of the Cre independent, AAV1.CAG.GFP virus (see figure 6). White arrow in I points at one of the few red cells labeled in the WT retinas, shown enlarged in the inset in J. D, F, H, and J represent higher magnifications of retinas shown in C, E, G, and J respectively. Left panels are tdTomato, red fluorescence, middle panels are GFP, green fluorescence, and right panels are merge channels. In all panels, red arrowheads point at tdTomato positive, green arrowheads at GFP positive, and yellow arrowheads at double positive cells. White arrowhead in F labels wandering axons characteristic of Brn3b null ($Brn3b^{Cre/Cre}$) retinas. White arrowhead in J points at the tdTomato positive cell in I, most likely a Müller Glia. Scale bars in I = 1 mm and J = 50 μm.

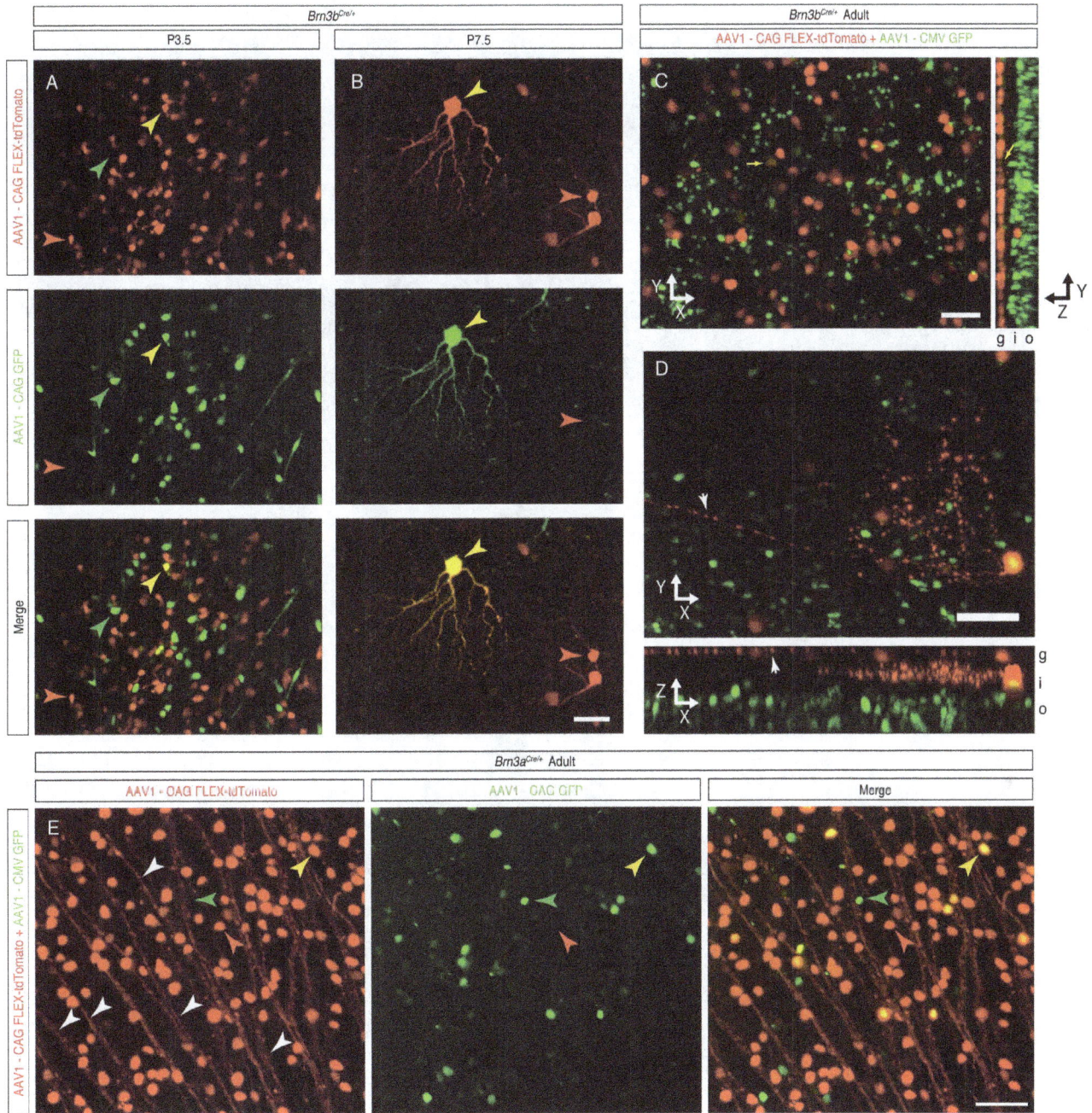

Figure 6. AAV1-FLEX-tdTomato infection at P0 results in RGC specific tdTomato expression, which is already detectable at P3.5 and P7.5. Retina samples are from P3.5 (A), P7.5 (B) pups or adults (C–E). Panels in A, B are tdTomato (top), GFP (middle), and merged (bottom) images. Panels A-D are from Brn3b$^{Cre/+}$, and E from from Brn3a$^{Cre/+}$ mice, infected at P0 after the protocol described in figure 5 A, B. Red, green and yellow arrowheads point at examples of red and green single positive or double positive cells. Note that the dendritic arbor of the double positive cell in B is clearly visible in both red and green channels. C, Projections along the z direction (left) and x direction (right) of a stack from a densely labeled Brn3b$^{Cre/+}$ retina, showing an overwhelming majority of red fluorescent cell bodies stratified in the GCL (g) layer, whereas abundant numbers of green fluorescent bodies are seen throughout the Inner and Outer Nuclear Layer (i and o). D, Projections along the z direction (top) and y direction (bottom) of a stack from a sparsely infected Brn3b$^{Cre/+}$ retina showing a displaced RGC, with its axon (white arrowheads) and dendritic arbor, seen both from the flat mount and transversal perspective. E, z projection of a stack from a densely labeled Brn3a$^{Cre/+}$ retina, showing single and double labeled cell bodies, and tdTomato labeled RGC axons (white arrowheads). Scale bars in B, C, D and E = 50 µm.

be used will be heterozygotes for the Brn3a or Brn3b loci. However, both Brn3a$^{Cre/WT}$ and Brn3b$^{Cre/WT}$ lines are viable and fertile, and it is known for a long time in the field that RGCs which are heterozygote for either Brn3a or Brn3b are normal in numbers, gene expression and morphology and the mice exhibit normal visual reflexes [8,16,63–65], (and Kretschmer and Badea, unpublished observations), perhaps because loss of Brn3 alleles tends to increase the expresson of the remaining alleles through the removal of an autoregulatory loop [66] However it is still

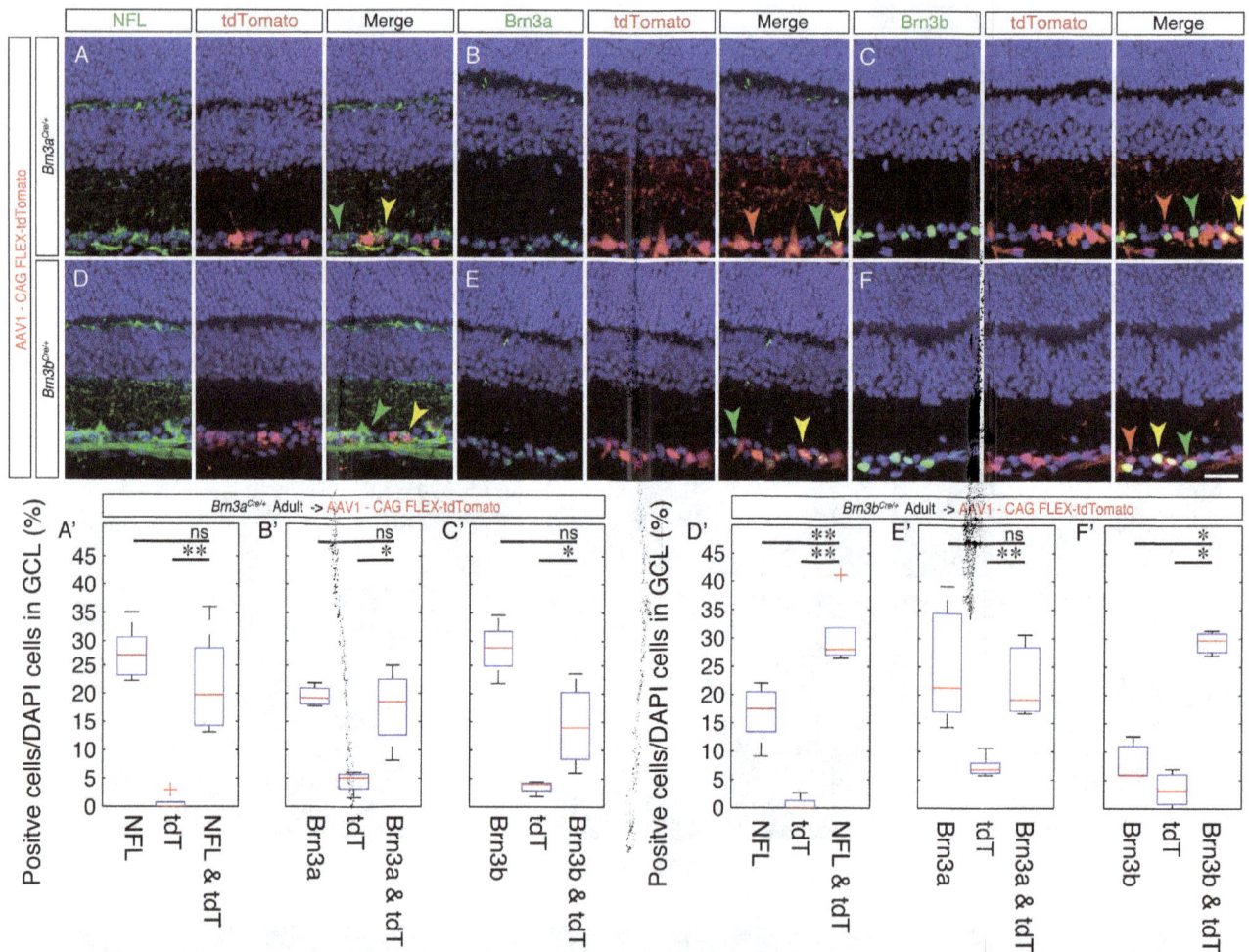

Figure 7. tdTomato positive cells in AAV1-FLEX-tdTomato infected *Brn3a*Cre/+ **and** *Brn3b*Cre/+ **retinas express the RGC markers Neurofilament Light Chain (NFL), Brn3a and Brn3b.** Eyes from *Brn3a*Cre/+ or *Brn3b*Cre/+ were infected with the Cre dependent AAV1-FLEX-tdTomato virus at P0, as shown in Figure 5. At p21, eye cups were prepared and briefly inspected under a fluorescence microscope for degree of infection. Highly infected eyes were then processed for sectioning and immunostaining. Sections from either *Brn3a*Cre/+ (A–C) or *Brn3b*Cre/+ (D–F) retinas were stained with antibodies against NFL (A, D), Brn3a (B, E) and Brn3b (C, F), shown in the left panels, in conjunction with endogenous tdTomato red fluorescence, shown in the middle pannels. Images were taken and quantitations performed on areas showing highest levels of infection, as judged by tdTomato fluorescence. Red, green and yellow arrowheads point at examples of red and green single positive or double positive cells. A'–F' Box-whisker plots representing quantitations of experiments shown in A–F. For each experiment, the numbers of marker, tdTomato, or double positive cells were normalized to the total number of DAPI positive cells in the GCL. Significance levels shown were calculated with the Kolmogorov – Smirnov test for comparisons of two unknown distributions: ns = not significant, *p < 0.05, **p < 0.01 For number of quantitated images, cells counted, averages, standard deviations and significance levels by Kolmogorov-Smirnov and Student T tests, see Tables S1 and S2. Scale bar in F = 25 μm.

conceivable that subtle differences, which have not yet been reported, persist between Brn3 heterozygote and wild type cells.

This opens the possibility for delivering a variety of virally encoded reporters or reagents to the about 14 RGC types which express either Brn3a or Brn3b, essentially throughout all stages of RGC development and adults. We chose to infect P0 pups and follow the dynamics of expression over several time points in early postnatal development, and found fairly high levels of expression of the fluorescent reporter as early as three days after infection, allowing us to image dendritic arbors of individual RGCs with fairly high accuracy. We feel that the window of Cre expression from the Brn3 alleles is uniquely useful for targeting broad populations of Retinal Ganglion Cells. Several other transgenic, BAC transgenic or knock-in lines that target populations of RGCs have been generated. Prior to Brn3 expression, RGC precursors are labeled by the transcription factor Atoh7 (Math5), and a

knock-in line as well as a BAC transgenic have been generated, both of which label a multitude of other cell types besides RGCs [67,68], implying that the transcription factor might be needed but not sufficient for the RGC lineage. Additionally several knock-in, transgenic and BAC transgenic lines using elements of the JamB, Melanopsin, Parvalbumin, and P2 genes [49,69–72] have been successfully used to label more or less restricted RGC subpopulations, typically at later onsets of development. Nevertheless, we believe that given the broad and early expression pattern, the *Brn3a*Cre and *Brn3b*Cre will be useful tools for delivering a variety of reporters and other tools to RGCs, beginning with early stages of development, as well as to other neuronal populations where Brn3s are expressed [45].

AAV1-FLEX Vectors in Retinal Ganglion Cells

In this study we have used intravitreal injection of a Cre dependent red fluorescent AAV1 (AAV1-FLEX-tdTomato-WPRE-bGH) in conjunction with a constitutive, Cre independent green fluorescent AAV1 (AAV1-CMV-eGFP-bGH) in mouse retinas at postnatal day 0. Both viruses have the AAV1 capsid, and ubiquitous promoters, but the Cre dependent AAV1-FLEX-tdTomato-WPRE-bGH was not expressed in wild type retinas, and expressed selectively in RGCs in the $Brn3a^{Cre}$ and $Brn3b^{Cre}$ retinas, whereas the Cre independent AAV1-CMV-eGFP-bGH was expressed in multiple layers including the Ganglion Cell Layer, both in wild type and $Brn3a^{Cre}$ and $Brn3b^{Cre}$ retinas. These observations encourage us to conclude that, when delivered in early postnatal (P0) retinas, the AAV1 capsid is successfully (and quite quantitatively) targeting RGCs. AAV vectors can be classified based on sequence of the capsid encoding *cap* gene into at least 11 serotypes, and it appears that each serotype is characterized by distinct efficiencies of expression and retinal cell population target selection [73–75]. Early characterizations of virus tropism amongst different cell populations suggested that the AAV2 capsid might be preferentially targeted to RGCs, however several newer studies suggest that, depending on subretinal or intravitreal delivery, and on the retinal developmental stage at which the virus is delivered, many other serotypes perform well in RGCs [49,76,77], and that AAV1 capsid vectors can effectively infect RGCs [49,77]. In these studies it is also shown that AAV1 positive cells tend to be seen much quicker (4–5 days after injection) compared to other serotypes [73]. We now report tdTomato positive cells in postnatal day 3.5 (P3.5) retinas infected at P0.5, however it is fair to assume that higher levels of expression of the reporter will accumulate over time. It is also not clear whether all cells of a retina are infected synchronously at the time of injection, and how many copies per cell are expressed, and hence how much protein will accumulate in each cell, determining when fluorescence would have reached detectable levels, however in most experimental scenarios we envision, the readout of the experiment would be performed in the adult stage, when hopefully expression would have saturated across samples.

The potential differential tropism of AAV capsids for different RGC types and/at different developmental stages, combined with the cell type distribution of Cre alleles expressed in broader or narrower RGC subpopulations, could be theoretically used to restrict AAV based RGC labeling and manipulation to specific cell types. Since infection efficiency will depend in a major fashion on the injection success, the absolute number of RGCs per retina infected by this strategy is less relevant, although the levels we observed are comparable to previously reported data, as referenced above. We find that, in sections from well infected areas of the retinal preparation about half of the RGCs are infected, as judged by double immunostaining with RGC markers. Since both Brn3b and Brn3a are expected to be expressed in about 75–85% of RGCs in the adult, this means that our strategy, under current optimal conditions, is missing about 25% of the adult

target population. Thus, most but not all $Brn3a^{Cre}$ or $Brn3b^{Cre}$ RGCs can be targeted by P0 intravitreal injections of AAV1. However, most of our experimental paradigms will require much lower levels of infection, which will allow the identification and characterization of individual RGC types by dendritic arbor morphology. The incomplete targeting could depend on variable injection technique, temporally dynamic expression pattern of $Brn3^{Cre}$ alleles in Brn3 RGCs during development, or changes in virus tropism for different cell types at different ages. These complex interactions between Cre alleles, and virus tropism should be seen as opportunities for generating combinatorial strategies for targeting specific cell types by choosing a certain Cre expressing allele, an AAV of a specific serotype and promoter and a specific time window of delivery. These strategies could allow us to map the developmental expression domain of Brn3a and Brn3b onto the adult distribution of RGC types, with relevance to the cell autonomous or non-autonomous roles of Brn3s in RGC development.

Supporting Information

Table S1 Quantitations for immunostaining experiments reported in Figure 7. Numbers of sections imaged, and cells counted for each image, together with average and standard deviation for each combination of genotype and marker (NFL, Brn3a and Brn3b) are provided.

Table S2 Statistical significance tests for comparisons presented in Figure 7. The Kolmogorov –Smirnov test for comparing two data samples of unknown distribution, and the Student T –test for comparing two distribution assumed to be normal were performed. For both tests, the null hypothesis states that the two data sets are drawn from the same distribution, or in other words, there are no differences between the compared conditions. A 0 indicates that the null hypothesis was not rejected, whereas a 1 indicates that the null hypothesis was rejected and the two samples are significantly different, with the corresponding p values.

Acknowledgments

The Authors acknowledge Seid Ali for help with mouse husbandry, Humphrey Yao for help with germ line expression of Brn3 genes and Zhijian Wu and Peter Colosi for their generous gift of AAV1-CAG-GFP expression virus.

Author Contributions

Conceived and designed the experiments: MS MG SS TB. Performed the experiments: MS MG SS PL LD NP TB. Analyzed the data: SS TB. Wrote the paper: TB OP.

References

1. Betley JN, Sternson SM (2011) Adeno-associated viral vectors for mapping, monitoring, and manipulating neural circuits. Hum Gene Ther 22: 669–677.
2. Knopfel T (2012) Genetically encoded optical indicators for the analysis of neuronal circuits. Nat Rev Neurosci 13: 687–700.
3. Szobota S, Isacoff EY (2010) Optical control of neuronal activity. Annu Rev Biophys 39: 329–348.
4. Hattar S, Liao HW, Takao M, Berson DM, Yau KW (2002) Melanopsin-containing retinal ganglion cells: architecture, projections, and intrinsic photosensitivity. Science 295: 1065–1070.
5. Huberman AD, Manu M, Koch SM, Susman MW, Lutz AB, et al. (2008) Architecture and activity-mediated refinement of axonal projections from a mosaic of genetically identified retinal ganglion cells. Neuron 59: 425–438.

6. Kim I-J, Zhang Y, Yamagata M, Meister M, Sanes JR (2008) Molecular identification of a retinal cell type that responds to upward motion. Nature 452: 478–482.
7. Yonehara K, Shintani T, Suzuki R, Sakuta H, Takeuchi Y, et al. (2008) Expression of SPIG1 reveals development of a retinal ganglion cell subtype projecting to the medial terminal nucleus in the mouse. PLoS ONE 3: e1533.
8. Badea TC, Cahill H, Ecker J, Hattar S, Nathans J (2009) Distinct roles of transcription factors brn3a and brn3b in controlling the development, morphology, and function of retinal ganglion cells. Neuron 61: 852–864.
9. Chen SK, Badea TC, Hattar S (2011) Photoentrainment and pupillary light reflex are mediated by distinct populations of ipRGCs. Nature 476: 92–95.
10. Fenno L, Yizhar O, Deisseroth K (2011) The development and application of optogenetics. Annu Rev Neurosci 34: 389–412.

11. Guler AD, Ecker JL, Lall GS, Haq S, Altimus CM, et al. (2008) Melanopsin cells are the principal conduits for rod-cone input to non-image-forming vision. Nature 453: 102–105.

12. Masland RH (2001) Neuronal diversity in the retina. Curr Opin Neurobiol 11: 431–436.

13. Masland RH (2012) The neuronal organization of the retina. Neuron 76: 266–280.

14. Troy JB, Shou T (2002) The receptive fields of cat retinal ganglion cells in physiological and pathological states: where we are after half a century of research. Prog Retin Eye Res 21: 263–302.

15. Wassle H (2004) Parallel processing in the mammalian retina. Nat Rev Neurosci 5: 747–757.

16. Badea TC, Nathans J (2011) Morphologies of mouse retinal ganglion cells expressing transcription factors Brn3a, Brn3b, and Brn3c: analysis of wild type and mutant cells using genetically-directed sparse labeling. Vision Res 51: 269–279.

17. Erkman L, McEvilly RJ, Luo L, Ryan AK, Hooshmand F, et al. (1996) Role of transcription factors Brn-3.1 and Brn-3.2 in auditory and visual system development. Nature 381: 603–606.

18. Gan L, Xiang M, Zhou L, Wagner DS, Klein WH, et al. (1996) POU domain factor Brn-3b is required for the development of a large set of retinal ganglion cells. Proc Natl Acad Sci U S A 93: 3920–3925.

19. Xiang M (1998) Requirement for Brn-3b in early differentiation of postmitotic retinal ganglion cell precursors. Dev Biol 197: 155–169.

20. Xiang M, Gan L, Li D, Chen ZY, Zhou L, et al. (1997) Essential role of POU-domain factor Brn-3c in auditory and vestibular hair cell development. Proc Natl Acad Sci U S A 94: 9445–9450.

21. Xiang M, Gan L, Zhou L, Klein WH, Nathans J (1996) Targeted deletion of the mouse POU domain gene Brn-3a causes selective loss of neurons in the brainstem and trigeminal ganglion, uncoordinated limb movement, and impaired suckling. Proc Natl Acad Sci U S A 93: 11950–11955.

22. Adams DJ, van der Weyden L (2008) Contemporary approaches for modifying the mouse genome. Physiol Genomics 34: 225–238.

23. Gama Sosa MA, De Gasperi R, Elder GA (2010) Animal transgenesis: an overview. Brain Struct Funct 214: 91–109.

24. Liu C (2013) Strategies for designing transgenic DNA constructs. Methods Mol Biol 1027: 183–201.

25. Rossant J, Nutter LM, Gertsenstein M (2011) Engineering the embryo. Proc Natl Acad Sci U S A 108: 7659–7660.

26. Anastassiadis K, Glaser S, Kranz A, Berhardt K, Stewart AF (2010) A practical summary of site-specific recombination, conditional mutagenesis, and tamoxifen induction of CreERT2. Methods Enzymol 477: 109–123.

27. Sauer B, Henderson N (1988) Site-specific DNA recombination in mammalian cells by the Cre recombinase of bacteriophage P1. Proc Natl Acad Sci U S A 85: 5166–5170.

28. Livet J, Weissman TA, Kang H, Draft RW, Lu J, et al. (2007) Transgenic strategies for combinatorial expression of fluorescent proteins in the nervous system. Nature 450: 56–62.

29. Schnutgen F, Doerflinger N, Calleja C, Wendling O, Chambon P, et al. (2003) A directional strategy for monitoring Cre-mediated recombination at the cellular level in the mouse. Nat Biotechnol 21: 562–565.

30. Lee G, Saito I (1998) Role of nucleotide sequences of loxP spacer region in Cre-mediated recombination. Gene 216: 55–65.

31. Buchholz F, Ringrose L, Angrand PO, Rossi F, Stewart AF (1996) Different thermostabilities of FLP and Cre recombinases: implications for applied site-specific recombination. Nucleic Acids Res 24: 4256–4262.

32. McLeod M, Craft S, Broach JR (1986) Identification of the crossover site during FLP-mediated recombination in the Saccharomyces cerevisiae plasmid 2 microns circle. Mol Cell Biol 6: 3357–3367.

33. Dymecki SM (1996) Flp recombinase promotes site-specific DNA recombination in embryonic stem cells and transgenic mice. Proc Natl Acad Sci U S A 93: 6191–6196.

34. Belteki G, Gertsenstein M, Ow DW, Nagy A (2003) Site-specific cassette exchange and germline transmission with mouse ES cells expressing phiC31 integrase. Nat Biotechnol 21: 321–324.

35. Groth AC, Olivares EC, Thyagarajan B, Calos MP (2000) A phage integrase directs efficient site-specific integration in human cells. Proc Natl Acad Sci U S A 97: 5995–6000.

36. Raymond CS, Soriano P (2007) High-efficiency FLP and PhiC31 site-specific recombination in mammalian cells. PLoS One 2: e162.

37. Nern A, Pfeiffer BD, Svoboda K, Rubin GM (2011) Multiple new site-specific recombinases for use in manipulating animal genomes. Proc Natl Acad Sci U S A 108: 14198–14203.

38. Tasic B, Hippenmeyer S, Wang C, Gamboa M, Zong H, et al. (2011) Site-specific integrase-mediated transgenesis in mice via pronuclear injection. Proc Natl Acad Sci U S A 108: 7902–7907.

39. Suzuki E, Nakayama M (2011) VCre/VloxP and SCre/SloxP: new site-specific recombination systems for genome engineering. Nucleic Acids Res 39: e49.

40. Sauer B, McDermott J (2004) DNA recombination with a heterospecific Cre homolog identified from comparison of the pac-c1 regions of P1-related phages. Nucleic Acids Res 32: 6086–6095.

41. Anastassiadis K, Fu J, Patsch C, Hu S, Weidlich S, et al. (2009) Dre recombinase, like Cre, is a highly efficient site-specific recombinase in E. coli, mammalian cells and mice. Dis Model Mech 2: 508–515.

42. Badea TC, Hua ZL, Smallwood PM, Williams J, Rotolo T, et al. (2009) New mouse lines for the analysis of neuronal morphology using CreER(T)/loxP-directed sparse labeling. PLoS One 4: e7859.

43. Soriano P (1999) Generalized lacZ expression with the ROSA26 Cre reporter strain. Nat Genet 21: 70–71.

44. Badea TC, Wang Y, Nathans J (2003) A noninvasive genetic/pharmacologic strategy for visualizing cell morphology and clonal relationships in the mouse. J Neurosci 23: 2314–2322.

45. Badea TC, Williams J, Smallwood P, Shi M, Motajo O, et al. (2012) Combinatorial expression of Brn3 transcription factors in somatosensory neurons: genetic and morphologic analysis. J Neurosci 32: 995–1007.

46. Xiang M, Zhou L, Macke JP, Yoshioka T, Hendry SH, et al. (1995) The Brn-3 family of POU-domain factors: primary structure, binding specificity, and expression in subsets of retinal ganglion cells and somatosensory neurons. J Neurosci 15: 4762–4785.

47. Budhram-Mahadeo V, Moore A, Morris PJ, Ward T, Weber B, et al. (2001) The closely related POU family transcription factors Brn-3a and Brn-3b are expressed in distinct cell types in the testis. Int J Biochem Cell Biol 33: 1027–1039.

48. Ohinata Y, Sano M, Shigeta M, Yamanaka K, Saitou M (2008) A comprehensive, non-invasive visualization of primordial germ cell development in mice by the Prdm1-mVenus and Dppa3-ECFP double transgenic reporter. Reproduction 136: 503–514.

49. Borghuis BG, Tian L, Xu Y, Nikonov SS, Vardi N, et al. (2011) Imaging light responses of targeted neuron populations in the rodent retina. J Neurosci 31: 2855–2867.

50. Shaner NC, Campbell RE, Steinbach PA, Giepmans BN, Palmer AE, et al. (2004) Improved monomeric red, orange and yellow fluorescent proteins derived from Discosoma sp. red fluorescent protein. Nat Biotechnol 22: 1567–1572.

51. Atasoy D, Aponte Y, Su HH, Sternson SM (2008) A FLEX switch targets Channelrhodopsin-2 to multiple cell types for imaging and long-range circuit mapping. J Neurosci 28: 7025–7030.

52. Zufferey R, Donello JE, Trono D, Hope TJ (1999) Woodchuck hepatitis virus posttranscriptional regulatory element enhances expression of transgenes delivered by retroviral vectors. J Virol 73: 2886–2892.

53. Harris JA, Oh SW, Zeng H (2012) Adeno-associated viral vectors for anterograde axonal tracing with fluorescent proteins in nontransgenic and cre driver mice. Curr Protoc Neurosci Chapter 1: Unit 1 20 21–18.

54. Huang CC, Sugino K, Shima Y, Guo C, Bai S, et al. (2013) Convergence of pontine and proprioceptive streams onto multimodal cerebellar granule cells. Elife 2: e00400.

55. Xiang M, Gan L, Li D, Zhou L, Chen ZY, et al. (1997) Role of the Brn-3 family of POU-domain genes in the development of the auditory/vestibular, somatosensory, and visual systems. Cold Spring Harb Symp Quant Biol 62: 325–336.

56. Qiu F, Jiang H, Xiang M (2008) A comprehensive negative regulatory program controlled by Brn3b to ensure ganglion cell specification from multipotential retinal precursors. J Neurosci 28: 3392–3403.

57. Quina LA, Pak W, Lanier J, Banwait P, Gratwick K, et al. (2005) Brn3a-expressing retinal ganglion cells project specifically to thalamocortical and collicular visual pathways. J Neurosci 25: 11595–11604.

58. Farago AF, Awatramani RB, Dymecki SM (2006) Assembly of the brainstem cochlear nuclear complex is revealed by intersectional and subtractive genetic fate maps. Neuron 50: 205–218.

59. Hirsch MR, d'Autreaux F, Dymecki SM, Brunet JF, Goridis C (2013) A Phox2b:FLPo transgenic mouse line suitable for intersectional genetics. Genesis 51: 506–514.

60. Harding SD, Armit C, Armstrong J, Brennan J, Cheng Y, et al. (2011) The GUDMAP database–an online resource for genitourinary research. Development 138: 2845–2853.

61. Davies JA, Little MH, Aronow B, Armstrong J, Brennan J, et al. (2012) Access and use of the GUDMAP database of genitourinary development. Methods Mol Biol 886: 185–201.

62. Jameson SA, Natarajan A, Cool J, DeFalco T, Maatouk DM, et al. (2012) Temporal transcriptional profiling of somatic and germ cells reveals biased lineage priming of sexual fate in the fetal mouse gonad. PLoS Genet 8: e1002575.

63. Wang SW, Gan L, Martin SE, Klein WH (2000) Abnormal polarization and axon outgrowth in retinal ganglion cells lacking the POU-domain transcription factor Brn-3b. Mol Cell Neurosci 16: 141–156.

64. Wang SW, Mu X, Bowers WJ, Kim D-S, Plas DJ, et al. (2002) Brn3b/Brn3c double knockout mice reveal an unsuspected role for Brn3c in retinal ganglion cell axon outgrowth. Development 129: 467–477.

65. Mu X, Beremand PD, Zhao S, Pershad R, Sun H, et al. (2004) Discrete gene sets depend on POU domain transcription factor Brn3b/Brn-3.2/POU4f2 for their expression in the mouse embryonic retina. Development 131: 1197–1210.

66. Eng SR, Gratwick K, Rhee JM, Fedtsova N, Gan L, et al. (2001) Defects in sensory axon growth precede neuronal death in Brn3a-deficient mice. J Neurosci 21: 541–549.

67. Brzezinski JAt, Prasov L, Glaser T (2012) Math5 defines the ganglion cell competence state in a subpopulation of retinal progenitor cells exiting the cell cycle. Dev Biol 365: 395–413.

68. Yang Z, Ding K, Pan L, Deng M, Gan L (2003) Math5 determines the competence state of retinal ganglion cell progenitors. Dev Biol 264: 240–254.

69. Münch TA, da Silveira RA, Siegert S, Viney TJ, Awatramani GB, et al. (2009) Approach sensitivity in the retina processed by a multifunctional neural circuit. Nat Neurosci 12: 1308–1316.

70. Ivanova E, Lee P, Pan ZH (2013) Characterization of multiple bistratified retinal ganglion cells in a purkinje cell protein 2-Cre transgenic mouse line. J Comp Neurol 521: 2165–2180.

71. Kim IJ, Zhang Y, Yamagata M, Meister M, Sanes JR (2008) Molecular identification of a retinal cell type that responds to upward motion. Nature 452: 478–482.

72. Ecker JL, Dumitrescu ON, Wong KY, Alam NM, Chen S-K, et al. (2010) Melanopsin-expressing retinal ganglion-cell photoreceptors: cellular diversity and role in pattern vision. Neuron 67: 49–60.

73. Auricchio A, Kobinger G, Anand V, Hildinger M, O'Connor E, et al. (2001) Exchange of surface proteins impacts on viral vector cellular specificity and transduction characteristics: the retina as a model. Hum Mol Genet 10: 3075–3081.

74. Lebherz C, Maguire A, Tang W, Bennett J, Wilson JM (2008) Novel AAV serotypes for improved ocular gene transfer. J Gene Med 10: 375–382.

75. Rabinowitz JE, Rolling F, Li C, Conrath H, Xiao W, et al. (2002) Cross-packaging of a single adeno-associated virus (AAV) type 2 vector genome into multiple AAV serotypes enables transduction with broad specificity. J Virol 76: 791–801.

76. Pang JJ, Lauramore A, Deng WT, Li Q, Doyle TJ, et al. (2008) Comparative analysis of in vivo and in vitro AAV vector transduction in the neonatal mouse retina: effects of serotype and site of administration. Vision Res 48: 377–385.

77. Watanabe S, Sanuki R, Ueno S, Koyasu T, Hasegawa T, et al. (2013) Tropisms of AAV for subretinal delivery to the neonatal mouse retina and its application for in vivo rescue of developmental photoreceptor disorders. PLoS One 8: e54146.

18

Conditional Reverse Tet-Transactivator Mouse Strains for the Efficient Induction of TRE-Regulated Transgenes in Mice

Lukas E. Dow[1], Zeina Nasr[2], Michael Saborowski[1], Saya H. Ebbesen[1,3], Eusebio Manchado[1], Nilgun Tasdemir[1,3], Teresa Lee[2], Jerry Pelletier[2,4]*, Scott W. Lowe[1,5]*

1 Cancer Biology and Genetics Program, Memorial Sloan Kettering Cancer Center, New York, New York, United States of America, **2** Department of Biochemistry, McGill University, Montreal, Quebec, Canada, **3** Watson School of Biological Sciences, Cold Spring Harbor Laboratory, Cold Spring Harbor, New York, United States of America, **4** The Rosalind and Morris Goodman Cancer Research Center, McGill University, Montreal, Quebec, Canada, **5** Howard Hughes Medical Institute, Memorial Sloan Kettering Cancer Center, New York, New York, United States of America

Abstract

Tetracycline or doxycycline (dox)-regulated control of genetic elements allows inducible, reversible and tissue specific regulation of gene expression in mice. This approach provides a means to investigate protein function in specific cell lineages and at defined periods of development and disease. Efficient and stable regulation of cDNAs or non-coding elements (e.g. shRNAs) downstream of the tetracycline-regulated element (TRE) requires the robust expression of a tet-transactivator protein, commonly the reverse tet-transactivator, rtTA. Most rtTA strains rely on tissue specific promoters that often do not provide sufficient rtTA levels for optimal inducible expression. Here we describe the generation of two mouse strains that enable Cre-dependent, robust expression of rtTA3, providing tissue-restricted and consistent induction of *TRE*-controlled transgenes. We show that these transgenic strains can be effectively combined with established mouse models of disease, including both Cre/LoxP-based approaches and non Cre-dependent disease models. The integration of these new tools with established mouse models promises the development of more flexible genetic systems to uncover the mechanisms of development and disease pathogenesis.

Editor: John P. Lydon, Baylor college of Medicine, United States of America

Funding: This work was supported by grants from the Cancer Target Discovery Development (CTDD) Consortium from the National Cancer Institute (NCI - CA168409) and a grant from the Canadian Institutes of Health Research (CIHR - MOP-106530). LED was supported by a National Health and Medical Research Council (NHMRC) Overseas Biomedical Fellowship and SWL is an investigator of the Howard Hughes Medical Institute. The funders had no role in study design, data collection and analysis decision to publish, or preparation of the manuscript.

Competing Interests: SWL and LED are members of the Scientific Advisory Board and hold equity in Mirimus Inc., a company that has licensed some of the technology reported in this manuscript.

* E-mail: jerry.pelletier@mcgill.ca (JP); lowes@mskcc.org (SWL)

Introduction

Genetically engineered mouse models (GEMMs) are an invaluable tool to investigate the biology of development and disease in a mammalian organism. Since the development of the first knockout mouse almost 25 years ago, a wide variety of knockout, knock-in and conditional mutant strains have been developed to interrogate gene function [1]. In recent years, use of inducible promoters to control genetic elements in adult mice has become increasingly valuable, allowing regulated control of cDNAs and, more recently, shRNAs. In concept, these systems enable the timed expression or silencing of any gene, in any tissue and at any stage of development or disease progression. The application of such experimental tools promises a detailed understanding of the temporal requirements for gene function in specific tissues and provides an opportunity to genetically validate proposed therapeutic targets prior to drug development.

The tetracycline (tet) system is, by far, the most widely used inducible model in mice. It consists of two essential components: a tetracycline-responsive element (*TRE*) that regulates gene or shRNA expression, and a *trans*-acting, tet-sensitive, tet-transactivator (tTA) or reverse tet-transactivator (rtTA) protein [2]. tTA promotes gene expression from *TRE* promoters, but is inhibited in the presence of tet, or its more common analog, doxycycline (dox). Conversely, rtTA promotes dox-dependent gene induction. Early versions of the rtTA protein showed 'leaky' gene expression in the absence of dox, but newer variants such as rtTA^{M2} and rtTA3 [3,4] show almost no dox-independent activity, and in the case of rtTA3, high sensitivity to low levels of dox. To date, more than 150 tTA/rtTA transgenic and knock-in strains have been developed to enable regulated gene expression in embryonic and adult tissues (http://www.tetsystems.com/fileadmin/tettransgenicrodents.pdf). As expression of the *TRE*-regulated cassette is dependent on both the presence of tet-transactivator and dox, induction can be ubiquitous or tissue specific (by controlling *tTA/rtTA* expression), inducible and reversible (by controlling dox exposure). Tissue specific *TRE* gene regulation is usually achieved by restricting the expression of tTA or rtTA to defined cell lineages using a tissue-specific promoter. Though convenient, this approach is absolutely dependent on robust expression of the tTA/rtTA. Moreover, cellular response downstream of transgene/shRNA induction may alter cell fate and compromise sustained *TRE*-regulated control. We recently described a new reverse tet-transactivator strain, *CAGs-rtTA3*, that shows stronger and more ubiquitous induction of *TRE*-regulated elements than any other existing strains we have tested [5]. Here we set out to develop a more flexible transgenic approach exploiting the strength of *TRE*-induction seen with *CAGs-rtTA3* and enabling tissue specific *TRE*-regulation.

Results

Conditional rtTA3 ES cell lines

Conditional gene modification in mice is most commonly achieved using tissue specific Cre recombinase strains and hundreds of such mice have already been developed and tested; More than 300 Cre strains are available from the Jackson Laboratory. We reasoned that Cre-dependent expression of rtTA3 from the *CAGs* promoter would enable robust and tissue-specific *TRE*-transgene induction in almost any cell lineage. Thus, we modified our *CAGs-rtTA3* construct to contain a *LoxP*-flanked polyadenylation signal or LoxP-Stop-LoxP (LSL) cassette (*CAGs-LSL-rtTA3*). In addition, we cloned a variant that also carried the mKate2 far-red fluorescent gene [6] downstream of an *Internal ribosomal entry site* (IRES) (*CAGs-LSL-rtTA3-IRES-mKate2* – *CAGs-LSL-RIK*) (Fig. 1A). In this context, mKate2 fluorescence serves as a reporter of Cre recombinase activity and rtTA3 expression. We cloned each construct into the *pRosa26-1* targeting backbone and transfected C10 ES cells [7]. Southern blot analysis of puromycin-selected clones identified correctly targeted clones for each construct (Fig. S1). In the case of *CAGs-LSL-RIK*, we identified one clone (designated D34), which showed homozygous targeting to the *Rosa26* locus. We further confirmed targeting on both copies of chromosome 6 in this ESC line by fluorescence in-situ hybridization (FISH) (Fig. S1C) and later, by breeding founder animals, which transmitted the allele to 100% of their progeny (not shown).

We first sought to confirm that both the *CAGs-LSL-rtTA3* (designated Y1) and *CAGs-LSL-RIK* (D34) cells showed robust expression of rtTA3 protein and therefore strong induction of *TRE*-regulated transgenes. To do this we took advantage of a recombinase-mediated cassette exchange (RMCE) 'landing pad' downstream of the *col1a1* locus in C10 ESCs [7]. This knock-in cassette allows efficient targeting of *TRE*-regulated (and other) transgenes, including cDNAs [8] and shRNAs [5]. We transfected Y1 and D34 cells with a *col1a1* RMCE targeting vector we previously described that expresses GFP and an shRNA directed against Renilla luciferase (*Ren.713*) downstream of *TRE* (*TG-Ren.713*). Throughout this study we use the *TG-Ren.713* transgene as a neutral fluorescent marker of dox-mediated induction. Following hygromycin selection, individual clones were expanded and transduced with a limiting titer of adenovirus expressing Cre recombinase to achieve recombination in 10–25% of cells. As expected, adenovirus transduced cells treated with doxycycline showed strong induction of GFP (Fig. S2). To confirm that single Cre-recombined clones showed uniform GFP induction we plated adenoviral treated Y1 cells at low density and isolated individual clones. 2/24 clones picked showed consistent and uniform, dox-dependent induction of GFP, whereas untreated cells showed no detectable GFP signal by flow cytometry (Fig. 1B).

To confirm the quality of Y1 and D34 ESCs for animal production, we generated wholly ESC-derived mice by tetraploid embryo complementation and bred multiple founder animals. Each cell line produced numerous viable and fertile mice that showed expected Mendelian transmission of the *Rosa26*-targeted allele (Table 1). Moreover, we also generated mice from both lines following *col1a1* re-targeting by RMCE (not shown), demonstrating that both Y1 and D34 are robust ESC lines that can serve as a platform for the production of conditional, TRE-inducible mice for analysis.

CAGs-rtTA3 and CAGs-RIK enable robust and widespread TRE-induction *in vivo*

We have previously noted that limited expression of rtTA from the endogenous *Rosa26* promoter (*R26-rtTA*) results in restricted induction of TRE-driven GFP in adult mouse tissues [5]. To test TRE-mediated induction from *CAGs-rtTA3* transgenes targeted to the *Rosa26* locus, we crossed each strain to a *CAGs-Cre* 'deletor' mouse that induces *LoxP* recombination at or before the two-cell stage [9]. Excision of the LSL cassette in the F1 progeny from each strain was confirmed by PCR and each of the recombined alleles could be propagated through breeding at Mendelian ratios, indicating there was no appreciable toxicity from the *CAGs*-driven expression of rtTA3 or mKate2 *in vivo* (Table 1).

We next bred *CAGs-rtTA3* and *CAGs-RIK* mice to *TG-Ren.713* mice generated previously [5], allowing GFP fluorescence to serve as a neutral marker of TRE induction. Similar to what was observed in our transgenic *CAGs-rtTA3* strain, *Rosa26*-targeted *CAGs-rtTA3* and *CAGs-RIK* enabled robust TRE-driven GFP expression in most tissues, including both solid organs and hematopoietic cells (Figs. 2, 3, 4 and Fig. S3). *CAGs-rtTA3* and *CAGs-RIK* showed significantly higher GFP induction in skin, liver, kidney, pancreas, compared to *R26-rtTA* (Fig. 2), while all animals showed high-level induction in intestine and T cells (Figs. 3A and 4), as previously reported [5,10]. Immunofluorescent staining for GFP and mKate2 revealed uniform and consistent staining in multiple cell types in each organ analyzed, compared to *R26-rtTA* that often showed absent or patchy expression in organs including pancreas and liver (Fig. 3B, Fig. S3B). As expected, *CAGs-RIK* mice showed uniform expression of mKate2 in all tissues examined. Interestingly, splenic B cells showed poor induction of the *TRE-GFP* transgene in all rtTA strains analyzed, including *CAGs-RIK*, despite relatively uniform mKate2. As mKate2 is expressed from the same polycistronic transcript as rtTA3, it is likely that these cells, which fail to induce high levels of GFP, contain abundant rtTA3 protein. To confirm this, we sorted mKate2+/GFP^{dim} and mKate2+/GFP^{hi} splenic cells from *CAGs-RIK/+;TG-Ren.713/+* mice treated with dox for one week, and assayed rtTA3 expression. Both QPCR and western blot analysis showed comparable levels of rtTA3 mRNA and protein in GFP^{hi} and GFP^{dim} cells (Fig. 4C,D), implying that the failure to induce GFP in some mKate2 positive cells is not due to reduced or absent rtTA3 expression, in contrast to previous observations with an independent CAGs-rtTA3 strain [10]. Although the mechanism underlying this observation is not known, it is possible that the *col1a1*-targeted, *TRE*-driven transgene is silenced or inaccessible in some cell types.

Importantly, both CAGs-driven strains showed almost identical GFP expression indicating that the presence of mKate2 in the *CAGs-RIK* strain did not alter rtTA3 levels. Together this data confirms that the CAGs promoter (at the *Rosa26* locus) can drive strong and widespread expression of rtTA3, allowing dox-dependent *TRE* induction in almost all tissues. Importantly, this implies that in combination with an appropriate Cre driver, both *CAGs-LSL-rtTA3* and *CAGs-LSL-RIK* strains can provide robust, tissue and/or cell-type specific TRE-mediated gene expression.

A

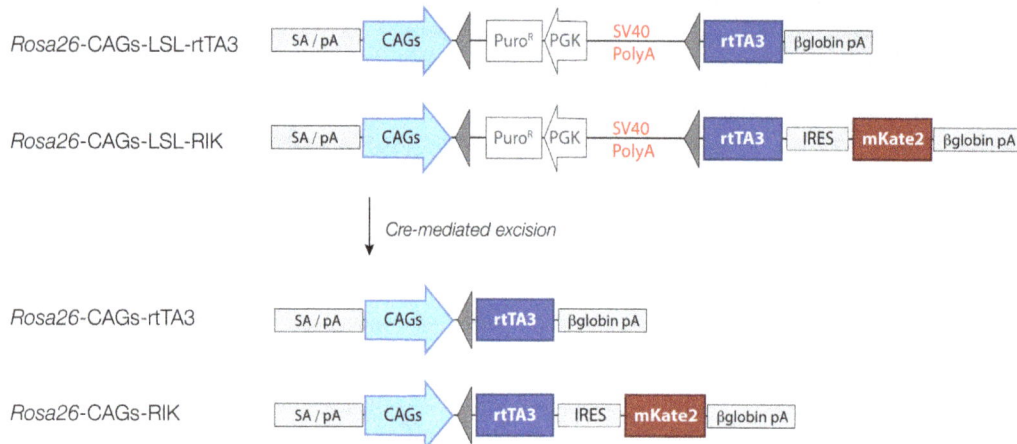

B

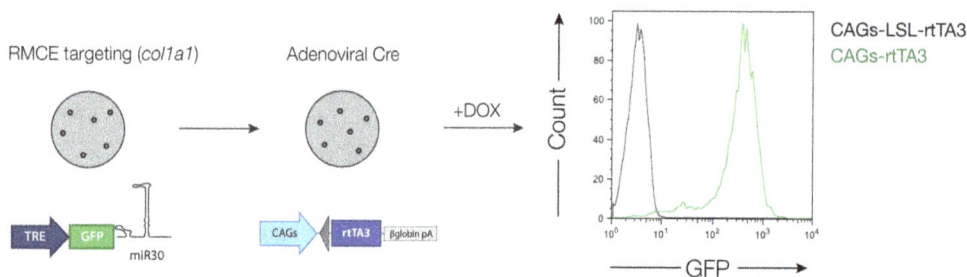

Figure 1. Generation of CAGs-LSL-rtTA3 and CAGs-LSL-RIK strains. A. Schematic representation of *CAGs-LSL-rtTA3* and *CAGs-LSL-rtTA3-IRES-mKate2 (RIK)* alleles targeted to the Rosa26 locus, prior to and following Cre-mediated recombination. **B**. Correctly targeted *CAGs-LSL-rtTA3* ESCs (Y1), were retargeted by recombinase mediated cassette exchange (RMCE) to introduce a *TRE-GFP-miR30* (TGM) construct to the *col1a1* recipient locus. Targeted cells were transduced with adenovirus expressing Cre and plated at low density to isolate individual clones. Clones were treated with doxycycline (1 ug/ml) for 2 days and analyzed by flow cytometry. Graph represents GFP fluorescence of TGM containing, dox-treated *CAGs-LSL-rtTA3* (black line) and recombined *CAGs-rtTA3* (green line) clones.

Mosaic TRE-induction through adenoviral Cre delivery

Tissue restricted Cre recombinase expression can be achieved in mice through the delivery of virus (adeno- or lentivirus) to specific organs. Intravenous (tail-vein) injection of adenovirus results in almost exclusive transduction of liver hepatocytes. As a first step to evaluate Cre-mediated, tissue specific TRE-induction we injected *CAGs-LSL-rtTA3* and *CAGs-LSL-RIK* mice (also carrying *TG-Ren.713*) with Adenoviral-Cre (AdenoCre) and treated mice with dox for one week. As expected, AdenoCre, dox-treated *CAGs-LSL-rtTA3* and *CAGs-LSL-RIK* mice showed mosaic expression of GFP (and mKate2 in *CAGs-LSL-RIK* animals) in the liver (Fig. 5A).

Importantly, we have never observed GFP signal in the absence of mKate2, suggesting mKate2 is a reliable indicator of rtTA3 expression, and no GFP was detected in any of the organs from dox-treated animals that were not exposed to Cre, indicating very low or no leaky expression of the rtTA3 transgene in adult mice.

Restricted delivery of AdenoCre to the mouse trachea allows mosaic Cre-mediated recombination in the lung epithelium [11]. We have previously shown that the Clara-cell Secretory Protein (CCSP)-rtTA transgenic strain can drive lung-specific expression of GFP-linked shRNAs in combination with AdenoCre-induced Kras[G12D] [5]. To assess whether our Cre-dependent rtTA3 alleles

Table 1. Mendelian transmission of Rosa26-targted CAGs-rtTA3 transgenes.

Genotype	CAGs-LSL-rtTA3	CAGs-rtTA3	CAGs-LSL-RIK	CAGs-RIK
Transgene	219 (217)*	110 (124)	84 (85)	130 (137.5)
Wildtype	215 (217)	138 (124)	86 (85)	145 (137.5)
p-value (two-tail, binomial test)	0.89	0.09	0.94	0.40

*Numbers represent: observed (expected) from heterzygote x wild-type crosses.

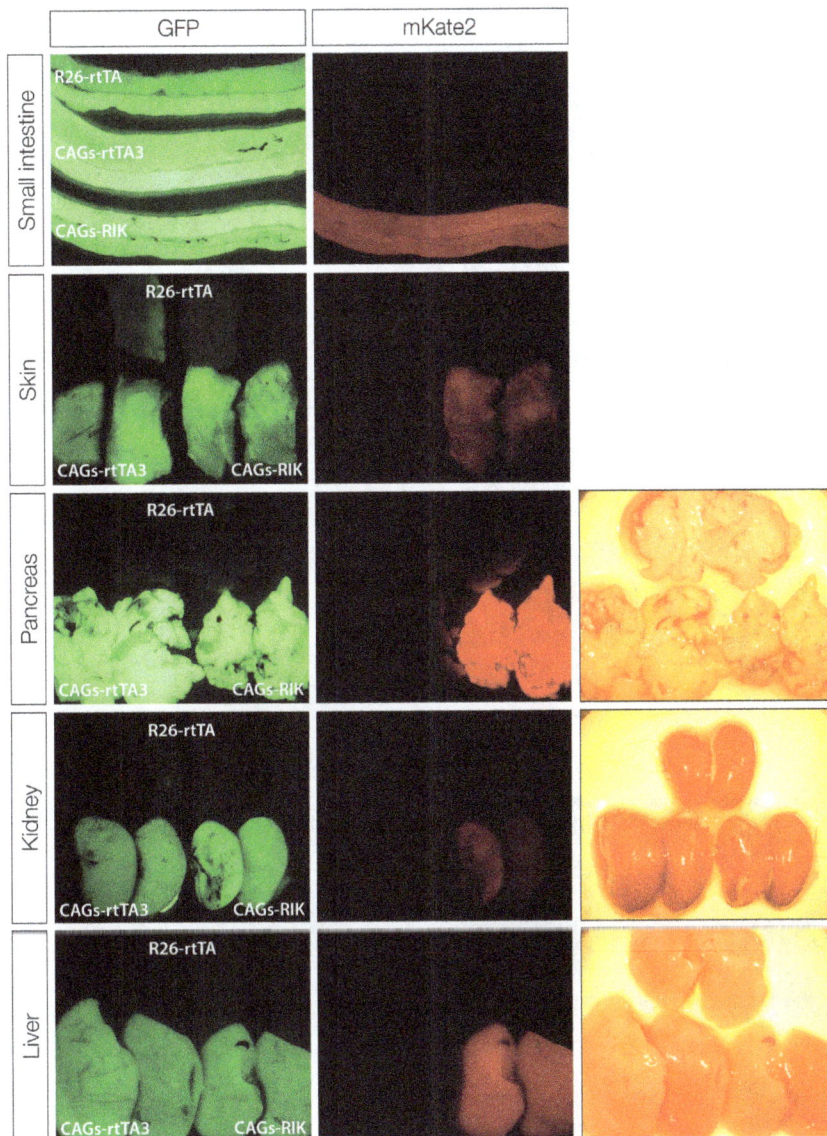

Figure 2. CAGs-rtTA3 and CAGs-RIK show strong expression in adult tissues. Whole mount epifluorescence images of small intestine, skin, pancreas kidney and liver from *R26-rtTA*, *CAGs-rtTA3* and *CAGs-RIK* transgenic animals (all containing *TG-Ren.713*). *R26-rtTA* shows strong expression in intestine and skin but weak or patchy expression in most other solid organs. *CAGs-rtTA3* and *CAGs-RIK* show almost identical expression patterns in adult mice. *CAGs-RIK* mice show strong and consistent expression of mKate2.

would similarly allow dox-dependent shRNA induction in Kras[G12D] expressing lung epithelium we treated *LSL-Kras*[G12D/+]; *TG-Ren.713/+;CAGs-LSL-RIK/+* mice with AdenoCre (5×10^6 PFU) via intratracheal injection and treated with dox for one week prior to analysis. Three months following AdenoCre treatment animals showed small Kras[G12D]-driven adenomas throughout the lung epithelium (Fig. 5B, Fig. S4) consistent with previous reports [12]. In almost all cases, Kras[G12D]-driven adenomas were GFP and mKate2 positive, indicating strong expression of rtTA3 and TRE-regulated shRNAs in these lesions. We observed only a very small proportion of individual mKate2 positive cells that did not express GFP (Fig. 5B, white arrows). In addition, in some lesions we noted some small areas of adenomas that were mKate2 negative (Fig. S4), implying activation of *LSL-Kras*[G12D] without *CAGs-LSL-RIK* recombination. It was not possible to determine whether there were also cases of *CAGs-LSL-RIK* recombination

without Kras[G12D] activation, as these events would not expand into adenomas. In all, this work demonstrates that both *CAGs-LSL-rtTA3* and *CAGs-LSL-RIK* allow robust and tissue-specific expression of rtTA3 in adult mice following restricted Cre exposure.

Tissue specific TRE-induction in transgenic Cre tumor models

In principle, integration of the Cre-conditional rtTA approach described above into existing disease models would enable a more precise investigation of gene function in complex genetic backgrounds. To test this we incorporated the *CAGs-LSL-RIK* into two well-established models, specifically, a Cre-dependent, *LSL-Kras*[G12D]-driven pancreatic cancer model [13] and Cre-independent, *MMTV-ErbB2* driven mammary cancer model [14].

Kras (codon 12 or 13) mutation is considered an initiating and driving event in most pancreatic ductal adenocarcinomas. In the

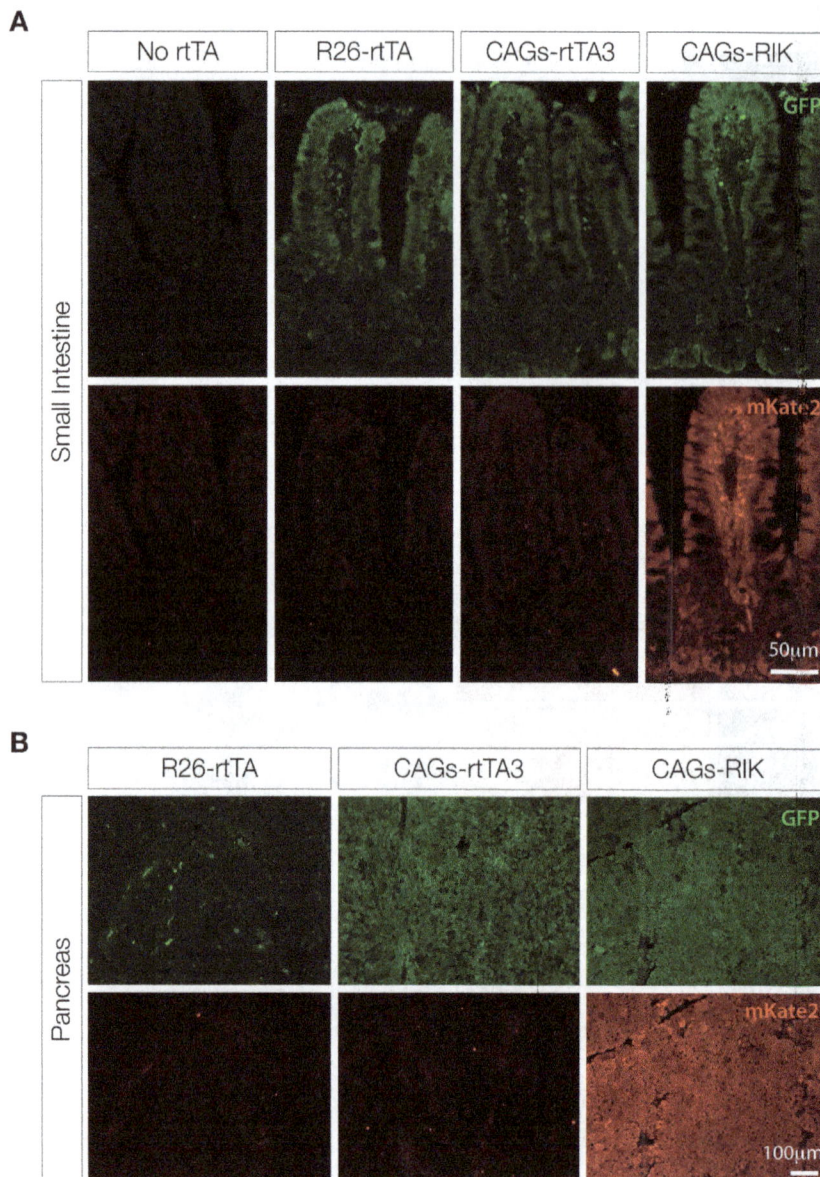

Figure 3. GFP induction and mKate2 expression is uniform in most organs of *CAGs-rtTA3* and *CAGs-RIK* mice. Immunofluorescence stains for GFP and mKate2 in the small intestine and pancreas of 'no rtTA', *R26-rtTA*, *CAGs-rtTA3* and *CAGs-RIK* mice following 1 week of doxycycline treatment. All rtTA strains show strong GFP induction in small intestine (**A**), but only *CAGs-rtTA3* and *CAGs-RIK* show robust and uniform GFP expression (and mKate2 for *RIK*) in the pancreatic acinar tissue (**B**).

mouse, pancreatic restricted expression of Kras^{G12D}, via a pancreas specific Cre such as *Pdx1-Cre*, leads to preneoplastic PanIN lesions [13]. We generated quadruple transgenic mice carrying *LSL-Kras$^{G12D/+}$;Pdx1-Cre/+;CAGs-LSL-RIK/+;TG-Ren.713/+* and treated them with doxycycline until 3 months of age. At this time the pancreas showed mosaic expression of mKate2 and GFP due to mosaic expression of the *Pdx1-Cre* allele. As expected, histology of the pancreas in 3-month old transgenic animals showed both normal acinar tissue and the development of acinar-to-ductal metaplasia and early PanIN lesions, as previously described [13]. The majority of PanIN lesions showed strong GFP and mKate2 expression confirming robust induction of rtTA3 and the GFP-linked shRNA. As noted following AdenoCre treatment in the lung, we observed some small regions of ADM/PanIN lesions that did not express mKate2 or GFP (Fig. 6A, white

arrows). We have now used the combination of Cre-driven pancreatic lesions with inducible shRNAs to investigate the genetic requirements underlying the initiation and progression of Kras-driven pancreas cancer [15].

Overexpression of HER2/Neu (encoded by the *ERBB2* gene) is a common feature of human breast cancer. In the mouse, this event can be mimicked via expression of a mutant rat ortholog *neu^{V664E}* allele downstream of a mouse mammary tumor virus (MMTV) promoter [14]. The *MMTV-neu^{V664E}* model has been used extensively to investigate the genetics and physiology of breast cancer progression. Expression of neu^{V664E} in this model is not dependent on the activity of Cre recombinase, therefore we asked whether our Cre-dependent *CAGs-LSL-RIK* allele could be used effectively in combination with *MMTV-neu^{V664E}* to express inducible shRNAs in the mammary gland and drive tumorigen-

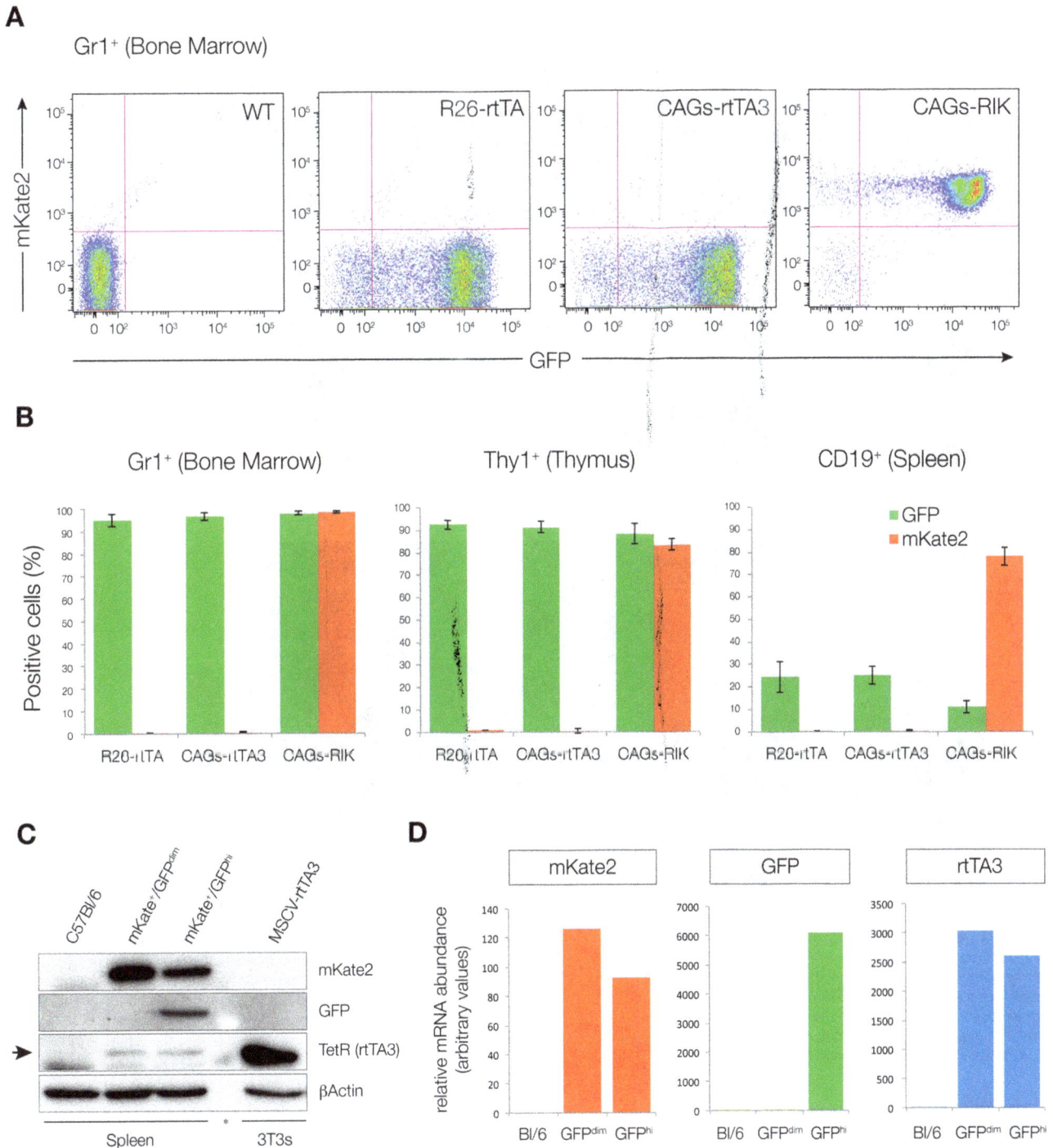

Figure 4. *CAGs-rtTA3* and *CAGs-RIK* enable GFP induction in myeloid and T lymphocyte lineages. A. Scatter plots representing GFP and mKate2 expression in Gr1 positive cells in the bone marrow of double transgenic (*rtTA/TGM*) animals following 1 week of doxycycline treatment (625 mg/kg in chow). **B**. Quantitation of GFP and mKate2 positive cells in Gr1, Thy1 and CD19 positive populations from the bone marrow, thymus and spleen respectively. Bars represent the mean percentage of GFP or mKate2 positive cells in each tissue, in 3 independent animals (per genotype) +/− SEM. **C**. Western blot of lysates from control (c57Bl/6), GFP negative and GFP positive splenocytes, indicating rtTA3, GFP and mKate2 expression in each population. Retrovirally transduced 3T3 cells serve as the positive control for rtTA3 expression. **D**. Graphs represent mRNA abundance in control (C57Bl/6), GFP negative and GFP positive splenocytes.

esis. For this, we generated female *MMTV-neu*$^{V664E/+}$;*CAGs-LSL-RIK/+;TG-Ren.713/+* animals also carrying the murine *whey acidic protein* gene promoter (WAP)-driven Cre transgene which responds to lactogenic hormones [16]. Parous females nursed litters for 3

weeks to induce expression of *WAP-Cre* and were treated with doxycycline to induce shRNA expression. These mice showed luminal epithelial expression of mKate2 and GFP (Fig. 6B, top panel), and developed neu^{V664E}-driven GFP and mKate2 positive

A

B

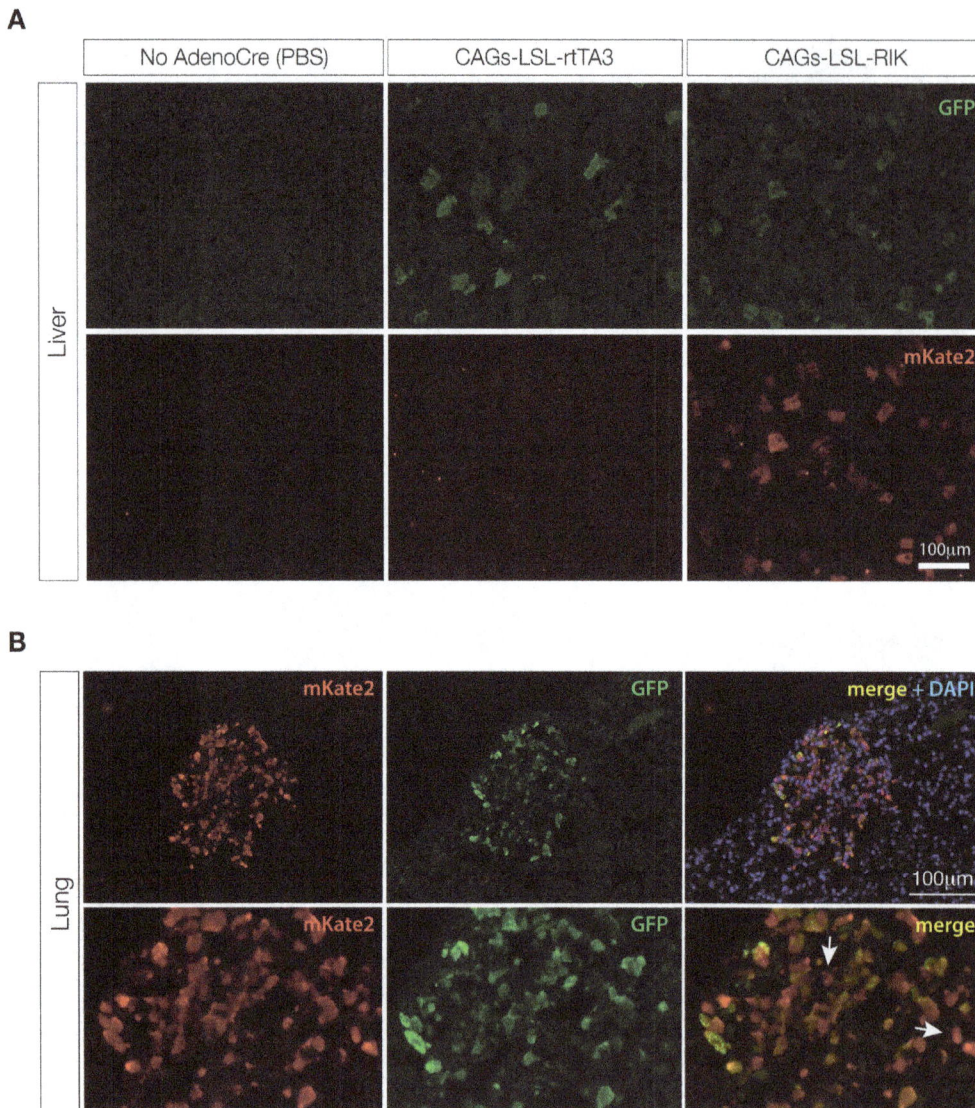

Figure 5. Adenoviral Cre induces mosaic activation of rtTA and GFP induction in *CAGs-LSL-rtTA3* and *CAGs-LSL-RIK* animals. A. Immunofluorescent stains for GFP and mKate2 in liver sections of *TG-Ren.713;CAGs-LSL-rtTA3* and *TG-Ren.713;CAGs-LSL-RIK* mice 1 week following intravenous injection of Adenoviral Cre (5×10^8 PFU) or PBS (*CAGs-LSL-RIK* only – left panel) and dox treatment. Double transgenic mice exposed to AdenoCre show mosaic expression of GFP (*CAGs-LSL-rtTA3*) or GFP and mKate2 (*CAGs-LSL-RIK*). No GFP of mKate2 expression was observed in animals not exposed to Cre. **B.** Immunofluorescent stains for GFP and mKate2 in lung sections of triple transgenic mice (*CAGs-LSL-rtTA3 or RIK;TG-Ren.713;LSL-Kras^{G12D}*). Kras^{G12D}-induced lung adenomas show strong expression of GFP and mKate2. Lowe panel: higher magnification of the lesion. White arrows indicate rare cells that show mKate2, but not GFP expression.

tumors at a median latency of approximately 160 days post partum (Fig. 6B, lower panel). Although not all tumors from *TG-Ren.713* mice showed mKate2/GFP expression due to an incomplete overlap of expression between the MMTV and *WAP* promoters, this novel multi-allelic system has proven a powerful platform for the study of tumor suppressor gene function in the context of HER2-driven breast cancer (SHE and SWL, *unpublished data*).

Discussion

Spatial and temporal control of gene expression provides a means to understand the contribution of genetic disruptions to disease progression and offers a setting to interrogate the role of individual genes in disease maintenance. Here we describe the

generation and characterization of two novel ESC lines and mouse strains that enable Cre-dependent, robust expression of the reverse tet-transactivator (rtTA3) and thus, tissue-restricted induction of TRE-controlled transgenes. Further, we show that these transgenic strains can be effectively combined with established mouse models of disease, including Cre/LoxP-based approaches and Cre-independent model systems. The integration of established models of disease with the flexibility of inducible and reversible gene regulation will allow a more detailed interrogation of the underpinnings of disease pathogenesis and evolution. For instance, model systems that incorporate constitutive and inducible genetic alterations with inducible and reversible gene silencing (or overexpression) offer the unique opportunity to investigate how the temporal order of events determines disease progression and whether those events are required for disease maintenance. Such

A

B

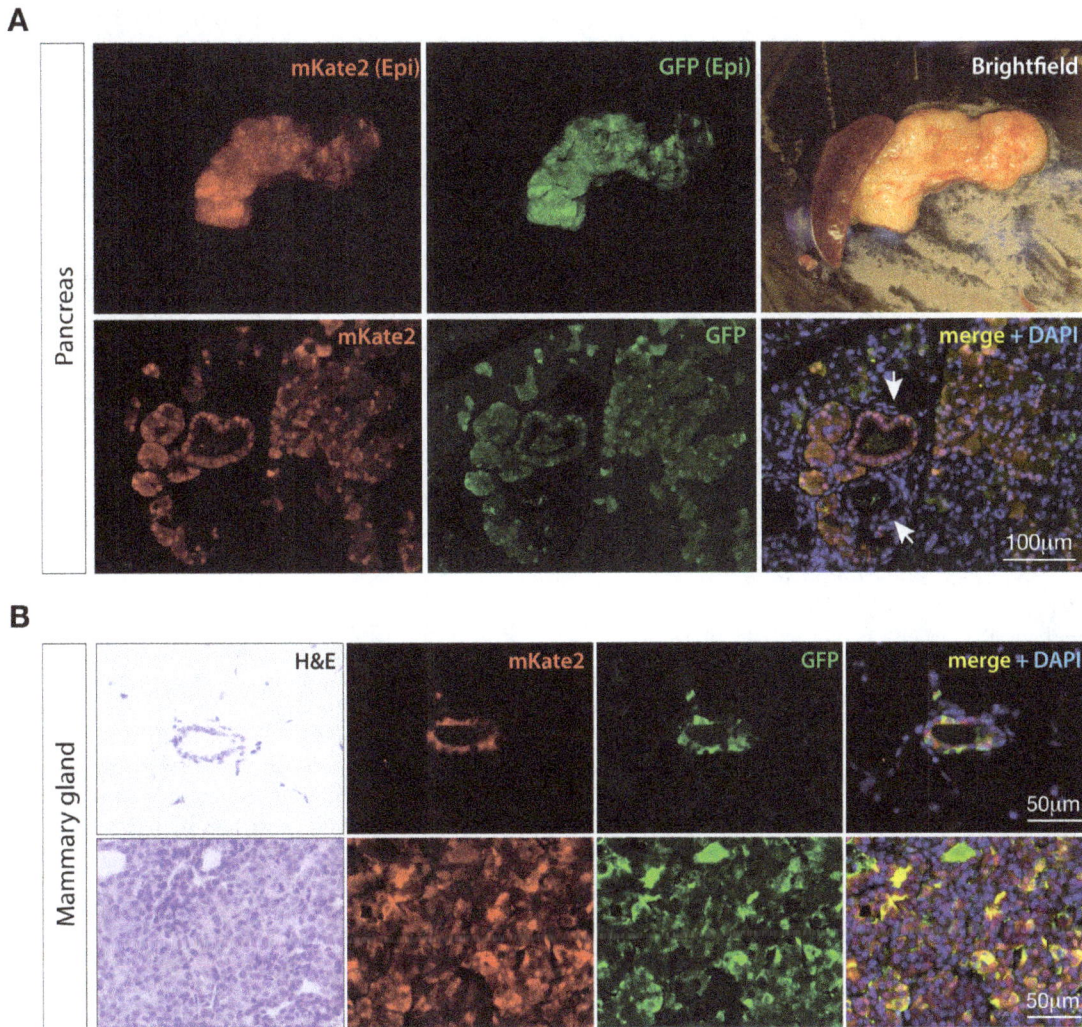

Figure 6. *CAGs-LSL-RIK* **enables tissue-restricted expression of** *TRE***-transgenes in transgenic models of disease. A.** Whole mount epifluorescence (top panel) and immunofluorescence images from a quadruple transgenic (*CAGs-LSL-RIK;TG-Ren.713;LSL-Kras^{G12D};Pdx1-Cre*) animal, showing induction of GFP and mKate2 in both normal acinar tissue and pre-neoplastic, Kras^{G12D}-induced PanIN lesions (top arrow). As observed in AdenoCre treated lungs, some PanIN lesions did not show GFP or mKate2 staining suggesting incomplete LSL excision in a small proportion of cells. **B.** Immunofluorescent stains for GFP and mKate2 in mammary tissue of *CAGs-LSL-RIK;TG-Ren.713;MMTV-Neu;WAP-Cre* transgenic mice treated with dox.

work will ultimately lead to better understanding of disease and the development of more accurate and effective preclinical models.

In contrast to strategies that rely on tet-transactivators driven by tissue specific promoters (http://www.tetsystems.com/fileadmin/tettransgenicrodents.pdf) our approach integrates the use of established tissue-restricted Cre recombinase strains to initiate *TRE* induction. Previous studies have reported the generation of Cre-dependent tTA and rtTA strains driven by the endogenous *Rosa26* promoter [17–19]. We have previously shown that *Rosa26* promoter activity can vary significantly in different tissues of adult mice, resulting in sub-optimal *TRE* induction in a range of cell types [5]. In contrast, the synthetic, ubiquitous *CAGs* promoter provides robust *TRE* induction in most adult tissues (Fig. 2). Thus, given an appropriate Cre driver, *CAGs-LSL-rtTA3* and *CAGs-LSL-RIK* provide potent, doxycycline-dependent transgene/shRNA expression in most tissues accessible to doxycycline. In addition, as the *CAGs* promoter drives robust expression of rtTA3 in many different cell lineages and at different stages of differentiation, *CAGs-LSL-rtTA3* strains offer a significant advantage over *tTA/*

rtTA strains that depend on lineage specific promoters, such as those expressed only in stem or progenitor cells. In this regard, we have observed that transient induction of Cre recombinase activity in stem cells of the intestine using *Lgr5-GFP-IRES-CreER*, promotes long-lived rtTA3 expression (and GFP induction) in individual crypts and villi of the small and large intestine (LED and SWL, unpublished data).

During our analysis of multi-allelic animals carrying *LSL-Kras^{G12D}* and *CAGs-LSL-RIK*, we noted some areas of tissue that showed characteristic Kras^{G12D}-induced changes (lung adenomas or PanINs), which did not show expression of mKate2, suggesting Cre-induced recombination of only a subset of 'floxed' genes in the genome. Because Kras^{G12D}-driven phenotypes are not 100% penetrant in either tissue, we have not been able to measure the frequency of cells that show recombination only at the *CAG-LSL-RIK* allele and not *Kras^{G12D}*. The reasons behind this incomplete recombination are not clear but may reflect the regions surrounding the *LoxP* sites. Of note, we have observed increased recombination efficiency of the *CAGs-LSL-rtTA3* allele compared to the *CAGs-LSL-RIK* strain in cases of low or transient Cre expression; the two alleles vary slightly in sequences close to the 3′ LoxP site due to alternate cloning strategies. It is likely that incomplete *LoxP* recombination is a feature of many complex, Cre-dependent models, but it goes undetected due to a lack of reporter-based approaches. The presence of mKate2 as a Cre reporter in *CAGs-LSL-RIK* provides a means to clearly identify cells and tissues that express rtTA3 and are capable of inducing shRNA or transgene expression, irrespective of fluorescent tags linked to such transgenes (e.g. GFP). Thus, it enables tracking and/or prospective isolation of Cre-recombined cells prior to induction of, and post-withdrawal of, *TRE*-driven transgene/shRNA expression.

We and others recently described a conceptually new approach to complex mouse modeling, based on the derivation and manipulation of conditional, multi-allelic embryonic stem cells (ESCs), which we use to generate tailored genetic models for analysis [1,5,20]. Such ESC-GEMMs provide a means to rapidly interrogate gene function in genetically complex animal models in a fraction of the time required for traditional breeding. During the genesis of the two transgenic strains described here we generated ESC lines carrying *CAGs-LSL-rtTA3* (Y1) or *CAGs-LSL-RIK* (D34) as well as the *col1a1* homing cassette for RMCE. Thus Y1 and D34 cells could be employed by investigators wanting to fast track analysis of a gene or genes in a setting where Cre recombinase is delivered extrinsically (i.e. intratracheal or intravenous injection) or used as a base ESC line for the introduction of additional genetic manipulations, such as the incorporation of tissue specific Cre knock-in alleles. Alternately these alleles could be incorporated into ESC-GEMMs through re-derivation of new ESC lines. In fact, we have recently validated the use of this approach by producing a number of Kras[G12D]-based pancreatic cancer models, using *CAGs-LSL-RIK* to drive pancreas-specific expression of positive and negative regulators of tumor initiation and progression [15].

Together, the ESCs and mouse strains described here bolster the already impressive arsenal of *in vivo*, tet-based systems for manipulation of gene expression by providing robust tools for tissue-restricted induction of transgenes and shRNAs. Integrating these new strains with existing mouse modeling platforms promises to provide a wealth of new discoveries by unearthing the details of gene function in all stages of development and throughout the pathogenesis of disease.

Materials and Methods

ES cell targeting

All ES cells were maintained on irradiated feeders in M15 media containing LIF as previously outlined [21]. Targeting vectors were linearized using a unique PmeI site introduced downstream of the Diphtheria Toxin A (DTA) expression cassette. C2 ES cells (1×10^7) were electroporated with 50 ug linearized targeting vector using a BioRad Gene Pulser and plated in M15 media as previously described [21]. Two days following transfections cells were treated with media containing 1 ug/ml puromycin and individual surviving clones were picked after 9–10 days of selection. Two days after clones were picked puromycin was removed from the media and cells were cultures in standard M15 thereafter. For Southern blots, genomic DNA from individual ES cell clones was digested overnight in either EcoRI or EcoRV/BglII (See Fig. S1).

Animal husbandry

ES cell-derived mice were produced by tetraploid complementation as has been described elsewhere (Zhao, Nat Prot 2010). For removal of the 'LSL cassette' *in vivo*, CAGs-LSL-rtTA3 and CAGs-LSL-RIK (*R26Sor[tm1(CAGs-LSL-rtTA3)Slo]* and *R26Sor[tm2(CAGs-LSL-RIK)Slo]*) mice were crossed to the CAGs-Cre transgenic strain and F1 progeny were genotyped for LoxP recombination using specific primers (see Table S1). TG-Ren.713 [5], WAP-Cre (Wagner et al, 1997), MMTV-Neu (Muller et al, 1988), LSL-KrasG12D [12] and Pdx1-Cre [13] mice have all been previously described. For mammary gland experiments, female mice were bred at 7 weeks and doxycycline feed was administered from date of litter birth. Litters were nursed for 3 weeks to induce WAP promoter activity. Parous mice monitored weekly for tumor formation by physical palpation. See Table S1 for genotyping information on CAGs-LSL-rtTA3 and CAGs-LSL-RIK strains.

Ethics statement

Production of mice and all treatments described were approved by the Institutional Animal Care and Use Committee (IACUC) at McGill University (Montreal, Canada) or Memorial Sloan Kettering Cancer Center (NY) under protocol numbers: 2001–4751 (McGill), 11-06-012, 11-06-015 and 12-04-006 (MSKCC).

Treatments

Doxycycline was administered via food pellets (625 mg/kg) (Harlan Teklad). Adenovirus expressing Cre recombinase (AdenoCre) was purchased from The University of Iowa Gene transfer Core. For adenoviral delivery to the liver: 5×10^8 PFU AdenoCre/mouse was injected intravenously via the tail vein. For adenoviral delivery to the lung: 6–10 week-old mice were anesthetized by i.p. injection of ketamine 80 mg/kg, xylazine 10 mg/kg [11] and treated once by intratracheal instillation of 5×10^6 PFU AdenoCre/mouse. Three months following AdenoCre treatment, lungs were collected, fixed and analyzed by immunofluorescence or immunohistochemistry.

Protein and RNA analysis

Protein lysates were prepared in RIPA buffer and quantified by Lowry assay (BioRad). Western blots were probed with antibodies against: TetR (rtTA3) (1:1000, mouse monoclonal Clone 9G9, Clontech #631131), GFP (1:2000, chicken polyclonal, Abcam #ab13970), tRFP (mKate2) (1:2000, rabbit polyclonal, Evrogen #B00201) and β-Actin-HRP (1:5000, mouse monoclonal AC15 clone, Sigma #A3854). RNA was prepared from sorted cells by Trizol extraction and column purification. cDNA was prepared from 1 μg total RNA using Taqman reverse transcription kit (Applied Biosystems, #N808-0234) with random hexamers. Quantitative PCR detection was performed using SYBR green reagents (Applied Biosystems) using primers specific to **rtTA3**: F: 5′-CAATGGTGTCGGTATCGAAG-3′, R: 5′-CTTGTTCT-TCACGTGCCAGT-3′; **mKate2**: F: GGTGAGCGAGCTGA-TTAAGG-3′ and R: 5′-TTTTGCTGCCGTACATGAAG-3′; and **GFP**: F: 5′-ATCGACTTCAAGGAGGACGGCA-3′ and R: 5′-CGTTCTTCTGCTTGTCGGCCAT-3′.

Immunophenotyping

Immunostaining and FACS analysis for blood lineages were performed as previously described [22]. Briefly, single cell suspensions from whole bone marrow, spleen and thymus were immunostained for CD45.2 (APC-conjugate, BD #559864) and cell lineage markers: Gr1 (Pacific blue, BioLegend #108430), CD19 (Pacific Blue, BioLegend #115526) or Thy1 (Pacific blue, BioLegend #105324) and the percentage of GFP and/or mKate2 expressing cells were calculated within these specific lineages populations. Data was collected on an LSR-II flow cytometer (BD BioSciences) and analyzed using FlowJo software (Tree Star).

Immunohistochemistry

Tissue, fixed in 10% neutral buffered formalin for 24 hours, was embedded in paraffin and sectioned by IDEXX RADIL (Columbia, MO). Sections were rehydrated and unmasked (antigen retrieval) by heat treatment for 5 mins in a pressure cooker in 10 mM Tris/1 mM EDTA buffer (pH 9) containing 0.05% Tween 20. For immunohistochemistry, sections were treated with 3% H_2O_2 for 10 mins and blocked in TBS containing 1% BSA. For immunofluorescence, sections were not treated with peroxidase. Primary antibodies, incubated at 4C overnight in blocking buffer, were: chicken anti-GFP (1:500, #ab13970), rabbit anti-tRFP (1:2000, Evrogen, #AB232) and rabbit anti-ki67 (1:100, Sp6 clone, Abcam #ab16667). For immunohistochemistry, sections were incubated with anti-rabbit ImmPRESS reagent (Vector Laboratories, #MP7401) and developed using ImmPACT DAB (Vector Laboratories, #SK4105) according to the manufacturer instructions. For immunofluorescent stains, secondary antibodies were applied for 1 hour at room temp in TBS in the dark, washed twice with TBS, counterstained for 5 mins with DAPI and mounted in ProLong Gold (Life Technologies, #P36930). Secondary antibodies used were: anti-chicken 488 (1:500, DyLight IgG, #ab96947) and anti-rabbit 568 (1:500, Molecular Probes, #a11036). Images of fluorescent and IHC stained sections were acquired on a Zeiss Axioscope Imager Z.1 using a 10x (Zeiss NA 0.3) or 20x (Zeiss NA 0.17) objective and an ORCA/ER CCD camera (Hamamatsu Photonics, Hamamatsu, Japan). Raw.tif files were processed using Photoshop CS5 software (Adobe Systems Inc., San Jose, CA) to adjust levels and/or apply false coloring.

Supporting Information

Figure S1 Targeting *CAGs-LSL-rtTA3* to the *Rosa26* locus. A. Schematic of the *Rosa26* locus before and after recombination of the *CAGs-LSL-rtTA3* or *CAGs-LSL-RIK* targeting vector. Key restriction sites used for clone identification by Southern blot are indicated. Sizes of each predicted fragment are also shown and a solid black line highlights the position of the Southern probe. **B**. Southern blot images showing identification of Y1 (2.3 kb band) and D34 (4.8 kb band) clones, following EcoRV/BglII and EcoRI digests, respectively. **C**. Fluorescence in situ hybridization on a metaphase spread from D34 ES cells using the CAGs-LSL-RIK fragment as a probe, showing homozygous targeting of *CAGs-LSL-RIK* to Chromosome 6.

Figure S2 GFP induction following Adenoviral Cre transduction in targeted Y1 and D34 ESCs. Y1 and D34 ESCs carrying *TG-Ren.713* at the *col1a1* locus were transduced with adenovirus expressing Cre (Cre, green line) or not transduced (no Cre, black line), treated with doxycycline (1 ug/ml) for 2 days and analyzed by flow cytometry. Graphs represent bulk population of transduced cells (not single clones). Bulk populations were single cell cloned to assess the uniformity of GFP induction in the presence of constitutive rtTA3 expression (see Fig. 1B).

Figure S3 GFP induction and mKate2 expression in large intestine and liver. Immunofluorescence stains for GFP and mKate2 in the large intestine and liver of 'no rtTA', *R26-rtTA*, *CAGs-rtTA3* and *CAGs-RIK* mice following 1 week of doxycycline treatment. All rtTA strains show strong GFP induction in large intestine (**A**), but only *CAGs-rtTA3* and *CAGs-RIK* show robust and uniform GFP expression (and mKate2 for *RIK*) in the liver tissue (**B**).

Figure S4 Mosaic mKate2 expression in a proportion of lung adenomas. Immunohistochemical stains for mKate2 and Ki67 in lung sections of double transgenic mice (*CAGs-LSL-RIK;LSL-Kras^{G12D}*) treated with intratracheal Adenoviral Cre (AdenoCre) or vehicle (Tris-HCl). 12 weeks following Cre delivery LSL-KrasG12D mice show small, moderately proliferative adenomas. Some adenomas show uniform mKate2 staining (top panel of 'AdenoCre'), while a subset showed both positive and negative mKate2 cells (arrows) suggesting Cre-driven activated Kras^{G12D} but not rtTA3-IRES-mKate2. Adenoma area highlighted by dotted line.

Table S1 Genotyping primers. Primer sequences and expected PCR product sizes for genotyping *Rosa26*-targeted CAGs-rtTA3 strains.

Acknowledgments

We thank Janelle Simon, Danielle Grace, Sha Tian and Jacqueline Cappellani for technical assistance with animal colonies, Kevin O'Rourke for editorial advice and other members of the Lowe and Pelletier laboratories for advice and discussions.

Author Contributions

Conceived and designed the experiments: LED ZN MS SHE NT JP SWL. Performed the experiments: LED ZN MS SHE EM NT TL JP. Analyzed the data: LED ZN JP. Contributed reagents/materials/analysis tools: LED MS SHE TL JP. Wrote the paper: LED JP SWL.

References

1. Dow LE, Lowe SW (2012) Life in the fast lane: mammalian disease models in the genomics era. Cell 148: 1099–1109.
2. Gossen M, Bujard H (1992) Tight control of gene expression in mammalian cells by tetracycline-responsive promoters. Proc Natl Acad Sci U S A 89: 5547–5551.
3. Das AT, Zhou X, Vink M, Klaver B, Verhoef K, et al. (2004) Viral evolution as a tool to improve the tetracycline-regulated gene expression system. The Journal of biological chemistry 279: 18776–18782.
4. Urlinger S, Baron U, Thellmann M, Hasan MT, Bujard H, et al. (2000) Exploring the sequence space for tetracycline-dependent transcriptional activators: novel mutations yield expanded range and sensitivity. Proceedings of the National Academy of Sciences of the United States of America 97: 7963–7968.
5. Premsrirut PK, Dow LE, Kim SY, Camiolo M, Malone CD, et al. (2011) A rapid and scalable system for studying gene function in mice using conditional RNA interference. Cell 145: 145–158.
6. Shcherbo D, Murphy CS, Ermakova GV, Solovieva EA, Chepurnykh TV, et al. (2009) Far-red fluorescent tags for protein imaging in living tissues. Biochem J 418: 567–574.
7. Beard C, Hochedlinger K, Plath K, Wutz A, Jaenisch R (2006) Efficient method to generate single-copy transgenic mice by site-specific integration in embryonic stem cells. Genesis 44: 23–28.
8. Hochedlinger K, Yamada Y, Beard C, Jaenisch R (2005) Ectopic expression of Oct-4 blocks progenitor-cell differentiation and causes dysplasia in epithelial tissues. Cell 121: 465–477.
9. Sakai K, Miyazaki J (1997) A transgenic mouse line that retains Cre recombinase activity in mature oocytes irrespective of the cre transgene transmission. Biochemical and biophysical research communications 237: 318–324.
10. Takiguchi M, Dow LE, Prier JE, Carmichael CL, Kile BT, et al. (2013) Variability of inducible expression across the hematopoietic system of tetracycline transactivator transgenic mice. PLoS ONE 8: e54009.

11. DuPage M, Dooley AL, Jacks T (2009) Conditional mouse lung cancer models using adenoviral or lentiviral delivery of Cre recombinase. Nature protocols 4: 1064–1072.

12. Jackson EL, Willis N, Mercer K, Bronson RT, Crowley D, et al. (2001) Analysis of lung tumor initiation and progression using conditional expression of oncogenic K-ras. Genes Dev 15: 3243–3248.

13. Hingorani SR, Petricoin EF, Maitra A, Rajapakse V, King C, et al. (2003) Preinvasive and invasive ductal pancreatic cancer and its early detection in the mouse. Cancer Cell 4: 437–450.

14. Muller WJ, Sinn E, Pattengale PK, Wallace R, Leder P (1988) Single-step induction of mammary adenocarcinoma in transgenic mice bearing the activated c-neu oncogene. Cell 54: 105–115.

15. Saborowski M, Saborowski A, Morris JPt, Bosbach B, Dow LE, et al. (2014) A modular and flexible ESC-based mouse model of pancreatic cancer. Genes & development 28: 85–97.

16. Wagner KU, Wall RJ, St-Onge L, Gruss P, Wynshaw-Boris A, et al. (1997) Cre-mediated gene deletion in the mammary gland. Nucleic acids research 25: 4323–4330.

17. Belteki G, Haigh J, Kabacs N, Haigh K, Sison K, et al. (2005) Conditional and inducible transgene expression in mice through the combinatorial use of Cre-mediated recombination and tetracycline induction. Nucleic acids research 33: e51.

18. Wang L, Sharma K, Deng HX, Siddique T, Grisotti G, et al. (2008) Restricted expression of mutant SOD1 in spinal motor neurons and interneurons induces motor neuron pathology. Neurobiology of Disease 29: 400–408.

19. Yu H-MI, Liu B, Chiu S-Y, Costantini F, Hsu W (2005) Development of a unique system for spatiotemporal and lineage-specific gene expression in mice. Proceedings of the National Academy of Sciences of the United States of America 102: 8615–8620.

20. Huijbers IJ, Krimpenfort P, Berns A, Jonkers J (2011) Rapid validation of cancer genes in chimeras derived from established genetically engineered mouse models. BioEssays: news and reviews in molecular, cellular and developmental biology 33: 701–710.

21. Dow LE, Premsrirut PK, Zuber J, Fellmann C, McJunkin K, et al. (2012) A pipeline for the generation of shRNA transgenic mice. Nature protocols.

22. Zuber J, Rappaport AR, Luo W, Wang E, Chen C, et al. (2011) An integrated approach to dissecting oncogene addiction implicates a Myb-coordinated self-renewal program as essential for leukemia maintenance. Genes Dev 25: 1628–1640.

19

BMPRIA Mediated Signaling Is Essential for Temporomandibular Joint Development in Mice

Shuping Gu[1,9,¤a], Weijie Wu[1,2,9], Chao Liu[1,¤b], Ling Yang[1,3], Cheng Sun[1], Wenduo Ye[1], Xihai Li[1,4], Jianquan Chen[5], Fanxin Long[5], YiPing Chen[1]*

1 Department of Cell and Molecular Biology, Tulane University, New Orleans, Louisiana, United States of America, **2** Department of Dentistry, ZhongShan Hospital, FuDan University, Shanghai, P.R. China, **3** Guanghua School of Stomatology, Sun Yat-sen University, Guangzhou, Guangdong, P.R. China, **4** Academy of Integrative Medicine, Fujian University of Traditional Chinese Medicine, Fuzhou, Fujian, P.R. China, **5** Department of Internal Medicine, Washington University School of Medicine, St. Louis, Missouri, United States of America

Abstract

The central importance of BMP signaling in the development and homeostasis of synovial joint of appendicular skeleton has been well documented, but its role in the development of temporomandibular joint (TMJ), also classified as a synovial joint, remains completely unknown. In this study, we investigated the function of BMPRIA mediated signaling in TMJ development in mice by transgenic loss-of- and gain-of-function approaches. We found that BMPRIA is expressed in the cranial neural crest (CNC)-derived developing condyle and glenoid fossa, major components of TMJ, as well as the interzone mesenchymal cells. *Wnt1-Cre* mediated tissue specific inactivation of *Bmpria* in CNC lineage led to defective TMJ development, including failure of articular disc separation from a hypoplastic condyle, persistence of interzone cells, and failed formation of a functional fibrocartilage layer on the articular surface of the glenoid fossa and condyle, which could be at least partially attributed to the down-regulation of *Ihh* in the developing condyle and inhibition of apoptosis in the interzone. On the other hand, augmented BMPRIA signaling by *Wnt1-Cre* driven expression of a constitutively active form of *Bmpria* (*caBmpria*) inhibited osteogenesis of the glenoid fossa and converted the condylar primordium from secondary cartilage to primary cartilage associated with ectopic activation of Smad-dependent pathway but inhibition of JNK pathway, leading to TMJ agenesis. Our results present unambiguous evidence for an essential role of finely tuned BMPRIA mediated signaling in TMJ development.

Editor: Songtao Shi, University of Southern California, United States of America

Funding: WW was supported by a fellowship from ZhongShan Hospital, FuDan University. LY was supported by a scholarship from the China Scholarship Council (No. 201208440191). This work was supported by the National Institutes of Health grants (R01 DE14044, DE17792) to YC. The funders had no role in this design, data collection and analysis, decision to publish, or preparation of the manuscript.

Competing Interests: The authors have declared that no competing interests exist.

* Email: ychen@tulane.edu

¤a Current address: Shanghai Research Center for Model Organisms, Pudong, Shanghai, P.R. China
¤b Current address: Department of Biomedical Sciences, Baylor College of Dentistry, Texas A&M University, Dallas, Texas, United States of America

9 These authors contributed equally to this work.

Introduction

As an evolutionary creature, the temporomandibular joint (TMJ) is a unique synovial joint generated only in mammals and is involved in food capture and intake, speech, as well as maturation of the facial contour [1]. It is made of specific components originated from the skull base and the low jaw including the glenoid fossa, condyle, articular disc, ligaments, and joint capsule. Although defined as a synovial joint, the developmental process of TMJ differs significantly from the joints of appendicular skeletons that are generated by cleavage or segmentation within a single skeletal condensation [2]. The TMJ develops from two distinct mesenchymal condensations, the glenoid fossa blastema that ossifies primarily through intramembranous bone formation, and the condylar blastema that undergoes endochondral ossification. These two primordia are initially separated widely by intervening mesenchyme that was thought to later contribute to the articular disc and capsule, as well as the synovial lining of joint cavity [3,4] Subsequently, the condylar primordium, arising from the periosteum of the mandibular bone and therefore classified as secondary cartilage [5,6], grows rapidly towards the glenoid fossa, and meanwhile, the articular disc forming from a condensed stripe flanking the apex of the condyle and subsequently separating from the latter, divides the interzone into the upper and lower joint cavities [7]. In mice, the mesenchymal condensation of condyle appears at embryonic day 13.5 (E13.5) and the glenoid fossa at E14.5 [8]. At E15.5, the shape of glenoid fossa and condyle has been established, and at E16.5, the upper synovial cavity becomes discernible with a disc beginning to form. Subsequently at E17.5, the lower joint cavity appears as a definite articular disc separates from the apex of the condyle. This intricate multi-step developmental process is regulated by intrinsic and extrinsic factors.

As for intrinsic constituents, the significance of genetic factors has attracted the attention of the field. Gene targeting studies have revealed essential roles for a number of transcription factors and growth factors in TMJ development, as evidenced by the absence of condylar cartilage in mice carrying mutations in *Sox9*, *Runx2*, or *Tgfbr2*, and by the abnormal development of mice carrying mutations in *Shox2* or *Spry1* and *Spry2* [8–14]. Ihh, which plays a pivotal role in long bone development and digit joint formation [15], has been implicated in TMJ development by initiating the formation of articular disc and instructing the disc to undergo proper morphogenesis and to separate from the condyle, as well as in maintaining proper structure and function of the TMJ after it forms [16–18]. Lack of Ihh or its downstream effector Gli2 results in missing of a distinct disc in the TMJ [16,17]. In addition, extrinsic factors such as biomechanical force also contribute to TMJ development [2].

Bone morphogenetic proteins (BMPs) exert diverse biological functions during development and postnatal homeostasis. BMP signals are transduced into cells through the type I and type II transmembrane serine/threonine kinase complexes by activating Smad-dependent (canonical) pathway, as well as Smad-independent (non-canonical) pathway via activation of the mitogen-activated protein kinase (MAPK) signaling [19]. Extensive studies have established critical roles for BMP signaling in skeletal development and joint morphogenesis, particularly in joint formation of long bones. Joint formation in the appendage skeletons begins with the formation of a condensed cell stripe known as interzone in the developing cartilage template [20]. Cells in the edges of the interzone give rise to the articular cartilage that covers the ends of the adjacent skeletal elements, while cells in the middle of the interzone undergo programmed cell death, leading to physical separation of the contiguous cartilage element and formation of joint cavity [21]. Several members of BMP family, including *Bmp2*, *Bmp4*, *Gdf5*, *Gdf6*, *Gdf7*, are expressed in the interzone along with BMP antagonists *Chondin* and *Noggin* [22–26]. Mice carrying mutations in *Gdf5* or *Gdf6* exhibit lack of joint formation at specific locations [25,27], demonstrating a direct action of BMP signaling in joint morphogenesis. On the other hand, elevated BMP signaling also blocks joint formation, as manifested by failure in joint formation in the limbs of *Noggin* mutant mice [28]. These loss-of- and gain-of-function studies indicate an essential role for tightly regulated BMP activity in synovial joint formation. Furthermore, BMP signaling is also involved in postnatal joint homeostasis and tissue remodeling [26,29].

Being one of the two primary BMP type I receptors (BMPRIA and BMPRIB), BMPRIA plays crucial roles in skeleton patterning and development. In developing limb skeletons, *BmprIa* is expressed in the joint interzone, perichondrium, periarticular cartilage, and hypertrophic chondrocytes [26,30–32]. Tissue specific deletion of *BmprIa* in cartilage lineage leads to chondrodysplasia attributed at least partially to the defective cell proliferation as well as the premature hypertrophy of chondrocytes associated with down-regulation of *Ihh* [33,34]. While joint defect was not identified in mice carrying cartilage specific inactivation of *BmprIa*, mice carrying tissue specific inactivation of *BmprIa* in the interzone indeed exhibited missed joints in the ankles [26], indicating a requirement of BMPRIA mediated signaling in joint formation.

Despite a wealth of documents on BMP signaling in bone and joint formation of appendicular skeletons, little is known about its role in TMJ development. Thus far, the only line of evidence implicating a possible involvement of BMP signaling in TMJ formation is that *Bmp2* and *Bmp7* were found to be expressed in

the developing condyle [17,35]. To gain an insight into BMP signaling in TMJ development, in this study, we used transgenic loss-of- and gain-of-function approaches to investigate the function of BMPRIA mediated signaling in TMJ development.

Materials and Methods

Ethics statement

Experiments that involved use of animals in this study was approved by the Institutional Animal Care and Use Committee (IACUC) of Tulane University (protocol number: 0367R) and was in strict accordance with the recommendations in the Guide for Care and Use of Laboratory Animals of the National Institutes of Health.

Animal and sample collection

The generation and identification of transgenic and gene-targeted animals, including *Wnt1-Cre*, *BmprIa^{f/f}*, and *pMes-caBmprIa* that carries a conditional constitutively active form (with Gln203 to Asp change) of *BmprIa* transgenic allele, have been described previously [15,31,36,37]. *Wnt1-Cre;BmprI^{f/f}* embryos were obtained by crossing *Wnt1-Cre;BmprIa^{f/+}* mice with *BmprIa^{f/f}* line. *Wnt1-Cre;pMes-caBmprIa* embryos were generated by mating *Wnt1-Cre* mice with *pMes-caBmprIa* transgenic line. *Ihh* null mutants were harvested from intercross of *Ihh* heterozygous mice. Embryos with *BmprIa* deficiency in their neural crest cells (*Wnt1-Cre;BmprIa^{f/f}*) die around E12.5 due to norepinephrine depletion [38,39]. To prevent early embryonic lethality, pregnant females were administrated with the β-adrenergic receptor agonist from 7.5 postcoitum (dpc) on by supplementing drinking water of dams with 200 µg/ml isoproterenol, which would allow *Wnt1-Cre;BmprIa^{f/f}* embryos to survive to term [40,41]. Embryos were collected from the timed pregnant females, and head samples were dissected in ice cold PBS, fixed individually in 4% paraformaldehyde (PFA) or z-fix (ANATECH Ltd; #170) overnight at 4°C, and tail clip from each embryo was used for PCR-based genotyping, respectively. Mutant and control heads were positioned for serial coronal sections through the TMJ. Comparable sections through the apex of the condyle were picked up for histological, in situ hybridization, immunostaining, and cell apoptosis analyses.

Histology, in situ hybridization, and immunohistochemistry, and Tunel assay

For histological study, paraffin sections were made at 6 µm and subjected for standard Hematoxylin/Eosin staining or Azoncarmine G/Aniline blue staining, as described [42]. Five mutant samples at each stage examined were used to ensure consistency of the phenotype. For in situ hybridization analyses, sections were cut at 10 µm and pretreated with proteinase K and hybridized with appropriate probes. Transcripts were detected by color reaction using BM purple (Roche) as described previously [43]. For immunohistochemical staining, frozen sections, made at 8 µm, were blocked with 4% goat serum and then incubated with primary antibodies against BMPRIA (Abcam; ab38560), Lubricin (Santa Cruz; sc-9854), pSmad1/5, pJNK, pERK, and p-p38 (from Cell Signaling; #9516, #9255, #4370, and #9211), respectively, at 4°C overnight. After washing, samples were incubated with secondary antibodies (Alexa Fluor488 goat anti-rabbit IgG from Invitrogen; #A-11034), counterstained with DAPI, and visualized under fluorescent microscope. Negative controls without primary antibodies were included in parallel. At least three samples of each genotype were used for histology, in situ hybridization, and immunohistochemistry analyses. Terminal deoxynucleotidyl transferase dUTP nick end labeling (Tunel) assay was applied to detect apoptotic cells using In Situ Cell Death Detection Kit (Roche), as described previously [44,45]. Three samples of each genotype were subjected to Tunel assays. Tunel-positive cells within the interzone were counted and presented as percentage of total cells within arbitrarily defined areas. Student's t-test was used to determine the significance of difference between wild type controls and mutants, and the results were presented as P value.

Results

Inactivation of Bmpr1a in CNC lineage leads to defective TMJ formation

To investigate the role of BmprIa in TMJ development, we began with examination of BMPRIA expression by immunohistochemistry. At E14.5 when both the primordial condyle and glenoid fossa become discernible, BMPRIA was found present in the developing condyle and glenoid fossa as well as the interzone, with a relatively low level in the condylar cartilage (Fig. 1A). At E15.5, BMPRIA expression retained in the condyle, glenoid fossa, and interzone, with an increased level in chondrocytes undergoing hypertrophy (Fig. 1C). This expression pattern is similar to that in the developing limb skeleton including the joints [26,30–32], suggesting a role for BMPRIA in TMJ morphogenesis. Since the condyle, glenoid fossa, and interzone cells are all derived from CNCs [8], we inactivated BmprIa in CNC lineage using the Wnt1-Cre transgenic allele. Immunohistochemistry confirmed the absence of BMPRIA in the developing TMJ of Wnt1-Cre;Bmpr1a$^{f/f}$ mice (Fig. 1B).

Histological analyses showed that the initial condensation of the condylar anlage in mutant mice appeared comparable to littermate controls at E13.5 (Fig. 2A, 2B). At E15.5, the morphology of the glenoid fossa did not exhibit an obvious difference between mutants and controls, but the size of the mutant condyle was reduced as compared to controls (Fig. 2C,

2D). At E18.5, the control TMJ displayed distinct structures, including a definite articular disc, the upper and lower synovial cavities, and the fibrocartilage/synovial membrane on the articular surface of the glenoid fossa and condyle (Fig. 2E, 2G). However, at this stage, the mutant TMJ exhibited a number of severe defects, including a hypoplastic condyle, lack of a definite disc, failed formation of an upper joint cavity evidenced by the existence of loose connective tissue in the interzone, as well as the absence of the fibrocartilage/synovial membrane of the glenoid fossa (Fig. 2F). Close examination of the mutant condyle at E18.5 revealed the formation of a disc-like structure that failed to separate from the apex of the condyle, leading to an absence of the lower joint cavity (Fig. 2F, 2H). The lack of a synovial joint cavity in the Wnt1-Cre;Bmpr1a$^{f/f}$ TMJ was further confirmed by the absent expression of Lubricin, a key component of joint fluids [46,47], as compared to its abundant expression in controls (Fig. 2I, 2J).

Delayed chondrocyte maturation in the Wnt1-Cre;Bmpr1a$^{f/f}$ condyle

Despite being a secondary cartilage, the growth of condylar cartilage takes the similar chondrogenesis and endochondral ossification process as that in long bone formation. Since BMPRIA mediated signaling is known to regulate primary cartilage differentiation [33,34], we set to examine chondrogenic differentiation process in the Wnt1-Cre;Bmpr1a$^{f/f}$ condyle. In the developing condyle, mesenchymal condensation appears at E13.5, and chondrogenic differentiation occurs at E14.5, and hypertrophy initiates at E15.5 [8]. We found that the timing of initial condensation of the condylar anlagen, as indicated by the expression of Sox9, and chondrogenic differentiation, determined by Col II expression, was comparable between wild type controls and Wnt1-Cre;Bmpr1a$^{f/f}$ mice (Fig. 3A–D). However, the mutant condyle exhibited a delayed terminal hypertrophy of chondrocytes, as assessed by the delayed Col X expression (Fig. 3E–H), and the longer distance between the apex and the beginning of hypertrophic zone in the mutant condyle at E18.5 (Fig. 3G, 3H). This phenotype differs from that in long bones where inactivation of BmprIa in condrocytes causes premature chondrocyte differentiation [34], likely due to different properties of primary v.s secondary cartilage.

Down-regulation of Ihh and inhibition of apoptosis in the Wnt1-Cre;Bmpr1a$^{f/f}$ TMJ

The similar TMJ phenotype between Wnt1-Cre;Bmpr1a$^{f/f}$ mice and the mice carrying mutations in Ihh or its downstream effectors [16,17], particularly the failure of disc separation, persistent interzone cells, and lack of fibrocartilaginous articular surface layer of the glenoid fossa in both mutants (Fig. 2F, 2H; Fig. 4B, 4D), and the overlapped expression pattern of BmprIa with Ihh in the developing condyle (Fig. 1) [8,17], prompted us to examine Ihh expression in the developing condyle of Wnt1-Cre;Bmpr1a$^{f/f}$ mice. In situ hybridization assay revealed a dramatic down-regulation of Ihh expression in the developing condyle of the mutants at E14.5 and E15.5, as compared to controls (Fig. 4G–J), consistent with the role of BMPRIA as a positive regulator of Ihh expression [34,48]. BmprIa thus likely acts through Ihh to regulate TMJ formation.

Because BmprIa is also expressed in the interzone mesenchymal cells that contribute to the articular disc and synovial membrane of the TMJ, the lack of fibrocartilage layer of the glenoid fossa and the persistence of interzone cells in the Wnt1-Cre;Bmpr1a$^{f/f}$ TMJ could be attributed either directly to the lack of BmprIa in these

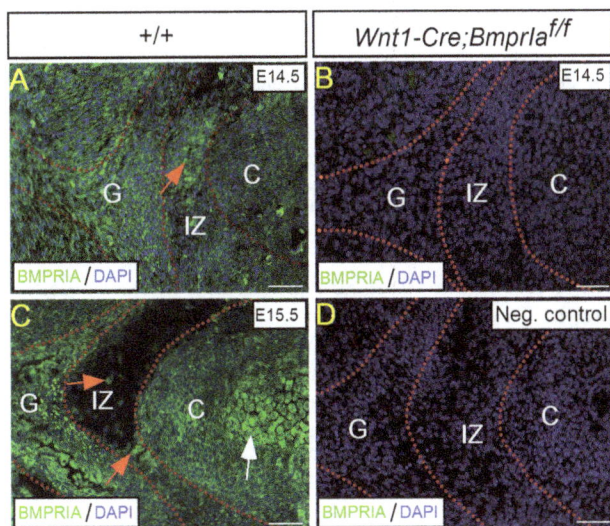

Figure 1. Expression of BMPRIA in the developing TMJ. (A–D) Immunohistochemistry shows expression of BMPRIA in the condylar cartilage, interzone, and glenoid fossa of E14.5 (A) and E15.5 (C) wild type animals, but a lack of staining on the Wnt1-Cre;Bmpr1a$^{f/f}$ TMJ (B) and on the negative control (D). Red arrows point to positive staining in the interzone, and white arrow points to the hypertrophic region where strong expression is detected. Abbreviation: C, condyle; G, glenoid fossa; IZ, interzone. Scale bar = 100 μm.

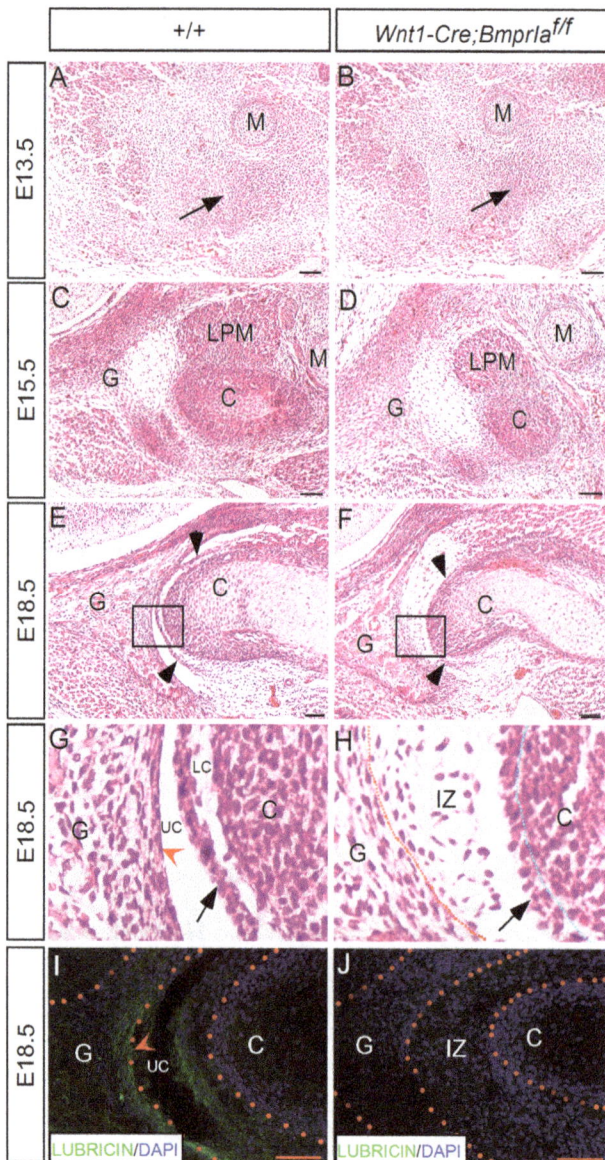

Figure 2. Wnt1-Cre;Bmprla^{f/f} mice display TMJ defects. (A–H) H&E staining shows histology of the developing TMJ of wild type controls (A, C, E, G) and mutants (B, D, F, H). Note that the initial condensation of the condylar anlagen at E13.5 (A, B) and the morphology of the glenoid fossa at E15.5 (C, D) appear comparable between the controls and mutants. However, the size of the mutant condyle is reduced at E15.5 (D). At E18.5, distinct structures including a definite disc, the upper and lower joint cavities, and the articular surface of the glenod fossa are well present in the control TMJ (E, G). However, in mutants, while a disc-like compact layer could be identified closely associated with the apex of the condyle, it fails to separate to form a distinct disc. In addition, the interzone cells persist, and a fibrocartilage layer fails to form on the articular surface of the glenoid fossa (F, H). (I, J) Immunnohistochemistry reveals expression of Lubricin in the synovial membrane of the control TMJ (I), and the complete absence of Lubricin in the mutant TMJ (J). Arrows in (A, B) point to the condylar condensation, and in (G, H) point to the disc. Arrowheads in (E, F) point to the disc. Red arrowhead points to the articular surface in (G) and the synovial membrane in (I). Abbreviation: C, condyle; G, glenoid fossa; M, Meckel's cartilage; IZ, interzone; LC, lower cavity; UC, upper cavity; LPM, lateral pterygoid muscle. Scale bar = 50 μm.

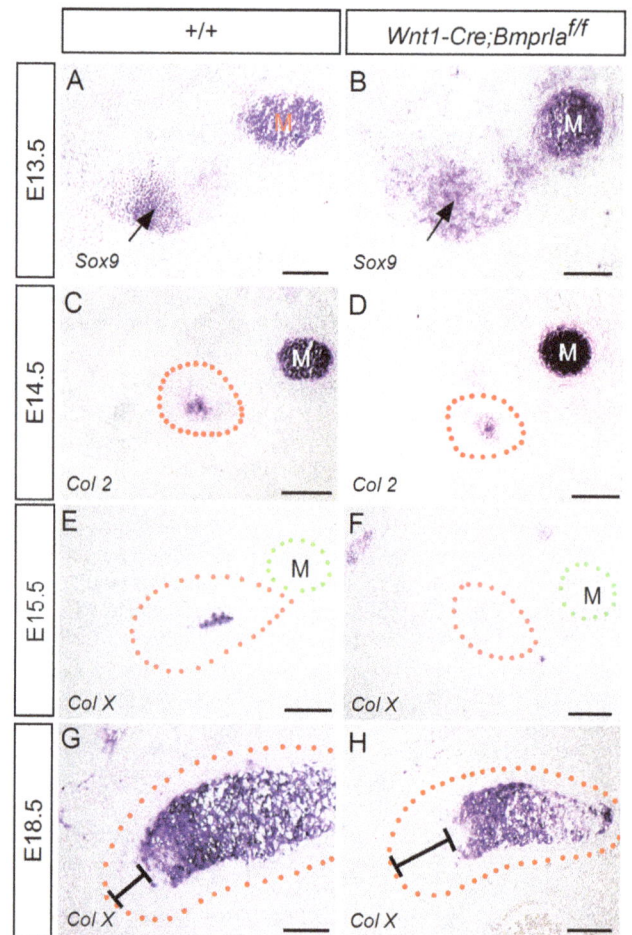

Figure 3. Delayed hypertrophic differentiation in the Wnt1-Cre;Bmprla^{f/f} condylar cartilage. (A, B) In situ hybridization shows Sox9 expression in the condylar condensation (arrow) of wild type (A) and mutant (B) at E13.5. (C, D) Col II expression exhibits comparable pattern in the condyle of wild type (C) and mutant (D) at E14.5. (E–H) In situ hybridization reveals Col X expression in the wild type condyle at E15.5 (E) and E18.5 (G). However, Col X expression is not detected in the mutant condyle at E15.5 (F), but is seen at E18.5 (H). The distance between the apex of the condyle and the beginning of hypertrophic zone is longer than that in the control condyle (G, H). Abbreviation: M, Meckel's cartilage. Scale bar = 100 μm.

cells or indirectly to the significantly reduced level of Ihh in the condyle. To distinguish these alternatives, we conducted immunohistochemistry to examine BMPRIA expression in the Ihh^{-/-} TMJ at E14.5. We found that although BMPRIA expression appeared comparable in the glenoid fossa of controls and mutants, its expression level was significantly reduced in the interzone and the condyle of the Ihh mutant TMJ (Fig. 4E–4F').

While the mechanism of cavitation in TMJ development remains to be addressed, programmed cell death in the interzone is regarded as a critical cellular mechanism for joint cavity formation in long bones [21]. Since BMPRIA mediated signaling is required for programmed cell death in the limb, particularly in the interdigital region [26,49], we wondered if the persistence of the loose connective tissue in the interzone of the Wnt1-Cre;Bmprla^{f/f} TMJ is a consequence of reduced level of apoptosis. Tunel assay indeed revealed abundant apoptotic cells specifically in the interzone of the control TMJ at E15.5 (Fig. 4K, 4K'). In contrast, Tunel assay detected a significantly reduced level of

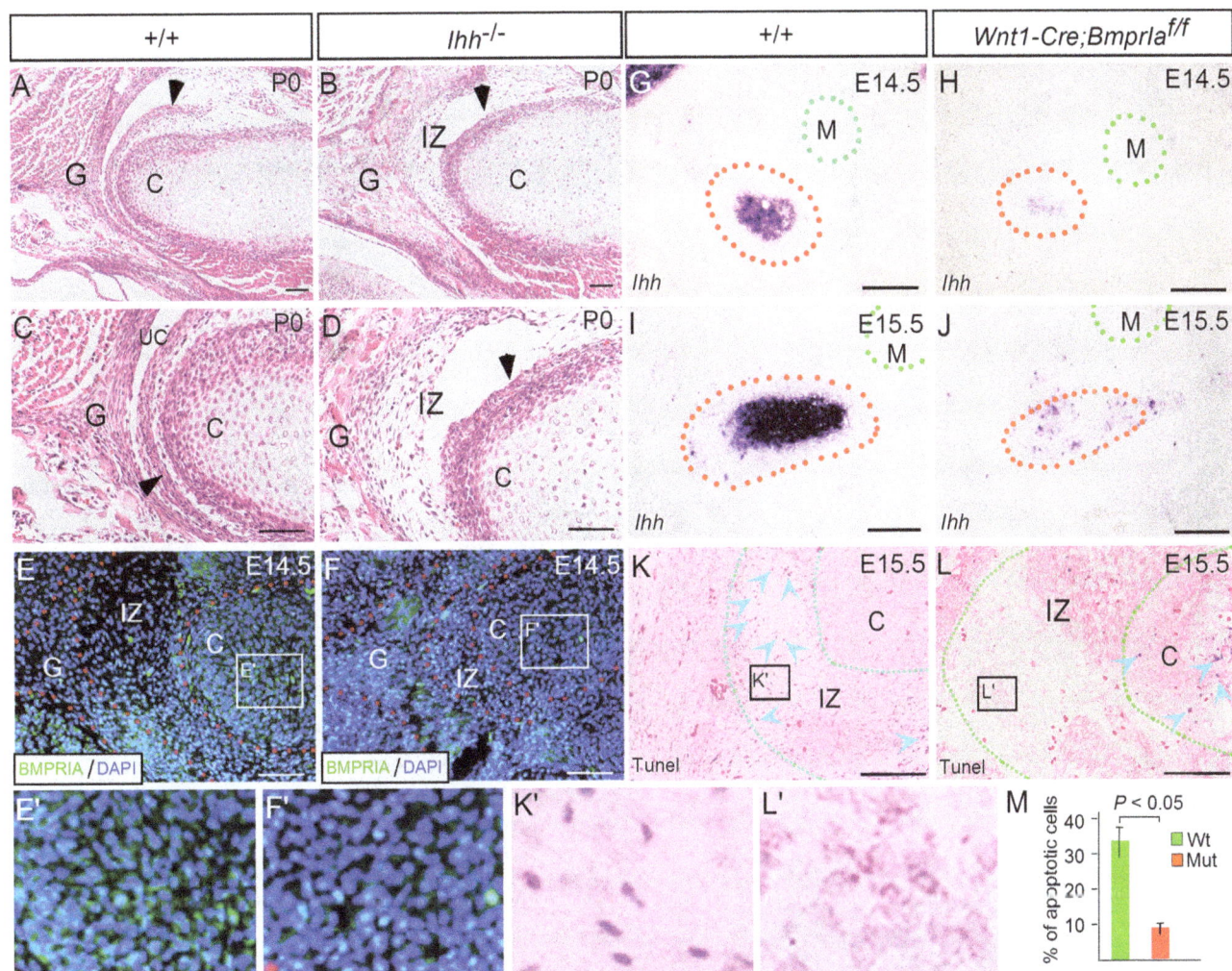

Figure 4. TMJ defects in *Ihh* mutants and reduced *Ihh* expression in the *Wnt1-Cre;Bmprla^{f/f}* condyle. (A–D) H&E staining reveals TMJ defects in the *Ihh* mutant TMJ (B, D), as compared to control (A, C) at P0. In mutant, a disc-like structure (arrowhead) forms but fails to separate from the apex of the condyle, the fibrocartilage layer fails to form on the articular surface of the glenoid fossa, and the interzone cells persist (B, D), as compared to the formation of distinct TMJ structures, including the disc (arrowhead), the fibrocartilginous articular surface, and the clear upper joint cavity, in controls (A, C). (E, E', F, F') Immunohistochemistry reveals significantly down-regulated BMPRIA expression in the condylar cartilage and interzone of *Ihh* mutant at E14.5 (F, F'), as compared to littermate control (E, E'). (G–J) In situ hybridization shows a dramatic down-regulation of *Ihh* in the condylar cartilage of *Wnt1-Cre;Bmprla^{f/f}* mice at both E14.5 (H) and E15.5 (J), as compared to controls (G, I). Tunel assay reveals numerous apoptotic cells (arrowheads) in the interzone of wild type control at E15.5 (K, K'), but very few apoptotic cells in the interzone of the *Wnt1-Cre;Bmprla^{f/f}* TMJ at the same age (L, L'). In contrast, some apoptotic cells (arrowheads) were observed in the mutant condyle (L). (M) Comparison of the percentage of apoptotic cells in the interzone of controls and mutants. Standard deviation values were indicated as the error bars, and the Student's t-test was used to determine the significance of difference between control and mutant, as presented as P value. Abbreviation: C, condyle; G, glenoid fossa; M, Meckel's cartilage; IZ, interzone; UC, upper cavity. Scale bar = 100 μm.

apoptotic cells in the interzone as well as some apoptotic cells in the condyle of the mutant TMJ (Fig. 4L, 4L', 4M). These observations suggest that similar to joint cavity formation in long bones, programmed cell death in the interzone also represents a critical cellular mechanism for joint cavitation during TMJ formation.

Augmented BMPRIA signaling in CNC lineage leads to TMJ agenesis

To further investigate the role of BMPRIA signaling in TMJ morphogenesis, we took a gain-of-function approach by transgenic expression of a constitutively active form of *BmprIa* (*pMes-caBmprIa*) [37] in CNC cells using the *Wnt1-Cre* allele. In situ hybridization revealed expression of *BmprIa* in the condensing

condylar blastema, the forming site of glenoid fossa, and cells between them, but not in Meckel's cartilage of wild type embryo at E13.5 (Fig. 5A). In *Wnt1-Cre;pMes-caBmprIa* mice at the same stage, strong and wide spread expression of *BmprIa* was found in the TMJ forming region and its surrounding tissues including Meckel's cartilage, indicating successful transgenic expression of *BmprIa* in CNC lineage (Fig. 5B). We and others have shown previously that elevated BMPRIA mediated signaling in CNC cells leads to a spectrum of craniofacial bone defects, including cleft secondary palate, ectopic cartilage formation, and craniosynostosis [50,51]. Histological analysis of the developing TMJ of *Wnt1-Cre;pMes-caBmprIa* mice identified unique TMJ developmental defects. Although the condensation of the condylar blastema occurred similarly to controls at E13.5 (Fig. 5C, 5D), the size of

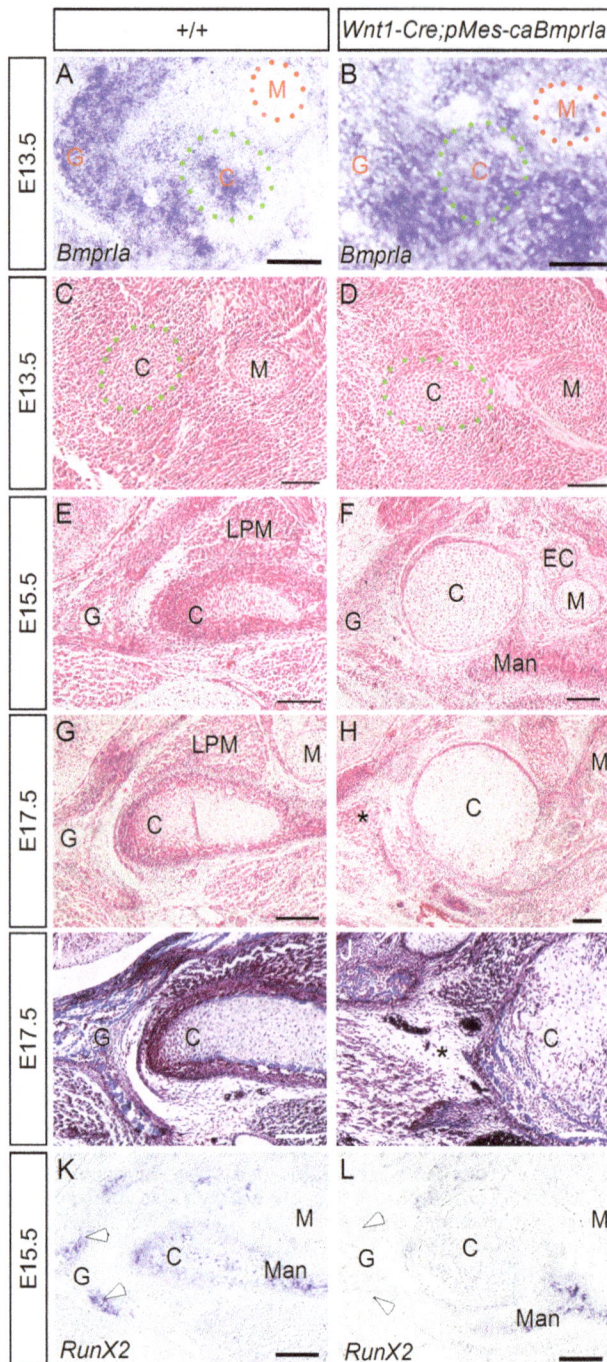

Figure 5. Augmented BMPRIA signaling in CNC cells leads to TMJ agenesis. (A, B) In situ hybridization shows expression of *Bmprla* in the condylar condensation, the future glenoid fossa forming site, and the interzone region, but not in Meckel's cartilage of an E13.5 wild type embryo (A), and an enhanced *Bmprla* expression in the TMJ forming site as well as the surrounding tissues including Meckel's cartilage in an E13.5 *Wnt1-Cre;pMes-cBmprla* embryo (B). (C–H) H&E staining reveals initial condensation of condylar anlagen in control and transgenic animals at E13.5 (C, D), growth and differentiation into primary cartilage of the condylar cartilage and lack of osteogenesis in the glenoid fossa in the transgenic TMJ at E15.5 (F) and E17.5 (H). (I, J) Azocarmine G/Aniline blue staining reveals glenoid fossa degeneration, evidenced by lack of bone formation, in transgenic mouse (J), as compared to control (I). (K, L) In situ hybridization assay shows expression of *Runx2* in the forming gelnoid fossa, perichondral region of the condyle, and mandibular bone of an E15.5 wild type control (K), but an absent expression of *Runx2* in

the glenoid fossa and a reduced expression in the condylar cartilage of an E15.5 transgenic animal (L). Note retention of *Runx2* expression in the mandibular bone of transgenic mouse (L). Asterisk in (H, J) indicates the site of glenoid fossa degeneration. Open arrowheads in (K, L) point to *Runx2* expression sites in the glenoid fossa. Abbreviation: C, condylar cartilage; G, glenoid fossa; M, Meckel's cartilage; EC, ectopic cartilage; LMP, lateral pterygoid muscle; Man, mandibular bone. Scale bar = 100 μm.

the condylar cartilage in transgenic animals became noticeably enlarged and the entire condylar cartilage appeared to become hypertrophic at E15.5, as compared to controls (Fig. 5E, 5F). Furthermore, unlike in controls that osteogenesis has begun in the glenoid fossa at this stage, the transgenic glenoid fossa failed to take osteogenic differentiation. At E17.5, the glenoid fossa became degenerated in transgenic animal (Fig. 5H). The failure of osteogenic differentiation in the glenoid fossa was further confirmed by the lack of bone formation in the glenoid fossa, assessed by Azocarmine G/Aniline blue staining, and by the absent expression of *Runx2*, a molecular marker for osteoblasts (Fig. 5I–L). Additionally, despite its expression in the mandibular bone, *Runx2* expression is also down-regulated in the condylar cartilage of *Wnt1-Cre;pMes-caBmprIa* mice (Fig. 5L). By E18.5, both the condyle and glenoid fossa degenerated and became unrecognizable (data not shown).

Enhanced BMPRIA signaling converts the condylar primordium from secondary cartilage to primary cartilage by ectopic activation of canonical signaling and inhibition of JNK signaling

As a secondary cartilage, the condylar cartilage expresses type I collagen (Col I), making it distinct from the primary cartilage [52]. Because of its aberrant differentiation, we wondered if the condylar cartilaginous element of *Wnt1-Cre;pMes-caBmprIa* mice retained its secondary cartilage characteristics. In situ hybridization assay revealed *Col I* expression in the control condylar cartilage and mandibular bone, but the absence of *Col I* in the transgenic condylar cartilaginous element despite its expression in the mandibular bone at E14.5 (Fig. 6A, 6B). However, the expression of *Col II* and *Col X* in the transgenic condylar cartilage confirmed its cartilage fate (Fig. 6C–F). We thus conclude that the *Wnt1-Cre;pMes-caBmprIa* condylar primordium adopts a fate of primary cartilage in response to an augmented BMPRIA-mediated signaling. Moreover, Tunel assay revealed an extensive apoptotic event in the transgenic condylar cartilage, beginning at E15.5, as compared to the lack of apoptosis in the control condyle at the same stage (Fig. 6G, 6H), which apparently contributes to the degeneration and disappearance of the condylar cartilage in transgenic animals.

We have shown previously that the expression of *caBmprIa* in CNC lineage induces ectopic activation of Smad1/5/8 signaling as well as p38 signaling in the developing palatal shelves [51]. We therefore set to examine alterations in BMP canonical and non-canonical signaling pathways in the condylar cartilage of *Wnt1-Cre;pMes-caBmprIa* mice by immunohistochemistry. Interestingly, we detected no activation of Smad-dependent as well as p38 and Erk1/2 pathways in the control condyle, as assessed by the lack of pSmad1/5, p-p38, and p-Erk1/2, but observed activity of p-JNK signaling (Fig. 7A, 7C,7E, 7G). In contrast, the condylar cartilage of *Wnt1-Cre;pMes-caBmprIa* mice exhibited an ectopic activation of pSmad1/5, but an absence of pJNK signaling, along with unaltered p38 and pEek1/2 pathways (Fig. 7B, 7D; 7F, 7H). These observations indicate that the switch between BMP canonical and non-canonical signaling pathways likely underlies the fate conversion from the secondary to primary cartilage, with the Smad-dependent signaling favoring the primary cartilage fate.

Discussion

Compared to synovial joint formation in the appendicular skeletons, TMJ development and the underlying molecular mechanisms are relatively under-studied. While the critical roles of BMP signaling in long bone joint development and homeostasis have been well documented [21,26,29], its role in TMJ formation remained completely unknown. In this study, we present evidence that BMPRIA mediated signaling is essential for TMJ morphogenesis, and overly activated BMPRIA signaling is detrimental to TMJ formation. Our results also reveal apoptosis in the interzone as a potential cellular mechanism for cavitation of the TMJ, similar to that in long bone joint formation.

A BMPRIA-Ihh positive regulatory pathway regulates TMJ development

It has been well established that BMP and Ihh signaling interact to regulate chondrocyte proliferation and hypertrophic differentiation [34,53–55]. In the developing limb, BMP signaling, particularly the BMPRIA mediated pathway, positively regulates *Ihh* expression that could also activate in the perichondrium the expression of several BMP ligands, forming a BMP-Ihh positive feedback loop [34,53,54]. Although there is no evidence for an interaction of BMP and Ihh signaling in joint development, the fact that several BMP ligands and receptor are expressed in the interzone and that mutations in either *Noggin* or *Ihh* lead to joint defects including joint ablation in limbs implies the existence of such interaction [15,21,28]. Indeed, in the current study, we found that the ablation of *BmprIa* in CNC lineage produces TMJ defects resembling that in *Ihh* mutant. In both mutants, a functional TMJ failed to form, evidenced by the absent Lubricin expression (Fig. 2) [16]. In addition, both mutants displayed a lack of a distinct articular disc due to failed disc separation from the condyle, persistence of the interzone cells, as well as absent synovial membrane on the articular surface of the glenoid fossa. Consistent with Ihh function in disc formation and separation during TMJ morphogenesis [16,17], we found a dramatic down-regulation of *Ihh* expression in the developing condyle of *Wnt1-Cre;BmprIa^{f/f}* mice. This result also indicates that similar to its role in developing appendicular skeletons, BMPRIA mediated signaling also acts as a positive regulator of *Ihh* expression in the condylar cartilage. On the other hand, *BmprIa* expression was significantly reduced in the condyle and interzone of the *Ihh^{-/-}* TMJ, suggesting the existence of a BMPRIA-Ihh positive feedback loop in the developing TMJ. However, the delayed hypertrophic differentiation observed in the condylar cartilage of *Wnt1-Cre;BmprIa^{f/f}* mice appears to be opposite to the premature hypertrophic differentiation defect seen in the *Ihh^{-/-}* condyle as well as in the long bones of mice carrying *BmprIa* deletion in chondrocyte lineage [16,34]. Although the underlying mechanism is currently unknown, the discrepancy would likely be attributed to the residual *Ihh* expression in the *Wnt1-Cre;BmprIa^{f/f}* condyle as well as the condyle's property as secondary cartilage.

Nevertheless, based on above mentioned observations and the established roles for BMP and Ihh signaling in limb development, we propose a model to summarize the function of the BMPRIA-Ihh regulatory pathway in regulating distinct steps during TMJ morphogenesis (Fig. 8). In this model, BMPRIA and Ihh regulate the expression of each other to coordinate chondrocyte proliferation and differentiation in the developing condyle. Meanwhile,

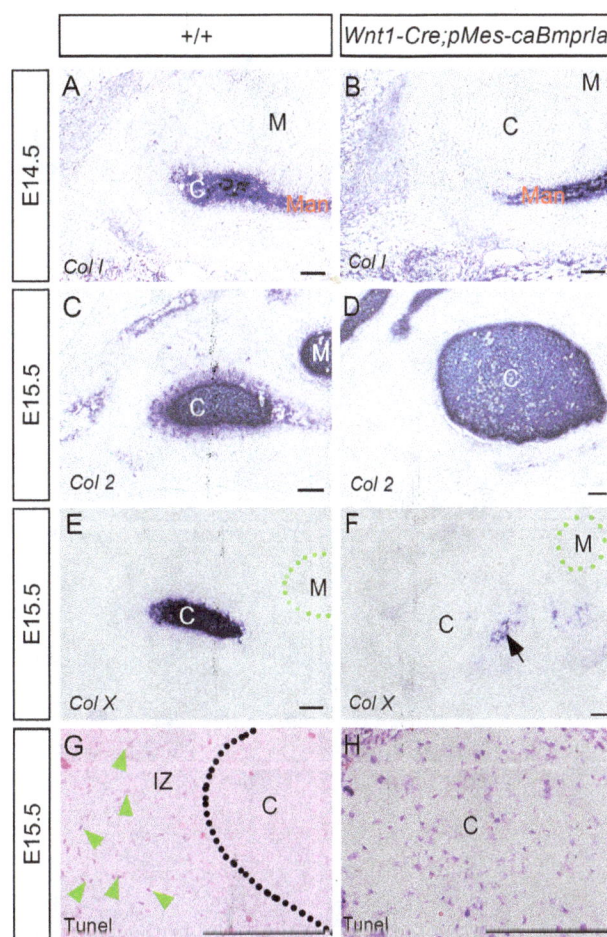

Figure 6. Elevated BMPRIA signaling converts secondary cartilage of the condylar primordium to primary cartilage and induces extensive cell death. (A–F) In situ hybridization detects *Col I* expression in the condylar cartilage and mandibular bone of an E14.5 control embryo (A), and in the mandibular bone of an E14.5 *Wnt1-Cre,pMes-caBmprIa* mice (B). However, *Col I* expression is not detected in the condylar cartilage of transgenic animal (B). *Col II* (C, D) and *Col X* (E, F) expression is observed in the condylar cartilage of control (C, E) and transgenic embryo (D, F) at E15.5. (G, H) Tunel assay reveals numerous apoptotic cells in the interzone but not in the condyle of the E15.5 wild type TMJ, but extensive cell death in the condylar cartilage of E15.5 transgenic embryo (H). Arrow in (F) points to *Col X* expression domain, and arrowheads in (G) point to apoptotic cells. Abbreviation: C, condylar cartilage; M, Meckel's cartilage; IZ, interzone. Scale bar = 100 μm.

Ihh, produced in the condyle, diffuses into the interzone to regulate disc separation and to maintain the expression of BMPRIA that in turn acts in a cell autonomous manner to regulate synovial membrane formation and to trigger apoptosis.

Augmented BMPRIA signaling in CNCs converts the secondary cartilage of the condylar primordium to primary cartilage

Despite being a secondary cartilage, the condyle shares many similarities with primary cartilage in development, including the expression of genes known to be important for cartilage growth and differentiation such as *Bmp2, Bmp7, Sox9, Runx2, Osterix, Ihh, Pthrp, Vegf, Col II* and *Col X* [8,16,17,35,56,57]. However, the condyle also differs from primary cartilage by its expression of Col I and Col II simultaneously and its capability of differentiating

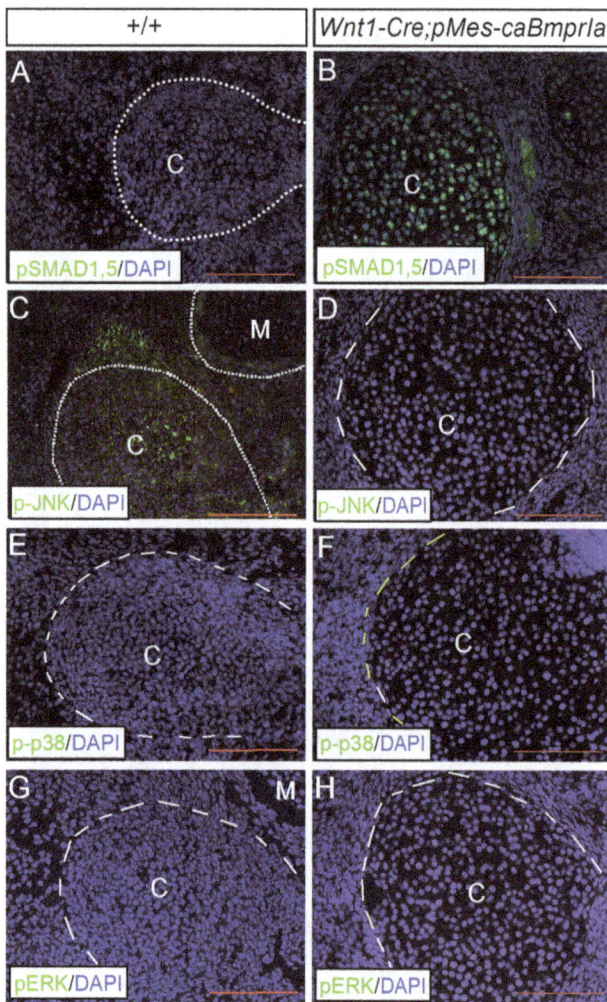

Figure 7. Enhanced BMPRIA signaling activates Smad-dependent pathway but inhibits JNK signaling pathway in the condylar cartilage. Immunohistochemistry shows absent pSmad1/5, p-p38, and pERK, but the presence of pJNK in the E14.5 control condyle (A, C, E, G), and the presence of pSmad1/5, but absent pJNK as well as p-p38 and pERK in the transgenic condylar cartilage (B, D, F, H). Abbreviation: C, condylar cartilage; G, glenoid fossa; M, Meckel's cartilage. Scale bar = 100 µm.

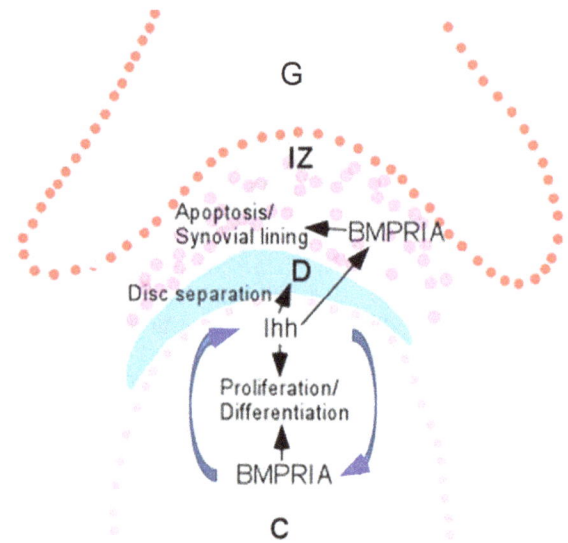

Figure 8. A model illustrating the interaction between BMPRIA and Ihh and their functions in regulating distinct steps during TMJ morphogenesis. Abbreviation: C, condyle; D, disc; G, glenoid fossa; IZ, interzone.

further support this notion. Furthermore, consistent with the role of BMPRIA signaling in apoptosis in the developing limb [26,32], overly activated BMPRIA signaling causes extensive apoptosis in the condylar cartilage and leads to condylar cartilage degeneration, indicating a detrimental effect on chondrocyte survival.

Despite an essential role for *BmprIa* in chondrogenesis and endochondral bone formation, in our current study, we found that inactivation of *BmprIa* did not affect glenoid fossa osteogenesis, suggesting that *BmprIa* may not be essential for intramembranous bone formation. However, elevated BMPRIA signaling instead inhibits osteogenesis in the glenoid fossa. Thus, although BMP signaling is generally accepted as a positive regulator of osteogenesis, elevated BMP signaling could have an opposite effect, depending on the tissue and cell types. Since a normal developing condyle is required to sustain the development of the glenoid fossa [9], the degeneration of the glenoid fossa in *Wnt1Cre;pMes-caBmprIa* mice could be the consequence of failed osteogenesis, or result from an abnormal condylar cartilage with altered property, or both.

Apoptosis as a cellular mechanism of TMJ cavitation

Cell death in the middle of the interzone is considered the cellular mechanism for physical separation of the contiguous cartilage elements during joint formation in long bones [21,58–61]. In the developing TMJ, although the primordial condyle and glenoid fossa form independently and become approximately through condylar growth, disappearance of the interzone cells is necessary for the formation of a joint cavity. The interzone mesenchymal cells are believed to contribute to the articular disc, capsule, and the synovial membrane of the joint cavity [3,4]. However, if apoptosis occurs in the interzone of the TMJ remains arguably. It was reported previously that in the rat TMJ at late developmental stage, apoptotic cells were found only at the subsurface of the condyle and in the region at which the lateral pterygoid muscle attached to the condyle, suggesting that apoptosis may be associated with the lower joint cavity formation of the TMJ [62]. However, in our studies, we found extensive cell death in the interzone of the control TMJ at E15.5, right before

into either chondrocytes or osteoblasts [52]. Although the underlying mechanism for fate determination of primary v.s. secondary cartilage remains elusive, our results that augmented BMPRIA signaling is able to convert the condylar cartilage to a primary cartilage, evidenced by the lack of *Col I* expression, implicate BMP signaling in such fate decision. *BmprIa* is expressed in the cartilage condensations of both long bone and condyle (this study) [33], suggesting its role in the fate decision of both primary and secondary cartilages. It appears that a tightly tuned BMPRIA signaling is essential for fate determination of secondary cartilage. Accompanied with this fate conversion is the switch of the downstream BMP signaling pathway from the non-canonical JNK signaling to the Smad-dependent pathway in the condylar cartilage, suggesting that higher activity of BMPRIA signaling preferentially activates the Smad-dependent pathway, which favors primary cartilage formation. Indeed, the lack of Smad-dependent signaling in the condyle (this study) and the strong expression of pSmad1/5/8 in the limb cartilage condensation [33] the upper cavity becomes discernibly at E16.5. However, in the *Wnt1-Cre;Bmpr1a^{f/f}* TMJ, such extensive apoptosis was not observed, consistent with the pro-apoptotic role of BMPRIA mediated signaling. The discrepancy between our results and that by Matsuda and colleagues [62] could be attributed to the stage difference. Nevertheless, the lack of apoptosis appears to contribute to the persistence of interzone cells in the mutant TMJ. In addition, BMPRIA signaling is also required for organization of some interzone cells to become synovial lining layer, which could also contribute to the cavitation of the TMJ. We thus propose that the upper joint cavity of the TMJ is formed by organization of the interzone cells into synovial lining layer and capsule, and by removal of excessive cells via apoptosis.

In conclusion, our studies using transgenic loss-of- and gain-of-function approaches reveal the importance of BMPRIA mediated signaling in TMJ morphogenesis and establish a BMPRIA-Ihh positive regulatory pathway in controlling disc separation, synovial membrane formation, as well as joint cavity formation.

Acknowledgments

The authors thank Dr. Yuji Mishina of the University of Michigan for sharing *Bmpr1a^{f/f}* mice.

Author Contributions

Conceived and designed the experiments: YC SG WW. Performed the experiments: SG WW CL LY CS WY XL JC. Analyzed the data: SG WW YC. Contributed reagents/materials/analysis tools: JC FL. Contributed to the writing of the manuscript: SG WW FL YC.

References

1. Kermack KA (1972) The origin of mammals and the evolution of the temporomandibular joint. Proc R Soc Med 65: 389–392.
2. Gu S, Chen Y (2013) Temporomandibular joint development. In: Huang, G.T.-J., Thesleff, I. (Eds.), Stem Cells in Craniofacial Development and Regeneration. Wiley-Blackwell; John Wily & Sons, Inc. p71–85.
3. Dixon AD (1997) Formation of the cranial base and craniofacial joints. In: Dixon, A.D., Hoyete, D.A.N., Rönning, O. (Eds.), Fundamentals of Craniofacial Growth. CRC Press LLC. p99–136.
4. Sperber GH (2001) Craniofacial Development. Ontario, BC Decker Inc. Hamilton.
5. Miyake T, Cameron AM, Hall BK (1997) Stage-specific expression patterns of alkaline phosphatase during development of the first arch skeleton in inbred *C57BL/6* mouse embryos. J Anat 190: 239–260.
6. Shibata S, Suda N, Suzuki S, Fukuoka H, Yamashita Y (2006) An in situ hybridization study of *Runx2*, *Osterix*, and *Sox9* at the onset of condylar cartilage formation in fetal mouse mandible. J Anat 208: 169–177.
7. Frommer J (1964) Prenatal development of the mandibular joint in mice. Anat Rec 150: 449–461.
8. Gu S, Wei N, Yu L, Fei J, Chen Y (2008) *Shox2*-deficiency leads to dysplasia and ankylosis of the temporomandibular joint in mice. Mech Dev 125: 729–742.
9. Wang Y, Liu C, Rohr J, Liu H, He F, et al. (2011) Tissue interaction is required for glenoid fossa development during temporomandibular joint formation. Dev Dyn 240: 2466–2473.
10. Purcell P, Jheon A, Vivero MP, Rahimi H, Joo A, et al. (2012) Spry1 and Spry2 are essential for development of the temporomandibular joint. J Dent Res 91: 387–393.
11. Mori-Akiyama Y, Akiyama H, Rowitch DH, de Crombrugghe B (2003) Sox9 is required for determination of the chondrogenic cell lineage in the cranial neural crest. Proc Natl Acad Sci USA 100: 9360–9365.
12. Shibata S, Suda N, Yoda S, Fukuoka H, Ohyama K, et al. (2004) *Runx2*-deficient mice lack mandibular condylar cartilage and have deformed Meckel's cartilate. Anat Embryol 208: 273–280.
13. Oka K, Oka S, Sasaki T, Ito Y, Bringas P Jr, et al. (2007) The role of TGF-beta signaling in regulating chondrogenesis and osteogenesis during mandibular development. Dev Biol 303: 391–404.
14. Oka K, Oka S, Hosokawa R, Bringas P Jr, Brockhoff HC 2nd, et al. (2008) TGF-beta mediated Dlx5 signaling plays a crucial role in osteo-chondroprogenitor cell lineage determination during mandible development. Dev Biol 321: 303–309.
15. St-Jacques B, Hammerschmidt M, McMahon AP (1999) Indian hedgehog signaling regulates proliferation and differentiation of chondrocytes and is essential for bone formation. Genes Dev 13: 2072–2086.
16. Shibukawa Y, Young B, Wu C, Yamada S, Long F, et al. (2007) Temporomandibular joint formation and condyle growth require Indian hedgehog signaling. Dev Dyn 236: 426–434.
17. Purcell P, Joo BW, Hu JK, Tran PV, Calicchio ML, et al. (2009) Temporomandibular joint formation requires two distinct hedgehog-dependent steps. Proc Natl Acad Sci USA 106: 18297–18302.
18. Ochiai T, Shibukawa Y, Nagayama M, Mundy C, Yasuda T, et al. (2010) Indian hedgehog roles in post-natal TMJ development and organization. J Dent Res 89: 349–354.
19. Massagué J (2012) TGFβ signaling in context. Nat Rev Mol Cell Biol 13: 616–630.
20. Mitrovic DR (1977) Development of the metatarsophalangeal joint of the chick embryo: morphological, ultrastructural and histochemical studies. Am J Anat 150: 333–347.
21. Pacifici M, Koyama E, Iwamoto M (2005) Mechanisms of synovial joint and articular cartilage formation: recent advances, but many lingering mysteries. Birth Defects Res C 75: 237–248.
22. Storm EE, Kingsley DM (1996) Joint patterning defects caused by single and double mutations in members of bone morphogenetic protein (BMP) family. Development 122: 3969–3979.
23. Wolfman NM, Hattersley G, Cox K, Celeste AJ, Nelson R, et al. (1997) Ectopic induction of tendon and ligament in rats by growth and differentiation factors 5, 6, and 7, members of the TGF-beta gene family. J Clin Invest 100: 321–330.
24. Francis-West PH, Parish J, Lee K, Archer CW (1999) BMP/DGF-signaling interactions during synovial joint development. Cell Tissue Res 296: 111–119.
25. Settle SH Jr, Rountree RB, Sinha A, Thacker A, Higgins K, et al. (2003) Multiple joint and skeletal patterning defects caused by single and double mutations in the mouse Gdf6 and Gdf5 genes. Dev Biol 254: 116–130.
26. Rountree R, Schoor M, Chen H, Marks M, Harley V, et al. (2004) BMP receptor signaling is required for postnatal maintenance of articular cartilage. PloS Biol 11: e255.
27. Storm EE, Huynh TV, Copeland NG, Jenkins NA, Kingsley DM, et al. (1999) Limb alterations in brachypodism mice due to mutations in a new member of the TGF beta-superfamily. Nature 368: 639–643.
28. Brunet LJ, McMahon JA, McMahon AP, Harland RM (1998) Noggin, cartilage morphogenesis, and joint formation in the mammalian skeleton. Science 280: 1455–1457.
29. Lories RJU, Luyten FP (2005) Bone morphogenetic protein signaling in joint hoemostasis and disease. Cytok Growth Fact Rev 16: 287–298.
30. Dewulf N, Verschueren K, Lonnoy O, Moren A, Grimsby S, et al. (1995) Distinct spatial and temporal expression patterns of two type I receptors for bone morphogenetic proteins during mouse embryogenesis. Endocrinology 136: 2652–2663.
31. Mishina Y, Suzuki A, Ueno N, Behringer RR (1995) *Bmpr* encodes a type I bone morphogenetic protein receptor that is essential for gastrulation during mouse embryogenesis. Genes Dev 9: 3027–3037.
32. Zou H, Wieser R, Massagué J, Niswander L (1997) Distinct roles of type I bone morphogenetic protein receptors in the formation and differentiation of cartilage. Genes Dev 11: 2191–2203.
33. Yoon BS, Ovchinnikov DA, Yoshii I, Mishina Y, Behringer RR, et al. (2005) *Bmpr1a* and *Bmpr1b* have overlapping functions and are essential for chondrogenesis in vivo. Proc Natl Acad Sci USA 102: 5062–5067.
34. Yoon BS, Pogue R, Ovchinikov DA, Yoshii I, Mishina Y, et al. (2006) BMPs regulate multiple aspects of growth-plate chondrogenesis through opposing actions on FGF pathways. Development 133: 4667–4678.
35. Fukuoka H, Shibata S, Suda N, Yamashita Y, Komori T (2007) Bone morphogenetic protein rescues the lack of secondary cartilage in *Runx2*-deficient mice. J Anat 211: 8–15.
36. Danielian PS, Puccino D, Rowitch DH, Michael SK, McMahon AP (1998) Modification of gene activity in mouse embryos in utero by a tamoxifen-inducible form of Cre recombinase. Curr Biol 8: 1323–1326.
37. He F, Xiong W, Wang Y, Matsui M, Yu X, et al. (2010) Modulation of BMP signaling by Noggin is required for the maintenance of palatal epithelial integrity during palatogenesis. Dev Biol 347: 109–121.
38. Stottmann RW, Choi M, Mishina Y, Meyers EN, Klingensmith J (2004) BMP receptor IA is required in mammalian neural crest cells for development of the cardiac outflow tract and ventricular myocardium. Development 131: 2205–2218.
39. Morikawa Y, Zehir A, Maska E, Deng C, Schneider M, et al. (2009) BMP signaling regulates sympathetic nervous system development through Smad-dependent and -independent pathways. Development 136: 3575–3584.
40. Morikawa Y, Cserjesi P (2008) Cardiac neural crest expression of Hand2 regulates outflow and second heart field development. Circ Res 103: 1422–1429.
41. Li L, Lin M, Wang Y, Cserjesi P, Chen Z, et al. (2011) *Bmpr1a* is required in mesenchymal tissue and has limited redundant function with *Bmpr1b* in tooth and palate development. Dev Biol 349: 451–461.
42. Presnell JK, Schreibman MP (1997) Humason's Animal Tissue Techniques. Fifth edition. Baltimore, MD, The Johns Hopkins University Press.

43. St Amand TR, Zhang Y, Semina EV, Zhao X, Hu YP, et al. (2000) Antagonistic signals between BMP4 and FGF8 define the expression of *Pitx1* and *Pitx2* in mouse tooth-forming anlage. Dev Biol 217: 323–332.

44. Zhang Z, Song Y, Zhao X, Zhang X, Fermin C, et al. (2002) Rescue of cleft palate in *Msx1*-deficient mice by transgenic *Bmp4* reveals a network of BMP and Shh signaling in the regulation of mammalian palatogenesis. Development 129: 4135–4146.

45. Alappat SR, Zhang Z, Suzuki K, Zhang X, Liu H, et al. (2005) The cellular and molecular etiology of the cleft secondary palate in *Fgf10* mutant mice. Dev Biol 277: 102–113.

46. Swann DA, Silver FH, Slayter HS, Stafford W, Shore E (1985) The molecular structure and lubricating activity of lubricin isolated from bovine and human synovial fluids. Biochem J 225: 195–201.

47. Marcelino J, Carpten JD, Suwairi WM, Gutierrez OM, Schwartz S, et al. (1999) CACP, encoding a secreted proteoglycan, is mutated in camptodactyly-arthropathy-coxa vara-pericarditis syndrome. Nat Genet 23: 319–322.

48. Seki K, Hata A (2004) Indian hedgehog gene is a target of the bone morphogenetic protein signaling pathway. J Biol Chem 279: 18544–18549.

49. Zou H, Niswander L (1996) Requirement for BMP signaling in interdigital apoptosis and scale formation. Science 272: 738–741.

50. Komatsu Y, Yu PB, Kamiya N, Pan H, Fukuda T, et al. (2012) Augmentation of Smad-dependent BMP signaling in neural crest cells causes craniosynostosis in mice. J Bone Miner Res 28: 1422–1433.

51. Li L, Wang Y, Lin M, Yuan G, Yang G, et al. (2013) Augmented BMPRIA-mediated BMP signaling in cranial neural crest lineage leads to cleft palate formation and delayed tooth differentiation. PLoS ONE 8(6): e66107.

52. Hall BK (2005) Bones and Cartilage: Developmental and Evolutionary Skeletal Biology. San Diego, CA, Elsevier Academic Press.

53. Pathi S, Rutenberg JB, Johnson RL, Vortkamp A (1999) Interaction of Ihh and BMP/Noggin signaling during cartilage differentiation. Dev Biol 209: 239–253.

54. Minina E, Wenzel HM, Kreschel C, Karp S, Gaffield W, et al. (2001) BMP and Ihh/PTHrP signaling interact to coordinate chondrocyte proliferation and differentiation. Development 128: 4523–4534.

55. Minina E, Kreschel C, Naski MC, Ornitz DM, Vortkamp A (2002) Interaction of FGF, Ihh/PThIh, and BMP signaling integrates chondrocyte proliferation and hypertrophic differentiation. Dev Cell 3: 1–20.

56. Fukada K, Shibata S, Suzuki S, Ohya K, Kuroda T (1999) In situ hybridization study of type I, II, X collagens and aggrecan mRNAs in the developing condylar cartilage of fetal mouse mandible. J Anat 95: 321–329.

57. Kuboki T, Kanyama M, Nakanishi T, Akiyama K, Nawachi K, et al. (2003) *Cbfa1/Runx2* gene expression in articular chondrocytes of the mice temporo-mandibular and knee joints in vivo. Arch Oral Biol 48: 519–525.

58. Mitrovic DR (1978) Development of the diathrodial joints in the rat embryo. Am J Anat 151: 475–485.

59. Nalin AM, Greenlee TK, Sandell LJ (1995) Collagen gene expression during development of avian synovial joints: transient expression of type II and X collagen genes in the joint capsule. Dev Dyn 203: 352–362.

60. Kimura S, Shiota K (1996) Sequential changes of programmed cell death in developing fetal mouse limbs and its possible role in limb morphogenesis. J Morph 229: 337–346.

61. Abu-Hijleh G, Reid O, Scothorne RJ (1997) Cell death in the developing chick knee joint. I. Spatial and temporal patterns. Clin Anat 10: 183–200.

62. Matsuda S, Mishima K, Yoshimura Y, Hatta T, Otani H (1997) Apoptosis in the development of the temporomandibular joint. Anat Embryo 196: 383–391.

20

Diffuse Glomerular Nodular Lesions in Diabetic Pigs Carrying a Dominant-Negative Mutant Hepatocyte Nuclear Factor 1-Alpha, an Inheritant Diabetic Gene in Humans

Satoshi Hara[1,4], Kazuhiro Umeyama[2], Takashi Yokoo[3], Hiroshi Nagashima[2], Michio Nagata[1]*

1 Department of Kidney and Vascular Pathology, University of Tsukuba, Tsukuba, Japan, **2** Meiji University International Institute for Bio-Resource Research, Kawasaki, Japan, **3** Division of Nephrology and Hypertension, Department of Internal Medicine, The Jikei University School of Medicine, Tokyo, Japan, **4** Division of Rheumatology, Department of Internal Medicine, Kanazawa University of Graduate School of Medicine, Kanazawa, Japan

Abstract

Glomerular nodular lesions, known as Kimmelstiel-Wilson nodules, are a pathological hallmark of progressive human diabetic nephropathy. We have induced severe diabetes in pigs carrying a dominant-negative mutant hepatocyte nuclear factor 1-alpha (HNF1α) P291fsinsC, a maturity-onset diabetes of the young type-3 (MODY3) gene in humans. In this model, glomerular pathology revealed that formation of diffuse glomerular nodules commenced as young as 1 month of age and increased in size and incidence until the age of 10 months, the end of the study period. Immunohistochemistry showed that the nodules consisted of various collagen types (I, III, IV, V and VI) with advanced glycation end-product (AGE) and N^ε-carboxymethyl-lysine (CML) deposition, similar to those in human diabetic nodules, except for collagen type I. Transforming growth factor-beta (TGF-β) was also expressed exclusively in the nodules. The ultrastructure of the nodules comprised predominant interstitial-type collagen deposition arising from the mesangial matrices. Curiously, these nodules were found predominantly in the deep cortex. However, diabetic pigs failed to show any of the features characteristic of human diabetic nephropathy; e.g., proteinuria, glomerular basement membrane thickening, exudative lesions, mesangiolysis, tubular atrophy, interstitial fibrosis, and vascular hyalinosis. The pigs showed only Armanni-Ebstein lesions, a characteristic tubular manifestation in human diabetes. RT-PCR analysis showed that glomeruli in wild-type pigs did not express endogenous HNF1α and HNF1β, indicating that mutant HNF1α did not directly contribute to glomerular nodular formation in diabetic pigs. In conclusion, pigs harboring the dominant-negative mutant human MODY3 gene showed reproducible and distinct glomerular nodules, possibly due to AGE- and CML-based collagen accumulation. Although the pathology differed in several respects from that of human glomerular nodular lesions, the somewhat acute and constitutive formation of nodules in this mammalian model might provide information facilitating identification of the principal mechanism underlying diabetic nodular sclerosis.

Editor: Leighton R. James, University of Florida, United States of America

Funding: This work was partially supported by Grants-in-Aid for Scientific Research (C) (#23500505) from the Ministry of Education, Culture, Sports, Science and Technology/Japan Society for the Promotion of Science (HN). The funders had no role in study design, data collection and analysis, decision to publish, or preparation of the manuscript.

Competing Interests: The authors have declared that no competing interests exist.

* E-mail: nagatam@md.tsukuba.ac.jp

Introduction

Diabetic nephropathy is the leading cause of end-stage renal disease [1,2]. Glomerular nodular lesions, known as Kimmelstiel-Wilson nodules, are a pathological hallmark of human diabetic nephropathy. In 1936 Kimmelstiel and Wilson first described intercapillary glomerulosclerosis as a sign of advanced diabetic glomerular changes [3]. The presence of glomerular nodular lesions is known to be associated with poor renal outcome [4,5].

Although investigated extensively, the morphogenesis of diabetic glomerular nodules remains to be determined. One major reason for this is a lack of animal models that accurately represent the nodules typically present in humans. Some rodent models of diabetes show segmental mesangial expansion and glomerular basement membrane (GBM) thickening, but few exhibit distinct

glomerular nodular lesions [6]. To date, the four representative diabetic rodent models with glomerular nodules are: endothelial nitric oxide synthase (eNOS) knockout db/db mice [7], receptor for advanced glycation end products (RAGE)/megsin/inducible nitric oxide synthase (iNOS) overexpressing transgenic mice [8], monocrotaline-treated Otsuka Long-Evans Tokushima Fatty (OLETF) rats [9] and BTBR ob/ob mice [10]. eNOS knockout db/db mice developed focal nodular glomerulosclerosis at 26 weeks of age [7]. RAGE/megsin/iNOS overexpressing transgenic mice also showed nodular-like lesions in 30–40% of glomeruli at 16 weeks of age [8]. Monocrotaline-treated OLETF rats showed a few nodular-like lesions at 50 weeks of age [9]. BTBR ob/ob mice showed diffuse but rare nodular mesangial sclerosis at 20 weeks of age [10]. These rodent models suggest that diabetic conditions in rodents do not lead to reproducible formation of diffuse glomerular nodular lesions. In addition, although two diabetic pig models—streptozotocin-induced diabetic pigs and LNS^{C94Y} transgenic pigs—were created, both failed to reproduce diabetic kidney manifestations [11,12,13]. Thus, it may be more appropriate to create a diabetic mammalian model with the same genetic mutations present in human diabetes that exhibits diabetic renal complications similar to those in humans.

In humans, several forms of diabetes are associated with genetic mutations. Maturity-onset diabetes of the young type-3 (MODY3) is an early onset, non-insulin-dependent form of diabetes characterized by autosomal-dominant inheritance [14]. Those suffering from MODY3 have insufficient insulin secretion, resulting in a similar pathophysiology to that seen in human type-2 diabetes [14,15]. Hepatocyte nuclear factor 1-alpha (HNF1α) is the transcription factor believed to be responsible for MODY3 [14,15]. It is expressed in the liver, pancreas, proximal tubules, stomach, and small intestine [15,16,17]. The most common mutation in the HNF1α gene is the result of a cytosine (C) nucleotide insertion into a poly-C tract around codon 291 (designated as P291fsinsC), which causes frameshift-mutation-mediated deletion of the transactivation domain [14,15].

We have successfully created diabetic pigs carrying the dominant-negative mutant HNF1αP291fsinsC gene that is responsible for severe hyperglycemia with decreasing numbers of pancreatic beta cells [18]. Using these transgenic animals, in the present study we investigated the sequence of morphological events that leads to glomerular nodular lesions in diabetic nephropathy based on the human MODY3 gene. We expected the components and processes of glomerular nodular lesions in diabetic pigs to resemble those in human diabetic nephropathy.

Materials and Methods

Animals

All animal experiments were approved by the Institutional Animal Care and Use Committee of Meiji University (IACUC-09-006). As described previously, focus was on the use of transgenic pigs carrying a dominant-negative mutant HNF1α gene [18]. In short, a transgenic pig carrying an expression vector for the mutant human HNF1α cDNA (HNF1αP291fsinsC) was used. The transgene construct consisted of the enhancer for an immediate-early gene of human cytomegalovirus, followed by a porcine insulin promoter, the human HNF1αP291fsinsC cDNA, a SV40 poly-adenylation signal and a chicken β-globin insulator. Transgenic pigs carrying this cDNA were produced as reported elsewhere [19].

Study protocol

One transgenic and three wild-type pigs were used for biochemical and histological analyses through kidney biopsy. Tests were conducted at monthly intervals until the animals were 10 months of age. For histological analyses, autopsy of additional three transgenic and three wild-type pigs was conducted at 19 weeks of age.

Biochemical analysis

Serum and urine were collected each month after birth until completion of the study. The following biochemical parameters were measured: blood urea nitrogen, creatinine, plasma glucose, total protein, total cholesterol, triglycerides, aspartate aminotransferase, alanine aminotransferase and 1,5-anhydroglucitol. Urine was also analyzed in terms of total protein/creatinine and albumin/creatinine.

Histochemistry of renal sections

For kidney biopsy, the animals were anesthetized by an intramuscular injection of ketamine (11 mg/kg, Fujita Pharmaceutical Co., Ltd., Tokyo, Japan), with isoflurane (DS Pharma Animal Health Co., Ltd., Osaka, Japan) inhalation for maintenance. After the kidney location was confirmed using an ultrasonic pulse-echo technique, specimens were obtained using a Bard Monopty disposable biopsy needle (18 G×20 cm, Bard Biopsy Systems, Tempe, AZ, USA). Kidney specimens were fixed with 4% paraformaldehyde for paraffin sections or 2% glutaraldehyde for electron microscopy. For kidney autopsy, the animals were anesthetized by an intramuscular injection of 1% mafoprazine (0.5 mg/kg, DS Pharma animal Health Co., Ltd.) and intravenous injection of pentobarbital (Kyoritsu Seiyaku Corporation, Tokyo, Japan). After the animals were sacrificed by exsanguination through cutting cervical artery under anesthesia, kidney tissues were dissected and fixed with 4% paraformaldehyde for paraffin sections.

Paraffin sections were processed for periodic acid–Schiff (PAS) staining, periodic acid–methenamine-silver (PAM) staining, Masson's trichrome (MT) staining and immunostaining. Specific primary antibodies were as follows: mouse anti-collagen I antibody (1:50; Abcam, Cambridge, UK), rabbit anti-collagen III antibody (1:400; Abcam), rabbit anti-collagen IV antibody (1:50; Abcam), mouse anti-collagen V antibody (1:50; Abcam), rabbit anti-collagen VI antibody (1:50; Abcam), rabbit anti-advanced glycation end products (AGE) antibody (1:250; Abcam), mouse anti-N^ε-carboxymethyl-lysine (CML) antibody (1:500; TransGenic Inc., Ltd., Kumamoto, Japan) and rabbit anti-transforming growth factor beta-1 (TGF-β1) (V) antibody (1:100; Santa Cruz Biotechnology Inc., Santa Cruz, CA, USA). For immunostaining, antigen retrieval was performed using a microwave (10 mM citrate buffer; pH 6.0) (collagen I) or 100 µg/mL proteinase K (collagen III, IV, V, and VI). Thereafter, primary antibodies were incubated in an EnVision labeled polymer-HRP (Dako, Glostrup, Denmark) or Histofine kit (Nichirei Bioscience Inc., Tokyo, Japan) followed by reaction with peroxidase-conjugated streptavidin (Nichirei). Peroxidase activity was visualized using a liquid diaminobenzidine substrate (Dako). Hematoxylin was used to stain nuclei.

Distribution of glomeruli with nodular lesions

To estimate the prevalence of glomeruli with nodular lesions between the superficial and deep cortexes, sections representing the entire depth of the cortex were subdivided into three zones of equal width: the superficial, middle and deep cortex. The proportion of glomeruli with nodules in each sample was calculated and compared between the superficial and deep cortexes in autopsy specimens of transgenic and wild-type pigs at 19 weeks of age (60–240 glomeruli per kidney per animal).

Measurement of glomerular tuft area

To estimate glomerular tuft area between the superficial and deep cortexes, sections representing the entire depth of the cortex were subdivided into three zones of equal width: the superficial, middle and deep cortex. The glomerular tuft area in each sample was calculated using NanoZoomer 2.0-RS (Hamamatsu Photonics K.K., Hamamatsu, Japan) and compared between the superficial and deep cortexes in autopsy specimens of transgenic and wild-type pigs at 19 weeks of age (69–393 glomeruli per kidney per animal).

Table 1. Analysis of biochemical parameters in transgenic (Tg) and wild-type (WT) pigs at age 1, 5 and 10 months.

	1 month old		5 months old		10 months old	
	Tg (n = 1)	WT (n = 3)	Tg (n = 1)	WT (n = 3)	Tg (n = 1)	WT (n = 1)
Blood urea nitrogen (mmol/l)	13.6	2.71±0.43	10.8	5.07±0.43	9.35	4.53
Plasma glucose (mmol/l)	33.3	6.11±0.03	>33.3	5.87±0.40	26.0	5.51
Creatinine (µmol/l)	53.0	61.9±0.0	35.4	88.4±8.8	26.5	115
Total protein (g/l)	54.0	48.0±1.0	63.0	62.0±0.0	72.0	68.0
Total cholesterol (mmol/l)	11.6	2.95±0.18	1.86	2.00±0.05	1.09	1.66
Triglycerides (mmol/l)	1.69	0.26±0.03	>5.60	0.50±0.10	4.35	0.17
Asperate aminotransferase (IU/l)	23.0	43.0±6.4	128	21.7±1.0	51.0	21.0
Alanine aminotransferase (IU/l)	38.0	31.0±2.3	68.0	33.3±0.7	62.0	32.0
1,5-anhydroglucitol (µg/ml)	2.8	8.8±0.3	1.1	9.6±0.7	2.5	6.7
Urinary protein/creatinine (g/gCr)	<0.20	0.24*	0.72	0.47±0.31	0.45	<0.20
Urinary albumin/creatinine (g/gCr)	<0.10	0.16*	1.41	0.77±0.59	0.23	<0.1

Aberrations: Tg = transgenic pigs; WT = wild-type pigs; *: n = 1.

Thickness of the glomerular basement membrane

In biopsy specimens of pigs at 4 weeks and 5 months of age, 2% glutaraldehyde-fixed kidney cortex tissue was visualized by transmission electron microscopy. GBM thickness was estimated by measurements at five random capillaries in one glomerulus per animal. In each capillary, a series of five photographs were taken at 12,000× magnification, a grid was overlaid on the photograph, and GBM thickness was measured at the points intersecting the grid, with the exception of paramesangial areas. This method is a modified version of that of Hudkins, et al. [10].

Glomerular isolation, RNA isolation and reverse transcription PCR (RT-PCR)

To evaluate HNF1α or HNF1β expression, RT-PCR was performed using isolated glomeruli from one wild-type pig at 4 weeks of age. The animal was anesthetized using isoflurane (DS Pharma Animal Health Co., Ltd.) and perfused with phosphate-buffered saline (PBS). The kidneys, liver and heart were then removed. Using the renal artery, the kidneys were perfused with a 1 mg/mL iron powder in PBS. They were then minced into 1-mm^3 pieces and passed through a 100-µm cell strainer. Finally, glomeruli containing the iron powder were isolated using a magnetic particle concentrator. Total RNA was extracted from the isolated glomeruli, liver, and heart using the RNeasy Mini Kit (Qiagen, Hilden, Germany). RNA was quantified using a Nanodrop 1000 spectrophotometer (Thermo Fisher Scientific K.K., Rockford, IL). Total RNA (1 µg) was reverse-transcribed using the Thermoscript RT-PCR System (Life Technologies Corporation, Carlsbad, CA, USA) into first-strand cDNA. Then, 10 ng of cDNA template and 0.25 mmol/l of sequence-specific primers were used to perform RT-PCR. Primer sequences (5' to 3') were as follows: HNF1α forward: CACAGTCTGCTGAG-CACAGA

HNF1α reverse: TTGGTGGTGTCGGTGATGAG
HNF1β forward: AGAGGGAGGCCTTAGTGGAG
HNF1β reverse: GAGAGGGGCGTCATGATGAG

The liver and heart were used as positive and negative controls, respectively.

Statistical analysis

Mann-Whitney U tests using StatView-J 5.0 (Adept Scientific, Acton, MA, USA) were performed for comparison of the glomerular nodular distribution and glomerular tuft area. Data are shown as means ± standard errors (SE). P-values were calculated from the data. Statistical significance was considered at p-values < 0.05.

Results

Transgenic pigs carrying a dominant-negative mutant HNF1α gene showed severe diabetic mellitus

The biochemical parameters of a single transgenic pig were compared with those of three wild-type pigs over a 10-month period (Table 1). Body weight was lower in transgenic pigs than in wild-type pigs (Figure S1A). In transgenic pigs, the plasma glucose levels were elevated to 22.2–33.3 mmol/L as early as 11 days after birth. This hyperglycemia persisted until 10 months of age (Figure S1B). 1,5-Anhydroglucitol, which reflects the increase in plasma glucose levels during the past several days, was at low levels, indicating severe diabetes mellitus (Figure S1C). In 1-month-old pigs, total cholesterol was high, but decreased after 2 months of age. In contrast, triglycerides were elevated throughout the lifespan of the pigs, which is a symptom also observed in humans with diabetic mellitus. However, serum creatinine levels were within the normal range and no proteinuria was detected in transgenic pigs until 10 months of age.

Transgenic pigs exhibited characteristic diffuse glomerular nodular lesions

Kidney autopsy revealed distinct glomerular nodular lesions at age 19 weeks in all three transgenic pigs (Figure 1A). These nodules were diffuse and acellular, consisting of abnormal matrices. Matrices were slightly evident by PAS staining, strongly by PAM staining, and appeared as a distinct blue color by MT staining. This staining pattern points to the abundance of collagen fibers in the nodules. Numerous nodules formed within an individual glomerulus and were distributed throughout with no discernible pattern. However, more were present in the deep cortex than in the superficial cortex (86.6±7.73 vs. 30.6±12.2%) (p = 0.0495; Figure 1B). Additionally, the glomerular tuft area in

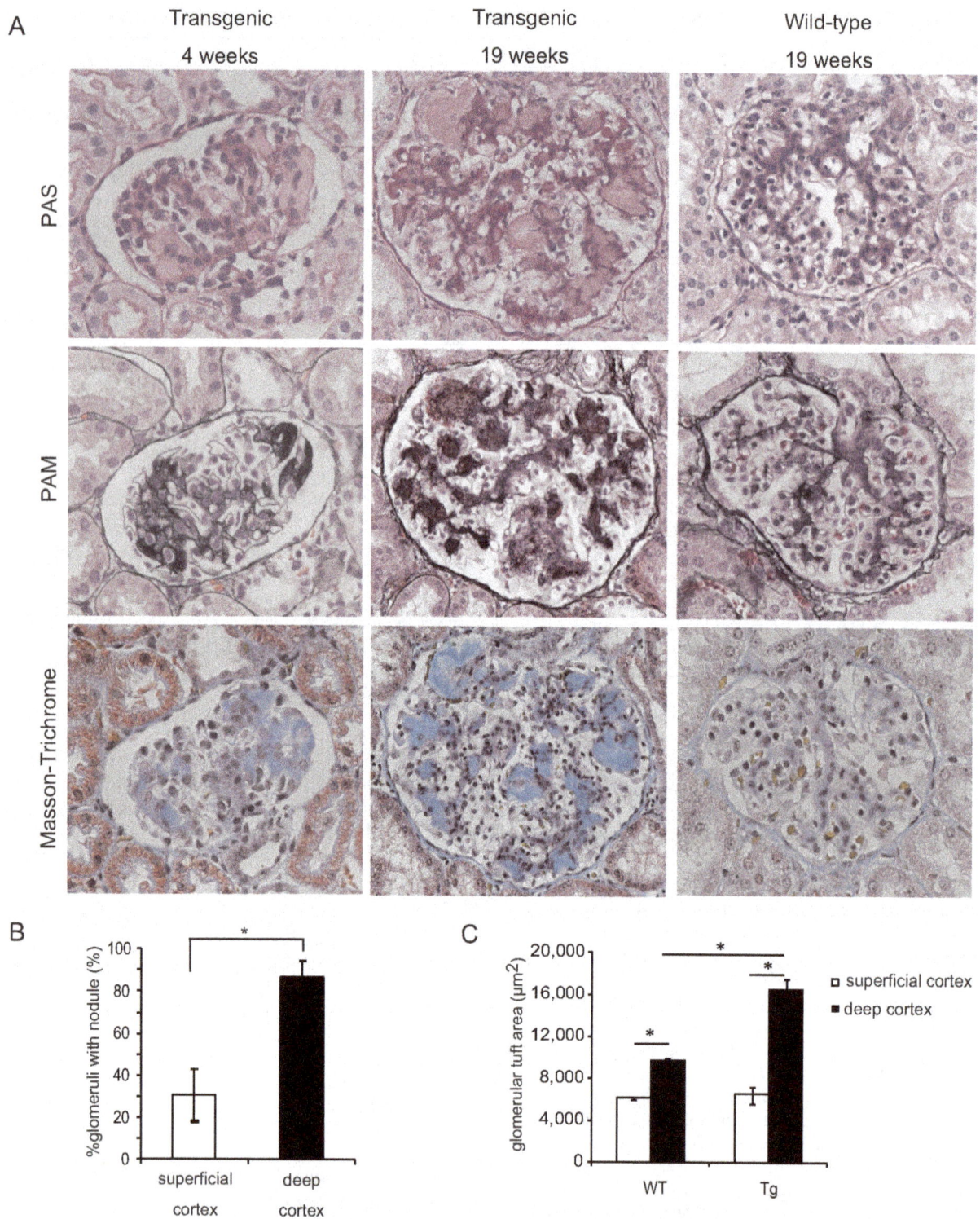

Figure 1. Renal pathological findings at age 4 and 19 weeks in transgenic pigs. A) In transgenic pigs, mesangial expansion commenced as early as 4 weeks. At 19 weeks, distinct glomerular nodules had formed. Magnification: 400×. **B)** The number of glomeruli with nodules as a fraction of the total number was compared between the superficial cortex and deep cortex. **C)** Glomerular tuft area in superficial and deep cortexes was compared between wild-type pigs and transgenic pigs. Transgenic pigs; n = 3, wild-type pigs; n = 3. *$P<0.05$. WT = wild-type pigs; Tg = transgenic pigs.

the deep cortex was significantly larger in transgenic pigs than in wild-type pigs ($16,566\pm983$ vs. $9,694\pm224$ μm^2; $p=0.0495$), but was not significantly different in the superficial cortex ($6,616\pm588$ vs. $6,166\pm80$ μm^2; $p=0.8273$; Figure 1C). This unique distribution of nodules and glomerular tuft size suggested that glomerular

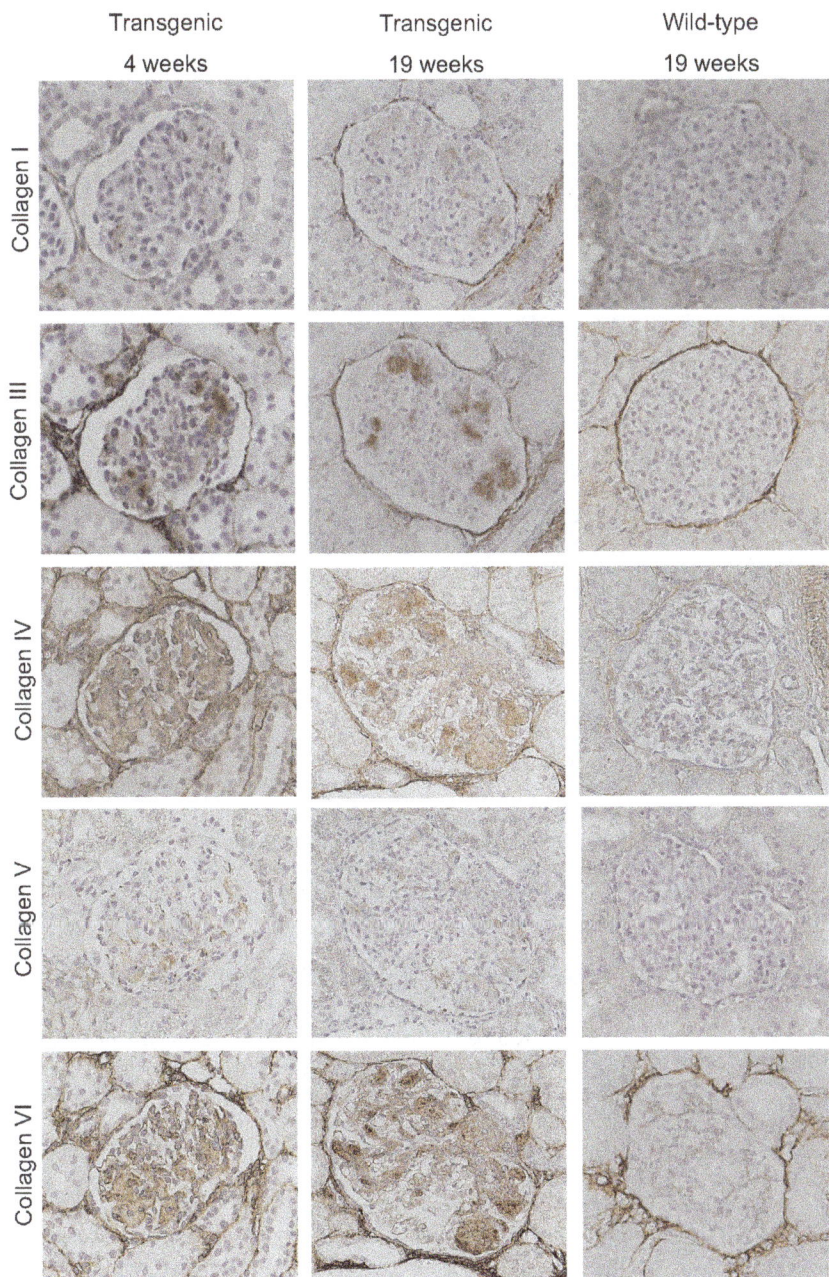

Figure 2. Immunostaining for collagen types I, III, IV, V and VI at age 4 weeks (left) and 19 weeks (middle) in transgenic pigs, and 19 weeks in wild-type pigs (right). In transgenic pigs, collagen types I, III, IV, V and VI were accumulated in the nodules as early as 4 weeks. Collagen types III, IV and VI were strongly positive. Magnification: 400×.

hyperfiltration might contribute to formation of nodules in transgenic pigs.

Immunostaining revealed that the nodules consisted of various types of collagen, including types I, III, IV, V and VI (Figure 2). Collagen types III, IV and VI were present at high concentrations, whereas collagen types I and V were relatively less abundant. AGE, CML and TGF-β1 were also detected in the nodules (Figure 3). AGE tended to be found at the margins of the nodule. CML and TGF-β1 were localized in the nodules in the same patterns as seen in human diabetic nephropathy [20,21,22].

To monitor the sequence of nodular formation, monthly kidney biopsies were performed until the age of 10 months. Mesangial

expansions were formed as early as 4 weeks of age and contained the abnormal matrices similar to those seen in transgenic pigs at 19 weeks of age (Figure 1A). Thereafter, the matrices expanded gradually with age. Collagen fibers and AGE deposition were exclusively associated from the early evolution to the end of the study period (Figures 2 and 3). Glomerular nodular lesions did not lead to segmental glomerulosclerosis or active adhesion.

Another major histological development was the vacuolization of the cytoplasm of epithelial cells in the proximal tubules, resembling Armanni-Ebstein lesions (Figure S2) [23]. The frequency of mesangiolysis and exudative lesions was low (~1 per 200 glomeruli). Other diabetic changes normally seen in

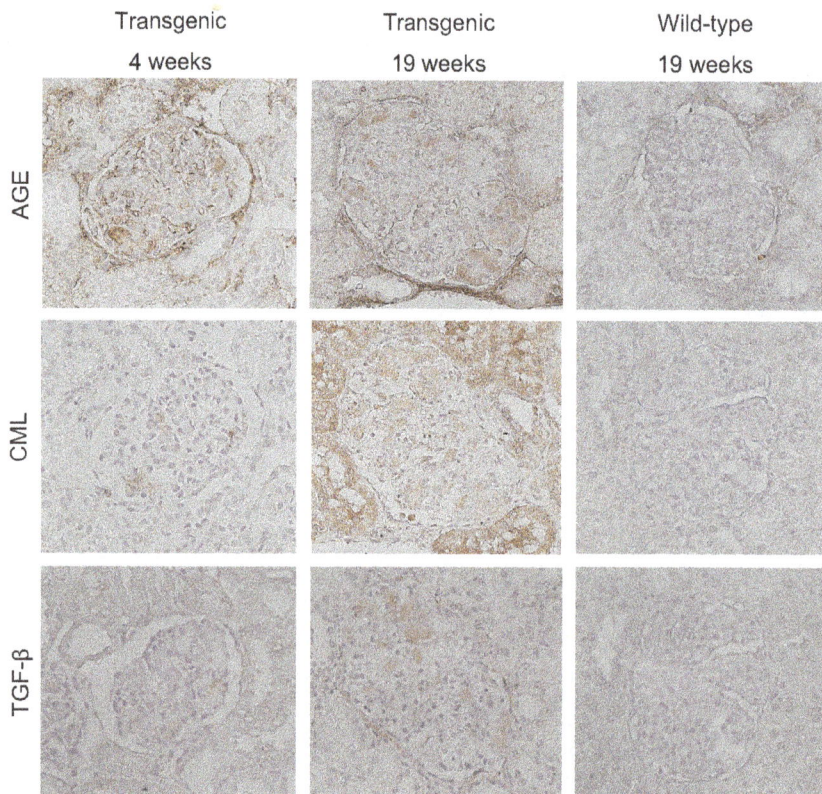

Figure 3. Immunostaining for AGE, CML and TGF-β at age 4 weeks (left) and 19 weeks (middle) in transgenic pigs, and 19 weeks in wild-type pigs (right). In transgenic pigs, AGE, CML and TGF-β accumulated in the glomerular nodules as early as 4 weeks. Magnification: 400×. AGE = advanced glycation end product; CML = N^6-carboxymethyl-lysine; TGF-β = transforming growth factor-beta.

humans were absent from the pig models, including tubular atrophy, interstitial fibrosis and arteriolar hyalinosis.

Glomerular nodular lesions consisted of interstitial forms of fibril collagen

To determine whether glomerular nodular lesions were associated with the typical diabetic changes found in humans, biopsy specimens from animals at 4 weeks and 5 months of age were visualized by electron microscopy. At 4 weeks, bright fibers began to appear in the mesangial matrices (Figure 4A and B), accompanied by lipid particles and cell debris. At a high magnification, the fibers were seen to closely resemble interstitial types of collagen, being of 46-nm diameter with a 50-nm cross-striation cycle (Figure 4E). These collagens were found predominantly around mesangial cells, suggesting that this was their point of origin (Figure 4A). Within 5 months the fibers had accumulated in the mesangium and had expanded to nodular formations (Figure 4C). This nodule expansion encroached upon capillary lumens and caused them to become occluded. A subendothelial widening, accompanied by a loss of endothelial fenestration and occasional mesangial interposition, was also noted (Figure 4D). The GBM thickness of the transgenic pigs was not different from that of wild-type pigs at both 4 weeks and 5 months of age (4 weeks: 163 nm in transgenic pigs $vs.$ 186±10.3 nm in wild-type pigs, 5 months: 194 nm in transgenic pigs $vs.$ 181±5.2 nm in wild-type pigs) (Figure 4F).

Endogenous HNF1α and HNF1β were absent from the glomeruli of wild-type pigs

RT-PCR for HNF1α and HNF1β in the glomeruli of wild-type pigs at 4 weeks of age was performed to determine whether insertion of the dominant-negative mutant HNF1αP291fsinsC gene contributed to the development of glomerular nodular lesions by inhibiting endogenous HNF1α or HNF1β function in glomerular cells. Both HNF1α and HNF1β were absent from the isolated glomeruli, but were expressed in the positive control liver tissue (Figure 5A and B). Therefore, the dominant-negative mutant HNF1αP291fsinsC gene insertion did not contribute directly to the glomerular nodular formation in transgenic pigs.

Discussion and Conclusions

Our pig model carrying a dominant-negative human MODY3 gene is the first to show reproducible diffuse glomerular nodular lesions in a mammalian model of diabetes. The ability to perform repeat kidney biopsies was a great advantage in terms of understanding the in $vivo$ morphological events involved in glomerular nodular formation.

Glomerular nodular lesions in our diabetic pigs were characterized by monotonous accumulation of interstitial forms of collagen fibrils in the mesangium. Initially, small nodules were detected as early as 1 month of age and developed diffusely until 10 months of age. Notably, these were basically acellular round nodules without mesangial proliferation, inflammatory infiltrates or mesangiolysis, (cold nodule); this differs from human diabetic nodules. Immunostaining for various collagens revealed predominantly collagen type III, IV, V and VI in our model, similar to in

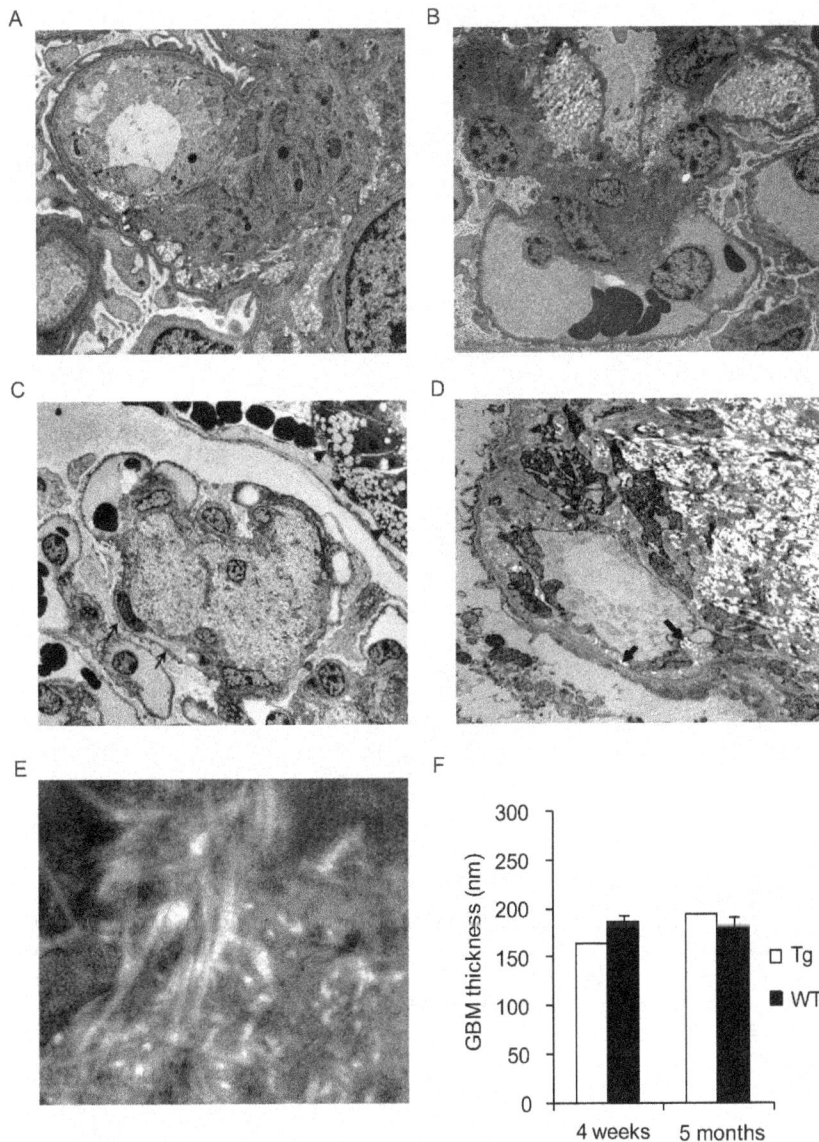

Figure 4. Transmission electron microscopy at age 4 weeks (A,B,E) and 5 months (C,D) in transgenic pigs. A) In 4-week-old transgenic pigs, mesangial widening is associated with fiber deposition in the mesangial matrices. Magnification: 2,000×. **B**) Fibers accumulated at mesangial areas, forming early lesion. Magnification: 500×. **C**) At 5 months, established glomerular nodules showed that mesangial areas and capillary lumens are filled with bright fibers (arrows). Vacuolations of proximal tubules were also seen (arrowheads). Magnification: 300×. **D**) Subendothelial widening with loss of endothelial fenestration and mesangial interposition are shown. Note that collagen is also found in the subendothelial spaces (arrows). Magnification: 1,500×. **E**) The nodules consist of fibril collagens with cross striation, indicating interstitial-type forms of collagen fibrils. Magnification: 10,000×. **F**) Thickness of glomerular basement membranes in transgenic pigs was no different from those in wild-type pigs at 4 weeks and 5 months old. Transgenic pigs; n = 1, wild-type pigs; n = 3. Tg = transgenic pigs; WT = wild-type pigs; GBM = glomerular basement membrane.

human diabetic nodules [24,25]. However, our diabetic nodules also exhibited collagen type I deposition, which is unusual in human diabetic nephropathy [24,25,26]. Electron microscopy showed a distinct interstitial collagen type, which appeared to be a mixture of types I, III and V collagen, as the main component. This was synthesized in the mesangial cells in the early stage, and tended to expand toward the corresponding capillary lumina, finally resulting in nodular sclerosis.

Although the detailed sequence of events leading to nodular formation, and the structure of the nodules, in this model may not be identical to that in humans with type-2 diabetes, the nodules expressed AGEs from a young age. AGEs are produced by non-enzymatic glycation under hyperglycemia, and glomerular AGE

deposition is an important characteristic of nodular morphogenesis in human diabetes [21,27,28]. Specifically, CML is the major AGE accumulated in nodular lesions [20,21]. Glomerular AGEs stimulate extracellular matrix production by mesangial cells through reactive oxygen species (ROS)-promoted TGF-β expression [27,28,29,30]. Glomerular ROS production was caused by AGE-mediated RAGE upregulation or glucose metabolism [27,28,30,31]. In this regard, early onset exclusive AGE deposition and TGF-β1 expression in the nodules of diabetic pigs suggest AGE-mediated collagen synthesis in mesangial cells under a persistent hyperglycemic condition. The differences in nodular morphogenesis and its collagen composition between our model and human diabetic nephropathy suggest that the mesangial

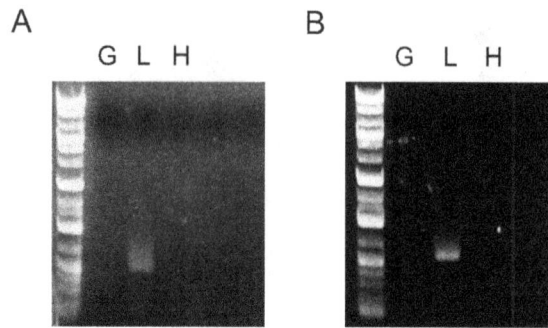

Figure 5. Reverse transcription-PCR for endogenous HNF1α and HNF1β in wild-type pigs at 4 weeks of age. Both HNF1α (**A**) and HNF1β (**B**) were negative in isolated glomeruli. Liver was used as a positive control, and heart as a negative control. G = isolated glomeruli; L = liver; H = heart.

cellular response in these species is different under diabetic conditions. Several reports of nodular sclerosis in the diabetic rodent model support this explanation.

In addition to the mesangial changes under hyperglycemia, our model suggests the involvement of unique hemodynamic factors in nodular morphogenesis. Glomerular hyperfiltration or hypertrophy promotes diabetic nephropathy; however, whether glomerular hemodynamic effects accelerate the formation of diabetic nodules in humans remains controversial. Accordingly, the present study showed that glomerular nodular lesions in diabetic pigs were localized predominantly in the deep cortex. Notably, glomeruli were significantly larger in the deep cortex of diabetic pigs compared to in that of wild-type pigs, but were unchanged in the superficial cortex of both groups. These observations suggest that glomerular hemodynamic effects also promote formation of glomerular nodular lesions in diabetic pigs. Similarly, diabetic nephropathy was accelerated in eNOS-knockout mice, attenuated by improvement of eNOS activity in *db/db* mice [32,33], and antihypertensive therapy alone significantly suppressed the development of nodular lesions and mesangiolysis in diabetic eNOS-knockout mice [34]. Based on these reports and our current findings, our results suggest the involvement of glomerular hypertension in nodular lesion formation in diabetes. Therefore, prominent glomerular hyperfiltration and hypertrophy may be the basis of glomerular nodule development in our diabetic pig model.

The inserted dominant-negative human MODY gene might have stimulated mesangial matrix synthesis by inhibiting endogenous HNF1α or HNF1β function, regardless of the diabetic milieu. Typically, HNF1α functions as a homodimer or a heterodimer with the structurally related protein HNF1β [14,15,35]. Thus, a dominant-negative mutant HNF1αP291fsinsC should inhibit HNF1α or HNF1β by forming an inactive heterodimer at the site of endogenous HNF1α or HNF1β

expression. The RT-PCR study confirmed that mutant HNF1αP291fsinsC could not interact with endogenous HNF1α and HNF1β in the glomeruli of transgenic pigs. This supports the notion that the diabetic milieu, but not the genetic alteration, promotes glomerular nodular formation in the pig model.

Our diabetic pig model lacks several diabetic renal features characteristic of human diabetic nephropathy; e.g., proteinuria, GBM thickening, exudative lesions, tubular atrophy, interstitial fibrosis and arteriolar hyalinosis. Furthermore, the glomeruli did not undergo glomerulosclerosis. These results suggest that our model does not accurately reproduce human diabetic nephropathy, even in pigs carrying the human MODY3 gene. In addition to the species difference in the cellular response to hyperglycemia, a possible explanation for this discrepancy is that we were unable to monitor the histology for a sufficiently long period because due to the relatively short lifespan of the pigs. In addition, the mechanism of nodular formation is considered to be different from that of other diabetic kidney lesions. Nevertheless, this pig model indicated that glomerular nodules could form independently of diabetic complications.

In conclusion, this was the first report of distinct and reproducible glomerular nodular lesions in transgenic pigs carrying a dominant-negative HNF1α mutation of the human MODY3 gene. Although there were several differences compared to the pathology of human glomerular nodular lesions, the somewhat acute and constitutive formation of nodules in the mammalian models might provide information that will facilitate identification of the principal mechanism underlying glomerular nodular formation.

Supporting Information

Figure S1 Body weight and diabetic parameter changes over time. A) Body weight was lower in transgenic pigs than in wild-type pigs. **B**) Plasma glucose was at a high level in transgenic pigs. **C**) 1,5-Anhydroglucitol was at a low level in transgenic pigs. Tg = transgenic pigs (n = 1); WT = wild-type pigs (up to 6 months of age, n = 3; 6–10 months of age, n = 1).

Figure S2 Armanni-Ebstein lesions in diabetic pigs at 19 weeks of age. Transgenic pigs revealed vacuolation of proximal tubules known as Armanni-Ebstein lesions. Note that distal tubules and the collecting duct are intact. **A**) Magnification: 100×. **B**) Magnification: 400×.

Author Contributions

Conceived and designed the experiments: KU HN. Performed the experiments: KU TY SH. Analyzed the data: SH TY MN. Contributed reagents/materials/analysis tools: SH TY KU HN MN. Wrote the paper: SH KU MN.

References

1. Nakai S, Iseki K, Itami N, Ogata S, Kazama JJ, et al. (2012) An overview of regular dialysis treatment in Japan (as of 31 December 2010). Ther Apher Dial 16: 483–521.
2. White SL, Chadban SJ, Jan S, Chapman JR, Cass A (2008) How can we achieve equity in provision of renal replacement therapy? Bull World Health Organ 86: 229–237.
3. Kimmelstiel P, Wilson C (1936) Intercapillary Lesions in the Glomeruli of the Kidney. Am J Pathol 12: 83–98.7.
4. Hong D, Zheng T, Jia-qing S, Jian W, Zhi-hong L, et al. (2007) Nodular glomerular lesion: a later stage of diabetic nephropathy? Diabetes Res Clin Pract 78: 189–195.

5. Heaf JG, Løkkegaard H, Larsen S (2001) The relative prognosis of nodular and diffuse diabetic nephropathy. Scand J Urol Nephrol 35: 233–238.
6. Brosius FC 3rd, Alpers CE, Bottinger EP, Breyer MD, Coffman TM, et al. (2009) Mouse models of diabetic nephropathy. J Am Soc Nephrol 20: 2503–2512.
7. Zhao HJ, Wang S, Cheng H, Zhang MZ, Takahashi T, et al. (2006) Endothelial nitric oxide synthase deficiency produces accelerated nephropathy in diabetic mice. J Am Soc Nephrol 17: 2664–2669.
8. Inagi R, Yamamoto Y, Nangaku M, Usuda N, Okamoto H, et al. (2006) A severe diabetic nephropathy model with early development of nodule-like lesions induced by megsin overexpression in RAGE/iNOS transgenic mice. Diabetes 55: 356–366.

9. Furuichi K, Hisada Y, Shimizu M, Okumura T, Kitagawa K, et al. (2011) Matrix metalloproteinase-2 (MMP-2) and membrane-type 1 MMP (MT1-MMP) affect the remodeling of glomerulosclerosis in diabetic OLETF rats. Nephrol Dial Transplant 26: 3124–3131.

10. Hudkins KL, Pichaiwong W, Wietecha T, Kowalewska J, Banas M, et al. (2010) BTBR ob/ob mutant mice model progressive diabetic nephropathy. J Am Soc Nephrol 21: 1533–1542.

11. Larsen MO, Wilken M, Gotfredsen CF, Carr RD, Svendsen O, et al. (2002) Mild streptozotocin diabetes in the Göttingen minipig. A novel model of moderate insulin deficiency and diabetes. Am J Physiol Endocrinol Metab 282: E1342–1351.

12. Bellinger DA, Merricks EP, Nichols TC (2006) Swine models of type 2 diabetes mellitus: insulin resistance, glucose tolerance, and cardiovascular complications. ILAR J 47: 243–258.

13. Renner S, Braun-Reichhart C, Blutke A, Herbach N, Emrich D, et al. (2013) Permanent neonatal diabetes in INS(C94Y) transgenic pigs. Diabetes 62: 1505–1511.

14. Yamagata K, Oda N, Kaisaki PJ, Menzel S, Furuta H, et al. (1996) Mutations in the hepatocyte nuclear factor-1α gene in maturity-onset diabetes of the young (MODY3). Nature 384: 455–458.

15. Yamagata K (2003) Regulation of pancreatic β-cell function by the HNF transcription network: lessons from maturity-onset diabetes of the young (MODY). Endocr J 50: 491–499.

16. Pontoglio M, Barra J, Hadchouel M, Doyen A, Kress C, et al. (1996) Hepatocyte nuclear factor 1 inactivation results in hepatic dysfunction, phenylketonuria, and renal Fanconi syndrome. Cell 84: 575–585.

17. Pontoglio M (2000) Hepatocyte nuclear factor 1, a transcription factor at the crossroads of glucose homeostasis. J Am Soc Nephrol 11: S140–S143.

18. Umeyama K, Watanabe M, Saito H, Kurome M, Tohi S, et al. (2009) Dominant-negative mutant hepatocyte nuclear factor 1alpha induces diabetes in transgenic-cloned pigs. Transgenic Res 18: 697–706.

19. Umeyama K, Honda K, Matsunari H, Nakano K, Hidaka T, et al. (2013) Production of diabetic offspring using cryopreserved epididymal sperm by in vitro fertilization and intrafallopian insemination techniques in transgenic pigs. J Reprod Dev 59: 599–603.

20. Tanji N, Markowitz GS, Fu C, Kislinger T, Taguchi A, et al. (2000) Expression of advanced glycation end products and their cellular receptor RAGE in diabetic nephropathy and nondiabetic renal disease. J Am Soc Nephrol 11: 1656–1666.

21. Horie K, Miyata T, Maeda K, Miyata S, Sugiyama S, et al. (1997) Immunohistochemical colocalization of glycoxidation products and lipid peroxidation products in diabetic renal glomerular lesions. Implication for glycoxidative stress in the pathogenesis of diabetic nephropathy. J Clin Invest 100: 2995–3004.

22. Yamamoto T, Nakamura T, Noble NA, Ruoslahti E, Border WA (1993) Expression of transforming growth factor beta is elevated in human and experimental diabetic nephropathy. Proc Natl Acad Sci U S A 90: 1814–1818.

23. Ritchie S, Waugh D (1957) The pathology of Armanni-Ebstein diabetic nephropathy. Am J Pathol 33: 1035–1057.

24. Nerlich A, Schleicher E (1991) Immunohistochemical localization of extracellular matrix components in human diabetic glomerular lesions. Am J Pathol 139: 889–899.

25. Makino H, Shikata K, Wieslander J, Wada J, Kashihara N, et al. (1994) Localization of fibril/microfibril and basement membrane collagens in diabetic glomerulosclerosis in type 2 diabetes. Diabet Med 11: 304–311.

26. Glick AD, Jacobson HR, Haralson MA (1992) Mesangial deposition of type I collagen in human glomerulosclerosis. Hum Pathol 23: 1373–1379.

27. Renner S, Braun-Reichhart C, Blutke A, Herbach N, Emrich D, et al. (2001) Biochemistry and molecular cell biology of diabetic complications. Nature 414: 813–820.

28. Yamagishi S, Matsui T (2010) Advanced glycation end products, oxidative stress and diabetic nephropathy. Oxid Med Cell Longev 3: 101–108.

29. Fukami K, Ueda S, Yamagishi S, Kato S, Inagaki Y, et al. (2004) AGEs activate mesangial TGF-β-Smad signaling via an angiotensin II type I receptor interaction. Kidney Int 66: 2137–2147.

30. Mason RM, Wahab NA (2003) Extracellular matrix metabolism in diabetic nephropathy. J Am Soc Nephrol 14: 1358–1373.

31. Yan SD, Schmidt AM, Anderson GM, Zhang J, Brett J, et al. (1994) Enhanced cellular oxidant stress by the interaction of advanced glycation end products with their receptors/binding proteins. J Biol Chem 269: 9889–9897.

32. Kanetsuna Y, Takahashi K, Nagata M, Gannon MA, Breyer MD, et al. (2007) Deficiency of endothelial nitric-oxide synthase confers susceptibility to diabetic nephropathy in nephropathy-resistant inbred mice. Am J Pathol 170: 1473–1484.

33. Cheng H, Wang H, Fan X, Paueksakon P, Harris RC (2012) Improvement of endothelial nitric oxide synthase activity retards the progression of diabetic nephropathy in db/db mice. Kidney Int 82: 1176–1183.

34. Kosugi T, Heinig M, Nakayama T, Connor T, Yuzawa Y, et al. (2009) Lowering blood pressure blocks mesangiolysis and mesangial nodules, but not tubulointerstitial injury, in diabetic eNOS knockout mice. Am J Pathol 174: 1221–1229.

35. Mendel DB, Hansen LP, Graves MK, Conley PB, Crabtree GR (1991) HNF-1 alpha and HNF-1 beta (vHNF-1) share dimerization and homeo domains, but not activation domains, and form heterodimers in vitro. Genes Dev 5: 1042–1056.

21

Diverse Functions of mRNA Metabolism Factors in Stress Defense and Aging of *Caenorhabditis elegans*

Aris Rousakis[1,2☯], Anna Vlanti[1☯], Fivos Borbolis[1,3], Fani Roumelioti[1,3¤a], Marianna Kapetanou[1,4¤b], Popi Syntichaki[1]*

1 Biomedical Research Foundation of the Academy of Athens, Center of Basic Research II, Athens, Greece, 2 Faculty of Medicine, University of Athens, Athens, Greece, 3 Faculty of Biology, School of Science, University of Athens, Athens, Greece, 4 Department of Biology, School of Science and Engineering, University of Crete, Heraklio, Crete, Greece

Abstract

Processing bodies (PBs) and stress granules (SGs) are related, cytoplasmic RNA-protein complexes that contribute to post-transcriptional gene regulation in all eukaryotic cells. Both structures contain translationally repressed mRNAs and several proteins involved in silencing, stabilization or degradation of mRNAs, especially under environmental stress. Here, we monitored the dynamic formation of PBs and SGs, in somatic cells of adult worms, using fluorescently tagged protein markers of each complex. Both complexes were accumulated in response to various stress conditions, but distinct modes of SG formation were induced, depending on the insult. We also observed an age-dependent accumulation of PBs but not of SGs. We further showed that direct alterations in PB-related genes can influence aging and normal stress responses, beyond their developmental role. In addition, disruption of SG-related genes had diverse effects on development, fertility, lifespan and stress resistance of worms. Our work therefore underlines the important roles of mRNA metabolism factors in several vital cellular processes and provides insight into their diverse functions in a multicellular organism.

Editor: Dimitris L. Kontoyiannis, BSRC 'Alexander FLEMING', Greece

Funding: The work was supported by European Research Council under the European Union's Seventh Framework Program (FP/2007–2013)/ERC Grant Agreement n. [201975]. The funder had no role in study design, data collection and analysis, decision to publish, or preparation of the manuscript.

Competing Interests: The authors have declared that no competing interests exist.

* Email: synticha@bioacademy.gr

☯ These authors contributed equally to this work.

¤a Current address: Institute of Immunology, BSRC "Alexander Fleming", Athens, Greece
¤b Current address: Institute of Biology, Medicinal Chemistry and Biotechnology, National Hellenic Research Foundation, Athens, Greece

Introduction

A plethora of evidence in all eukaryotes has established that aging is a multifactorial process that can be modulated by several means, ranging from genes to lifestyle and pharmacological interventions. Genetic and genome-wide studies in many organisms shed light on the basic molecular mechanisms of aging and identified conserved signaling pathways as master modulators of lifespan [1]. Fundamental cellular functions that primarily alter metabolism or preserve homeostasis can influence the rate of aging and age-related disability/degeneration. Such cellular functions comprise the activation of stress response and repair mechanisms, the enhancement of catabolic processes and the reduction of anabolic processes. Protein synthesis is an anabolic process that, when reduced by gene mutations, drugs, hormonal or stress signals, increases lifespan in diverse species. The underlying mechanisms could involve energy conservation and/or reprogramming of gene expression that is essential for cell survival [2]. In accordance with this, several studies demonstrate that protein synthesis is tightly regulated by environmental stress in eukaryotes. Protein synthesis rates are mainly controlled at the level of translation initiation but post-transcriptional regulation of mRNAs, including processing, export, decay and localization, can also affect translation. Accordingly, the impact of post-transcriptional regulation of gene expression by RNA-binding proteins and cytoplasmic RNA granules is an emerging field in regulators of aging and age-related diseases [3–5].

Processing bodies (PBs) and stress granules (SGs) are major cytoplasmic RNA-protein complexes that regulate, in part, the abundance and translation of mRNAs in all eukaryotic cells. PBs contain non-translatable mRNAs and a variety of proteins involved in translational repression, 5' to 3' mRNA decay and RNA-induced silencing [6–8]. SGs function in mRNA storage following environmental stress and contain stalled 43S pre-initiation complexes and many RNA-binding proteins that regulate the stability, storage or translation of mRNAs [9]. Both PBs and SGs are highly dynamic structures that share specific components and can spatially interact to determine the fate of mRNAs in the cytoplasm, in equilibrium with translating polysomes [10–14]. However, the physiological role of these complexes has not yet been fully elucidated; the processes of translation inhibition and mRNA decay can occur normally in yeast, *Drosophila* and mammalian cells that are defective in PB or SG formation, suggesting that their accumulation is a consequence of rather than a perquisite for mRNA metabolism [15–19]. Furthermore, under some conditions in yeast, decapping and 5' to 3' decay occur while mRNAs are still associated with polyribosomes [20]. Nevertheless, the evolutionary conservation of both

RNA granules supports that they function to impart specificity in molecule interactions and efficiency in translation repression and mRNA degradation.

Both PBs and SGs participate in the transient translational silencing and global reprogramming of gene expression that occur in response to cellular stress [8]. They accumulate within minutes of stress and dissolve within a few hours of stress recovery. The aggregation of non-translatable mRNAs into PBs and SGs could either promote or inhibit mRNA decay and might be crucial for the adaptation or recovery of cells e.g. through the stabilization of specific transcripts. These include either stress-responsive transcripts with short half-lives [21] or energy-costly mRNAs, as those encoding the abundant cytoplasmic ribosomal proteins [22]. In addition, SGs can affect key signaling pathways involved in the global stress response, such the MAPK, TOR and HIF-1 pathways [23–26]. Studies in yeast and mammalian cells have shown that stress-induced granules can differ in composition and assembly rules, depending on the type of stress insult [11,27].

In *C. elegans* several types of RNA granules have been described during the processes of oogenesis and embryogenesis. P granules are a type of germ granule, required for germ cell development, with a primary role in post-transcriptional regulation of maternal mRNAs in the gonad [28–31]. Other RNA granules accumulate within the *C. elegans* gonad in response to developmental or environmental signals and differ in composition and function [32–34]. All these diverse RNA granules participate in proper mRNA control during early development or in aged/stressed oocytes and share some components with somatic PBs and SGs in other organisms[32–37]. Such components include the worm homologues of proteins implicated in mRNA decapping or translational repression, as the decapping enzyme DCAP-2 (Dcp2 in yeast and mammals) and its cofactor DCAP-1 (Dcp1), the translation regulator CGH-1 (Dhh1/RCK), the decapping activators PATR-1 (Pat1) and LSM-1 (Lsm1) or the 5′ to 3′ exonuclease XRN-1 (Xrn1). In addition, they harbor RNA-binding proteins, such as the worm homologues of human ataxin-2 (ATX-2), the poly(A) binding protein PABP (PAB-1) and TIA-1/TIAR proteins (TIAR-1, -2) that are SG components in mammalian cells. Although the structure, localization and function of these germline and early embryonic RNA granules have been thoroughly characterized, somatic PBs and SGs have not yet been well investigated in *C. elegans*. The developmental consequences of loss of several P granule components have been nicely described [34–36,38] but the effect of somatic PB and SG components on survival of the organism during stress conditions or aging are largely unknown.

Here, we present data on worm's stress survival and lifespan modulation by factors involved in mRNA metabolism. We first provide *in vivo* insights into the dynamic formation of somatic PBs and SGs in intact adult animals, in response to various environmental stresses. Moreover, an age-dependent accumulation of PBs, but not SGs, was observed in somatic cells of worms. Aggregation of PBs in young adults is also triggered in response to alterations in 5′ to 3′ mRNA degradation but not to reduced mRNA translation. Furthermore, we show that direct alterations in PB-related genes, by mutations or RNA interference (RNAi), can influence lifespan and stress resistance, beyond their developmental role. We reveal that stress conditions induce distinct modes of SG formation and deletion of SG-related genes has diverse effects on development, lifespan and stress response of worms. Our work provides genetic characteristics of proteins localized to PBs and SGs and offers new insights into the pattern and function of these RNA granules at the organismal level.

Materials and Methods

C. elegans strains and culture

Standard methods of culturing and handling worms were used [39]. Worms were raised on NGM plates seeded with *Escherichia coli* OP-50 or HT115 (DE3) for RNAi experiments, at the indicated temperature. Wild-type Bristol N2 and some mutant strains were provided by the *Caenorhabditis* Genetics Center (CGC, University of Minnesota), which is funded by NIH Office of Research Infrastacture Programs (P40 OD010440). Other mutant strains were provided by the Mitani Lab through the National Bio-Resource Project of the MEXT, Japan. Table S1 shows all strains used in this study. All single mutants were outcrossed 3–5 times with N2 and relevant mutations were tracked in F2 progeny by PCR (Table S2). Double mutants were made by crossing the relevant strains and PCR-based selection in F2 progeny. Transgenic animals were generated by microinjection of plasmid DNAs into the gonad of N2 young adults, using *rol-6(su1006)* as co-transformation marker [40]. Multiple lines were generated for each genotype and screened for the representative expression pattern. Transgenic mutants were generated by crossing N2 transgenic hermaphrodites with males of the desired mutant background.

Constructs

RNAi plasmids were constructed by inserting gene-specific PCR product, amplified from genomic DNA using the appropriate primers (Table S2), into the L4440 feeding vector (pPD129.36) [41]. For the double *dcap-1/-2(RNAi)* construct both gene fragments were cloned into a single L4440 vector [42]. See Table S3 for RNAi construct details. The presumptive promoter of *dcap-1* (~500-kb) was obtained by PCR (Table S2) and inserted into the Fire Lab vector pPD95.77 for the $P_{dcap-1}::gfp$ transcriptional fusion. For the *dcap-1::gfp* transgene, the *gfp* sequence was inserted between the end of *dcap-1* coding region and its 3′ UTR in the plasmid FLAG::*dcap-1*, kindly provided by Dr. Min Han [43]. The *tiar-3::gfp* transgene was constructed by cloning the promoter (1969-bp) and coding region in pPD95.77. For *gfp::tiar-1* and *gfp::tiar-2* a modified version of pPD95.77 was used, allowing insertion of the promoter regions of *tiar-1* (1421-bp) or *tiar-2* (1614-bp) upstream of *gfp* with subsequent in-frame insertion of their corresponding coding regions followed by their own 3′ UTRs (489-bp for *tiar-1* and 680-bp for *tiar-2*). The intermediate constructs with the promoter only sequences fused to *gfp* of pPD95.77 were used as transcriptional reporters. The tagRFP fusions were constructed by replacing the GFP by tagRFP, obtaining the relevant sequence by PCR amplification from pHb9::tagRFP (a gift from Dr. Ivo Lieberam).

RNA interference

RNAi experiments were carried out by adding synchronized L4s or eggs to plates seeded with HT115(DE3) bacteria that express dsRNA for the indicated gene. HT115 bacteria transformed with the relevant RNAi vectors were grown at 37°C in LB medium with ampicillin (50 µg/ml) and tetracycline (10 µg/ml). On the following day, fresh cultures with ampicillin were induced with 1 mM isopropylb-D-thiogalactopyranoside (IPTG) and seeded on RNAi plates [44]. Bacteria carrying the empty vector (pL4440) and treated likewise were used as control cultures (Control(RNAi)).

Microscopy

The expression pattern of transgenic worms was monitored by mounting levamizole-treated animals on 3% agarose pads, on glass microscope slides. Animals (~20 animals per condition in at least three experiments) were imaged using a Leica TCS SP5 confocal imaging system. For imaging assays of transgenic worms at different ages, adults from age-synchronized cultures, raised at the indicated temperatures from hatching, were picked and monitored under the same microscopy settings. For heat-shock assays 1-day adults were shifted at 35°C for the indicated time and were immediately monitored. For recovery assays the heat-shocked animals were incubated for 0.5–3 h at 20°C before mounting. For oxidative stress 1-day adults were transferred on NGM plates seeded with UV-killed OP-50 bacteria containing 10–15 mM sodium arsenite and were visualized after 2–3 h. For osmotic stress 1-day adults were transferred on NGM plates containing 350 mM NaCl and were visualized after 0.5–1 h. All images were taken at 20× magnification, keeping the same microscopy settings for each strain per condition. Images shown from confocal are 2D maximal projections of z-stacks, or optical sections where is indicated, processed in Photoshop CS3.

The quantification of PBs formed under HS was performed as follows: three random squares of 50 μm×50 μm were drawn along a worm; the number of aggregates within each square was measured and their mean value, representing the number of granules per 2500 μm^2 per worm, was calculated. This procedure was repeated for a total of 20 worms from at least two independent experiments and the average of the mean values is presented in Table S4. The same method was applied to measure SGs visualized using the P$_{myo-3}$::GFP::TIAR-2 marker. In the case of age-induced PBs and SGs we measured aggregates formed in the area of the head (from the pharynx to the tip of the worm) to avoid the intestinal autofluorescence. The same method was used to measure TIAR-1 granules as well as SA-induced PBs. When using the GFP::TIAR-2 marker, we measured the number of granules formed in 100 μm length of excretory cell. Again a total of 15–20 worms were counted in each case from at least two independent experiments. All measurements were performed using the count tool in Photoshop CS3. Statistical analysis (unpaired t-test) was performed using GraphPad Prism version 5 (GraphPad Software, San Diego, California USA).

RNA isolation and quantitative reverse transcription PCR (qRT-PCR)

Total RNA was prepared from frozen worm pellets, of the indicated genetic backgrounds and developmental stages, using a NucleoSpin RNA XS kit (Macherey-Nagel) and measured by Quant-iT RNA Assay Kit (Invitrogen). Total RNA was reverse transcribed with iScriptTM cDNA Synthesis Kit (Biorad) and quantitative PCR was performed using the SsoFastTM EvaGreen Supermix (BioRad) in the MJ MiniOpticon system (BioRad). The relative amounts of mRNA were determined using the Comparative Ct method for quantification and gene expression data are presented as the fold change relative to control. qRT-PCR was performed in triplicates and each sample was independently normalized to endogenous reference ama-1. The mean ± the standard deviation (SD) of at least two independent experiments is presented. All statistical comparisons were performed by Student's t-test for unpaired samples in GraphPad Prism 5. The primer sequences for qRT-PCR are available upon request.

Western Blotting

About 1500–2000 synchronized worms of each strain, grown on OP50 plates at 25°C, were subjected to heat-shock (35°C for 3 h), in the first day of adulthood and were collected in M9 buffer, washed 2–3 to remove bacteria and frozen in ethanol dry ice. Control worms (no heat-shocked) of the same age were treated similarly. Likewise, ~1500 L4-synchronized worms of each strain at 25°C, were transferred on OP50 plates containing 50 μM 5-fluoro-2'-deoxyuridine (FUDR) to prevent progeny production and collected as either 1-day or 5-day adults in M9 buffer. After 2–3 washes to remove bacteria they were frozen in ethanol dry ice. In all cases, before loading onto SDS-PAGE, worm pellets were boiled in 2× SDS-sample buffer for 10 min. Worm lysates were resolved on a 10% SDS-PAGE, western blotted and analyzed with primary antibodies against to either GFP (1:10,000, BD Biosciences) or actin (1:5,000, Clone C4, Millipore). Secondary anti-rabbit and anti-mouse IgG antibodies (HRP) were used respectively for immunoblot signal detection with ECL (Thermo Scientific). Quantification of immunoblot signals was performed using ImageJ software. Ratio of DCAP-1::GFP to actin levels was measured in two independent experiments.

Lifespan assays

Lifespan analysis was conducted at 20°C or 25°C as described previously [45]. Briefly, eggs or mid-to-late L4 larvae of each strain (at least 70 animals per experiment) were transferred to NGM plates seeded with OP-50 or RNAi bacteria of interest and were moved to fresh plates every two days. Viability of the worms was daily scored and animals that failed to respond to stimulation by touch were referred as dead, whereas that bagged, ruptured or crawled off the plates were referred as censored in the analysis. For post-developmental assays, day 0 of adulthood was defined as the day that mid-to-late L4s were transferred to plates. Lifespan and statistical analysis were performed using GraphPad Prism version 5. Each population is compared with the appropriate control population and p-values were determined using the log-rank (Mantel-Cox) test.

Fertility assay

Worms of each genotype were grown at 20°C or 25°C and 5–15 L4 hermaphrodites were placed on individual NGM plates to lay eggs. Animals were transferred daily to fresh plates until egg-laying ceased and the hatched progeny were counted in each plate. The total number of progeny per worm (brood size) was counted and the average brood size (mean ± SD) of each strain was plotted. Unpaired t-test was used to calculate p-values in GraphPad Prism 5.

Stress resistance assays

For heat-shock assays, 1-day adult worms were shifted to 35°C for 6 h (8 h for daf-2 mutants). After 16 h of recovery at 20°C the percentage of worms surviving was determined. For UV resistance assays 5-day adults were irradiated on plates without bacteria at 0.2 J/cm^2 and then were transferred to plates with food at 20°C. Three days later the percentage of worms surviving was determined. For oxidative stress 1-day adults of each strain were transferred on plates seeded with UV-killed OP-50 containing 5 mM sodium arsenite and the percentage of worms surviving was determined after 24 h (for transgenic worms) or 48 h (non-transgenic worms). For osmotic stress 1-day adults were transferred on NGM plates containing 400 mM NaCl and scored for survival after 24 h. The average (mean ± SD) of at least three independent experiments with ~100 individuals for each strain per experiment was plotted. Unpaired t-test was used to calculate p-values in GraphPad Prism 5.

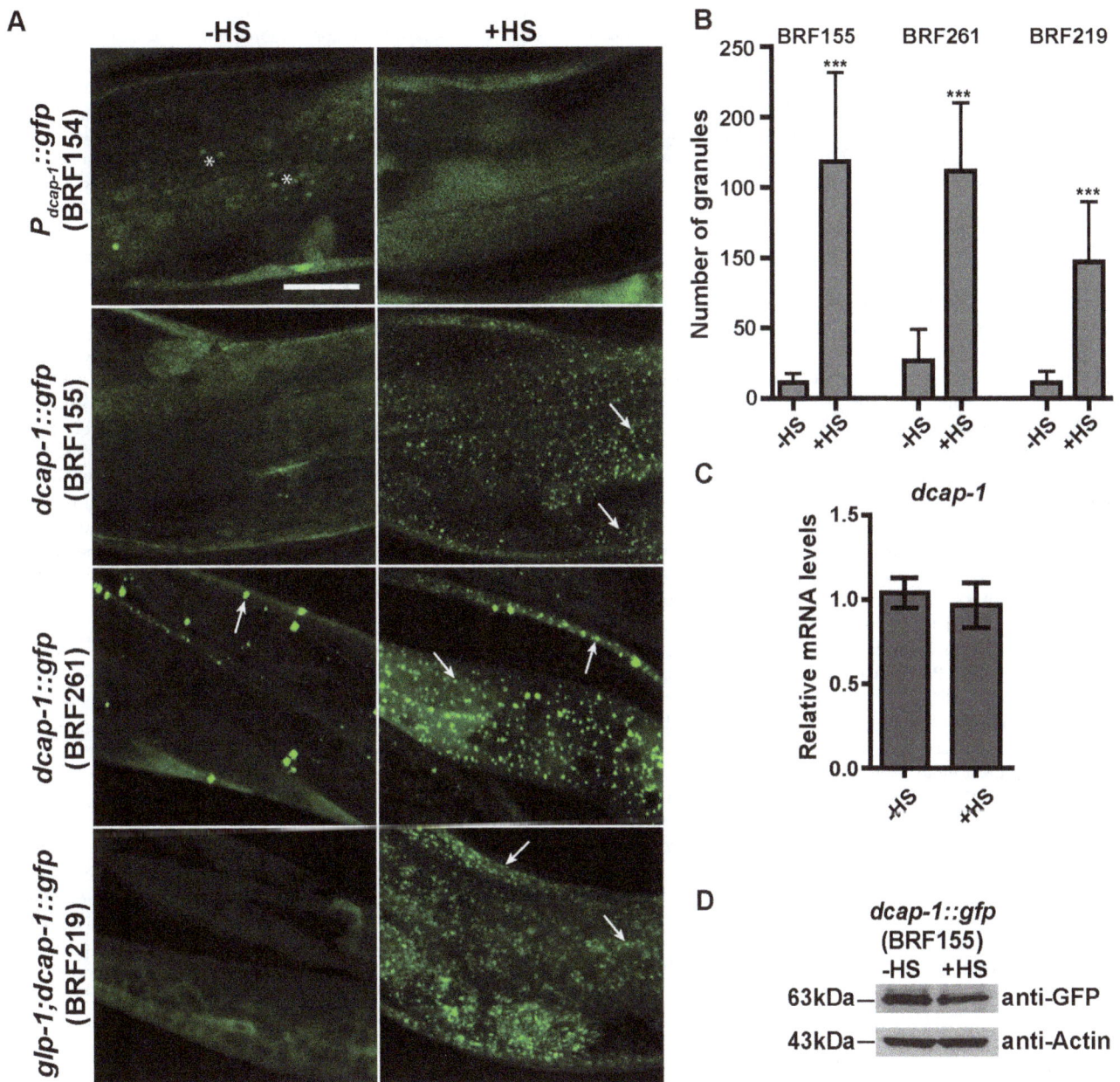

Figure 1. Aggregation of DCAP-1 protein in response to heat-shock. (A) Representative confocal images of somatic tissues in 1-day adult worms expressing the transcriptional fusion $P_{dcap-1}::gfp$ (BRF154) or the translational fusion $dcap-1::gfp$ in N2 (BRF155 and BRF261) and germline-deficient $glp-1(e2141)$ worms (BRF219), normally grown at 25°C (-HS) or transiently subjected to heat-shock (+HS, 35°C for 3 h). Arrows point to DCAP-1::GFP granules in body wall muscles that are highly induced upon stress. Asterisks denote the intestinal autofluorescence. Scale bar: 25 µm. (B) Quantification of data in (A). Values on Y axis show the number of granules per 2500 µm² per worm (mean±SD, see Materials and Methods and Table S4). (C) Endogenous $dcap-1$(mRNA) levels in 1-day adults N2, grown at 25°C before (-HS) or after heat shock (+HS, 35°C for 3 h) were measured by quantitative RT-PCR and normalized to endogenous $ama-1$(mRNA) levels. Error bars represent the standard deviation of the means of five independent experiments (p-value = 0.3466, calculated by Student's t-test). (D) Western blot analysis of DCAP-1::GFP protein expression in 1-day adult BRF155 transgenic animals, before (-HS) or after heat shock (+HS), using anti-GFP or anti-Actin (as a loading control) antibodies. Band intensity of DCAP-1::GFP normalized to actin gives a value of 0.83, showing similar protein levels in the two conditions.

Results

Induction of PBs in *C. elegans* somatic cells by environmental stress

A well-described marker of PBs in all organisms is the decapping subunit Dcp1, which is encoded by the *dcap-1* (Y55F3AM.12.1) gene in worms (http://www.wormbase.org). DCAP-1 protein is a component of P granules in germline and

PB-like structures in embryos [32,36–38]. We generated transgenic animals that carry either the transcriptional ($P_{dcap-1}::gfp$) or the translational ($P_{dcap-1}::dcap-1::gfp::3'$ UTR^{dcap-1} referred to as $dcap-1::gfp$ for simplicity) fluorescent reporter gene to monitor the expression pattern in somatic cells of adult wild-type (N2) worms, under normal growth conditions (Fig. S1A). Worms expressing the $P_{dcap-1}::gfp$ reporter (line BRF154, Fig. S1A) displayed a smooth and diffused fluorescent signal throughout the cytoplasm of most cells. Worms bearing the *dcap-1::gfp* fusion (BRF155, Fig. S1A) showed a similar ubiquitous expression pattern with occasional cytoplasmic puncta, varying among the worms. Although several protein components of PBs have an intrinsic capacity to aggregate [16,46–48], in mammalian cell lines PB components show a diffuse distribution in the cytoplasm and only few PBs are formed under normal conditions [49]. Also, high expression levels of fluorescent PB proteins in mammalian or yeast cells can further induce the aggregation of PBs and can alter the cellular stoichiometry of other PB components leading to aberrant structures [48–52]. Having established that the *dcap-1::gfp* fusion is functional (see text below and Fig. S5C), we showed by quantitative RT-PCR that the expression levels of *dcap-1* were not higher but underrepresented in BRF155 transgenic animals compared to the physiological levels of *dcap-1* in N2 (Fig. S1B). We interpreted that the transgene silencing effects resulted from the presence of vector backbone, as microinjection of the *dcap-1::gfp* reporter construct in linear, vector-free form [53] greatly improved its expression; a representative transgenic line (BRF261, Fig. S1A) had increased *dcap-1(mRNA)* levels (Fig. S1B) and more punctate pattern under normal conditions. Moreover, in a germline-deficient background (*glp-1*) the *dcap-1::gfp* transgene silencing was abolished (BRF219, Fig. S1A, B) confirming that such effects originate from the germline [54].

To validate that the DCAP-1::GFP-containing granules represent dynamic PBs in living animals we subjected all the above transgenic worms to heat-shock (+HS, Fig. 1A), a stress that generally enhances PB formation in other organisms, in a reversible manner [51,55]. Monitoring worms under confocal microscopy immediately after incubation at 35°C for 3 h, resulted in increased number and size of DCAP-1::GFP-containing cytoplasmic puncta in all transgenic lines (Fig. 1A, B). In contrast, there was no change in the expression pattern of worms carrying the transcriptional fusion transgene (BRF154, Fig. 1A), suggesting that the DCAP-1::GFP granules result from the accumulation of diffuse cytoplasmic protein rather than transcriptional or translational induction. This was verified by qRT-PCR for *dcap-1(mRNA)* levels and western blotting for DCAP-1::GFP protein levels, under normal or heat-shock conditions (Fig. 1C, D). The formation of DCAP-1::GFP aggregates in response to heat-shock was also reversible, with the majority of puncta dissolving within 3 h of recovery at 20°C (Fig. 2A, B). Similar enhanced formation of DCAP-1::GFP granules was observed in response to oxidative stress induced by sodium arsenite (BRF155, Fig. S2A). We further showed that RNAi-mediated knockdown of a core PB component, *cgh-1* (encoding a conserved DEAD-box RNA helicase) was sufficient to disrupt the formation of DCAP-1::GFP granules during heat (Fig. 2C, D) or oxidative stress (data not shown), in agreement with studies in yeast or human cells [8,56].

Another putative component of mammalian PBs is the mRNA-binding protein GW182 [57], which has two homologues in *C. elegans*, AIN-1 and AIN-2, both involved in miRNA-induced silencing function [58]. However, only AIN-1 interacts with components of the decapping complex and co-localizes with DCAP-1 in PB-like structures in *C. elegans*, under normal conditions [43]. Transgenic worms expressing the *ain-1::gfp* reporter have sporadic cytoplasmic puncta under normal conditions, which are further increased after heat-shock (Fig. S2B) without changes in *ain-1(mRNA)* levels (Fig. S2D), similar to *dcap-1::gfp*-expressing animals. We also tested whether the somatic PBs contain the translation initiation factor eIF4E, which localizes to PBs in unstressed mammalian cells but is found to both PBs and SGs upon exposure to stress [12,56]. In *C. elegans*, *ife-2* encodes the major somatic isoform of eIF4E [59]. Transgenic

animals expressing the transcriptional fusion $P_{ife-2}::gfp$ (BRF68, Fig. S2B) displayed a uniform fluorescence in all cells and did not show any aggregation under heat stress. In contrast, animals carrying the translational fusion *ife-2::gfp*, that drives expression of *ife-2* by its own promoter, had a similar diffused fluorescence pattern under normal growth conditions but rapidly formed numerous cytoplasmic granules in response to heat-shock (Fig. S2B) or oxidative stress (Fig. S2A). In co-localization experiments we monitored that several of heat-induced IFE-2::GFP granules also contain the PB marker DCAP-1::tagRFP (Fig. S2C). However, this partial co-localization could result from docking between SGs and PBs under stress, as it was shown to occur in yeast and mammalian cells [12,15,60]. The homogenous cytoplasmic fluorescence pattern of the strong *ife-2::gfp* reporter in unstressed animals argues that IFE-2 does not localize to constitutive PBs. Additionally, we found that *cgh-1(RNAi)*, which prevents stress-induced aggregation of PBs, had no significant effect on the formation and the relative number of IFE-2::GFP granules under heat-shock (Fig. S3A). This indicates that IFE-2 localizes mainly in SGs under stress conditions (see below Fig. S7A). We also found that deletion of *ife-2* did not affect the heat-induced accumulation of PBs (Fig. S3B). Thus, stress-induced PBs in the soma of intact adult worms resemble yeast and mammalian PBs in their pattern, formation and reversibility, albeit differ in localization of IFE-2/eIF4E factor.

Accumulation of somatic PBs with age

We next explored the pattern of PBs during the aging process, which is associated with marked alterations in protein synthesis and homeostasis [61,62]. Using the *dcap-1::gfp* reporter we demonstrated that the number and size of PBs gradually increased with age as, there was an increased punctate pattern in 5-day adults compared to 1-day adults, at 25°C (BRF155 and BRF261, Fig. 3A, B). This was not due to altered mRNA or protein levels with age, as verified by qRT-PCR for the endogenous *dcap-1(mRNA)* levels and western blotting for DCAP-1::GFP protein (Fig. 3C, D). The aggregation of PBs with age was also evident in worms grown at 20°C (Fig. S4A, B), albeit in later time-points compared to 25°C consistent with the slower aging process at lower temperatures. Similar age-dependent accumulation of PBs, was reported in worms expressing the DCAP-1::RFP fusion protein [63]. Cytoplasmic granules containing the AIN-1::GFP protein (MH2704, Fig. S2B) also accumulated in aged animals, without changes in *ain-1(mRNA)* levels (Fig. S2D), suggesting that aging can alter the profile of somatic PBs. Interestingly, we did not observe any age-dependent accumulation of the IFE-2::GFP fusion protein (BRF70, Fig. S2B), indicating that IFE-2/eIF4E factor is not part of the PB foci that accumulate with age. Taken together our observations led us to consider that the age-dependent formation of PBs in adult worms could not result from oxidative stress that takes place in aged cells and tissues [64]; this should have induced the assembly of SGs as well, something that we did not monitor by IFE-2 or other SG markers (see below).

We next considered the possibility that reduction of mRNA translation or degradation rates with age might cause or contribute to the accumulation of PBs in aged worms. Studies in yeast and mammalian cells have described the induction of PBs when either translation initiation or mRNA decay is inhibited, due to accumulation of non-translatable mRNAs into PBs [12,14]. Thus, we performed early in life RNAi of *ife-2* or other translation factors such as eIF2α (encoded by Y37E3.19), *ppp-1*/eIF2Bγ and *rsks-1*/S6K in N2 worms carrying the *dcap-1::gfp* transgene but we didn't observe any PB aggregation, in 1-day adults (Fig. 4A). We validated the efficiency of RNAi clones by qRT-PCR (Fig. 4C)

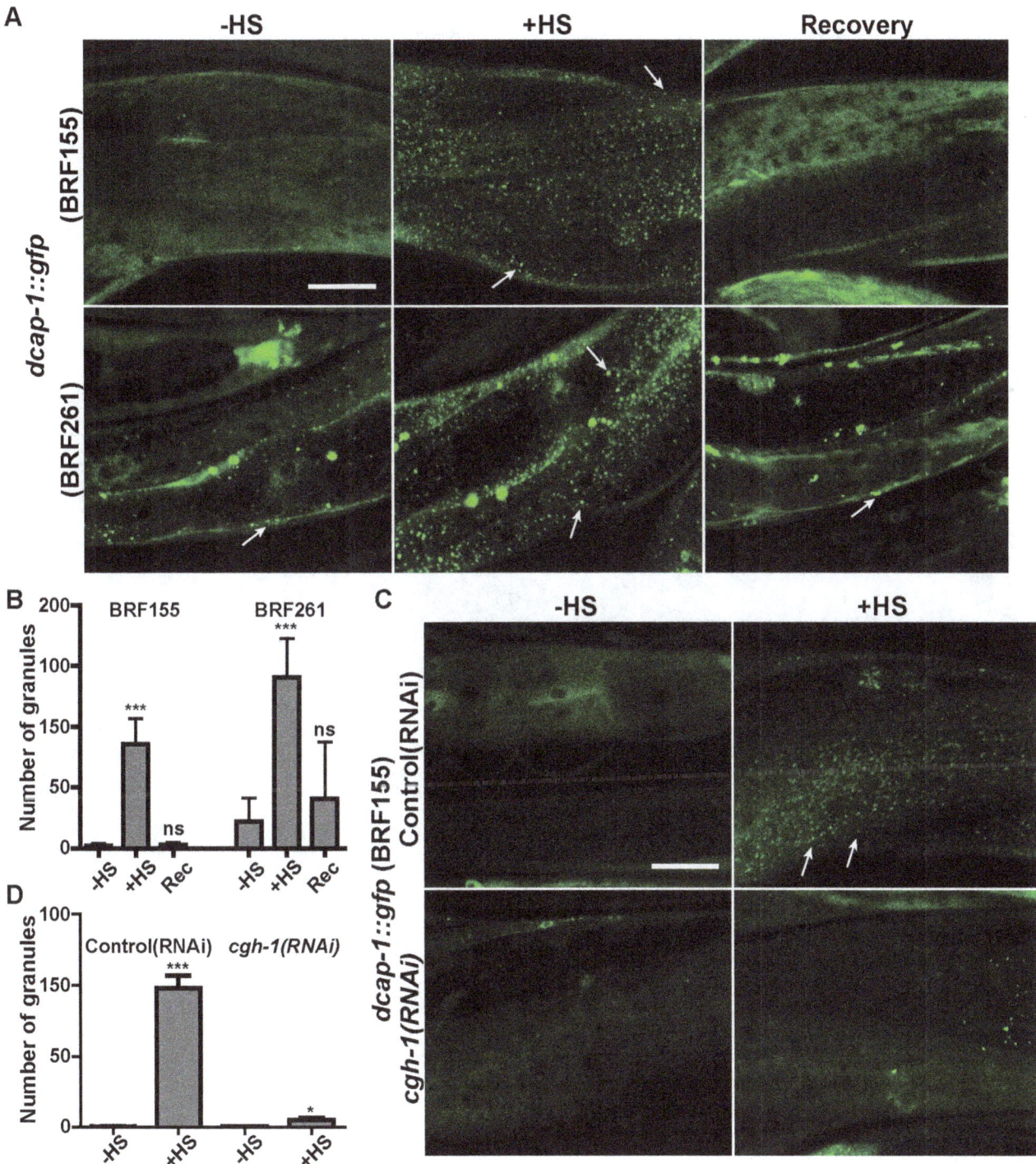

Figure 2. Accumulation of DCAP-1-containing granules under heat-shock is rapid, reversible and sensitive to *cgh-1(RNAi)*. (A) Representative confocal images of 1-day adult worms expressing *dcap-1::gfp* (BRF155 and BRF261 strains) normally grown at 25°C (-HS) or transiently subjected to heat-shock (+HS, 35°C for 3 h),following recovery for 3 h at 20°C (Recovery). (B) Quantification of data in (A). (C) Representative confocal images of 1-day adult worms expressing *dcap-1::gfp* (BRF155) fed from eggs with Control(RNAi) or *cgh-1(RNAi)* bacteria and grown at 25°C (-HS) or subjected to heat-shock at 35°C for 3 h (+HS). In all cases arrows point to DCAP-1::GFP granules in body wall muscles. (D). Quantification of data in (C). Values on Y axis show the number of granules per 2500 μm² per worm (mean±SD, see Materials and Methods and Table S4). Scale bar: 25 μm.

or by testing their effects in the treated adults and their progeny [45,65]. Consistent with *ife-2(RNAi)*, a strain carrying a null *ife-2(ok306)* mutation and exhibiting lower protein synthesis rates than N2 [45,65,66] did not form PBs in adult somatic tissues (*ife-2(ok306); dcap-1::gfp*, Fig. 4D). On the other hand, RNAi of *xrn-*

1, the major cytoplasmic 5′ to 3′ exonuclease, resulted in the accumulation of DCAP-1::GFP granules under the same conditions (Fig. 4A, B). Thus, the dynamics of PBs appear to be more prone to regulation by alterations in mRNA degradation rather than in translation rates. We suggest that an age-dependent

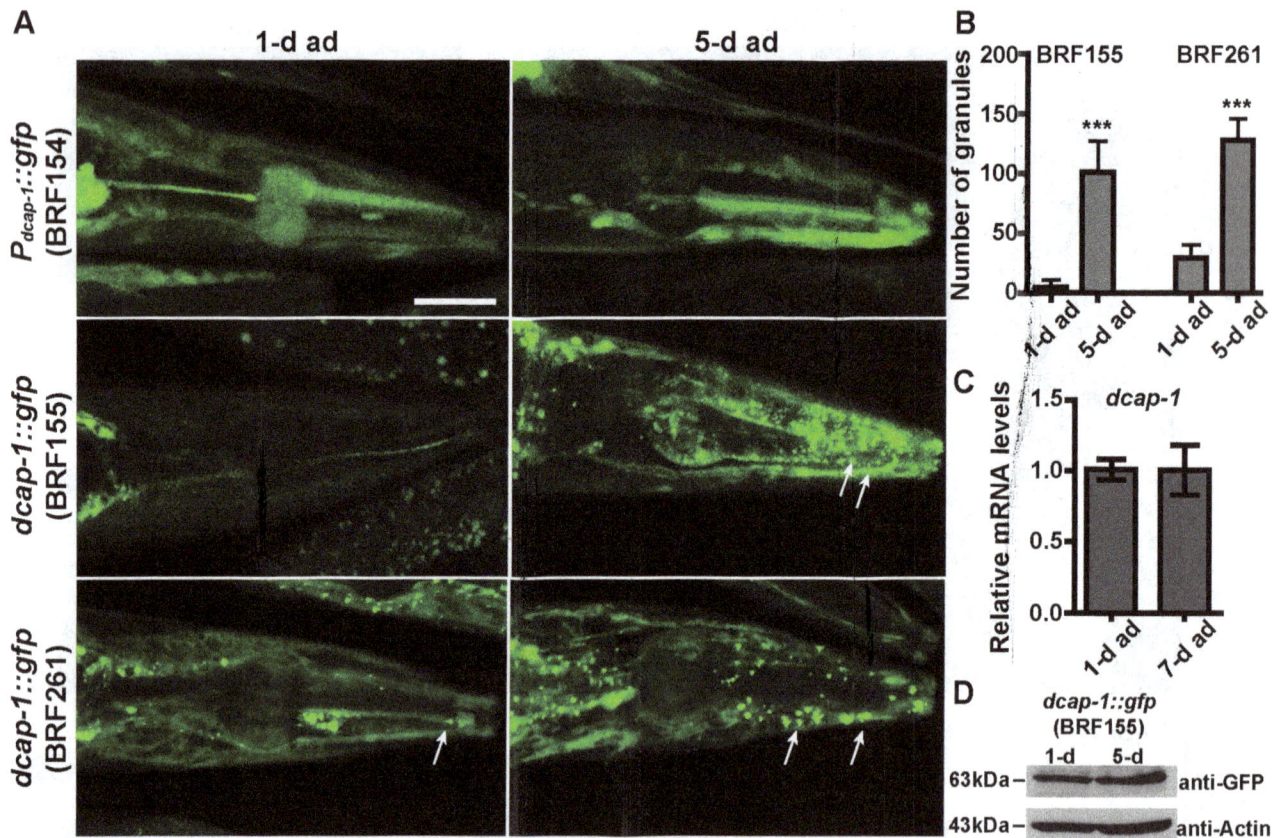

Figure 3. Accumulation of PBs with age. (A) Representative confocal images of 1-day and 5-day adults, grown at 25°C and expressing the transcriptional fusion P_{dcap-1}::gfp reporter (BRF154) or the translational fusions dcap-1::gfp (BRF155 and BRF261). Arrows point to age-induced DCAP-1::GFP granules. Scale bar: 25 μm. (B) Quantification of data in (A). Values on Y axis show the number of granules per head (mean±SD, see Materials and Methods and Table S4). (C) Endogenous dcap-1(mRNA) levels in 1-day and 7-day N2 adults grown at 25°C before (-HS) or after heat shock (+HS, 35°C for 3 h) were measured by quantitative RT-PCR and normalized to endogenous ama-1(mRNA) levels. Error bars represent the standard deviation of the means of six independent experiments (p-value = 0.9697, calculated by Student's t-test). (D) Western blot analysis of DCAP-1::GFP protein expression in 1-day and 5-day adult BRF155 transgenic animals, grown at 25°C, using anti-GFP or anti-β-Actin (as a loading control) antibodies. Band intensity of DCAP-1::GFP normalized to actin gives a value of 0.95, showing similar protein levels in the two conditions.

decline in mRNA decay process could be a causal factor in the onset of PBs aggregation in aged tissues but further experimental validation is needed. Interestingly, xrn-1(RNAi) did not cause aggregation of IFE-2::GFP, which remained cytoplasmic in worms that co-express DCAP-1::tagRFP and form distinct granules (Fig. 4E), showing that IFE-2 does not relocalize to PBs under impaired mRNA degradation conditions.

Disruption of PB-related genes affects development, lifespan and stress response of worms

We further set out to evaluate the importance of PB-related genes in the physiology of worms and whether disruption of their function could protect from or contribute to mortality. Worms bearing mutations in the decapping genes dcap-1(tm3163) and dcap-2(ok2023) showed a profound delay in development (Fig. S5A) and reduced brood size (Fig. S5B) compared to N2, at both growth temperatures (20 and 25°C). They also exhibited high levels (~50%) of matricidal death ("bagging" due to internal hatching of eggs) and uncoordinated locomotion. Both mutants had a significantly shorter lifespan than N2 (Fig. 5A and Table S5) and were more sensitive, as adults, in a range of environmental stressors (Fig. 5B). When we expressed the dcap-1::gfp reporter in dcap-1(tm3163) we rescued the growth and lifespan defects and

restored resistance to stresses (Table S5 and Fig. S5C). We noticed that the mutant phenotypes of dcap-1 were more severe at 25°C compared to 20°C due to a possible residual function of the mutated DCAP-1 protein; dcap-1(tm3163) harbors a 334-bp deletion that results in a truncated mRNA, generating a new amino-acid sequence at position 263 (E263) of the C'-terminus of wild-type protein (Fig. S5D). Since the C' end of DCAP-1 contains a conserved trimerization domain, which in metazoan is required for incorporation of Dcp1 into PBs and mRNA decapping in vivo [67], it is possible that the allele tm3163 doesn't completely remove gene function at 20°C, but at higher temperature (25°C) its phenotypes are greatly enhanced. Therefore, the lifespan of dcap-1 worms is ~25% shorter than N2 at 20°C and ~45% shorter than N2 at 25°C, resembling the lifespan of the null dcap-2 mutants (Table S5). Also, each gene disruption by RNAi from the time of egg hatching resulted in similarly decreased lifespan of N2 (Table S6).

We considered that the mortality of decapping mutants could emerge from their early developmental defects. To assess the effects of PB components on lifespan beyond development we performed RNAi post-developmentally, by initiating RNAi during late larval stages of N2. Although the effect of single decapping gene depletion was small we observed significantly decreased lifespan when dcap-1 and dcap-2 were simultaneously inactivated

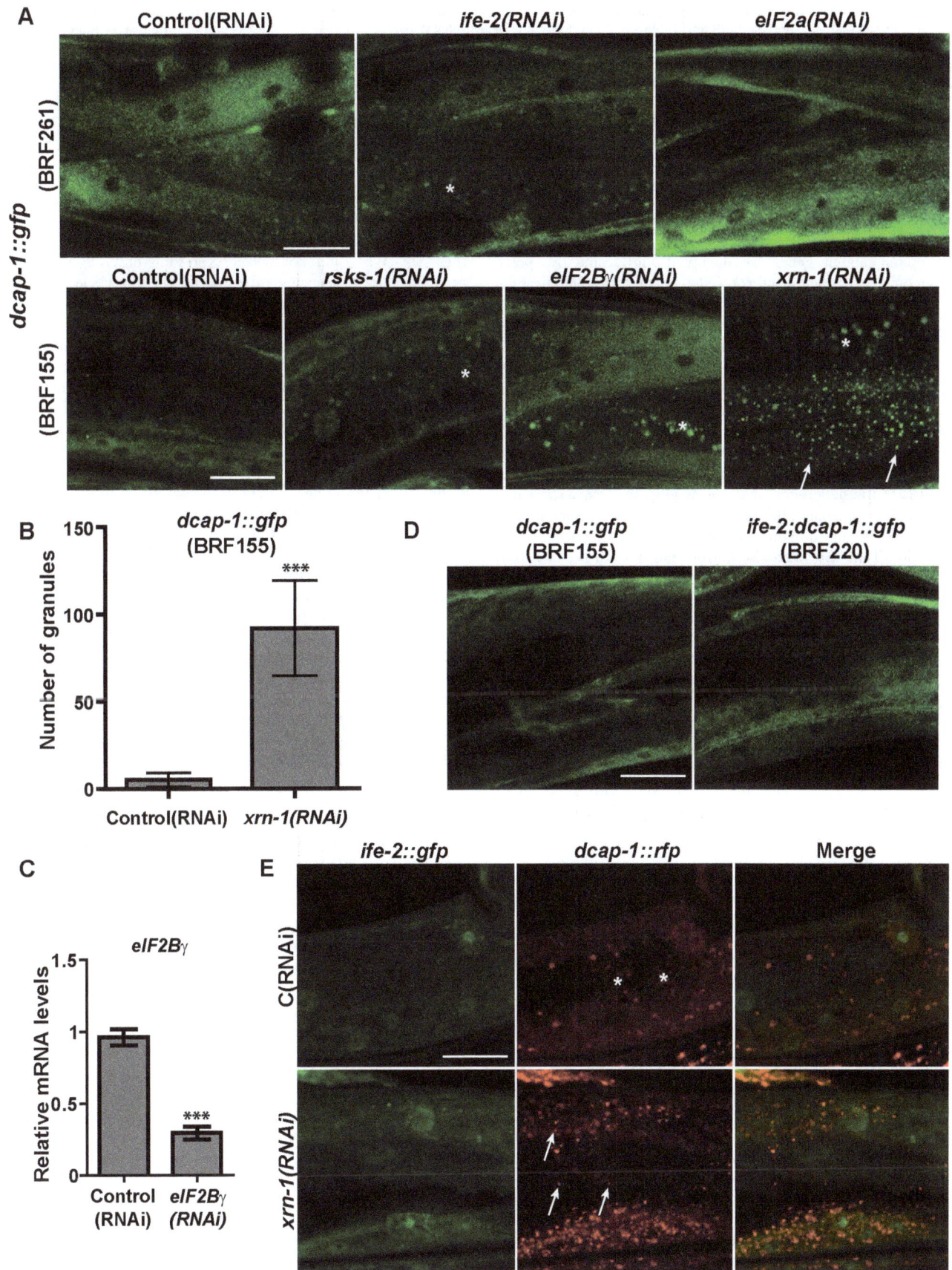

Figure 4. Disruption of mRNA degradation, but not of translation factors, triggers PBs formation that lack *ife-2*/eIF4E. (A) Representative confocal images of 1-day adults expressing *dcap-1::gfp* in N2 (BRF155 and BRF261) fed with *ife-2(RNAi)*, *eIF2a(RNAi)*, *rsks-1(RNAi)*,

elF2Bγ(RNAi) or *xrn-1(RNAi)* bacteria. (B) Quantification of data in (A). Values on Y axis show the number of granules per 2500 μm² per worm (mean±SD, see Materials and Methods and Table S4). (C) Endogenous *elF2Bγ(mRNA)* levels in 1-day adults BRF155 worms grown at 25°C and fed with *elF2Bγ(RNAi)* from eggs, were measured by quantitative RT-PCR and normalized to *ama-1(mRNA)* levels. Error bars represent the standard deviation of the means of three independent experiments (***p-value<0.0001, in unpaired *t*-test). (D) Representative confocal images of 1-day adults expressing *dcap-1::gfp* in N2 (BRF155) or *ife-2(ok306)* (BRF220) background. (E) Representative confocal images of 1-day adults co-expressing *dcap-1::rfp* and *ife-2::gfp* (BRF313), and treated with *xrn-1(RNAi)* to monitor the lack of co-localization of the two proteins. In all cases, targeted RNAi was initiated from eggs at 25°C in parallel with Control(RNAi) to assess the localization of DCAP-1::GFP into PBs. Asterisks show the intestinal autofluorescence. Scale bar: 25 μm.

(dcap-1/2(RNAi) compared to Control(RNAi)-treated worms (Fig. 5C and Table S6). This is consistent with the model of a more efficient disruption of decapping activity when both decapping subunit genes are disrupted [36]. Similarly, post-developmental depletion of *xrn-1*, *patr-1* or *cgh-1* significantly shortened lifespan of N2 (Fig. 5C and Table S6). Given that *xrn-1(RNAi)* induced the accumulation of PBs we postulate that this accumulation is not sufficient to confer longevity, probably due to impaired degradation of resident mRNAs. Furthermore, disruption of *dcap-1* or *dcap-2* had an impact in the lifespan of several long-lived mutants. Post-developmental RNAi-treatment of worms lacking the insulin/IGF-1-like receptor *daf-2* gene [68,69] resulted in shortening of their extreme long life (Table S6). Similarly, RNAi of decapping genes during adulthood in *ife-2(ok306)* or *rsks-1(ok1255)* translation-defective worms [45], in dietary-restricted *eat-2(ad465)* animals [70] as well as in somatic tissues of the germline-deficient mutant *glp-1(e2141)* [71], reduced their longevity (Table S6). Likewise, introducing the *dcap-1(tm3163)* or *dcap-2(ok2023)* alleles in the genome of these long-lived mutants resulted in significantly reduced lifespan and stress resistance compared to controls (Fig. 5D, E for *daf-2* and Table S5), suggesting that the decapping mutant phenotypes were not rescued by any longevity pathway. On the other hand, overexpression of *dcap-1* in somatic tissues of *glp-1* mutants had a positive effect on their ability to cope with stress during adulthood, as it significantly enhanced survival under heat and oxidative stress (Fig. 5F). Taken together, our data indicate that genes directly related to PBs modulate adult lifespan, separately from their developmental role and alterations in functions of PB components increase the mortality rate and the sensitivity of adult worms to various stressors.

Accumulation of SGs in somatic cells upon stress but not with age

PBs are functionally related to SGs that have a crucial role in cellular response to environmental stress. SG components in human cells include TIA-1 (T-cell intracellular antigen-1) and TIAR (TIA-1-related) RNA-binding proteins that regulate mRNA stability, in addition to other cellular functions [72]. In *C. elegans* genome (http://www.wormbase.org), three genes, named *tiar-1* (C18A3.5), *tiar-2* (Y46G5A.13) and *tiar-3* (C07A4.1) encode homologues of human TIA-1/TIAR family proteins. The predicted proteins TIAR-1 and TIAR-2 contain three RNA-recognition motifs (RRMs) at their N′-termini and prion-related domains at their C′-termini, whereas TIAR-3 has only two RRMs and no prion-related domain (Fig. 6A). The latter domain in human TIA-1/TIAR proteins is essential for the assembly of SGs under stress [73]. We generated animals expressing transcriptional or translational *gfp*-reporters of *tiar* genes and through observation of several transgenic lines we concluded that: (a) the promoter of both *tiar-1* and *tiar-2* is expressed in most, if not all, cells (P*tiar-1*::gfp and P*tiar-2*::gfp in Fig. S6); (b) the *gfp::tiar-1* and *gfp::tiar-2* N′-terminal fusion genes, driven by their own promoters, display a diffused fluorescent signal dispersed in both nucleus and cytoplasm

(Fig. S6); (c) *tiar-3::gfp*, a C′-terminal fusion driven by its own promoter, is exclusively nuclear, strongly expressed mainly in spermatheca, ventral nerve cord and some head and tail neurons (Fig. S6). We could not detect expression of GFP fusions in germline due to silencing of transgenes in this tissue [74] but previous studies have shown the expression of TIAR-1 [33,75] and TIAR-2 [32] in germ granules.

We next examined the pattern of the GFP fusion transgenes in adult worms subjected to heat or oxidative stress, conditions that elicit SG assembly in mammalian and yeast cells. In contrast to *tiar-3::gfp* protein, which constantly remained nuclear after each stress (data not shown), we observed the formation of numerous cytoplasmic aggregates containing *gfp::tiar-1* in response to heat-shock (HS) and sodium arsenite (SA) (Fig. 6B). In the case of *tiar-2*, the aggregates induced by heat-shock were more visible in the excretory cell of worms, but were also vigorously formed in muscles when we used a strong, muscle-specific promoter (P*myo-3*::gfp::tiar-2) to enhance the fluorescent signal and avoid the intestinal autofluorescence at the confocal settings of *gfp::tiar-2* (Fig. 6B). Under oxidative stress we could monitor granules only in muscle-expressed *gfp::tiar-2* (Fig. 6B). The relative mRNA levels of each transgenic line are shown in figure 6C. The heat-shock induced granules, monitored by *gfp::tiar-1* or *gfp::tiar-2*, were not the result of increased transcription of *tiar-1* or *tiar-2*, respectively (Fig. 6D) and were not formed in worms expressing the transcriptional *gfp* fusions (data not shown). In accordance with the dynamic nature of SGs, the above heat-induced granules were formed rapidly (within 45 min) and could be dissociated within 2 h of recovery at 20°C (Fig. 7). Deletion of *tiar-1* did not affect the aggregation of *tiar-2*-containing granules and *vice versa* (data not shown). Moreover, under heat-shock conditions several RFP::TIAR-1 granules contain the IFE-2::GFP and partially colocalise with DCAP-1::GFP granules, while GFP::TIAR-2 also co-localizes partially with DCAP-1::RFP (Fig. S7A–C), again consistent with the spatial overlapping of PBs and SGs [12,15,60]. As opposed to PBs, RNAi-mediated depletion of *cgh-1* did not affect the formation of SGs upon heat-shock and *xrn-1(RNAi)* did not lead to accumulation of SGs under normal conditions of growth (data not shown), similar to our observations with IFE-2::GFP marker. Finally, our data using IFE-2 reporter suggest that aging cannot induce the formation of SGs, in sharp contrast to PBs. Indeed, we did not observe an age-dependent accumulation of TIAR proteins in SGs, comparing 6-day to 1-day adult worms at 25°C (Fig. 6B), which show similar expression levels of *tiar-1* or *tiar-2* (data not shown). These data strongly suggest that neither the mild stress accompanying aging nor the age-related alterations in mRNA degradation is sufficient to induce SG formation.

In the process of monitoring the fluorescent *tiar* granules, we noticed a tissue-specific response in their formation under heat-shock, with *gfp::tiar-1* aggregating mostly in head neurons, muscles, intestine, vulval and hypodermal cells, whereas *gfp::tiar-2* granules were visible mainly in excretory cell and less in intestine, muscle and hypodermal cells. Furthermore, by generating transgenic animals co-expressing *rfp::tiar-1* and *gfp::tiar-2* we observed the same tissue-specific aggregates in

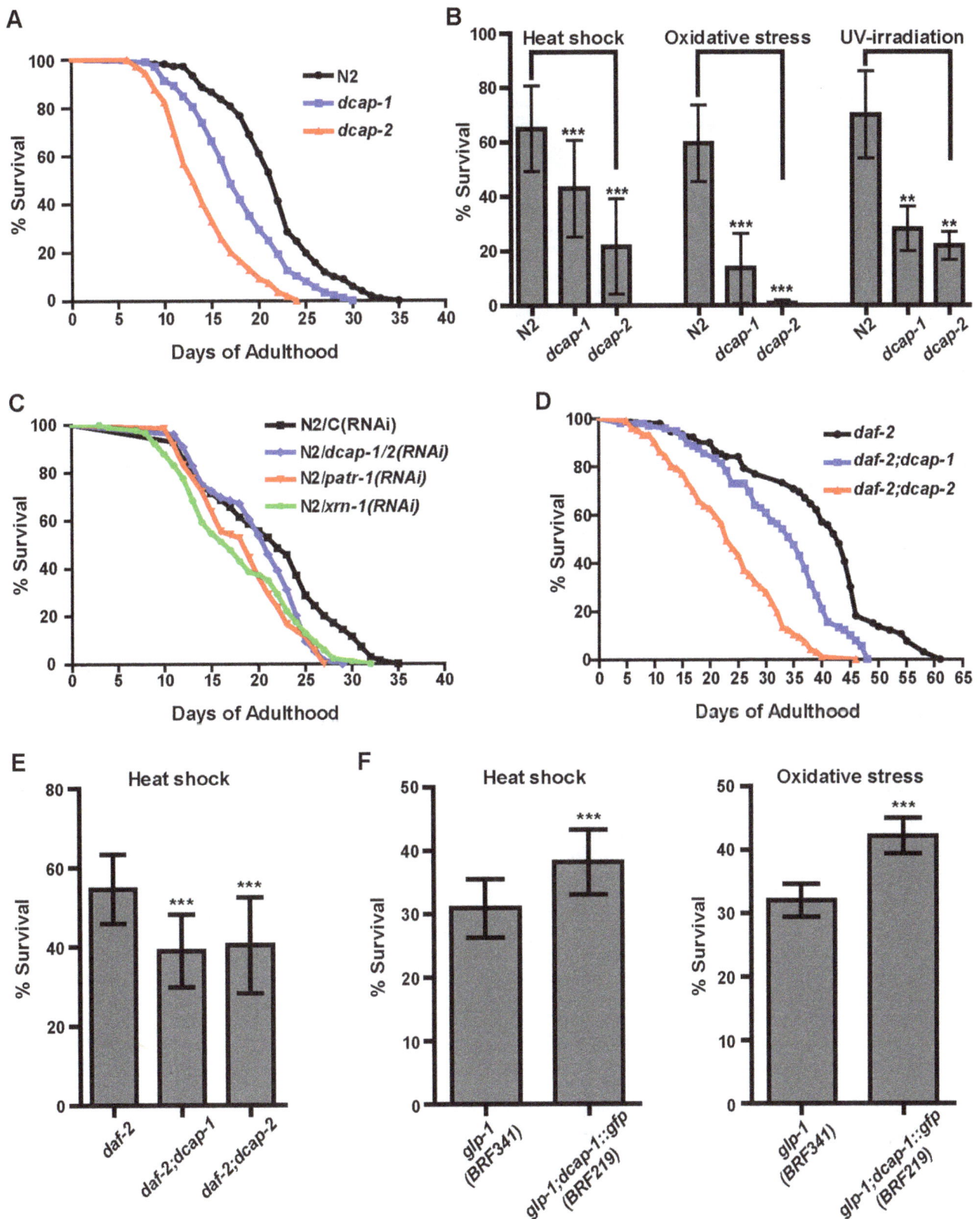

Figure 5. PB components are important for normal lifespan and stress response. (A) Survival curves of N2, *dcap-1* and *dcap-2* mutants at 20°C. (B) Survival of adults *dcap-1* and *dcap-2* mutants in heat-shock (1-day adults at 35°C for 6 h), oxidative stress (1-day adults in 5 mM sodium arsenite for 48 h) and UV-irradiation (5-day adults at 0.2 J/cm²), compared to N2. (C) Survival curves of N2 worms treated post-developmentally with *dcap-1/-2(RNAi)*, *patr-1(RNAi)* or *xrn-1(RNAi)* at 20°C. (D) Lifespan of *daf-2*, *daf-2; dcap-1* and *daf-2; dcap-2* mutants at 20°C and (E) survival of 1-day adults after heat-shock (35°C for 8 h). (F) Survival of 1-day adult *glp-1* germline-deficient mutants that overexpress *dcap-1::gfp* after heat-shock (at 35°C for 6 h) or oxidative stress (5 mM sodium arsenite for 48 h). In lifespan assays the p-values were determined using the log-rank test (see Tables

S5 and S6) and in stress assays the error bars show the SD in unpaired *t*-tests (see Materials and Methods). ** indicates very significant (p-value 0.001 to 0.01); *** indicates extremely significant (p<0.001).

response to heat-shock, with partial co-localization of the two fusion proteins only in muscles, hypodermal and intestinal cells (Fig. S7D). This implies some tissue-specific functions of each TIAR protein in cellular stress response. In support of this, only TIAR-2-containing granules were formed under osmotic stress (Fig. S8A). In addition, we observed a different requirement on the activity of GCN-2 kinase to initiate the formation of each type of SGs in response to oxidative stress. Having shown the conserved function of GCN-2 as an eIF2α kinase in worms [76] we tested whether SGs were formed in the *gcn-2(ok871)* loss-of-function mutant. While both TIAR-1 and TIAR-2 foci could assemble upon heat-shock in *gcn-2(ok871)*, only TIAR-2 aggregates were formed after treatment with sodium arsenite (Fig. S8B). Thus, TIAR-1 accumulation in response to oxidative stress requires phosphorylation of eIF2α by GCN-2 (and not by the other eIF2α kinase in worms, PEK-1, data not shown). Overall, we showed that the three TIAR proteins in *C. elegans* differ in cellular localization and only TIAR-1 and TIAR-2 participate in cytoplasmic SGs that have both overlapping and distinct patterns, with different assembly requirements depending on the stress stimulus.

Diverse function of SG-related genes in development, lifespan and stress response

Because of the differences in cellular responses of TIAR proteins we tested the impact of each *tiar* gene deletion on development, lifespan and stress survival. We used the mutant strains *tiar-1(tm361)*, *tiar-2(tm2923)* and *tiar-3(ok144)* that harbor internal deletions in the coding region of *tiar* genes and accessed their phenotypes. Mutant *tiar-1* worms exhibited delayed developmental rate and reduced brood size compared to N2, while *tiar-2* and *tiar-3* mutants had milder (at 25°C) or no defect (at 20°C) on both development and fecundity (Fig. 8A). Also, *tiar-1* mutants displayed early behavioral decline, with premature onset of uncoordinated locomotion at day 6 of adulthood (Movies S1 and S2), which was not observed in *tiar-2* and *tiar-3* mutants. Animals lacking both *tiar-1* and *tiar-2* genes (*tiar-1; tiar-2*, Fig. 8A) displayed even slower development than the single mutants, with high matricide deaths at 25°C, suggesting additive effects.

We also assessed differences in the lifespan of *tiar* mutants; deletion of each *tiar* gene shortened normal lifespan, but the effect was more dramatic for *tiar-1* allele (Table S5). Since both the developmental (data not shown) and lifespan (Table S5) defects of *tiar-1* were rescued by the functional GFP::TIAR-1 fusion, these defects are linked to *tm361* allele and not to a secondary mutation. For *tiar-2* allele, there was a temperature-dependent effect on lifespan, as in development, with significantly shorter lifespan at 25°C (Table S5). The double mutant *tiar-1; tiar-2* had an even shorter lifespan than each single mutant at any temperature (Fig. 8B and Table S5). We obtained similar results on N2 lifespan by RNAi of *tiar* genes post-developmentally (Fig. 8B and Table S6). This suggests that the lifespan reduction was not solely due to the developmental abnormalities of each mutant. In addition, by introducing each mutation to the germline-deficient *glp-1(e2141)* worms we observed that loss of *tiar-1* and, to a greater extent, *tiar-2* from somatic cells significantly reduced the long lifespan of *glp-1* worms (Table S5). Surprisingly, deletion of *tiar-3* significantly improved the mean lifespan of *glp-1* worms, contrary to N2 (Table S5). Thus, the impact of *tiar* deletions on lifespan can be influenced by temperature or germline signaling.

Finally, we measured diverse responses of *tiar* mutants against various insults, which are, in some cases, related to their expression pattern. *tiar-1* deletion reduced the survival of worms in sodium arsenite-induced oxidative stress, or after UV-irradiation but increased survival under heat or osmotic stress (Fig. 8C). The impaired oxidative stress response of *tiar-1(tm361)* was rescued by the *gfp::tiar-1* transgene (Fig. 8D). In the imaging analysis we observed TIAR-1-positive granules upon heat and oxidative stress but not under osmotic stress (Fig. 6B and Fig. S8A). We assume that deletion of TIAR-1 affects only specific functions of SGs related to oxidative stress response and not to heat-shock, although it localizes to SGs formed under both conditions. Alternatively, since TIAR-1 is required to induce germ cell apoptosis under various stresses [75], the increased resistance of *tiar-1* mutants to heat-shock might be related to the systemic stress response that is observed in several DNA-damage checkpoint mutants affecting programmed cell death in germline [77,78]. In sharp contrast to *tiar-1* worms, *tiar-2* or *tiar-3* mutants had increased mortality under heat and osmotic stress but survived as N2 under oxidative or UV-irradiation stress (Fig. 8C). Interestingly, the expression of *tiar-2* driven by its own promoter was strong in the excretory system of worms which has a role in osmoregulation, analogous to the 'kidney' of higher animals [79]. TIAR-2 granules formed in oxidative stress, (BRF255 in Fig. 6B), were hardly visible while TIAR-1 granules under the same conditions, were readily formed in all tissues (Fig. 6B), suggesting a more crucial role of TIAR-1 under this stress. The double mutant *tiar-1; tiar-2* survived similar to the single *tiar-1*, at least under heat and oxidative stress (Fig. 8C). While further studies are required to address these issues, our data show that the three *tiar* genes exhibit diverse developmental, lifespan and stress response outputs at the organismal level that are possibly reflected by their specific functions, which are differentially regulated under various cellular contexts and stressors.

Discussion

Post-transcriptional regulation of eukaryotic gene expression is attained through several mRNA-specific control mechanisms, such as processing, export, turnover and translation. PBs and SGs are cytoplasmic aggregates of RNP complexes that mediate the subcellular localization, translation or decay of bulk mRNA in eukaryotic cells [6,8,11]. PBs consist of mRNAs associated with translation repressors or the 5' to 3' mRNA decay machinery, whereas SGs contain a pool of mRNAs stalled in the process of translation initiation in response to stress. Both are dynamic structures that transiently interact and share many components [12]. Hence, PBs and SGs are considered as hubs of mRNP trafficking that determine the fate of mRNAs but also as regulators of stress-responsive signaling pathways that determine the survival of cells under stress [80]. Although the function of these granules provides another level for fine-tuning gene expression to maintain cellular and tissue homeostasis, studies addressing their impact or pattern during stress or aging, at the level of intact organism, are limited. In *C. elegans*, related RNA-protein particles function in maternal mRNA metabolism and have been well-studied in germline development and embryogenesis [30–32,34,36].

Here, we documented the formation of PBs and SGs in somatic cells of *C. elegans* by using fluorescent reporters of known components of these granules. These include the decapping

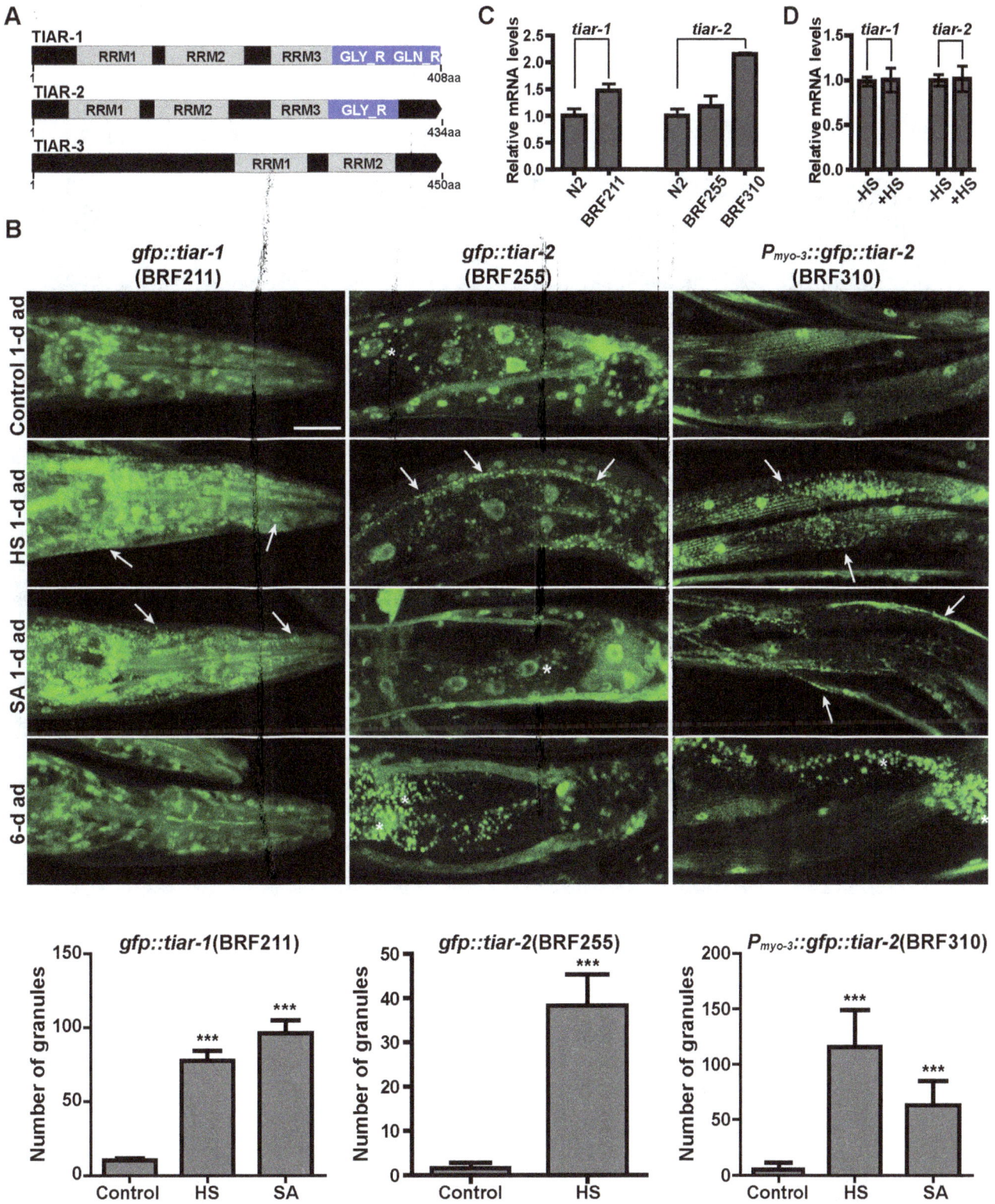

Figure 6. Aggregation of TIAR-1 and TIAR-2 proteins in response to stress but not to age. (A) The domain structure of TIAR proteins, designed using the Prosite MyDomains (http://prosite.expasy.org/mydomains/). The graphic was created using the Exon-Intron Graphic Maker (http://wormweb.org/exonintron). RRM: RNA-recognition motif, GLY_R: Glycine rich domain, GLN_R: Glutamine rich domain. (B) Representative confocal images of 1-day adults of the indicated transgenic strains grown at 25°C (Control 1-d ad), upon heat-shock (HS, 35°C for 3 h), oxidative stress (SA, 15 mM sodium arsenite for 3 h) or compared to 6-day adults under normal conditions. Arrows point to formed granules and quantification of the granule number is shown in the lower panel for each strain. Values on Y axis show the number of granules per head (BRF211), per 100 µM length of excretory cell (BRF255) or per 2500 µm^2 per worm (mean±SD, see Materials and Methods and Table S4). Asterisks show the intestinal auto-fluorescence. Scale bar: 25 µm. (C–D) Expression levels of *tiar-1* or *tiar-2* genes in (C) the indicated transgenic strains expressing *gfp::tiar-1* and *gfp::tiar-2* by their own promoter (BRF211 and BRF255, respectively) or the muscle specific P$_{myo-3}$::*gfp::tiar-2* (BRF310) and (D) N2 worms after heat

shock (35°C for 3 h). Quantification of each mRNA level, relative to *ama-1(mRNA)* levels, in 1-day adults and the mean $\pm$ SD of biological triplicates are shown (p>0.05 in Student's *t*-tests).

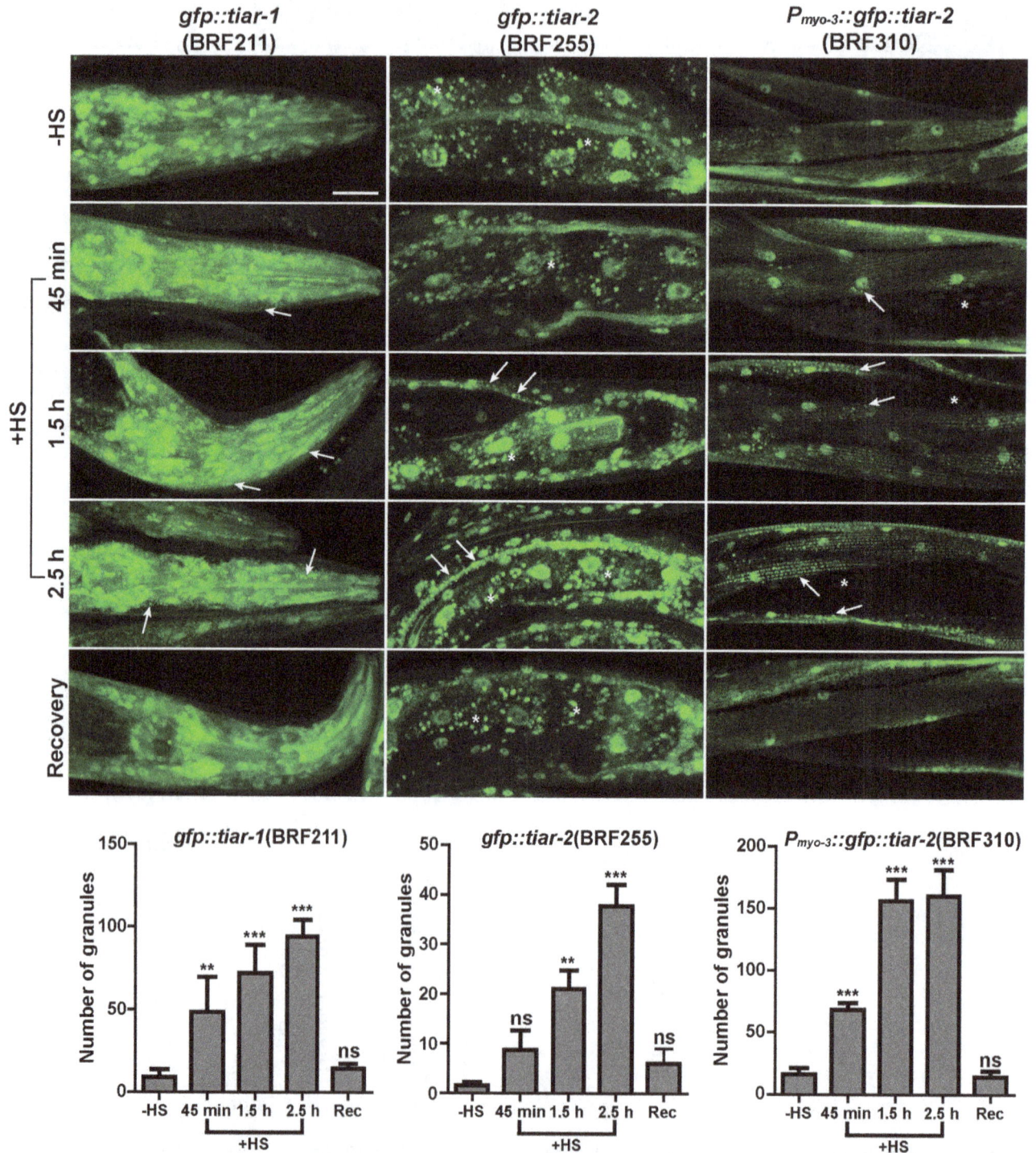

Figure 7. Rapid formation and dissociation of heat-induced SGs. Representative confocal images of 1-day adults expressing *gfp::tiar-1* and *gfp::tiar-2* by their own promoter (BRF211 and BRF255, respectively) or the muscle specific P_{myo-3}::*gfp::tiar-2* (BRF310) under normal growth conditions (-HS) and upon heat-shock (+HS, 35°C for 45 min to 2.5 h), following recovery at 20°C for 2 h. Arrows point to formed granules and quantification of the granule number is shown in the lower panel for each strain. Values on Y axis show the number of granules per head (BRF211), per 100 µM length of excretory cell (BRF255) or per 2500 µm² per worm (mean±SD, see Materials and Methods and Table S4). Asterisks show the intestinal auto-fluorescence. Scale bar: 25 µm.

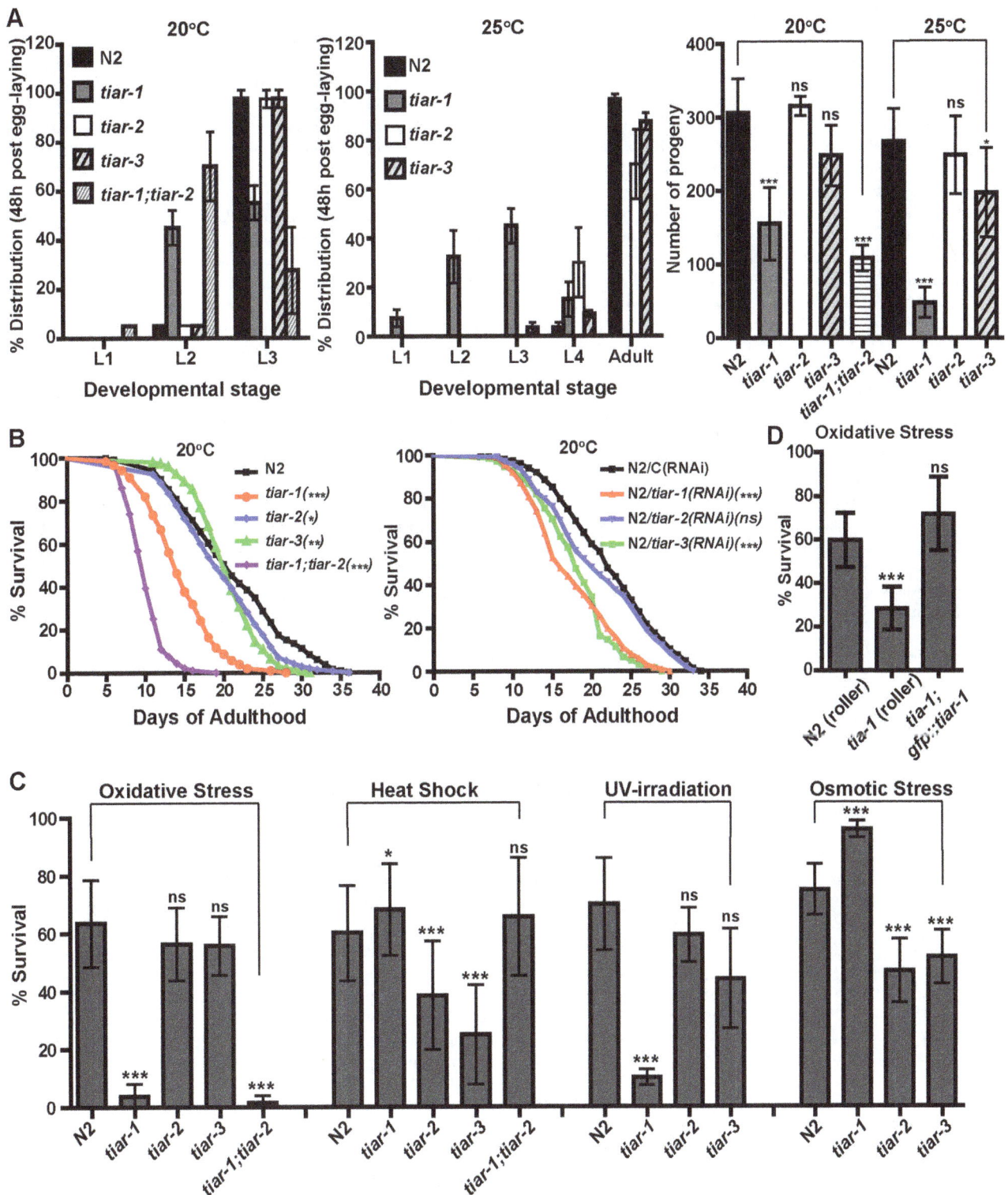

Figure 8. Effects of *tiar* genes deletion on development, fertility, lifespan and stress survival. (A) Developmental rate and brood size of *tiar-1*, *tiar-2*, *tiar-3* and *tiar-1;tiar-2* mutant worms, compared to N2. (B) Survival curves of the above mutants or of N2 worms treated post-developmentally with *tiar-1(RNAi)*, *tiar-2(RNAi)* or *tiar-3(RNAi)*. (C) Survival of adults of the above mutants in oxidative stress (1-day adults at 5 mM sodium arsenite for 48 h), heat-shock (1-day adults at 35°C for 6 h), UV-irradiation (5-day adults at 0.2 J/cm^2) or osmotic stress (1-day adults at 400 mM NaCl for 24 h, compared to N2. (D) Survival of *tiar-1* mutants expressing the *gfp::tiar-1* rescuing transgene in oxidative stress (1-day adults at 5 mM sodium arsenite for 24 h) compared to N2 or *tiar-1* animals carrying only the *rol-6(su1006)* roller marker. In lifespan assays the p-values were determined using the log-rank test (see Tables S5 and S6) and in stress assays the error bars show the SD in unpaired *t*-tests (see Materials and Methods). ns indicates not significant (p>0.05); * indicates significant (p-value 0.01 to 0.05); ** indicates very significant (p-value 0.001 to 0.01); *** indicates extremely significant (p<0.001).

complex subunit DCAP-1/DCP1 and the miRNA-binding protein AIN-1/GW182, as PB markers and the homologues of human TIA-1/TIAR RNA-binding proteins, TIAR-1-to-3, as SG markers. Of the three worm TIAR proteins, only TIAR-1 and TIAR-2 contain a QN-rich C'-terminal domain and are distributed in both nucleus and cytoplasm. As shown by live imaging in adult transgenic worms there were no SGs and few PBs under normal conditions but there was robust aggregation of PB and SG reporters in the cytoplasm of most cells/tissues, in response to heat or oxidative stress. The dynamic nature of these aggregates is reminiscent of yeast and mammalian PBs and SGs; their formation is rapid and reversible, whereas the partial co-localization between the heat-induced DCAP-1 and TIAR-1 or TIAR-2 foci indicates that PBs and SGs could overlap or dock to each other in worms [12,15,49,51,60]. Additionally, we showed that the translation initiation factor IFE-2/eIF4E is not a component of constitutive PBs and localizes mainly to SGs in response to stress. The heat-induced accumulation of PBs, as monitored by DCAP-1::GFP, was affected by alterations in genes influencing PB assembly in other organisms; it was induced by RNAi of the 5' to 3' exonuclease xrn-1 and prevented by depletion of the translation regulator cgh-1/RCK (Dhh1 in yeast). In C. elegans gonad, loss of cgh-1 induces the formation of aberrant sheet-like structures through the relocalization of various PB components [34,35,81,82]. Interestingly, we did not observe such structures in somatic tissues of cgh-1(RNAi)-treated animals; instead PBs aggregation was prevented even upon stress. Possible reasons for this discrepancy could be related to different tissues or markers that we used since it has been suggested that the function of CGH-1 helicase depends on cellular and developmental context [34] and cgh-1 loss induces relocalization of a subset of RNP factors (as CAR-1) into sheet-like structures but dissociates PATR-1 or PGL-1 [82,83]. Furthermore, the fact that knockdown of cgh-1 prevents PB accumulation without affecting SG formation under stress, suggests that SGs can form independently of PBs. This supports evidence in yeast and mammalian cells, indicating that assembly of PBs and SGs is regulated by different signaling pathways [12,27,60,84].

We also investigated the impact of aging on the pattern of somatic PBs and SGs in C. elegans. We demonstrated that only PBs accumulated with age, showing an increase in their number and size. In sharp contrast, we did not observe SG formation in aged animals, despite that in aged oocytes both PB and SG components co-localize to large RNP structures [33]. Since IFE-2 and TIAR-1/-2 were localized in cytoplasmic aggregates induced by sodium arsenite, we reasoned that the accumulation of PBs with age does not result from oxidative stress in aged tissues. The accumulation of non-translatable mRNAs into PBs, due to reduction of either translation or degradation rates with age, provides another possible explanation for the increased PB aggregation during aging. However, knockdown of several translation factors or deletion of ife-2 did not induce the formation of DCAP-1::GFP granules in adult worms, in contrast to the increased number and size of PBs caused by xrn-1(RNAi). Similarly to the age-induced aggregates, the xrn-1(RNAi)-induced PBs did not contain SG markers (IFE-2 or TIAR-1/-2). Thus, we presume that the increased formation of PBs in aged tissues could be an outcome of age-related alterations in decay rates rather than translation but further experimentation is required to support this hypothesis. Alternatively, the accumulation of PBs in older ages could be just a consequence of the aging process, which is associated with a large increase in protein insolubility and aggregation of diverse proteins [85,86]. Among these are translation factors, ribosomal subunits and proteins involved in

mitochondrial respiration, whose inhibition has been linked to longevity in many organisms [1]. Thus, we examined the effects of direct alterations in PB components on lifespan. In contrast to the above factors, post-developmental RNAi of genes related to PBs significantly shortened lifespan of N2 and several long-lived mutants. Moreover, increased expression of dcap-1 in somatic cells of young adults improved their resistance in both heat and oxidative stress.

The maintenance of protein homeostasis, through regulation of both translation and protein degradation is considered a common longevity assurance mechanism [87]. Our findings support that mRNA metabolism factors contribute to such mechanisms and impaired function of PBs can limit lifespan. Interestingly, studies in S. cerevisiae have demonstrated that PBs are required for the long-term survival of stationary phase cells [88]. PBs functions are also vital for normal development and stress management. Mutant worms for the decapping genes exhibited severe defects in development, growth, fecundity, movement and impaired stress responses to a variety of environmental insults. A role in the protection of the nascent germline from stress has been described for P granules in worms [89]. Noteworthy, we noticed that dcap-1 is highly expressed in the three pairs of coelomocytes, the scavenger cells that are considered as a primitive immune system in C. elegans. Moreover, we revealed diverse roles of worm TIAR proteins in the cellular stress response, as they displayed specific expression pattern and assembly requirements that can differ according to the stress stimulus. Consistent with such diverse roles, we measured differences in development, growth and lifespan of worms carrying mutations in any of the three tiar genes. The effects of tiar alleles on lifespan were influenced by temperature or germline signaling; loss of each tiar gene shortened normal lifespan, mainly at 25°C, whereas in the long-lived, germline-deficient mutant glp-1 loss of tiar-3 resulted in increased mean lifespan, in contrast to tiar-1 or tiar-2 deletion that reduced its longevity. The latter indicates a currently unknown role of the predicted RNA-binding protein TIAR-3 in germline-mediated longevity. We finally demonstrated diverse responses of tiar mutants against various stressors, suggesting different activities for these genes in worms. Such diverse cellular roles have been described for several RNA-binding proteins which regulate various mRNA subpopulations, often in a coordinated manner [90,91]. For example, several hundred mRNAs of various functions were identified as targets of the dsRNA binding protein STAU-1, the worm homologue of human and Drosophila Staufen proteins that are localized in neuronal RNPs and SGs under stress [92–94]. In conclusion, our work implicates factors related to PBs and SGs in the normal lifespan and stress responses, beyond their developmental roles in C. elegans. Our observations expand our knowledge on the formation, function and relationship between PBs and SGs in the somatic cells of worms and reveal their tissue- and stress-specific properties.

Supporting Information

Figure S1 Expression pattern and mRNA levels of dcap-1 in the used transgenic lines. (A) Representative confocal images of 1-day adult worms expressing the transcriptional fusion $P_{dcap-1::gfp}$ (BRF154) or the translational fusion dcap-1::gfp in N2 (BRF155 and BRF261) or germline-deficient glp-1(e2141) worms (BRF219), normally grown at 25°C. m: muscles, n: neurons, sp: spermatheca, exc: excretory cell, v: vulva, i: intestine, cc: coeloemocytes. Scale bar: 50 μm. (B) Expression levels of dcap-1 gene in N2, glp-1(e2141) or the indicated transgenic strains measured by quantitative RT-PCR and normalized to ama-

1(mRNA) levels. Error bars show the SD of the means of two independent experiments.

Figure S2 Monitoring of granule formation under oxidative stress, heat-shock or aging by using other reporter markers.

(A) Representative confocal images of 1-day adults expressing the translational fusion *dcap-1::gfp* (BRF155 in Table S1) or *ife-2::gfp* (BRF70 in Table S1) under normal conditions (-SA) or after exposure to 10 mM sodium arsenite for 3 h (+SA) at 25°C. (B) Representative confocal images of 1-day adults expressing the translational fusion *ain-1::gfp* (MH2704 in Table S1), the transcriptional fusion of *ife-2* putative promoter to *gfp* (BRF68 in Table S1) or the translational fusion *ife-2::gfp* (BRF70 in Table S1) under normal conditions (-HS) or after heat-shock at 35°C for 3 h (+HS). The same transgenic animals are shown at the day 5 of adulthood, grown under normal conditions, at 25°C. (C) Representative confocal images of 1-day adults co-expressing *ife-2::gfp* and *dcap-1::rfp* (BRF313 in Table S1). Arrows point to induced granules. Scale bar: 25 μm. (D) Expression levels of *ain-1* gene in N2, as 1-day or 5-day adults grown under normal conditions (-HS) or as 1-day adults exposed to heat-shock at 35°C for 3 h (+HS), measured by quantitative RT-PCR and normalized to *ama-1(mRNA)* levels. Error bars show the SD of the means of two independent experiments.

Figure S3 Effects of *cgh-1(RNAi)* and *ife-2* deletion in heat-induced granule formation.

(A) Representative confocal images of 1-day adults expressing the translational fusion *ife-2::gfp* (BRF70 in Table S1) fed from eggs either control Control(RNAi) or *cgh-1(RNAi)* bacteria, at 25°C, under normal conditions (-HS) or after heat-shock at 35°C for 3 h (+HS). (B) Representative confocal images of 1-day adults expressing the translational fusion *dcap-1::gfp* in N2 (BRF155 in Table S1) or in *ife-2(ok306)* mutant background (BRF220 in Table S1), grown under normal conditions (-HS) or after heat-shock at 35°C for 3 h (+HS). Arrows point to induced granules. Scale bar: 25 μm.

Figure S4 Accumulation of PBs with age at 20°C.

(A) Representative confocal images of 1-day, 5-day, 10-day and 15-day adults grown at 20°C, expressing the translational fusion *dcap-1::gfp* (BRF155). Arrows point to DCAP-1::GFP granules. Scale bar: 25 μm. (B) Quantification of data presented in (A). Values on Y axis show the number of granules per head.

Figure S5 Decapping genes are important for development, growth, fertility and stress response.

(A) Distribution of N2, *dcap-1* and *dcap-2* progeny in the indicated developmental stages (L1-L4 and adult) 48 h after egg-laying, at 20°C and 25°C. (B) Number of *dcap-1* and *dcap-2* progeny (brood size), compared to N2, at 20°C and 25°C. (C) Development and stress resistance of *dcap-1* mutants expressing the *dcap-1::gfp* rescuing transgene, compared to N2 and *dcap-1* worms carrying only the *rol-6(su1006)* roller marker, at 20°C. For stress assays see Materials and Methods. Error bars show the SD in unpaired t-tests. ns indicates not significant (p>0.05); ** indicates very significant (p-value 0.001 to 0.01); *** indicates extremely significant (p<0.001). (D) Reverse transcription (RT)-PCR of *dcap-1(mRNA)* in 1-day adult N2 and *dcap-1* mutants and alignment of the wild-type DCAP-1 protein and the one encoded by the *tm3163* allele using MultAlin (http://multalin.toulouse.inra.fr/multalin/multalin.html).

Figure S6 Expression pattern of *tiar* genes in the used transgenic lines.

(A) Representative confocal images of 1-day adult worms expressing the transcriptional fusion $P_{tiar-1}::gfp$ (BRF118 in Table S1), the translational fusion *gfp::tiar-1* (BRF211 in Table S1), the transcriptional fusion $P_{tiar-2}::gfp$ (BRF238 in Table S1), the translational fusion *gfp::tiar-2* (BRF255 in Table S1) or the translational fusion *tiar-3::gfp* (BRF120 in Table S1), normally grown at 20°C. m: muscles, n: neurons, sp: spermatheca, exc: excretory cell, v: vulva, i: intestine, cc: coeloemocytes. Scale bar: 50 μm.

Figure S7 Spatial overlapping of PB and SG components in somatic cells of worms.

Representative confocal images of 1-day adults, under normal conditions (-HS) or after heat-shock at 35°C for 3 h (+HS), co-expressing: (A) *rfp::tiar-1* and *ife-2::gfp* (BRF312 in Table S1), (B) *rfp::tiar-1* and *dcap-1::gfp* (BRF328 in Table S1), (C) *gfp::tiar-2* and *dcap-1::rfp* (BRF369 in Table S1), (D) *rfp::tiar-1* and *gfp::tiar-2* (BRF361 in Table S1). B and C are confocal optical sections. All fusions are driven by their own promoters. Arrows point to induced granules. Scale bar: 25 μm.

Figure S8 TIAR-1 and TIAR-2 granules show differences in their formation.

Representative confocal images of 1-day adults expressing (A) *gfp::tiar-1* (BRF211 in Table S1), *gfp::tiar-2* (BRF255 in Table S1) or $P_{myo-3}::gfp::tiar-2$ (BRF310 in Table S1) subjected to 350 mM NaCl and visualized after 0.5–1 h, (B) *gfp::tiar-1* or $P_{myo-3}::gfp::tiar-2$ in N2 (BRF211 or BRF310, respectively in Table S1) and *gcn-2(ok871)* background (BRF294 or BRF339, respectively in Table S1), subjected to heat-shock (HS, 35°C for 3 h) or sodium arsenite (SA, 15 mM for 3 h). Arrows point to induced granules. Scale bar: 25 μm.

Table S1 List of the strains used in this study.

Table S2 Primers used in this study.

Table S3 RNAi plasmids used in this study.

Table S4 Quantification of number of granules.

Table S5 Lifespan assays in OP-50 plates.

Table S6 Lifespan assays in RNAi plates.

Movie S1 Motility of 6-day adults of N2 on OP-50 plates, at 25°C. Movie was taken in a Leica M205 FA fluorescence stereoscope with a Leica DFC340 FX camera.

Movie S2 Motility of 6-day adults of *tiar-1(tm361)* on OP-50 plates, at 25°C. Movie was taken in a Leica M205 FA fluorescence stereoscope with a Leica DFC340 FX camera.

Acknowledgments

We thank the BIU of BRFAA for using the Confocal system; Dr. Min Han and Dr. Ivo Lieberam for providing us with constructs; Michael Fasseas and Giota Poirazi for suggestions on this manuscript; Dimitris Stravopodis for insightful discussions. Some strains were provided by the Caenorhabditis Genetic Center, which is funded by the National Institutes for Health

National Center for Research Resources (Minneapolis). Other strains were provided by the Mitani Lab through the National Bio-Resource Project of the MEXT, Japan.

Author Contributions

Conceived and designed the experiments: PS AR AV FB. Performed the experiments: AR AV FB FR MK PS. Analyzed the data: PS AR AV FB. Contributed reagents/materials/analysis tools: AR AV FB FR MK. Wrote the paper: PS.

References

1. Kenyon CJ (2010) The genetics of ageing. Nature 464: 504–512.
2. Spriggs KA, Bushell M, Willis AE (2010) Translational regulation of gene expression during conditions of cell stress. Mol Cell 40: 228–237.
3. Renoux AJ, Todd PK (2012) Neurodegeneration the RNA way. Prog Neurobiol 97: 173–189.
4. Wang W (2012) Regulatory RNA-binding proteins in senescence. Ageing Res Rev 11: 485–490.
5. Westmark CJ, Malter JS (2012) The regulation of AbetaPP expression by RNA-binding proteins. Ageing Res Rev 11: 450–459.
6. Eulalio A, Behm-Ansmant I, Izaurralde E (2007) P bodies: at the crossroads of post-transcriptional pathways. Nat Rev Mol Cell Biol 8: 9–22.
7. Filipowicz W, Bhattacharyya SN, Sonenberg N (2008) Mechanisms of post-transcriptional regulation by microRNAs: are the answers in sight? Nat Rev Genet 9: 102–114.
8. Parker R, Sheth U (2007) P bodies and the control of mRNA translation and degradation. Mol Cell 25: 635–646.
9. Anderson P, Kedersha N (2008) Stress granules: the Tao of RNA triage. Trends Biochem Sci 33: 141–150.
10. Brengues M, Teixeira D, Parker R (2005) Movement of eukaryotic mRNAs between polysomes and cytoplasmic processing bodies. Science 310: 486–489.
11. Buchan JR, Parker R (2009) Eukaryotic stress granules: the ins and outs of translation. Mol Cell 36: 932–941.
12. Kedersha N, Stoecklin G, Ayodele M, Yacono P, Lykke-Andersen J, et al. (2005) Stress granules and processing bodies are dynamically linked sites of mRNP remodeling. J Cell Biol 169: 871–884.
13. Souquere S, Mollet S, Kress M, Dautry F, Pierron G, et al. (2009) Unravelling the ultrastructure of stress granules and associated P-bodies in human cells. J Cell Sci 122: 3619–3626.
14. Teixeira D, Sheth U, Valencia-Sanchez MA, Brengues M, Parker R (2005) Processing bodies require RNA for assembly and contain nontranslating mRNAs. Rna 11: 371–382.
15. Buchan JR, Muhlrad D, Parker R (2008) P bodies promote stress granule assembly in Saccharomyces cerevisiae. J Cell Biol 183: 441–455.
16. Decker CJ, Teixeira D, Parker R (2007) Edc3p and a glutamine/asparagine-rich domain of Lsm4p function in processing body assembly in Saccharomyces cerevisiae. J Cell Biol 179: 437–449.
17. Eulalio A, Behm-Ansmant I, Schweizer D, Izaurralde E (2007) P-body formation is a consequence, not the cause, of RNA-mediated gene silencing. Mol Cell Biol 27: 3970–3981.
18. Loschi M, Leishman CC, Berardone N, Boccaccio GL (2009) Dynein and kinesin regulate stress-granule and P-body dynamics. J Cell Sci 122: 3973–3982.
19. Stoecklin G, Mayo T, Anderson P (2006) ARE-mRNA degradation requires the 5'-3' decay pathway. EMBO Rep 7: 72–77.
20. Hu W, Sweet TJ, Chamnongpol S, Baker KE, Coller J (2009) Co-translational mRNA decay in Saccharomyces cerevisiae. Nature 461: 225–229.
21. Romero-Santacreu L, Moreno J, Perez-Ortin JE, Alepuz P (2009) Specific and global regulation of mRNA stability during osmotic stress in Saccharomyces cerevisiae. Rna 15: 1110–1120.
22. Arribere JA, Doudna JA, Gilbert WV (2011) Reconsidering movement of eukaryotic mRNAs between polysomes and P bodies. Mol Cell 44: 745–758.
23. Arimoto K, Fukuda H, Imajoh-Ohmi S, Saito H, Takekawa M (2008) Formation of stress granules inhibits apoptosis by suppressing stress-responsive MAPK pathways. Nat Cell Biol 10: 1324–1332.
24. Gottschald OR, Malec V, Krasteva G, Hasan D, Kamlah F, et al. (2010) TIAR and TIA-1 mRNA-binding proteins co-aggregate under conditions of rapid oxygen decline and extreme hypoxia and suppress the HIF-1alpha pathway. J Mol Cell Biol 2: 345–356.
25. Takahara T, Maeda T (2012) Transient sequestration of TORC1 into stress granules during heat stress. Mol Cell 47: 242–252.
26. Wippich F, Bodenmiller B, Trajkovska MG, Wanka S, Aebersold R, et al. (2013) Dual specificity kinase DYRK3 couples stress granule condensation/dissolution to mTORC1 signaling. Cell 152: 791–805.
27. Buchan JR, Yoon JH, Parker R (2011) Stress-specific composition, assembly and kinetics of stress granules in Saccharomyces cerevisiae. J Cell Sci 124: 228–239.
28. Pitt JN, Schisa JA, Priess JR (2000) P granules in the germ cells of Caenorhabditis elegans adults are associated with clusters of nuclear pores and contain RNA. Dev Biol 219: 315–333.
29. Scheckel C, Gaidatzis D, Wright JE, Ciosk R (2012) Genome-wide analysis of GLD-1-mediated mRNA regulation suggests a role in mRNA storage. PLoS Genet 8: e1002742.
30. Sheth U, Pitt J, Dennis S, Priess JR (2010) Perinuclear P granules are the principal sites of mRNA export in adult C. elegans germ cells. Development 137: 1305–1314.
31. Updike D, Strome S (2010) P granule assembly and function in Caenorhabditis elegans germ cells. J Androl 31: 53–60.
32. Gallo CM, Munro E, Rasoloson D, Merritt C, Seydoux G (2008) Processing bodies and germ granules are distinct RNA granules that interact in C. elegans embryos. Dev Biol 323: 76–87.
33. Jud MC, Czerwinski MJ, Wood MP, Young RA, Gallo CM, et al. (2008) Large P body-like RNPs form in C. elegans oocytes in response to arrested ovulation, heat shock, osmotic stress, and anoxia and are regulated by the major sperm protein pathway. Dev Biol 318: 38–51.
34. Noble SL, Allen BL, Goh LK, Nordick K, Evans TC (2008) Maternal mRNAs are regulated by diverse P body-related mRNP granules during early Caenorhabditis elegans development. J Cell Biol 182: 559–572.
35. Boag PR, Atalay A, Robida S, Reinke V, Blackwell TK (2008) Protection of specific maternal messenger RNAs by the P body protein CGH-1 (Dhh1/RCK) during Caenorhabditis elegans oogenesis. J Cell Biol 182: 543–557.
36. Lall S, Piano F, Davis RE (2005) Caenorhabditis elegans decapping proteins: localization and functional analysis of Dcp1, Dcp2, and DcpS during embryogenesis. Mol Biol Cell 16: 5880–5890.
37. Navarro RE, Shim EY, Kohara Y, Singson A, Blackwell TK (2001) cgh-1, a conserved predicted RNA helicase required for gametogenesis and protection from physiological germline apoptosis in C. elegans. Development 128: 3221–3232.
38. Squirrell JM, Eggers ZT, Luedke N, Saari B, Grimson A, et al. (2006) CAR-1, a protein that localizes with the mRNA decapping component DCAP-1, is required for cytokinesis and ER organization in Caenorhabditis elegans embryos. Mol Biol Cell 17: 336–344.
39. Brenner S (1974) The genetics of Caenorhabditis elegans. Genetics 77: 71–94.
40. Mello C, Fire A (1995) DNA transformation. Methods Cell Biol 48: 451–482.
41. Timmons L, Fire A (1998) Specific interference by ingested dsRNA. Nature 395: 854.
42. Min K, Kang J, Lee J (2010) A modified feeding RNAi method for simultaneous knock-down of more than one gene in Caenorhabditis elegans. Biotechniques 48: 229–232.
43. Ding L, Spencer A, Morita K, Han M (2005) The developmental timing regulator AIN-1 interacts with miRISCs and may target the argonaute protein ALG-1 to cytoplasmic P bodies in C. elegans. Mol Cell 19: 437–447.
44. Kamath RS, Martinez-Campos M, Zipperlen P, Fraser AG, Ahringer J (2000) Effectiveness of specific RNA-mediated interference through ingested double-stranded RNA in Caenorhabditis elegans. Genome Biol 2: RESEARCH0002.
45. Syntichaki P, Troulinaki K, Tavernarakis N (2007) eIF4E function in somatic cells modulates ageing in Caenorhabditis elegans. Nature 445: 922–926.
46. Ozgur S, Chekulaeva M, Stoecklin G (2010) Human Pat1b connects deadenylation with mRNA decapping and controls the assembly of processing bodies. Mol Cell Biol 30: 4308–4323.
47. Yang Z, Jakymiw A, Wood MR, Eystathioy T, Rubin RL, et al. (2004) GW182 is critical for the stability of GW bodies expressed during the cell cycle and cell proliferation. J Cell Sci 117: 5567–5578.
48. Yu JH, Yang WH, Gulick T, Bloch KD, Bloch DB (2005) Ge-1 is a central component of the mammalian cytoplasmic mRNA processing body. Rna 11: 1795–1802.
49. Aizer A, Brody Y, Ler LW, Sonenberg N, Singer RH, et al. (2008) The dynamics of mammalian P body transport, assembly, and disassembly in vivo. Mol Biol Cell 19: 4154–4166.
50. Fenger-Gron M, Fillman C, Norrild B, Lykke-Andersen J (2005) Multiple processing body factors and the ARE binding protein TTP activate mRNA decapping. Mol Cell 20: 905–915.
51. Kedersha N, Tisdale S, Hickman T, Anderson P (2008) Real-time and quantitative imaging of mammalian stress granules and processing bodies. Methods Enzymol 448: 521–552.
52. Swisher KD, Parker R (2010) Localization to, and effects of Pbp1, Pbp4, Lsm12, Dhh1, and Pab1 on stress granules in Saccharomyces cerevisiae. PLoS One 5: e10006.
53. Etchberger JF, Hobert O (2008) Vector-free DNA constructs improve transgene expression in C. elegans. Nat Methods 5: 3.
54. Dernburg AF, Zalevsky J, Colaiacovo MP, Villeneuve AM (2000) Transgene-mediated cosuppression in the C. elegans germ line. Genes Dev 14: 1578–1583.
55. Teixeira D, Parker R (2007) Analysis of P-body assembly in Saccharomyces cerevisiae. Mol Biol Cell 18: 2274–2287.
56. Andrei MA, Ingelfinger D, Heintzmann R, Achsel T, Rivera-Pomar R, et al. (2005) A role for eIF4E and eIF4E-transporter in targeting mRNPs to mammalian processing bodies. Rna 11: 717–727.
57. Eystathioy T, Jakymiw A, Chan EK, Seraphin B, Cougot N, et al. (2003) The GW182 protein colocalizes with mRNA degradation associated proteins hDcp1 and hLsm4 in cytoplasmic GW bodies. Rna 9: 1171–1173.

58. Zhang L, Ding L, Cheung TH, Dong MQ, Chen J, et al. (2007) Systematic identification of C. elegans miRISC proteins, miRNAs, and mRNA targets by their interactions with GW182 proteins AIN-1 and AIN-2. Mol Cell 28: 598–613.

59. Keiper BD, Lamphear BJ, Deshpande AM, Jankowska-Anyszka M, Aamodt EJ, et al. (2000) Functional characterization of five eIF4E isoforms in Caenorhabditis elegans. J Biol Chem 275: 10590–10596.

60. Groušl T, Ivanov P, Frydlova I, Vasicova P, Janda F, et al. (2009) Robust heat shock induces eIF2alpha-phosphorylation-independent assembly of stress granules containing eIF3 and 40S ribosomal subunits in budding yeast, Saccharomyces cerevisiae. J Cell Sci 122: 2078–2088.

61. Koga H, Kaushik S, Cuervo AM (2011) Protein homeostasis and aging: The importance of exquisite quality control. Ageing Res Rev 10: 205–215.

62. Syntichaki P, Tavernarakis N (2006) Signaling pathways regulating protein synthesis during ageing. Exp Gerontol 41: 1020–1025.

63. Sun Y, Yang P, Zhang Y, Bao X, Li J, et al. (2011) A genome-wide RNAi screen identifies genes regulating the formation of P bodies in C. elegans and their functions in NMD and RNAi. Protein Cell 2: 918–939.

64. Squier TC (2001) Oxidative stress and protein aggregation during biological aging. Exp Gerontol 36: 1539–1550.

65. Pan KZ, Palter JE, Rogers AN, Olsen A, Chen D, et al. (2007) Inhibition of mRNA translation extends lifespan in Caenorhabditis elegans. Aging Cell 6: 111–119.

66. Hansen M, Taubert S, Crawford D, Libina N, Lee SJ, et al. (2007) Lifespan extension by conditions that inhibit translation in Caenorhabditis elegans. Aging Cell 6: 95–110.

67. Tritschler F, Braun JE, Motz C, Igreja C, Haas G, et al. (2009) DCP1 forms asymmetric trimers to assemble into active mRNA decapping complexes in metazoa. Proc Natl Acad Sci U S A 106: 21591–21596.

68. Kenyon C, Chang J, Gensch E, Rudner A, Tabtiang R (1993) A C. elegans mutant that lives twice as long as wild type. Nature 366: 461–464.

69. Kimura KD, Tissenbaum HA, Liu Y, Ruvkun G (1997) daf-2, an insulin receptor-like gene that regulates longevity and diapause in Caenorhabditis elegans. Science 277: 942–946.

70. Lakowski B, Hekimi S (1998) The genetics of caloric restriction in Caenorhabditis elegans. Proc Natl Acad Sci U S A 95: 13091–13096.

71. Arantes-Oliveira N, Apfeld J, Dillin A, Kenyon C (2002) Regulation of life-span by germ-line stem cells in Caenorhabditis elegans. Science 295: 502–505.

72. Kedersha NL, Gupta M, Li W, Miller I, Anderson P (1999) RNA-binding proteins TIA-1 and TIAR link the phosphorylation of eIF-2 alpha to the assembly of mammalian stress granules. J Cell Biol 147: 1431–1442.

73. Gilks N, Kedersha N, Ayodele M, Shen L, Stoecklin G, et al. (2004) Stress granule assembly is mediated by prion-like aggregation of TIA-1. Mol Biol Cell 15: 5383–5398.

74. Kelly WG, Xu S, Montgomery MK, Fire A (1997) Distinct requirements for somatic and germline expression of a generally expressed Caernorhabditis elegans gene. Genetics 146: 227–238.

75. Silva-Garcia C, Navarro R (2013) The C. elegans TIA-1/TIAR homolog TIAR-1 is required to induce germ cell apoptosis. Genesis 51: 690–707.

76. Rousakis A, Vlassis A, Vlanti A, Patera S, Thireos G, et al. (2013) The general control nonderepressible-2 kinase mediates stress response and longevity induced by target of rapamycin inactivation in Caenorhabditis elegans. Aging Cell 12: 742–751.

77. Ermolaeva MA, Segref A, Dakhovnik A, Ou HL, Schneider JI, et al. (2013) DNA damage in germ cells induces an innate immune response that triggers systemic stress resistance. Nature 501: 416–420.

78. Judy ME, Nakamura A, Huang A, Grant H, McCurdy H, et al. (2013) A shift to organismal stress resistance in programmed cell death mutants. PLoS Genet 9: e1003714.

79. Nelson FK, Albert PS, Riddle DL (1983) Fine structure of the Caenorhabditis elegans secretory-excretory system. J Ultrastruct Res 82: 156–171.

80. Stoecklin G, Kedersha N (2013) Relationship of GW/P-bodies with stress granules. Adv Exp Med Biol 768: 197–211.

81. Audhya A, Hyndman F, McLeod IX, Maddox AS, Yates JR, 3rd, et al. (2005) A complex containing the Sm protein CAR-1 and the RNA helicase CGH-1 is required for embryonic cytokinesis in Caenorhabditis elegans. J Cell Biol 171: 267–279.

82. Hubstenberger A, Noble SL, Cameron C, Evans TC (2013) Translation repressors, an RNA helicase, and developmental cues control RNP phase transitions during early development. Dev Cell 27: 161–173.

83. Updike DL, Strome S (2009) A genomewide RNAi screen for genes that affect the stability, distribution and function of P granules in Caenorhabditis elegans. Genetics 183: 1397–1419.

84. Mollet S, Cougot N, Wilczynska A, Dautry F, Kress M, et al. (2008) Translationally repressed mRNA transiently cycles through stress granules during stress. Mol Biol Cell 19: 4469–4479.

85. David DC, Ollikainen N, Trinidad JC, Cary MP, Burlingame AL, et al. (2010) Widespread protein aggregation as an inherent part of aging in C. elegans. PLoS Biol 8: e1000450.

86. Reis-Rodrigues P, Czerwieniec G, Peters TW, Evani US, Alavez S, et al. (2012) Proteomic analysis of age-dependent changes in protein solubility identifies genes that modulate lifespan. Aging Cell 11: 120–127.

87. Ben-Zvi A, Miller EA, Morimoto RI (2009) Collapse of proteostasis represents an early molecular event in Caenorhabditis elegans aging. Proc Natl Acad Sci U S A 106: 14914–14919.

88. Ramachandran V, Shah KH, Herman PK (2011) The cAMP-dependent protein kinase signaling pathway is a key regulator of P body foci formation. Mol Cell 43: 973–981.

89. Gallo CM, Wang JT, Motegi F, Seydoux G (2010) Cytoplasmic partitioning of P granule components is not required to specify the germline in C. elegans. Science 330: 1685–1689.

90. Gerber AP, Herschlag D, Brown PO (2004) Extensive association of functionally and cytotopically related mRNAs with Puf family RNA-binding proteins in yeast. PLoS Biol 2: E79.

91. Keene JD (2003) Posttranscriptional generation of macromolecular complexes. Mol Cell 12: 1347–1349.

92. LeGendre JB, Campbell ZT, Kroll-Conner P, Anderson P, Kimble J, et al. (2013) RNA targets and specificity of Staufen, a double-stranded RNA-binding protein in Caenorhabditis elegans. J Biol Chem 288: 2532–2545.

93. Micklem DR, Adams J, Grunert S, St Johnston D (2000) Distinct roles of two conserved Staufen domains in oskar mRNA localization and translation. Embo J 19: 1366–1377.

94. Thomas MG, Martinez Tosar LJ, Desbats MA, Leishman CC, Boccaccio GL (2009) Mammalian Staufen 1 is recruited to stress granules and impairs their assembly. J Cell Sci 122: 563–573.

22

Tubular Overexpression of Gremlin Induces Renal Damage Susceptibility in Mice

Alejandra Droguett[1], Paola Krall[1], M. Eugenia Burgos[1], Graciela Valderrama[1], Daniel Carpio[3], Leopoldo Ardiles[1], Raquel Rodriguez-Diez[4], Bredford Kerr[2], Katherina Walz[2], Marta Ruiz-Ortega[4], Jesus Egido[4], Sergio Mezzano[1]*

1 Division Nephrology, School of Medicine, Universidad Austral de Chile, Valdivia, Chile, **2** Centro de Estudios Científicos, Valdivia, Chile, **3** Hystopathology Division, School of Medicine, Universidad Austral de Chile, Valdivia, Chile, **4** Cellular Biology in Renal Diseases Laboratory, Universidad Autónoma Madrid, Madrid, Spain

Abstract

A growing number of patients are recognized worldwide to have chronic kidney disease. Glomerular and interstitial fibrosis are hallmarks of renal progression. However, fibrosis of the kidney remains an unresolved challenge, and its molecular mechanisms are still not fully understood. Gremlin is an embryogenic gene that has been shown to play a key role in nephrogenesis, and its expression is generally low in the normal adult kidney. However, gremlin expression is elevated in many human renal diseases, including diabetic nephropathy, pauci-immune glomerulonephritis and chronic allograft nephropathy. Several studies have proposed that gremlin may be involved in renal damage by acting as a downstream mediator of TGF-β. To examine the *in vivo* role of gremlin in kidney pathophysiology, we generated seven viable transgenic mouse lines expressing human gremlin (GREM1) specifically in renal proximal tubular epithelial cells under the control of an androgen-regulated promoter. These lines demonstrated 1.2- to 200-fold increased GREM1 expression. GREM1 transgenic mice presented a normal phenotype and were without proteinuria and renal function involvement. In response to the acute renal damage cause by folic acid nephrotoxicity, tubule-specific GREM1 transgenic mice developed increased proteinuria after 7 and 14 days compared with wild-type treated mice. At 14 days tubular lesions, such as dilatation, epithelium flattening and hyaline casts, with interstitial cell infiltration and mild fibrosis were significantly more prominent in transgenic mice than wild-type mice. Tubular GREM1 overexpression was correlated with the renal upregulation of profibrotic factors, such as TGF-β and αSMA, and with increased numbers of monocytes/macrophages and lymphocytes compared to wild-type mice. Taken together, our results suggest that GREM1-overexpressing mice have an increased susceptibility to renal damage, supporting the involvement of gremlin in renal damage progression. This transgenic mouse model could be used as a new tool for enhancing the knowledge of renal disease progression.

Editor: Jean-Claude Dussaule, INSERM, France

Funding: Supported by Ciberdem, Redinren, Fondecyt 1120480, Fondecyt 1080083, Fondecyt 1100821, Chile. The funders had no role in study design, data collection and analysis, decision to publish, or preparation of the manuscript.

Competing Interests: The authors have declared that no competing interests exist.

* Email: smezzano@uach.cl

Introduction

Kidney disease and renal failure are worldwide health problems and increasing research efforts are required to understand the molecular mechanisms underlying kidney injury to identify new therapeutic approaches.

Gremlin is a highly conserved secreted protein that is present both on the external cell surface and within the ER-Golgi compartment of different cell types [1,2], affecting diverse biological processes such as growth, differentiation and development [2]. At early stages of development, gremlin is expressed in the mesoderm and inhibits BMP (bone morphogenetic protein) signaling by binding to and blocking BMP activity [3]. Thus, gremlin is considered a BMP antagonist. The *Grem1*-null mouse was the first *in vivo* gremlin model. This model was neonatally lethal, and the embryos presented kidney and lung defects [4]. Gremlin is critical in nephrogenesis but is quiescent after birth and absent in the normal adult kidney [5]. The gremlin gene may be induced in human mesangial cell cultures that are exposed to high

glucose levels and is expressed in the kidneys of diabetic rats [6–8]. In biopsies obtained from patients with diabetic nephropathy, we have observed gremlin expression in areas with tubule-interstitial fibrosis, and it colocalizes with transforming growth factor-β (TGF-β) [9]. Moreover, gremlin is also expressed in cellular glomerular crescents and in the tubular and infiltrating interstitial cells of human biopsies of pauci-immune glomerulonephritis and chronic allograft nephropathy, broadening the range of activity to a more global role for gremlin in renal diseases [10,11]. We have recently shown that recombinant gremlin directly regulates profibrotic events in cultured tubulo-interstitial cells and acts as a mediator of TGF-β responses [12]. Blockade of gremlin in experimental models of diabetes, using heterozygous *grem1* mice or gene silencing, has been shown to ameliorate renal damage, including proteinuria and fibrosis, suggesting that gremlin contributes to renal damage progression [13,14]. Furthermore, gremlin overexpression in rat lungs results in a partly reversible lung fibrosis through the activation of alveolar epithelial cell proliferation [15]. Interestingly, there are no data regarding the direct effect of gremlin in the kidney.

With the interest to investigate the potential role of gremlin in the kidney in physiological and pathological conditions *in vivo*, we generated viable transgenic (TG) mice expressing the human gremlin gene (GREM1) in the renal proximal tubular cells under the control of a specific kidney androgen-regulated promoter (KAP) that can be used as a molecular "on-off" switch.

We found that GREM1-expressing mice presented normal renal function, but they were more susceptible to developing renal damage induced by folic acid (FA) administration (a known experimental model of acute renal injury), suggesting that gremlin plays a pathogenic role in renal damage. These mice could be used as an *in vivo* experimental model to study the role of gremlin in renal diseases, such as diabetic nephropathy.

Materials and Methods

GREM1 cloning

Because human and murine mRNAs and proteins for gremlin exhibit high homology (89% and 98%, respectively), all of the experiments were performed using the human sequence. GREM1 cDNA was purchased from the Mammalian Gene NIH Collection (Bethesda, Maryland USA). To facilitate the detection of human gremlin (as opposed to the endogenous mouse protein), we added a c-myc tag to the 3′ portion of GREM1 using PCR with the forward primer 5′AGTGCGGCGGCTGAGGACCC GCCGC-ACTGACAT-3′ and the reverse primer 5′-ATAGCCGCCGCT-TACAGATCCTCTTCTGAGATGAGTTTTTGTTCATCCA-AATCGATGGATATGC-3′. To add another signal to the transgene to facilitate detection in transfected cells, we inserted an e-GFP sequence downstream of human gremlin as follows. The IRES-eGFP sequence was obtained by PCR using a pIRES2-EGFP plasmid (Clontech Mountain View, CA, USA) as the template with the following primers: IRES-eGFP-F (5′-TACAT-TAATGGGCCCGGGATCCGCCCCTC-3′) and IRES-eGFP-R (5′-GGCCATATGCGCCTTAAGATACATTGATG-3′). The GREM1-c-myc and IRES-eGFP fragments were independently cloned into a pGEMT-Easy vector and then sequenced (Macrogen, Seoul, Korea) to confirm the modifications and absence of additional mutations. Next, both the GREM1-c-myc and IRES-eGFP fragments were subcloned into a modified pCDNA3 vector using the *Eco*RI and *Not*I restriction sites, respectively.

Transfection of EBNA293 and HK-2 cells

To determine whether the pCDNA3-GREM1-myc-IRES-eGFP generated stable proteins, EBNA293 and human renal proximal tubulo-epithelial (HK2 cell line, ATCC CRL-2190, Virginia, USA) cells were both grown in RPMI with 10% fetal bovine serum (FBS), 1% non-essential amino acids, 100 U/ml penicillin, 100 μg/ml streptomycin, Insulin Transferrin Selenium (5 μg/ml) and hydrocortisone (36 ng/ml) in 5% CO_2 at 37°C and were then transfected. EBNA293 cells were transfected with 0.6 μg of plasmid using LF2000 (Invitrogen, Carlsbad, CA, USA). Immunofluorescence was performed using a mouse anti-c-myc antibody (M4439, clone 9E10, Sigma, St. Louis, USA) at a 1/500 dilution followed by a donkey anti-mouse IgG conjugated to Cy3 antibody (Jackson ImmunoResearch, West Grove, PA, USA) at a 1/800 dilution. Images were captured using a Zeiss Axiovert 100M epifluorescence microscope. HK2 cells were grown in the aforementioned conditions and at 60–70% of confluence; cells were growth-arrested in serum-free medium for 24 hours before experiments. HK2 cells were transiently transfected for 48 hours using FuGENE (Roche, Basel Switzerland) and the pCDNA3-

GREM1-c-myc-IRES2-eGFP plasmid vector. To study the expression of GREM1 in renal-specific tubulo-epithelial cells, HK2 cells were transiently transfected with pKAP-GREM1-c-myc-IRES2-eGFP to confirm the promoter activity. Immunocytochemical studies were performed in cells grown on coverslips. The cells were then fixed in Merckofix (Merck, Darmstadt, Germany) and permeabilized with 0.2% Triton-X100 for 1 min. After blocking with 4% bovine serum albumin and 8% of the corresponding serum (secondary antibody) for 1 hour, the cells were incubated with primary antibodies overnight at 4°C, followed by an Alexa Fluor 633-conjugated antibody (Invitrogen) at a 1/300 dilution for 1 hour. Negative control samples were processed in the absence of primary antibody. The cells were mounted in Mowiol 40–88 (Sigma) and examined using a Leica DM-IRB confocal microscope.

Transgenesis, genotyping and colony expansion

The pKAP plasmid, which has been previously shown to be specific in males [16–20], was used as the backbone for transgene generation. The pKAP plasmid was modified by excising an exon of the human angiotensinogen gene and the poly (A) signal. First, the IRES-eGFP sequence was subcloned downstream of GREM1-c-myc into pGEMT-Easy. The resulting construct was digested with *Sph*I and *Nde*I, and the 2.3 kb fragment containing GREM1-myc IRES-eGFP was subcloned into the modified pKAP plasmid. The 4.1 kb-long transgene was isolated with *Ase*I and *Eco*RV and purified using the QIAEx II kit (QIAGEN, Valencia, CA, USA). Five hundred molecules were microinjected into hybrid C57BL/6J x CBA/J zygotes, which were then transplanted into 13 pseudopregnant mothers. The born mice were maintained in a specific pathogen-free mouse facility in a 12-hour light:dark cycle with access to food and water *ad libitum*. The sacrifice of animals was done with administration of anaesthesia and analgesia, following the protocols approved by the Committee on the Ethics of Animal Experiments of Universidad Austral de Chile (Permit Number:20.2011), and FONDECYT Ethics Committee, and according to the NIH Guidelines. The founders and pups were screened using PCR with the following primers: *GREM1* intron 1F (5′-GCCAGTA AGGAATTCTAATAGG-3′), *KAP* promoter F (5′-ATGAGGACTCTAA TGCGTACAT-3′) and *GREM1* exon 2R (5′-TCCAAATCGATGGATA TGCAAC-3′). The PCR reaction generated two differential products (820 bp endogenous; 1040 bp TG). Once the genotype of the founders was confirmed, the founders were mated with pure C57BL/6J mice to expand the colony. F1 mice were screened using PCR as previously described and used for further molecular and phenotypic characterization studies.

Mouse characterization

First, kidneys from 4 to 5 week-old mice were analyzed by indirect immunofluorescence using a rabbit anti-GFP antibody at a 1/100 dilution (Invitrogen) and immunohistochemistry (IHC) to detect c-myc. Following confirmation that the female transgenic mice did not express the transgene, these mice were used only for mating and to evaluate the off/on system by the administration of 2.5 mg of testosterone via an intraperitoneal (i.p.) injection over five consecutive days [17,18]. Male TG and wild-type (WT) mice of five selected lines were also analyzed using western blotting analyses. The kidneys were dissected from both WT and TG mice homogenized in lysis buffer (125 mM Tris pH 6.8 and 1% SDS) supplemented with 1X protease inhibitor cocktail (Sigma P8340). Twenty micrograms of protein was electrophoresed in a 4–12% SDS-PAGE gel and transferred onto a PVDF membrane (Bio-Rad, Dreieich, Germany). The membrane was incubated with

a

DAPI   eGFP   c-myc

b

eGFP   Gremlin   MERGE

c

pCDNA -eGFP   pCDNA-GREM1-c-myc-
                                    IRES-e GFP

E-cadherin

Vimentin

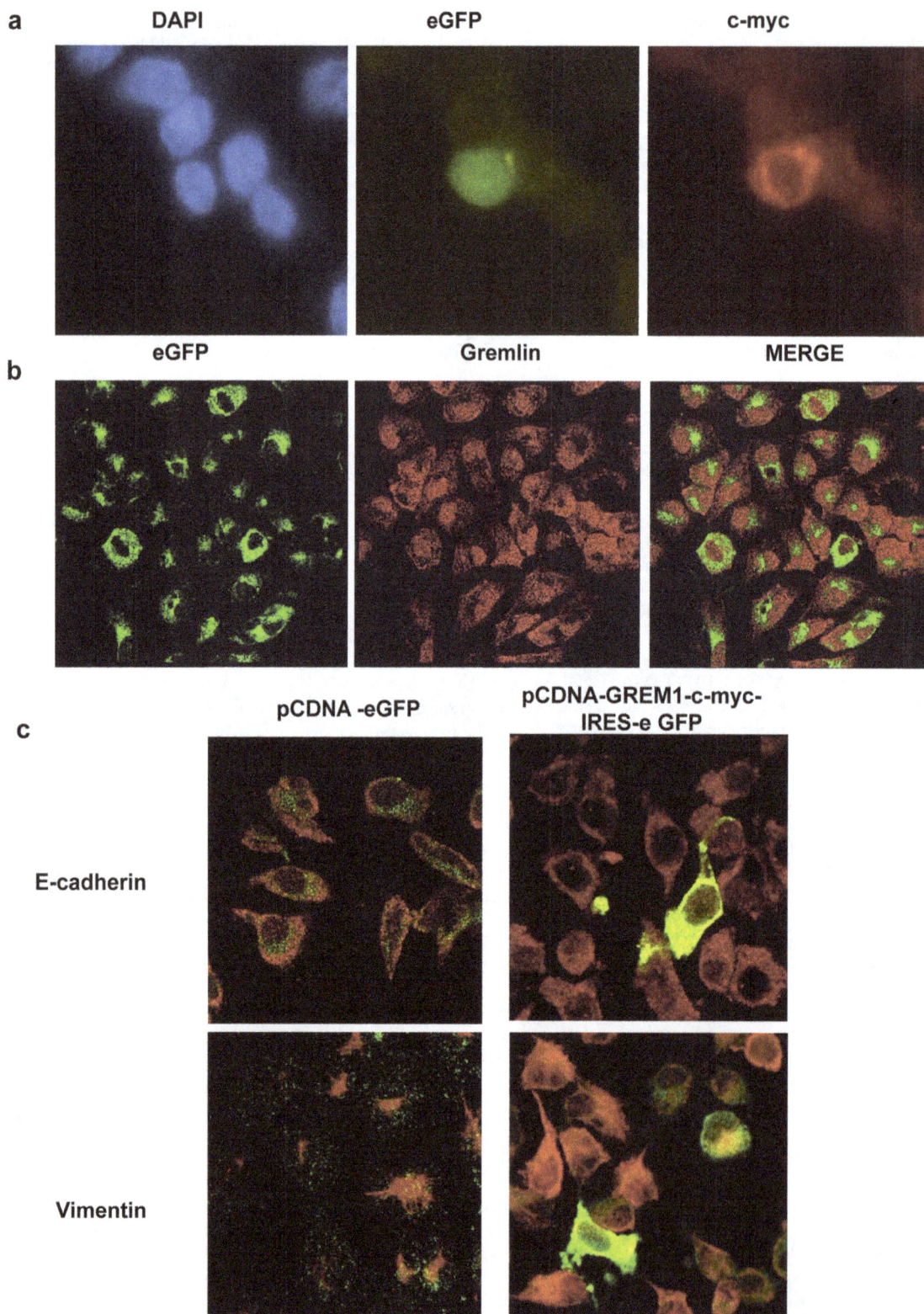

Figure 1. In vitro validation of the GREM1 plasmid. (a) EBNA293 cells were transfected with pCDNA3-GREM1-c-myc-IRES-eGFP, as described in the methods. Immunofluorescence shows that eGFP (green) and c-myc (red) are expressed in the same transfected EBNA293 cell. Nuclei were stained with DAPI (blue) (1000x). (b) In HK-2 cells transfected with pCDNA3-GREM1-c-myc-IRES-eGFP, GREM-1 expression was evaluated by immunocytochemistry using an antibody against GREM-1, followed by a secondary TRICT antibody (red staining). The figure shows eGFP and GREM-1 expressed in the same cell (800x). (c) Confocal immunofluorescence of HK-2 transfected cells, showing the loss of E-cadherin and induction of vimentin in eGFP-positive GREM-1-expressing cells (1600X). E-cadherin and vimentin immunostaining was detected with secondary anti-FITC antibodies (green).

anti-c-myc at 1/2500 or anti-actin at 1/5000 (Sigma). Densitometric analysis of the immunoreactive bands was performed using Quantity One software (Bio-Rad).

mRNA expression

Total RNA was extracted with TRIzol according to the manufacturers instructions and quantified using Qubit reagent (Invitrogen). RNA was treated with DNase I (Ambion, Austin TX, USA) to remove potential contamination and reverse transcribed using random primers and the ImProm-II kit (Promega) to synthesize double-stranded cDNA. qPCR was performed with the commercial reagent Maxima SYBR Green qPCR Master Mix (Promega, Madison WI, USA) to determine *GREM1, cyclophilin* and *GAPDH* mRNA expression levels using the following primers: human GREM1 F (5′-CCCGGGGAGGAGGTGCTGGAGT-3′); human GREM1 R (5′-CCGGATGTGCCTGGGGATGTA-GAA-3′); mouse cyclophilin1 F (5′-GCAGACATGGTCAACCC-CACCG-3′); mouse cyclophilin1 R (5′-GAAATTAGAGTTGT-CCACAGTCGG-3′); mouse GAPDH F(5′-TCCGCCCCTTCT-GCCGATG-3′); and mouse GAPDH R (5′-CACGGAAGGCC-ATGGCAGTGA-3′). PCR product specificity was verified by melting curve analysis, and all of the real-time PCR reactions were

performed in triplicate. The $2^{-\Delta\Delta CT}$ method was used to analyze the relative changes in gene expression levels [21].

Induction of acute renal damage in TG mice

The adult TG lines A and D and wild-type (WT) male littermates aged 4 to 5 months were used. TG and WT mice were injected i.p. with 250 mg/kg body weight of FA (Sigma F7876), dissolved in the vehicle 0.3 M sodium bicarbonate (veh). Control animals, both WT and TG, received 0.3 ml of veh. Additional studies were done in transgenic homozygous mice from line A, injected with FA or vehicle (used as control, because there were not wild type littermates). Spot urine and serum were collected on days 0, 7 and 14 from all of the animals and analyzed for proteinuria and creatininuria using Bradford assay (Bio-Rad) and a Creatinina Wiener Lab Kit (Wiener Laboratorios, Rosario, Argentina), respectively. Seven or 14 days after the injection, the animals were anesthetized with 2% 2,2,2-tribromethanol (Sigma) dissolved in 2-methyl-buthanol (Sigma). The kidneys were removed, decapsulated and cut along the sagittal plane. The left kidney was fixed in 4% formaldehyde, while the right kidney was immediately frozen in liquid nitrogen and processed for RNA and protein extraction. The specimens were embedded in paraffin and cut into 4 μm tissue sections for further histological (PAS/Masson)

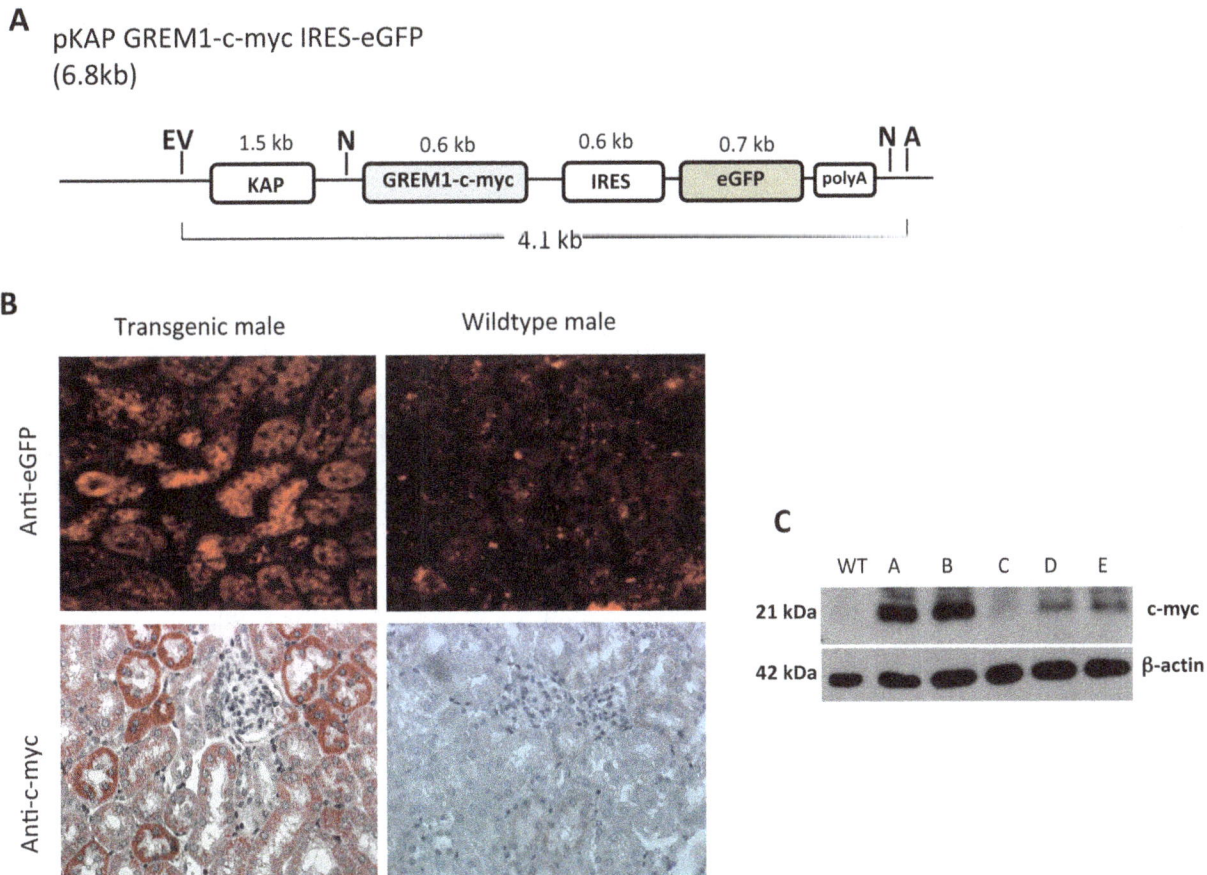

Figure 2. Generation and validation of transgenic mice with specific tubular GREM1 overexpression. (a) Illustration of the pKAP GREM1-c-myc-IRES-eGFP plasmid. Restriction sites used for transgene isolation are indicated with EV (*Eco*RV) and A (*Ase*l). (b) **eGFP and c-myc detection in renal tubular epithelial cells of transgenic mice.** Immunofluorescence against eGFP and immunohistochemistry for c-myc (peroxidase immunostaining) to detect these proteins in the kidney tissue of transgenic males from line A and WT mice (400x). (c) Kidneys were dissected from WT and transgenic male mice of lines A, B, C, D and E, and isolated proteins were subjected to western blotting using an antibody against c-myc (1:1000); anti β-actin (1:2500) was used as a loading control. GREM1 expression was determined by densitometric analysis of the c-myc/β-actin ratio and normalized to transgenic line C expression.

Table 1. Molecular characterization of transgenic mice.

Line	Copy number	Transmittance F1–F2 (%)	eGFP signal (IIF)	GREM1 levels (mRNA)	GREM1 levels (c-myc)	Male/Female (%)	Ectopicity
A	44	54	+++	190 ± 92	7.54	55/45	2
B	2–3	53	+	3.9 ± 0.4	6.38	52/48	1
C	2	50	++	2.2 ± 0.7	1.00	45/55	2
D	8–18	81***	++	10.5 ± 11.1	2.07	53/47	2
E	24	49	++	9.7 ± 7.2	2.4	48/52	1
F	11	46	+	54.6 ± 29.9	n.d	48/52	1

Qualitative and quantitative data from the transgenic lines are shown. Copy number and GREM1 expression levels were determined using real-time PCR. The eGFP signal was qualitatively analyzed. c-myc levels were quantified by densitometric analysis in each line and normalized to line C. Ectopic expression is indicated as the number of the eight analyzed extrarenal organs that demonstrated increased for GREM1 expression compared with wild-type expression. M, male; F, female; n.d., no data. *** p < 0.001.

and IHC studies using antibodies against gremlin and αSMA, F4/80, CD3 and PCNA.

Histological analysis and IHC

Tubular and interstitial lesions were graded from 0 to 4 and analyzed as previously described [22]. IHC for different markers was performed following heat-induced epitope retrieval (microwaving for 10 min in citrate buffer), and sections were incubated overnight with rabbit anti-human gremlin 1/300 (AP6133a, Abgent, San Diego, CA, USA), followed by incubation with Impress anti-rabbit reagent (Vector, Burlingame, CA, USA); or mouse anti-αSMA 1/100 (DAKO, Carpinteria, CA, USA), monoclonal anti-c-myc clone 9E10 1/300 (Thermo, Rockford IL, USA) or anti-PCNA (PC10, DAKO) followed by incubation with the M.O.M. Immunodetection kit (PK 2200 Vector). All of the tissue sections were developed using AMEC red chromogen (SK 4285, Vector) or DAB, and counterstained with hematoxylin. Interstitial infiltrating cells were detected by mean of F4/80 (monocytes/macrophages) and CD3 (T lymphocytes) antibodies. F4/80 was detected by using the MA1-91124 antibody (dilution: 1/100, THERMO, Rockford, IL, USA) followed by Immpress Reagent Kit (MP 7444, Vector, USA) and CD3 was detected using Trilogy epitope retrieval (Cell Marque, Rocklin, USA) and A 0452 (181–195) antibody (dilution: 1/200, DAKO, USA) followed by horseradish peroxidase streptavidin (dilution 1:500, SA-5004 Vector, USA), reveled with DAB, and counterstained with hematoxylin.

Murine Gremlin was detected using anti-murine Gremlin antibody (dilution: 1/20 AF 956, R&D Systems, Minneapolis, MN, USA) overnight at 4°C, and the reaction was developed with Impact NovaRed SK-4805 (Vector, USA).

Image analysis and quantification of the IHC signals were performed using the KS300 imaging system, version 3.0 (Zeiss). For each sample, the mean staining area was obtained by an analysis of 20 fields (20x). The staining score is expressed as the mm^2/dens

Statistical analysis

The results were expressed as the means ± SEM. Two-tailed chi-square tests were performed to determine the statistical significance of the viability of the transgene and the proportion of female:male pups born in the TG lines. A factorial ANOVA followed by the Tukey test was performed to compare proteinuria and *Grem1* mRNA expression. The Mann-Whitney U-test was performed to compare the tubular/interstitial lesions and Grem1, αSMA, F4/80 and CD3 IHC signals in the TG and WT mice injected with FA. A Spearman rank correlation was performed to determine the correlation between GREM1 and TGF-β, and Kruskal-Wallis analysis was performed to evaluate endogenous murine gremlin and PCNA expression. Values of p< 0.05 were considered significant.

Results

Generation of TG GREM1 mice

Validation of the GREM1 expression vector. Expression of pCDNA3-GREM1-c-myc-IRES-eGFP in EBNA293 cells revealed that this construct produced stable proteins of c-myc-GREM1 and eGFP, and both were expressed simultaneously in these cells (Figure 1a). To distinguish the human TG protein from the murine endogenous gremlin, a c-myc tag was fused to the GREM1 protein, and we used eGFP as reporter for transgene expression. *In vitro* experiments in EBNA293 cells were performed to detect the c-myc signal and showed an expression

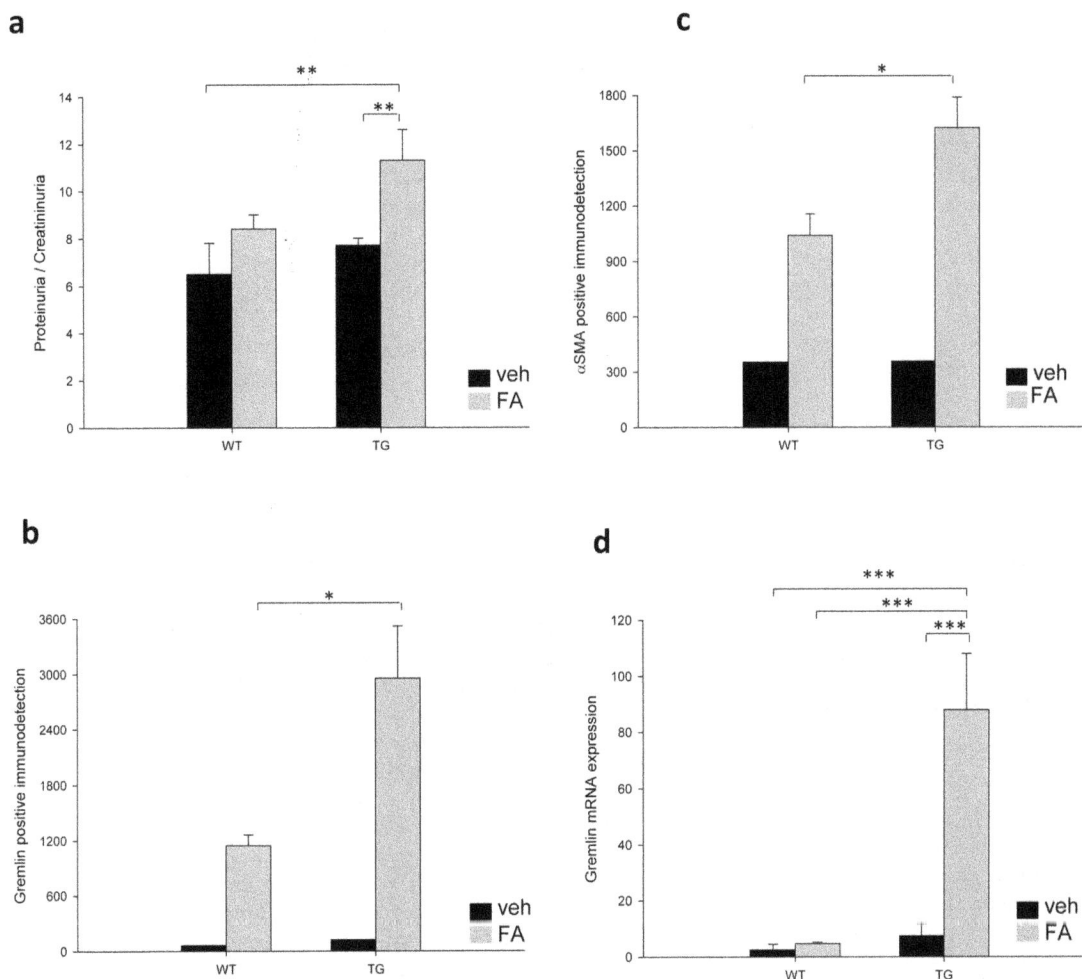

Figure 3. FA injection induces proteinuria and GREM1 and αSMA expression in transgenic line D. (a) The urinary protein to creatinine ratio (μg/mg) was examined in each experimental group. Positive IHC signals were quantified for (b) gremlin and (c) αSMA with KS300 image analyzer software. (d) Gremlin mRNA expression levels were determined using real-time PCR. The four parameters were significantly increased in GREM1-overexpressing transgenic mice. Data are shown as the mean ± SEM of 5-6 mice per group * p < 0.05; ** p < 0.01; *** p < 0.001. TG vs WT control

pattern consistent with GREM1 subcellular localization [1,2]; in a similar pattern to that found in human renal biopsies [9–11], where this protein is localized in cytoplasm and nucleus of the affected tubular epithelial cells. HK-2 cells transfected with pCDNA3-GREM1-c-myc-IRES-eGFP were positive for GREM1 (Figure 1b), which confirmed that eGFP could be used as a reporter for *in vivo* experiments and that c-myc could be used to differentiate human GREM1 in TG animals.

To validate our TG construction, exclude loss of function of the GREM1-c-myc-fused protein and determine whether GREM1 was functional, several experiments were performed in HK2 cells. We have previously demonstrated that stimulation with recombinant GREM1 protein in HK2 cells induces phenotypic changes related to epithelial to mesenchymal transition (EMT) [12]. Transient transfection of these cells with pCDNA3 GREM1-c-myc IRES-eGFP also induced characteristic EMT features, such as the downregulation of E-cadherin immunostaining and an induction of vimentin expression, as well as changes in the cell phenotype to a fibroblast-like morphology, confirming that our expression vector displayed similar effects to the GREM1 recombinant protein (Figure 1c).

Thus, to specifically induce the expression of GREM1 in proximal tubular renal cells, we used the promoter of kidney androgen-regulated protein (pKAP) to drive transgene expression because it is transcriptionally active only in these cells and its activity is testosterone dependent.

Transgene isolation was performed by digestion of the pKAP GREM1-c-myc IRES-eGFP plasmid (Figure 2a) with *Ase*I and *Eco*RV, which generated three fragments of 4.1, 1.5 and 1.2 kb. The 4.1 kb fragment was purified and quantified to obtain 500 molecules/picoliter and then microinjected into C57BL/6J x CBA/J hybrid zygotes. Ninety percent (232/259) of the microinjected zygotes were transferred to pseudopregnant mothers. Eight of the 57 (14%) pups born were confirmed to be TG using PCR. All of the founders reached the age of 3 months, and seven of the founders were subsequently mated with pure C57BL/6 to generate F1 mice.

Molecular and phenotypic characterization of GREM1 mice

TG lines (named in alphabetic order (A–G)) demonstrated normal fertility and litter sizes. Transgene transmission from all of the lines was verified using PCR analysis and appeared normal in

Homozygote mice line A treated with FA

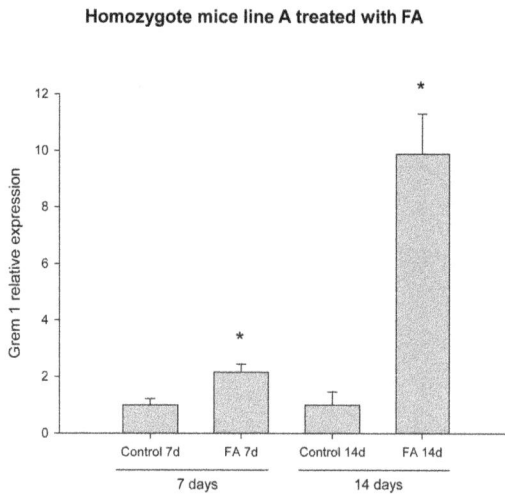

Figure 4. Effect of FA administration on GREM1 expression in transgenic line A homozygous mice. Gremlin expression in transgenic mice was examined 7 and 14 days after treatment with FA. The GREM1 relative expression in homozygous mice of transgenic line A was increased at 7 days after injection with FA and remained increased at 14 days. Data are shown as the mean ± SEM of 4-9 mice per group * p≤0.0049 TG-FA vs TG-Veh, used as control because no WT littermates of the transgenic homozygotes mice were available.

the first and second (F1 and F2) generations in all lines but was significantly increased in line D (p ≤ 0.001) (Table 1). In all TG lines, Mendelian ratios were observed between TG/WT males and females at birth, discarding an effect of transgene expression on male survival during gestation (Table 1).

Males and females were subjected to immunofluorescence analysis for eGFP and immunohistochemistry for c-myc. Males showed specific tubular epithelial expression of eGFP that was qualitatively variable in each TG line (Table 1). The eGFP and c-myc signals were not detected in WT males (Figure 2b). Moreover, this signal was not detected in TG females but could be induced in testosterone-treated TG females (data not shown), verifying the functionality of the promoter as previously described [18,20].

To confirm that the TG mice expressed a stable TG protein, western blotting analysis for c-myc-tagged GREM1 was performed. The assay detected a protein with the expected molecular weight (~21 kDa) in the kidneys of the TG lines, and this c-myc-tagged protein was absent in WT mice (Figure 2c). Densitometric analysis revealed that the GREM1 expression levels were variable in each TG line and ranged between 1 (line C exhibited the lowest expression) and 7.5 (line A exhibited the highest expression) (Table 1). Line G was not used in further experiments because the RT-PCR analysis detected ectopic expression of GREM1 in seven extra-renal tissues. The other six TG lines showed low extra-renal expression (cerebellum, data not shown). The levels of GREM1 mRNA in the kidney ranged between a 2- and 200-fold increase compared with the WT mice and were related to the transgene copy number (2 to 44 transgene copy number) (Table 1). Proteinuria (urine protein to creatinine ratio) was monitored in the first generation of males every 2 weeks until the age of 6 months in the three groups (WT and TG lines B and D), and no abnormal increases were observed compared with the WT mice, indicating that GREM1 expression alone was not sufficient to develop renal injury. Moreover, no histological lesions were observed at 6 months of age in any group.

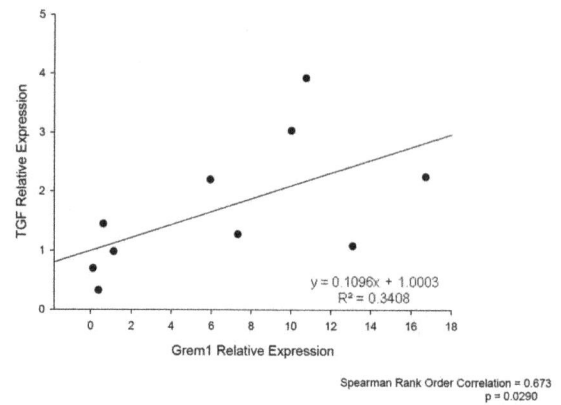

Figure 5. Correlation of TGF-β and GREM1 expression in transgenic line A homozygous mice. TGF-β gene expression was measured in FA-injected GREM1 transgenic mice. We observed a strongly positive correlation between TGF-β and GREM1 expression (p≤ 0.029, R = 0.67; 4-9 mice per group).

Renal injury induction in GREM1 TG mice

To determine the effect of GREM1 in renal damage *in vivo*, we selected mice from two lines and challenged these mice with a model of FA-mediated acute renal failure [23].

The line D, TG mice (8–18 transgene copies) were evaluated first. Seven days after FA injection, TG mice presented a significant increase in proteinuria compared with the treated WT mice (Figure 3a). However, the morphological lesions score found at 7 days, did not reach statistical significance between TG-FA compared with WT-FA (TG 9.0 ± 1.0 vs. WT 8.0 ± 2.1; p = 0.328, n = 5–6 mice per group). To further evaluate profibrotic markers, IHC for αSMA, as the first phenotypic marker of activated fibroblasts, was performed. In FA-TG mice, αSMA was markedly upregulated, showing a significant increase compared with FA-treated WT mice (Figure 3c).

Gremlin expression was also evaluated using real-time PCR. In FA-injected TG mice, (TG-FA) gremlin mRNA expression showed a nearly 10-fold increase compared with veh-injected mice and was significantly higher than FA-treated WT mice (WT-FA) (Figure 3d). Similar findings were observed by IHC, with a significant increase in gremlin staining after FA injection in TG mice compared with WT-FA mice (Figure 3b). Furthermore, we found a significant correlation between αSMA and gremlin IHC signals ($F_{1,10}$: 14.34; p < 0.0356; r = 0.5892), indicating that gremlin is associated with the variation in αSMA (r = 0.5892). Taken together, these results indicated that the effect observed *in vitro* [12] could be corroborated *in vivo*.

To further evaluate the effect of gremlin in FA-induced damage and determine if this was a transient or sustained effect, we performed additional studies using homozygous mice from TG line A and additionally evaluated its expression 14 days after treatment with FA. This line had the greatest number of copies of GREM1 (88 transgene copies) and gremlin mRNA. By real time PCR, we demonstrated that Grem1 was significantly increased at 7 and 14 days after FA injection (Figure 4) and by IHC gremlin renal staining was also increased at 7 and 14 days, with higher level at 14 days after treatment (score, expressed as mean of percentage/mm^2: TG-VEH; 14, TG-FA at 7 days; 343, TG-FA at 14 days; 862, p ≤ 0.0049), indicating that GREM1 expression in TG mice induces higher renal synthesis of gremlin following FA administration. Additionally, gremlin has been proposed as a downstream mediator of TGF-β [6], and we have previously

WT-FA

TG-FA

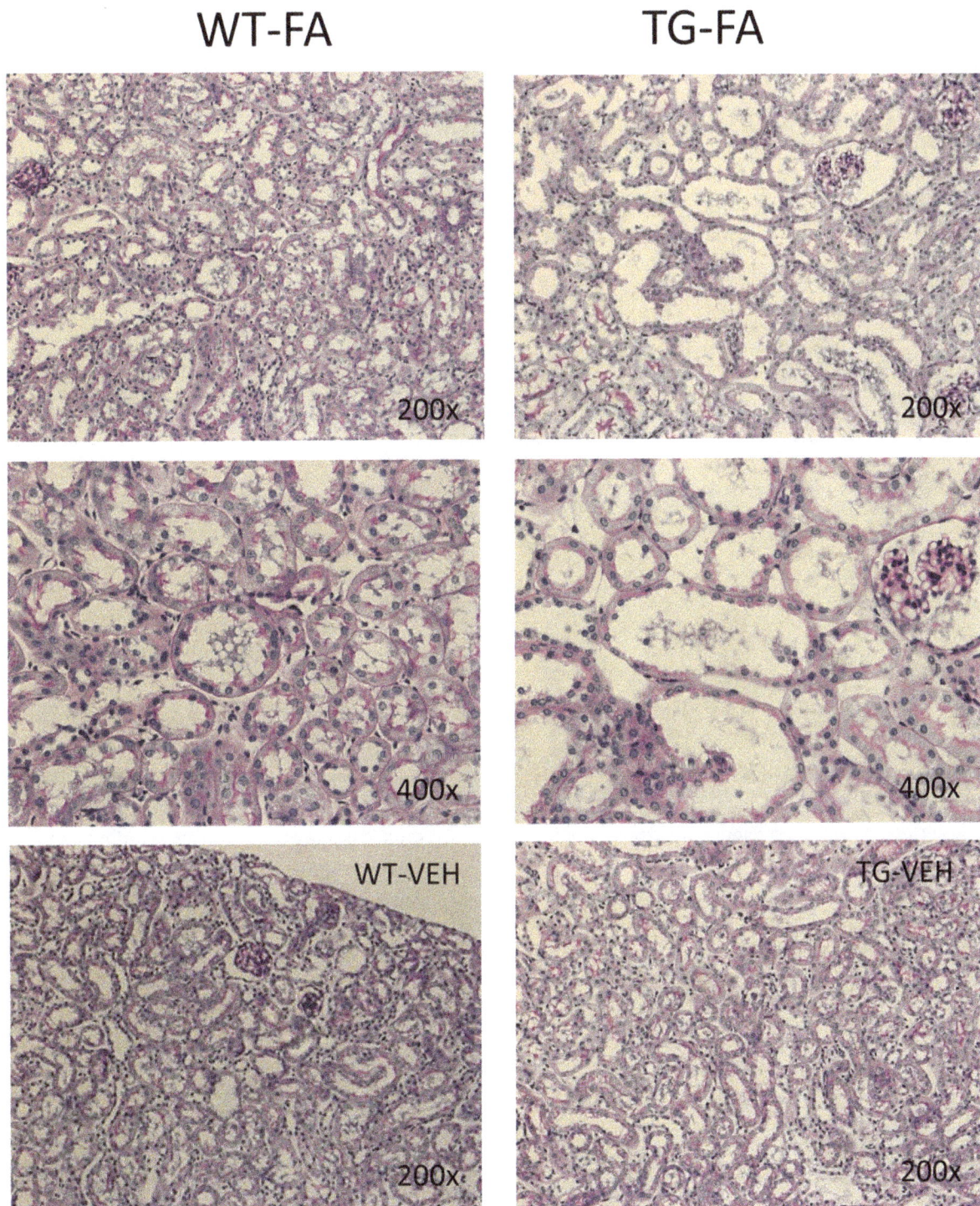

Figure 6. Histological analysis (PAS, Masson) of FA-injected mice in transgenic line A homozygous mice. TG-FA mice (right column) showed more severe morphological lesions (tubular dilatation, flattening of tubular epithelial cells, hyaline casts, interstitial infiltrating cells and mild interstitial fibrosis) compared with WT mice (left column). p < 0.05 (200x-400x). Control vehicle treated mice are shown at bottom. Figure shows representative mice of each group of 14–18 studied.

reported that TGF-β induced gremlin expression in renal tubular cells *in vitro* [12]. Therefore, to further evaluate profibrotic factors, TGF-β gene expression was measured. In FA-injected GREM1 TG mice, an increased relative renal expression of TGF-β at 7 and 14 days after treatment was observed compared with WT mice. Moreover, a strongly positive correlation was found between TGF-β and gremlin expression (p ≤ 0.029, R = 0.67) (Figure 5). Renal histological analysis at 14 days showed more severe morphological lesions (tubular dilatation, epithelium flattening, hyaline casts, interstitial cell infiltration and mild interstitial fibrosis) in TG-FA mice compared with WT-FA mice (p = 0.0037 Fisher's test, and Mann-Whitney p< 0.01) (Figure 6).

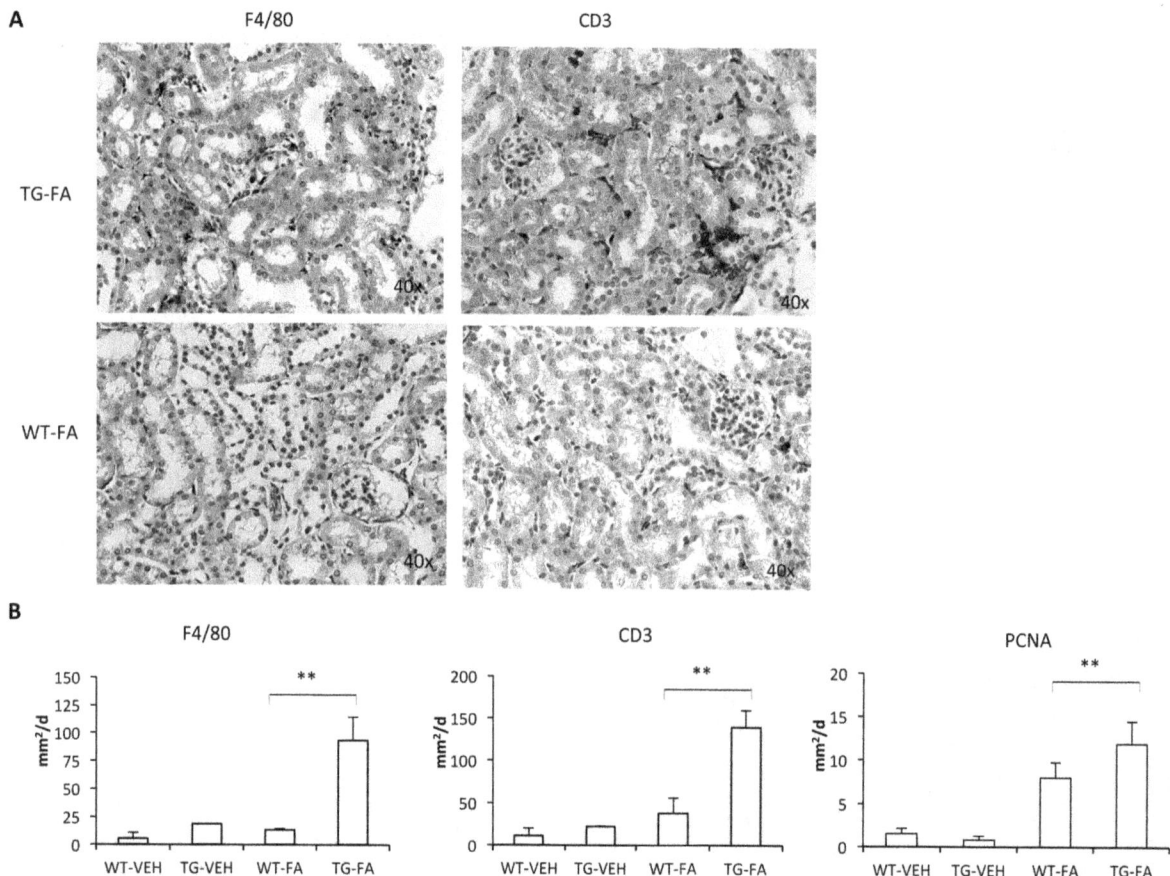

Figure 7. FA injection induces interstitial cell infiltration in transgenic line A homozygous mice. The inflammatory cell infiltration was characterized by immunohistochemistry with anti-F4/80 (monocytes/macrophages) and anti-CD3 (T cells) antibodies. (A and B). Representative immunostaining of one mouse from each group (x400 magnification). (C) Quantification of positive IHC signals were quantified for (a) F4/80, (b) CD3 and (c) PCNA using KS300 image analyzer software. All parameters were significantly increased in transgenic mice. Data are shown as the mean ± SEM of 14–18 mice per group * p < 0.05; ** p < 0.01 vs WT-FA.

To further evaluate inflammatory cell infiltration and proliferation, IHC for F4/80 (murine macrophages), CD3 (murine T cells), and PCNA was performed. In TG-FA mice, all markers were strongly upregulated, showing a significant increase compared with WT-FA mice (Figure 7). Also, a positive correlation was found between gremlin expression and PCNA at 14 days (R = 0.88, p ≤ 0.0019; data not shown)

Discussion

We reported here, for the first time, the generation of TG mice expressing GREM1 in a sex- and renal tubular cell-specific manner with no evident lethal effects and normal renal function and morphology. This mouse model was designed as a molecular tool to analyze the effect of gremlin expression in renal damage. Our results suggest that under normal conditions, GREM1 in adult tubular cells is not sufficient to cause renal damage. However, in response to acute renal injury caused by FA injection, GREM1 TG mice presented exacerbated renal damage, suggesting that gremlin could participate in renal damage progression *in vivo*. Gremlin is a developmental gene involved in renal morphogenesis due to its role as a BMP antagonist, but its function in the adult kidney is unknown. Several *in vitro* studies have evaluated the effect of gremlin in renal cells; however, the *in vivo* function has not been investigated. In tubular epithelial cells

in vitro, overexpression of this expression vector resembles the effect of recombinant GREM1, inducing phenotypic changes related to EMT [12].

These pKAP GREM1-overexpressing mice, which presented a specific GREM1 expression pattern in kidney tubular epithelial cells, demonstrated normal renal function and morphology. The specific pKAP promoter [20], motivated the specific cell type and hormone-regulated targeting of transgene expression. In addition, due to the developmental role of gremlin, expression driven by the pKAP promoter prevented any lethal effects at embryonic stages and generated a testosterone-dependent off/on switch in TG females.

Studies in TG mice overexpressing profibrotic factors, such as connective tissue growth factor (CTGF-CCN2) in different tissues have shown similar findings, as observed in our renal GREM1 TG mice. Although *in vitro* studies have shown that recombinant CCN2 increased extracellular matrix production [24], as observed with gremlin, several *in vivo* studies have shown that CCN2 alone is not sufficient to cause ongoing profibrotic changes. In the kidney, podocyte-specific CCN2-TG mice (in C57BL/6 background) exhibit no glomerular abnormalities, proteinuria or matrix accumulation [25]. In C57BL/6 mice, systemic CCN2 administration has been shown induce a transient overexpression of profibrotic genes at day 5, but it is not sufficient to induce progressive fibrosis [26], as observed following CCN2 overexpression in rat lungs [27]. In a mouse skin model, only coinjection of CCN2 and TGF-β1, not either cytokine alone, caused persistent fibrosis [28].

Several authors have suggested that gremlin could be considered as a mediator of renal injury in diabetic nephropathy, based on experimental studies showing a beneficial effect of gremlin inhibition [13,14]. In response to FA-induced acute renal damage, GREM1 TG mice developed higher proteinuria after 7 and 14 days than WT mice. Tubular GREM1 overexpression was associated with renal upregulation of profibrotic factors, such as TGF-β and αSMA, recruitment of F4/80 and CD3 positive cells, and increased cell proliferation in TG mice challenged with FA compared with WT mice. Furthermore, as hypothesized these GREM1 overexpressing mice developed more severe histological damage in response to FA injection at 14 days, particularly those mice with more transgenic copies and at the time of more gremlin expression.

In biopsies from patients with diabetic nephropathy, we have demonstrated that GREM1 is expressed in areas of tubular-interstitial fibrosis and that it colocalizes with profibrotic markers, including TGF-β, αSMA and vimentin. These changes also correlate directly with renal dysfunction, as shown by serum creatinine levels [9]. All these data suggest a role for gremlin in the pathogenesis of kidney damage.

Indeed, in the transgenic mice we found more acute tubular injury induced by folic acid than chronic damage progression. However it is important to note, than acute tubular injury is clearly associated with progression towards end stage renal disease [29], and on the other hand we found a significant interstitial cell infiltration that is commonly considered as the major initial mechanism leading to renal fibrosis.

Some evidence supports a potential interrelation between TGF-β1 and gremlin responses. We have previously demonstrated that *in vitro* blockade of endogenous gremlin by a specific siRNA inhibits TGF-β1-induced profibrotic gene overexpression and extracellular matrix production in renal fibroblasts. Moreover, gremlin blockade inhibits TGF-β1-mediated phenotypic-changes in tubular epithelial cells [12]. Even more, recently, it has been reported that gremlin likely induces endogenous TGF-β/Smad signaling, resulting in podocyte injury in mouse podocytes cultured in high glucose conditions [30]. Many data suggest that gremlin could be an important promoter of fibrosis in different pathologies, including liver fibrosis and lung diseases, particularly pulmonary hypertension, idiopathic pulmonary fibrosis and cancer invasion [31-35], as we have shown here in an experimental model of renal damage.

Our results suggest that GREM1-overexpressing mice have an increased susceptibility to renal damage, supporting the involvement of gremlin in renal damage progression.

Acknowledgments

The authors thank Dr. Juan Young for gifting the plasmid pMeCP2-flag IRES-eGFP (CECs, Chile) and Dr. Curt Sigmund from the University of Iowa, USA, for providing the pKAP2 plasmid. We thank Ms. Vanesa Marchant for her technical help.

Author Contributions

Conceived and designed the experiments: AD PK LA KW MRO JE SM. Performed the experiments: AD PK MEB GV RRD BK. Analyzed the data: AD PK MEB GV DC LA RRD BK KW MRO JE SM. Contributed reagents/materials/analysis tools: AD PK DC LA RRD BK KW MRO JE SM. Wrote the paper: AD PK BK MRO JE SM.

References

1. Topol LZ, Marx M, Laugier D, Bogdanova NN, Boubnov NV, et al. (1997) Identification of dmr, a novel gene whose expression is suppressed in transformed cells and which can inhibit growth of normal but not transformed cells in culture. Mol Cell Biol 17: 4801–4810.
2. Topol LZ, Bardot B, Zhang Q, Resau J, Huillard E, et al. (2000) Biosynthesis, post-translation modification, and functional characterization of Drm/Gremlin. J Biol Chem 275: 8785–8793.
3. Hsu DR, Economides AN, Wang X, Eimon PM, Harland RM (1998) The Xenopus dorsalizing factor Gremlin identifies a novel family of secreted proteins that antagonize BMP activities. Mol Cell1: 673–683.
4. Michos O, Panman L, Vintersten K, Beier K, Zeller R, et al. (2004) Gremlin-mediated BMP antagonism induces the epithelial-mesenchymal feedback signaling controlling metanephric kidney and limb organogenesis. Development 131: 3401–3410.
5. Roxburgh SA, Murphy M, Pollock CA, Brazil D (2006) Recapitulation of embryological programmes in renal fibrosis-the importance of epithelial cell plasticity and developmental genes. Nephron Physiol 103: 139–148.
6. McMahon R, Murphy M, Clarkson M, Taal M, Mackensie H, et al. (2000) IHG-2, a mesangial cell gene induced by high glucose, is human gremlin. Regulation by extracellular glucose concentration, cyclic mechanical strain, and transforming growth factor-beta1. J Biol Chem 275: 9901–9904.
7. Murphy M, Godson C, Cannon S, Kato S, Mackenzie HS, et al. (1999) Suppression subtractive hybridization identifies high glucose levels as a stimulus for expression of connective tissue growth factor and other genes in human mesangial cells. J Biol Chem 274: 5830–5834.
8. Lappin DW, McMahon R, Murphy M, Brady HR (2002) Gremlin: an example of the re-emergence of developmental programmes in diabetic nephropathy. Nephrol Dial Transplant 17s9: 65–67.
9. Dolan V, Murphy M, Sadlier D, Lappin D, Doran P, et al. (2005) Expression of gremlin, a bone morphogenetic protein antagonist, in human diabetic nephropathy. Am J Kidney Dis 45: 1034–1039.
10. Mezzano S, Droguett A, Burgos ME, Aros C, Ardiles L, et al. (2007) Expression of gremlin, a bone morphogenetic protein antagonist, in glomerular crescents of pauci-immune glomerulonephritis. Nephrol Dial Transplant 22: 1882–1890.
11. Carvajal G, Droguett A, Burgos ME, Aros C, Ardiles L, et al. (2008) Gremlin: a novel mediator of epithelial mesenchymal transition and fibrosis in chronic allograft nephropathy. Transplant Proc 40: 734–739.
12. Rodríguez Díez R, Lavoz C, Carvajal G, Rayego-Mateos S, Rodríguez Díez RR, et al. (2012) Gremlin is a downstream profibrotic mediator of transforming growth factor-beta in cultured renal cells. Nephron Exp Nephrol 122: 62–74.
13. Roxburgh SA, Kattla JJ, Curran SP, O'Meara YM, Pollock CA, et al. (2009) Allelic depletion of grem1 attenuates diabetic kidney disease. Diabetes 58: 1641–1650.
14. Zhang Q, Shi Y, Wada J, Malakauskas SM, Liu M, et al. (2010) In vivo delivery of Gremlin siRNA plasmid reveals therapeutic potential against diabetic nephropathy by recovering bone morphogenetic protein-7. PLoS One 5: e11709.
15. Farkas L, Farkas D, Gauldie J, Warburton D, Shi W, et al. (2011) Transient overexpression of Gremlin results in epithelial activation and reversible fibrosis in rat lungs. Am J Respir Cell Mol Biol 44: 870–878.
16. Ding Y, Davisson RL, Hardy DO, Zhu L, MerrilL DC, et al. (1997) The kidney androgen-regulated protein promoter confers renal proximal tubule cell-specific and highly androgen-responsive expression on the human angiotensinogen gene in transgenic mice. J Biol Chem 272: 28142–28148.
17. Ding Y, Sigmund C (2001) Androgen-dependent regulation of human angiotensinogen expression in KAP-hAGT transgenic mice. Am J Physiol Renal Physiol 280: F54–F60.
18. Lavoie JL, Lake-Bruse KD, Sigmund CD (2004) Increased blood pressure in transgenic mice expressing both human renin and angiotensinogen in the renal proximal tubule. Am J Physiol Renal Physiol 286: F965–F971.
19. Sachetelli S, Liu Q, Zhang SL, Liu F, Hsieh TJ, et al. (2006) RAS blockade decreases blood pressure and proteinuria in transgenic mice overexpressing rat angiotensinogen gene in the kidney. Kidney Int 69: 1016–1023.
20. Li H, Zhou X, Davis DR, Xu D, Sigmund CD (2008) An androgen-inducible proximal tubule-specific Cre recombinase transgenic model. Am J Physiol Renal Physiol 294: F1481–F1486.
21. Livak KJ, Schmittgen TD (2001) Analysis of relative gene expression data using real-time quantitative PCR and the 2(-Delta Delta C(T)) Method. Methods 25: 402–408.
22. Zoja C, Corna D, Camozzi D, Cattaneo D, Rottoli D, et al. (2002) How to fully protect the kidney in a severe model of progressive nephropathy: a multidrug approach. J Am Soc Nephrol 13: 2898–2908.
23. Ortega A, Rámila D, Ardura JA, Esteban V, Ruiz-Ortega M, et al. (2006) Role of parathyroid hormone-related protein in tubulointerstitial apoptosis and fibrosis after folic acid-induced nephrotoxicity. J Am Soc Nephrol 17: 1594–1603.
24. Phanish MK, Winn SK, Dockrell ME (2010) Connective tissue growth factor-(CTGF, CCN2)- a marker mediator and therapeutic target for renal fibrosis. Nephron Exp Nephrol 114: e83–92.

25. Yokoi H, Mukoyama M, Mori K, Kasahara M, Suganami T, et al. (2008) Overexpression of connective tissue growth factor in podocytes worsens diabetic nephropathy in mice. Kidney Int 73: 446–455.

26. Alfaro MP, Deskins DL, Wallus M, DasGupta J, Davidson JM, et al. (2013) A physiological role for connective tissue growth factor in early wound healing. Lab Invest 93: 81–95.

27. Bonniaud P, Margetts PJ, Kolb M, Haberberger T, Kelly M, et al. (2003) Adenoviral gene transfer of connective tissue growth factor in the lung induces transient fibrosis. Am J Respir Crit Care Med 168: 770–778.

28. Mori T, Kawara S, Shinozaki M, Hayashi N, Kakinuma T, et al. (1999) Role and interaction of connective tissue growth factor with transforming growth factor-beta in persistent fibrosis: a mouse fibrosis model. J Cell Physiol 181: 153–159.

29. Gentle ME, Shi S, Daehn I, Zhang T, Qi H, et al. (2013) Epithelial cell TGF-β signaling induces acute tubular injury and interstitial inflammation. J Am Soc Nephrol 24:787–799

30. Li G, Li Y, Liu S, Shi Y, Chi Y, et al. (2013) Gremlin aggravates hyperglycemia-induced podocyte injury by a TGFβ/Smad dependent signaling pathway. J Cell Biochem 114: 2101–2112.

31. Guimei M, Baddour N, Elkaffash D, Abdou L, Taher Y (2012) Gremlin in the pathogenesis of hepatocellular carcinoma complicating chronic hepatitis C: an immunohistochemical and PCR study of human liver biopsies. BMC Res Notes 5: 390.

32. Costello CM, Cahill E, Martin F, Gaine S, Mc Loughlin P (2010) Role of Gremlin in the lung: development and disease. Am J Respir Cell Mol Biol 42: 517–23.

33. Cahill E, Costello CM, Rowan SC, Harkin S, Howell K, et al. (2012) Gremlin plays a key role in the pathogenesis of pulmonary hypertension. Circulation 125: 920–930.

34. Koli K, Myllärniemi M, Vuorinen K, Salmenkivi K, Ryynänen MJ, et al. (2006) Bone morphogenetic protein-4 inhibitor gremlin is overexpressed in idiopathic pulmonary fibrosis. Am J Pathol 169: 61–71.

35. Karagiannis GS, Berk A, Dimitromanolakis A, Diamandis EP (2013) Enrichment map profiling of the cancer invasion front suggests regulation of colorectal cancer progression by the bone morphogenetic protein antagonist, gremlin-1. Mol Oncol7:826–839.

23

Generation of BAC Transgenic Epithelial Organoids

Gerald Schwank[1], Amanda Andersson-Rolf[1,2], Bon-Kyoung Koo[1,2], Nobuo Sasaki[1], Hans Clevers[1]*

1 Hubrecht Institute, KNAW and University Medical Center Utrecht, Utrecht, The Netherlands, **2** Wellcome Trust - Medical Research Council Stem Cell Institute, University of Cambridge, Cambridge, United Kingdom

Abstract

Under previously developed culture conditions, mouse and human intestinal epithelia can be cultured and expanded over long periods. These so-called organoids recapitulate the three-dimensional architecture of the gut epithelium, and consist of all major intestinal cell types. One key advantage of these ex vivo cultures is their accessibility to live imaging. So far the establishment of transgenic fluorescent reporter organoids has required the generation of transgenic mice, a laborious and time-consuming process, which cannot be extended to human cultures. Here we present a transfection protocol that enables the generation of recombinant mouse and human reporter organoids using BAC (bacterial artificial chromosome) technology.

Editor: Derya Unutmaz, New York University, United States of America

Funding: This work was funded by grants from the European Research Council (EU/232814-StemCeLLMark), the KNAW/3V-fund, the SNF fellowship for advanced researchers PA00P3 139732 (G.S.), the Human Frontiers in Science Program long-term fellowship LT000422/2012 (G.S.), and the National Research Foundation of Korea NRF-2011-357-C00093 (B.-K.K.). The funders had no role in study design, data collection and analysis, decision to publish, or preparation of the manuscript.

Competing Interests: The authors have declared that no competing interests exist.

* E-mail: h.clevers@hubrecht.eu

Introduction and Results

In the past decades, the mouse has been extensively studied to understand vertebrate development. Progress has been largely driven by the generation of genetic tools, which enabled to manipulate the mouse genome. The generation of transgenic mice is however a time-consuming procedure, and many tissues are poorly accessible for in vivo live imaging.

We recently developed a method that allows the culture of three-dimensional multi-cellular structures from single Lgr5+ intestinal stem cells [1]. These so- called 'miniguts' recapitulating most features of the normal gut epithelium. Lgr5+ stem cells and the niche supporting Paneth cells are located in a domain that resembles the crypt bottom, and enterocytes as well as goblet - and enteroendocrine cells move upwards to build a villus-like domain that lines the central lumen. The organoid cultures are grown ex vivo in matrigel supplemented with a defined growth medium, and can be expanded for over a year. Direct genetic manipulation of organoid cultures has been previously demonstrated using a retroviral transduction based method, enabling overexpression and shRNA-mediated downregulation of target genes [2]. However, due to size-limitations of viral vectors entire genes including their cis-regulatory regions cannot be integrated into the host genome [3], and expression of transgenes under their endogenous promoter is therefore not possible. Here we present a method enabling stable insertions of more than 100 kilobase large BACs into mouse and human intestinal organoids, relying on liposome-based transfection.

Due to their large size BACs are able to carry the entire genomic locus of genes, and therefore often ensure precise expression patterns [4]. BAC libraries covering the entire mouse and human genome have been established (www.chori.org), and BAC recombineering allowed the generation of libraries with fluorescently tagged genes [5,6]. Here we used previously established BAC reporters with enhanced green fluorescent (EGFP) tagged genes [7], and in addition generated recombineering cassettes that allow protein tagging with the red fluorescent protein tagRFP [8], the cyan fluorescent protein mTurquoise [9], and the yellow fluorescent protein mVenus [10] (Fig. 1A). These vectors carry a kanamycin-neomycin selection marker, which is flanked by loxP sites and therefore enables excision using Cre recombinase [11]. The tamoxifen inducible *CreERT2* can be delivered to mouse and human organoids by retroviral infection, or by deriving organoids from transgenic mice expressing the enzyme [2,12].

To explore methods for transfection of mouse intestinal organoids we delivered a 3.5 kb reporter plasmid expressing EGFP (pmax-GFP, Amaxa™) into organoids, using either electroporation based transfection (Nucleofector™) or liposome-mediated transfection (Lipofectamine®). We expanded mouse intestinal organoids in Wnt-conditioned media to enrich for stem cells, separated them from matrigel by pipetting, and trypsinized them in order to get a single cell suspension. Cells were then transfected as described in Materials and Methods, and analyzed for EGFP expression 48 hours later. Nucleofection lead to a transfection efficiency of 18%, and lipofectamine-mediated transfection resulted in 2.5% positive cells (Fig. 1B). We next transfected mouse organoids with a 120 kb BAC reporter containing the genomic locus of the core histone H2A with a C-terminal EGFP tag [7] (Fig. 1A). While nucleofection did not lead to successful BAC transfection, lipofection resulted on average in 3.8 (+/-1.3 STD) positive cells per well (Fig. 1C). Co-expression of an IRES-driven neomycin resistance gene enabled selection for stable expression. After two weeks organoids showed uniform nuclear GFP localization (Fig. 1D), and expression remained stable for more than 6 months. Similarly, we were able to transfect a BAC reporter containing the EGFP-tagged genomic locus of the

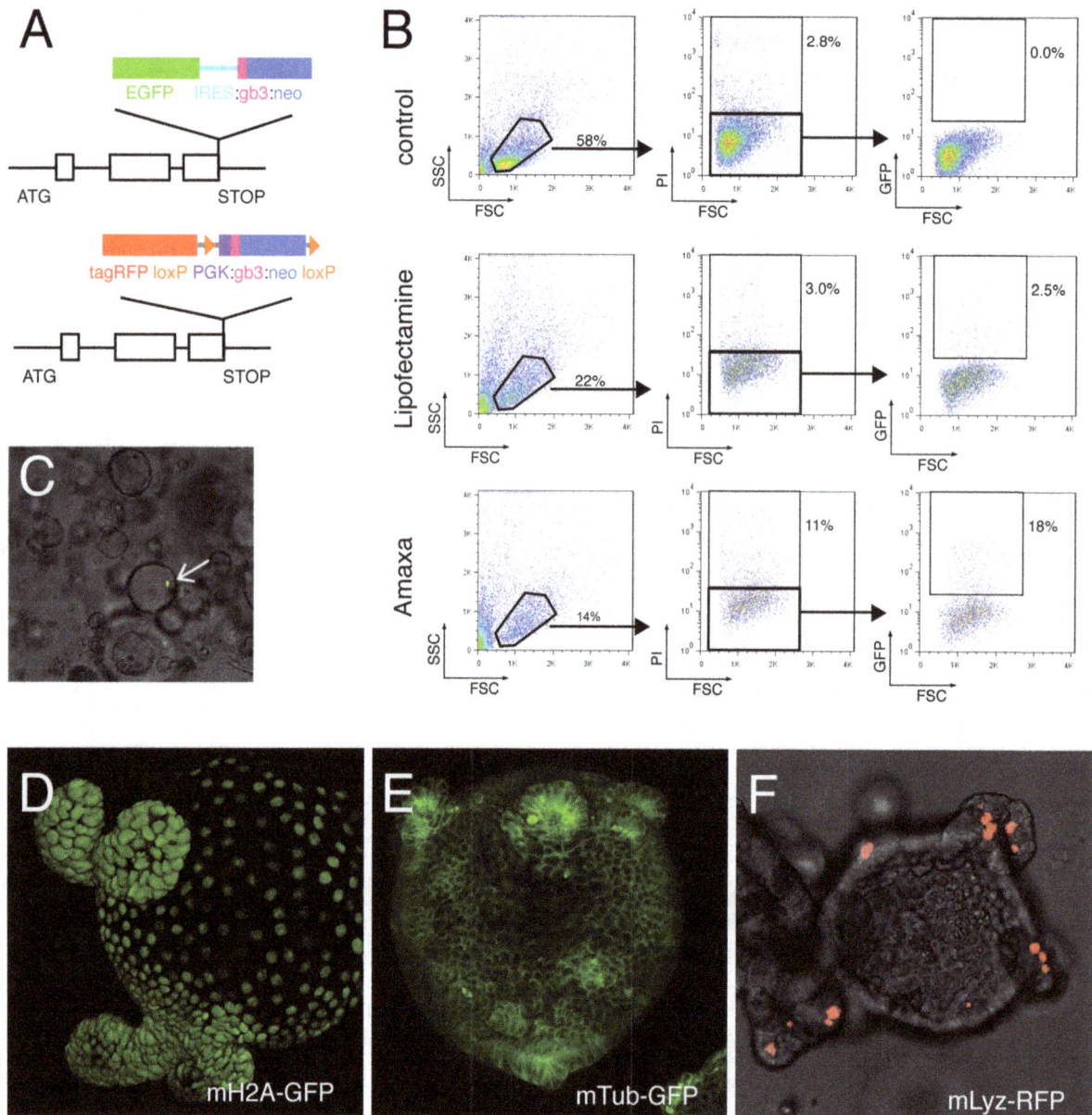

Figure 1. BAC transgenic mouse intestinal organoids. (A) Tagging cassettes for BAC recombineering. Upper panel illustrates the C-terminal GFP-tagging cassette with a neomycin resistance gene downstream of an IRES sequence, which was used in [7] to tag histone H2A and TUBB5. Lower panel shows the C-terminal tagRFP tagging cassette with a neomycine resistance gene downstream of a PGK promoter and flanked by loxP sites, which was used to tag the lysozyme gene. gb3: bacterial promoter, PGK: phosphoglycerate kinase promoter, IRES: internal ribosome entry site. (**B**) FACS analysis of mouse intestinal organoids 48 h after transfection with the pmax-GFP plasmid. Middle panel shows cell viability using propidium iodide (PI), right panel shows the percentage of GFP transfected cells. (**C**) Section of a well with mouse organoids 48 h after H2A-GFP BAC transfection using lipofectamine. Arrow points to a successfully transfected cell. Fluorescence images of a (**D**) H2A-GFP BAC transgenic organoid, (**E**) TUBB5-GFP BAC transgenic organoid, and (**F**) Lysozyme-tagRFP BAC transgenic organoid.

ß-Tubulin gene TUBB5, a major constituent of microtubules. After selection, organoids ubiquitously expressed the transgene and had a GFP labeled microtubule network, including the mitotic spindle of dividing cells (Fig. 1E).

Next, we tested if our method can be used to label specific cell lineages. We generated a BAC reporter with tagRFP labeled lysozyme, a specific marker for Paneth cells [13], and transfected organoids from *villin-creERT* transgenic mice [12]. IRES could not be used to drive the neomycin resistance gene, as the lack of lysozyme expression in stem cells would kill also successfully transfected organoids. Instead we used the constitutively active

PGK promoter (Fig. 1A). After selection and CreERT2-mediated removal of the neomycin selection cassette, recombinant organoids specifically expressed the tagRFP reporter in Paneth cells located at the base of the organoid crypts (Fig. 1F).

Modification of the original growth-factor composition has allowed us to also grow epithelial organoids derived from human intestine [14,15]. To test if our method can be applied to human organoid cultures, we optimized the protocol and transfected human small intestinal organoids with the pmax-GFP plasmid. Surprisingly, both transfection methods were more efficient in human organoids compared to mouse organoids; nucleofection lead to a transfection efficiency of 36%, and lipofectamine-mediated transfection resulted in 6.3% positive cells (Fig. 2A). We next used lipofectamine to transfect human organoids with the histone H2A-GFP BAC reporter, resulting on average in 36.6 (+/−6.3 STD) positive cells per well (Fig. 2B). Stable clones were selected using G418 and ubiquitously expressed nuclear H2A-GFP (Fig. 2C).

As a proof of concept we performed time lapse ex vivo imaging on the BAC transgenic organoid lines. We used conventional confocal and spinning disc microscopes to image organoids up to 77 hours ((Fig. 3), Movies S1–S3). The 3D reconstruction of the acquired images allowed us to visualize spatiotemporal processes such as crypt bud formation, cell division, and growth (Movies S1–S3).

Taken together, we present a method that allows the generation of BAC recombinant organoids. This technique can be used to study gene function ex vivo in organoid cultures under endogenous expression levels by live imaging, and is applicable to mouse and human organoid cultures. Thus, it will help to circumvent the time-consuming and costly process of generating transgenic mice,

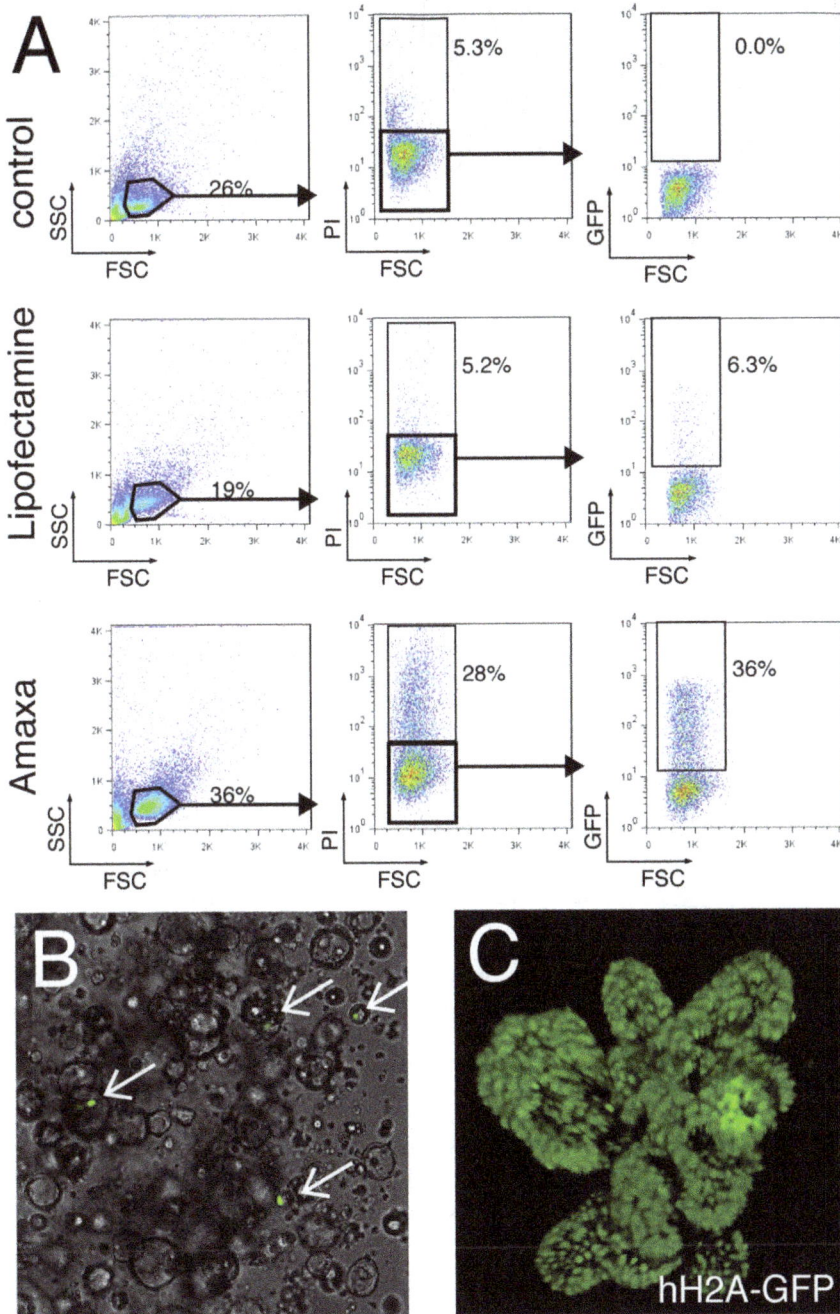

Figure 2. BAC transgenic human intestinal organoids. (**A**) FACS analysis of human intestinal organoids 48 h after transfection with the pmax-GFP plasmid. Middle panel shows cell viability using propidium iodide (PI), right panel shows the percentage of GFP transfected cells. (**B**) Section of a well with human organoids 48 h after H2A-GFP BAC transfection with lipofectamine. Arrows point to successfully transfected cells. (**C**) Fluorescence image of a H2A-GFP BAC transgenic human organoid.

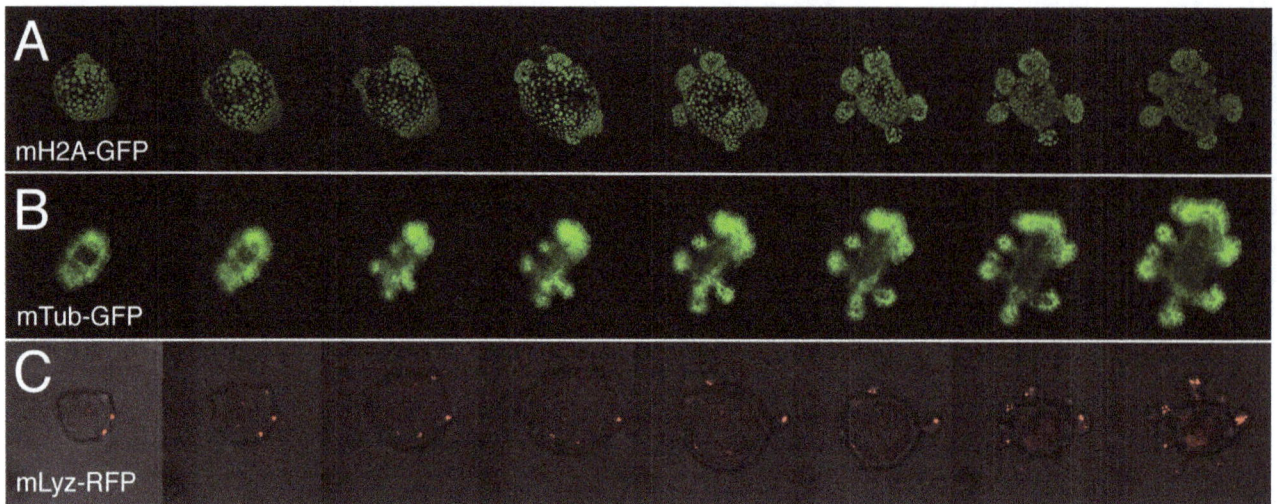

Figure 3. Snapshots of BAC transgenic organoid movies. (A) a 36 h movie (2.4 frames/hour) of a growing mouse H2A BAC transgenic organoid, **(B)** a 51 h movie (2.2 frames/hour) of a growing mouse TUBB5-GFP BAC transgenic organoid, and **(C)** a 77 h movie (1 frame/hour) of a growing mouse Lysozyme-tagRFP BAC transgenic organoid.

and enable to study the role of genes in human epithelial tissues, potentially opening a new avenue to examine genes involved in human diseases.

Materials and Methods

Human Samples

This study was approved by the ethical committee of the University Medical Centre Utrecht, and all samples were obtained with informed consent. The participants provided their informed consent to participate in this study in a written form.

Mouse Samples

All animal experiments have been conducted according to relevant national and international guidelines. Experimental setup was approved by the animal welfare committee (DEC) of the Royal Dutch academy of sciences (KNAW).

Organoid Culture

Crypts were isolated from mouse and human small intestines by incubating the tissue for 60 minutes with 2 mM EDTA in PBS at 4°C. Detached crypts were subsequently plated in 20 µl drops of matrigel, and after polymerization the previously described growth medium was added [16]. In short, mouse intestinal growth medium consists of advanced DMEM/F12 medium (Invitrogen) including the supplements B27 (Invitrogen), N2 (Invitrogen) and N-Acetylcysteine (Sigma-Aldrich) and the growth factors noggin (Peprotech), Rspo1 [17], and epidermal growth factor (Peprotech). Human intestinal growth medium additionally contains Wnt conditioned media, TGF-ß type I Receptor inhibitor A83-01 (Tocris), Nicotinamide (Sigma-Aldrich) and P38 inhibitor SB202190 (Sigma-Aldrich). Confluent organoids were mechanically dissociated using a fire polished glass pipette. Fragmented organoids were centrifuged at 1000 g for 5 minutes, and resuspended in cold matrigel in a 1:4 ratio. Rosa-CreERT2 mice were used to generate mouse small intestinal organoids. Isolation of human crypts was described elsewhere [15].

Vector Construction

BAC recombineering cassettes: For cloning we used the In-Fusion Advanced PCR cloning kit (Clontech). The *loxP-pgk:gb2:neo-loxP* cassette was PCR amplified (with the 5′ primer containing a NotI site and the 3′ primer containing a XhoI site) from the R6Kamp-hNGFP vector (kindly provided by Anthony Hyman, MPI Dresden), and inserted into the HspI site of the pMSCV puro vector (Clontech). Vectors containing the fluorescent proteins mTurquoise and mVenus were kindly provided by Joachim Goedhart (University of Amsterdam), and tagRFP was obtained by Evrogen. The sequences for the fluorescent proteins were PCR amplified with primers that contain a nuclear localization sequence after the start codon, and cloned upstream of the *loxPpgk:gb2:neoloxP* cassette into the NotI restriction site. The BGH polyA signal (Invitrogen) was PCR amplified and cloned downstream of the *loxP-pgk:gb2:neo-loxP* cassette into the XhoI restriction site. Additionally we also generated vectors with the SV40 polyA signal (PCR amplified from the pTagRFP-N vector (Evrogen)) between the fluorescent protein sequences and the *loxP-pgk:gb2:neo-loxP* cassette. For BAC recombineering a 360 bp homology region around the lysozyme stop codon was synthesized and cloned into pIDTsmart backbone (IDT). The sequence was flanked with SacII sites, and the regions upstream and downstream of the stop codon were separated by BamHI and XbaI sites. This design allowed to insert the *tagRFP;loxP-pgk:gb2:neo-loxP* cassette between the homology arms, and to linearize the entire construct for recombineering.

BAC Recombineering

The BAC clone RP11-1105J23 which contains the Lysozyme 2 locus (ENSMUSG00000069516) was obtained from the BACPAC resources center (Children's Hospital Oakland Research Institute in Oakland, California, USA). Recombineering was done using the Quick and Easy BAC Modification Kit (Gene Bridges), and the provided protocol was followed. The recombineering cassette was linearized using the SacII sites and purified by LiCl precipitation. 0.5 µg of DNA were used for the transfection. H2A-GFP (ENSMUSG00000037894) and tubb5-GFP (ENSMUSG00000001525) BACs were kindly provided by Anthony Hyman (MPI Dresden).

Preparing Organoids for Transfection

Before transfection mouse organoids were cultured for two generations in growth medium plus Nicotinamide and Wnt-conditioned medium. Under these conditions the cultures mainly consist of stem cells, which can form new organoids after seeding single cells. Stem cell enriched organoids were first mechanically dissociated (per transfection reaction organoids of approximately six 20 μl matrigel drops were used), transferred into 15 ml falcon tubes and centrifuged for 5 minutes at 1000 g. The pellet was resuspended in TriplE (Invitrogen) and trypsinized for 5 minutes at 37°C to obtain single cells. Human organoids were grown in expansion media, which already contains Wnt and Nicotinamide (see above). To obtain single cells, human organoids were trypsinized in TriplE for 10–15 minutes at 37°C, with short vortexing steps every 3 minutes.

Transfection using Electroporation

We used the Amaxa™ Mouse/Rat-Hepatocyte-Nucleofector™ kit. After trypsinization cells were spinned at 1000 g for 5 minutes and the pellet was resuspended in nucleofactor solution plus supplement and plasmid DNA. Electroporation was performed according to the standard Amaxa protocol. After electroporation cells were incubated for 15 minutes at room temperature in the nucleofactor solution. The cell suspension was transferred to an Eppendorf tube, spun at 1000 g for 5 minutes, resuspended in 100 μl cold matrigel, and split into 5 wells of a 48-well culture plate. After polymerization we added growth medium plus Nicotinamide, Wnt-conditioned medium, and the Rho kinase inhibitor Y-27632 to mouse organoids. To human organoids we added human expansion media plus Y-27632.

Transfection using Liposomes

After trypsinization cells were spun at 1000 g for 5 minutes, the supernatant was removed, and mouse cells were resuspended in 450 μl growth medium plus Nicotinamide, Wnt-conditioned media and the Rho kinase inhibitor Y-27632 (human cells were resuspended in human expansion media plus Y-27632). Cells were then plated in 48 well plates at high density (80–90% confluent). Nucleic acid-Lipofectamine® 2000 complexes were prepared according to the standard Lipofectamine protocol. In short, 4 μl of Lipofectamine® 2000 reagent and 1.5 μg plasmid DNA were each diluted in 50 μl Opti-MEM® medium. Both mixes were pooled and incubated for 5 minutes before the DNA-reagent complex was added to the cells (50 μl per well). We centrifuged the plate at 600 g at 32°C for 60 min, and then incubated the plate for additional 4 hours at 37°C. Cells were collected in eppendorf tubes, centrifuged at 1000 g, resuspended in 100 μl cold matrigel, and plated as described above in the electroporation protocol.

After Transfection

Two days after transfection we added 200 μg/ml G418 (Invitrogen) to the medium. When necessary organoids were split 1:3. After selection of stable organoids Wnt3a and Nicotinamid were removed from the mouse organoid media, and sphere-like organoids changed into budding organoids within 1–2 weeks. For the cell sorting experiments organoids were trypsinized 48 h after the transfection, and single cells were analyzed using a MoFlo (Dako Colorado, Inc.) FACS machine.

Live Imaging and Image Analysis

Images and movies of histone H2A-GFP and tubulin TUBB5-GFP organoids were taken with a PerkinElmer Ultraview VoX spinning disk microscope. For image analysis the Volocity 3D image analysis software (PerkinElmer) was used. Lysozyme-tagRFP organoids were imaged using a Leica Sp5 confocal microscope. For image analysis ImageJ was used.

Acknowledgments

We thank Ina Poser and Andrew Hyman for sharing reagents and advise during establishing the lipofectamine transfection protocol in organoids. Rosa-CreERT2 mice were donated by A. Smith (Welcome Trust Centre for Stem Cell Research, University of Cambridge), mTurquoise was donated by Joachim Goedhart (University of Amsterdam).

Author Contributions

Conceived and designed the experiments: GS B-KK HC. Performed the experiments: GS AA-R NS. Analyzed the data: GS AA-R B-KK. Contributed reagents/materials/analysis tools: GS AA-R B-KK. Wrote the paper: GS HC.

References

1. Sato T, Vries RG, Snippert HJ, van de Wetering M, Barker N, et al. (2009) Single Lgr5 stem cells build crypt-villus structures in vitro without a mesenchymal niche. Nature 459: 262–265.
2. Koo BK, Stange DE, Sato T, Karthaus W, Farin HF, et al. Controlled gene expression in primary Lgr5 organoid cultures. Nat Methods 9: 81–83.
3. Park F (2007) Lentiviral vectors: are they the future of animal transgenesis? Physiol Genomics 31: 159–173.
4. Giraldo P, Montoliu L (2001) Size matters: use of YACs, BACs and PACs in transgenic animals. Transgenic Res 10: 83–103.
5. Ciotta G, Hofemeister H, Maresca M, Fu J, Sarov M, et al. Recombineering BAC transgenes for protein tagging. Methods 53: 113–119.
6. Copeland NG, Jenkins NA, Court DL (2001) Recombineering: a powerful new tool for mouse functional genomics. Nat Rev Genet 2: 769–779.
7. Poser I, Sarov M, Hutchins JR, Heriche JK, Toyoda Y, et al. (2008) BAC TransgeneOmics: a high-throughput method for exploration of protein function in mammals. Nat Methods 5: 409–415.
8. Merzlyak EM, Goedhart J, Shcherbo D, Bulina ME, Shcheglov AS, et al. (2007) Bright monomeric red fluorescent protein with an extended fluorescence lifetime. Nat Methods 4: 555–557.
9. Goedhart J, van Weeren L, Hink MA, Vischer NO, Jalink K, et al. (2010) Bright cyan fluorescent protein variants identified by fluorescence lifetime screening. Nat Methods 7: 137–139.
10. Nagai T, Ibata K, Park ES, Kubota M, Mikoshiba K, et al. (2002) A variant of yellow fluorescent protein with fast and efficient maturation for cell-biological applications. Nat Biotechnol 20: 87–90.
11. Kaartinen V, Nagy A (2001) Removal of the floxed neo gene from a conditional knockout allele by the adenoviral Cre recombinase in vivo. Genesis 31: 126–129.
12. el Marjou F, Janssen KP, Chang BH, Li M, Hindie V, et al. (2004) Tissue-specific and inducible Cre-mediated recombination in the gut epithelium. Genesis 39: 186–193.
13. Porter EM, Bevins CL, Ghosh D, Ganz T (2002) The multifaceted Paneth cell. Cell Mol Life Sci 59: 156–170.
14. Jung P, Sato T, Merlos-Suarez A, Barriga FM, Iglesias M, et al. (2011) Isolation and in vitro expansion of human colonic stem cells. Nat Med 17: 1225–1227.
15. Sato T, Stange DE, Ferrante M, Vries RG, Van Es JH, et al. (2011) Long-term expansion of epithelial organoids from human colon, adenoma, adenocarcinoma, and Barrett's epithelium. Gastroenterology 141: 1762–1772.

16. Sato T, Clevers H (2013) Primary mouse small intestinal epithelial cell cultures. Methods Mol Biol 945: 319–328.

17. Kim KA, Kakitani M, Zhao J, Oshima T, Tang T, et al. (2005) Mitogenic influence of human R-spondin1 on the intestinal epithelium. Science 309: 1256–1259.

PERMISSIONS

LIST OF CONTRIBUTORS

Katja U. Beiser, Anne Glaser, Gudrun A. Rappold, Isabell Scholl and Ralph Röth
Department of Human Molecular Genetics, Heidelberg University Hospital, Heidelberg, Germany

Kerstin Kleinschmidt and Wiltrud Richter
Division of Experimental Orthopaedics, Orthopaedic University Hospital, Heidelberg, Germany

Li Li and Norbert Gretz
Medical Research Center (ZMF), Medical Faculty Mannheim at Heidelberg University, Mannheim, Germany

Gunhild Mechtersheimer
Institute of Pathology, Heidelberg University Hospital, Heidelberg, Germany

Marcel Karperien
Department of Developmental Bioengineering, University of Twente, Enschede, The Netherlands

Antonio Marchini
Department of Human Molecular Genetics, Heidelberg University Hospital, Heidelberg, Germany
German Cancer Research Center (DKFZ), Heidelberg, Germany

Andrew F. Teich, Mitesh Patel and Ottavio Arancio
Department of Pathology and Cell Biology, Taub Institute for Research on Alzheimer's Disease and the Aging Brain, Columbia University, New York, New York, United States of America

So-Hyeon Baek, Woon-Chul Shin, Chun-Sun Seo and Hyeon-Jung Kang
National Institute of Crop Science, Rural Development Administration, Iksan, Chonbuk, Korea

Hak-Seung Ryu, Dae-Woo Lee and Jong-Seong Jeon
Graduate School of Biotechnology, Kyung Hee University, Yongin, Gyeonggi, Korea
Crop Biotech Institute, Kyung Hee University, Yongin, Gyeonggi, Korea

Eunjung Moon and Eunson Hwang
Graduate School of East-West Medical Science, Kyung Hee University, Yongin, Gyeonggi, Korea

Hyun-Seo Lee, Mi-Hyun Ahn, Youngju Jeon and Seong-Tshool Hong
Laboratory of Genetics and Department of Microbiology, Chonbuk National University Medical School, Jeonju, Chonbuk, Korea

Sang-Won Lee
Crop Biotech Institute, Kyung Hee University, Yongin, Gyeonggi, Korea
Department of Plant Molecular Systems Biotechnology, Kyung Hee University, Yongin, Gyeonggi, Korea

Sun Yeou Kim
College of pharmacy, Gachon University, Incheon, Korea

Roshan D'Souza and Hyeon-Jin Kim
BDRD Research Institute, JINIS Biopharmaceuticals Inc., Wanju, Chonbuk, Korea

Derek Silvius, Rose Pitstick, Delisha Meishery, George A. Carlson and Teresa M. Gunn
McLaughlin Research Institute, Great Falls, Montana, United States of America

Misol Ahn, Stephen J. DeArmond and Abby Oehler
Institute for Neurodegenerative Diseases and Department of Pathology, University of California San Francisco, San Francisco, California, United States of America

Gregory S. Barsh
Departments of Genetics and Pediatrics, Stanford University, Stanford, California, United States of America

Yah-se K. Abada
Neuropharmacology, EVOTEC AG, Hamburg, Germany
Brain Research Institute Dept. of Neuropharmacology, University of Bremen – FB 2, Bremen, Germany

Huu Phuc Nguyen
Institute of Medical Genetics and Applied Genomics, University of Tübingen, Tübingen, Germany

Bart Ellenbroek
School of Psychology, Victoria University of Wellington, Wellington, New Zealand

Rudy Schreiber
Behavioral Physiology & Pharmacology, University of Groningen, Groningen, The Netherlands

Florian Krismer, Gregor K. Wenning, Werner Poewe and Nadia Stefanova
Division of Neurobiology, Department of Neurology, Innsbruck Medical University, Innsbruck, Austria

Yuntao Li
Division of Neurobiology, Department of Neurology, Innsbruck Medical University, Innsbruck, Austria
The Second School of Clinical Medicine, The Second Affiliated Hospital, Nanjing Medical University, Nanjing, China

Maurizio Pocchiari, Anna Poleggi, Maria Puopolo, Marco D'Alessandro and Dorina Tiple, Anna Ladogana
Department of Cell Biology and Neurosciences, Istituto Superiore di Sanità , Rome, Italy

Xiaojian Wang, Jizheng Wang, Ming Su, Changxin Wang, Jingzhou Chen, Hu Wang and Rutai Hui
Sino-German Laboratory for Molecular Medicine, State Key Laboratory of Cardiovascular Disease, FuWai Hospital & Cardiovascular Institute, Chinese Academy of Medical Sciences, Peking Union Medical College, Beijing, People's Republic of China

Lei Song and Yubao Zou
Department of Cardiology, State Key Laboratory of Cardiovascular Disease, FuWai Hospital & Cardiovascular Institute, Chinese Academy of Medical Sciences, Peking Union Medical College, Beijing, People's Republic of China

Lianfeng Zhang
Key Laboratory of Human Disease Comparative Medicine, Ministry of Health, Institute of Laboratory Animal Science, Chinese Academy of Medical Sciences and Comparative Medical Center, Peking Union Medical College, Beijing, People's Republic of China

Youyi Zhang
Institute of Vascular Medicine, Peking University Third Hospital, Beijing, People's Republic of China

Kwang-Hwan Choi, Jin-Kyu Park, Hye-Sun Kim, Kyung-Jun Uh, Dong-Chan Son and Chang-Kyu Lee
Department of Agricultural Biotechnology, Animal Biotechnology Major, and Research Institute for Agriculture and Life Science, Seoul National University, Seoul, Korea

Zhi-Yuan Zhang, Chaoyun Li, Caroline Zug and Hermann J. Schluesener
Division of Immunopathology of the Nervous System, Institute of Pathology and Neuropathology, University of Tuebingen, Tuebingen, Germany

Hongsheng Men and Elizabeth C. Bryda
Rat Resource and Research Center, Department of Veterinary Pathobiology, University of Missouri, Columbia, Missouri, United States of America

Thomas R. Whitesell, Regan M. Kennedy, Alyson D. Carter, Evvi-Lynn Rollins, Sonja Georgijevic and Sarah J. Childs
Department of Biochemistry and Molecular Biology, and Smooth Muscle Research Group, University of Calgary, Calgary, Alberta, Canada

Massimo M. Santoro
VIB Vesalius Research Center, University of Leuven (KU Leuven), Leuven, Belgium

Masashi Shin, Coralee E. Tye, Xiaomu Guan, Jerry V. Antone and John D. Bartlett
Department of Mineralized Tissue Biology and Harvard School of Dental Medicine, The Forsyth Institute, Cambridge Massachusetts, United States of America

Charles E. Smith
Department of Biologic and Materials Sciences, University of Michigan School of Dentistry, Ann Arbor, Michigan, United States of America
Facility for Electron Microscopy Research, Department of Anatomy & Cell Biology, and Faculty of Dentistry, McGill University, Montreal, QC, Canada

Yuanyuan Hu and James P. Simmer
Department of Biologic and Materials Sciences, University of Michigan School of Dentistry, Ann Arbor, Michigan, United States of America

Craig C. Deagle
Program in Endodontics, Harvard School of Dental Medicine, Boston Massachusetts, United States of America

Yuhong Zhang, Xiaojin Zhou, Rumei Chen and Wei Zhang
Biotechnology Research Institute, Chinese Academy of Agricultural Sciences, Beijing, P. R. China

Kun Meng, Huiying Luo, Bin Yao, Xiaolu Xu and Peilong Yang
Key Laboratory for Feed Biotechnology of the Ministry of Agriculture, Feed Research Institute, Chinese Academy of Agricultural Sciences, Beijing, P. R. China

Jianhua Yuan and Qingchang Meng
Institute of Food Crops, Jiangsu Academy of Agricultural Sciences, Nanjing, P. R. China

Soumee Bhattacharya, Christin Haertel and Dirk Montag
Neurogenetics Special Laboratory, Leibniz Institute for Neurobiology, Magdeburg, Germany

Alfred Maelicke
Galantos Pharma GmbH, Nieder-Olm, Germany

Chenxi Liu
Xinjiang Laboratory of Animal Biotechnology, Urumqi, Xinjiang, China
Key Laboratory of Genetics, Breeding and Reproduction of Grass Feeding Livestock, Ministry of Agriculture, Urumqi, Xinjiang, China
College of Life Science and Technology, Xinjiang University, Urumqi, Xinjiang, China

Liqin Wang, Wenrong Li, Xuemei Zhang, Yongzhi Tian, Ning Zhang, Sangang He, Tong Chen, Juncheng Huang and Mingjun Liu
Xinjiang Laboratory of Animal Biotechnology, Urumqi, Xinjiang, China
Key Laboratory of Genetics, Breeding and Reproduction of Grass Feeding Livestock, Ministry of Agriculture, Urumqi, Xinjiang, China
Animal Biotechnology Research Center, Xinjiang Academy of Animal Science, Urumqi, Xinjiang, China

Melody Shi, Pinghu Liu, Lijin Dong, Nadia Parmhans and Tudor Constantin Badea
National Eye Institute, NIH, Bethesda, Maryland, United States of America

Szilard Sajgo and Miruna Georgiana Ghinia
National Eye Institute, NIH, Bethesda, Maryland, United States of America
Biology Department, Babes-Bolyai University, Cluj-Napoca, Cluj, Romania

Octavian Popescu
Biology Department, Babes-Bolyai University, Cluj-Napoca, Cluj, Romania
Institute of Biology, Romanian Academy, Bucharest, Romania

Lukas E. Dow, Michael Saborowski and Eusebio Manchado
Cancer Biology and Genetics Program, Memorial Sloan Kettering Cancer Center, New York, New York, United States of America

Saya H. Ebbesen and Nilgun Tasdemir
Cancer Biology and Genetics Program, Memorial Sloan Kettering Cancer Center, New York, New York, United States of America
Watson School of Biological Sciences, Cold Spring Harbor Laboratory, Cold Spring Harbor, New York, United States of America

Scott W. Lowe
Cancer Biology and Genetics Program, Memorial Sloan Kettering Cancer Center, New York, New York, United States of America
Howard Hughes Medical Institute, Memorial Sloan Kettering Cancer Center, New York, New York, United States of America

Zeina Nasr and Teresa Lee
Department of Biochemistry, McGill University, Montreal, Quebec, Canada

Jerry Pelletier
Department of Biochemistry, McGill University, Montreal, Quebec, Canada
The Rosalind and Morris Goodman Cancer Research Center, McGill University, Montreal, Quebec, Canada

Shuping Gu, Chao Liu, Cheng Sun, Wenduo Ye and YiPing Chen
Department of Cell and Molecular Biology, Tulane University, New Orleans, Louisiana, United States of America

Weijie Wu
Department of Cell and Molecular Biology, Tulane University, New Orleans, Louisiana, United States of America
Department of Dentistry, ZhongShan Hospital, FuDan University, Shanghai, P.R. China

Ling Yang
Department of Cell and Molecular Biology, Tulane University, New Orleans, Louisiana, United States of America
Guanghua School of Stomatology, Sun Yat-sen University, Guangzhou, Guangdong, P.R. China

Xihai Li
Department of Cell and Molecular Biology, Tulane University, New Orleans, Louisiana, United States of America
Academy of Integrative Medicine, Fujian University of Traditional Chinese Medicine, Fuzhou, Fujian, P.R. China

Jianquan Chen and Fanxin Long
Department of Internal Medicine, Washington University School of Medicine, St. Louis, Missouri, United States of America

Michio Nagata
Department of Kidney and Vascular Pathology, University of Tsukuba, Tsukuba, Japan

Kazuhiro Umeyama and Hiroshi Nagashima
Meiji University International Institute for Bio-Resource Research, Kawasaki, Japan

Takashi Yokoo
Division of Nephrology and Hypertension, Department of Internal Medicine, The Jikei University School of Medicine, Tokyo, Japan

Satoshi Hara
Department of Kidney and Vascular Pathology, University of Tsukuba, Tsukuba, Japan
Division of Rheumatology,Department of Internal Medicine, Kanazawa University of Graduate School of Medicine, Kanazawa, Japan

Popi Syntichaki and Anna Vlanti
Biomedical Research Foundation of the Academy of Athens, Center of Basic Research II, Athens, Greece

Aris Rousakis
Biomedical Research Foundation of the Academy of Athens, Center of Basic Research II, Athens, Greece
Faculty of Medicine, University of Athens, Athens, Greece

Fivos Borbolis and Fani Roumelioti
Biomedical Research Foundation of the Academy of Athens, Center of Basic Research II, Athens, Greece
Faculty of Biology, School of Science, University of Athens, Athens, Greece

Marianna Kapetanou
Biomedical Research Foundation of the Academy of Athens, Center of Basic Research II, Athens, Greece
Department of Biology, School of Science and Engineering, University of Crete, Heraklio, Crete, Greece

Alejandra Droguett, Paola Krall, M. Eugenia Burgos, Graciela Valderrama, Leopoldo Ardiles and Sergio Mezzano
Division Nephrology, School of Medicine, Universidad Austral de Chile, Valdivia, Chile

Bredford Kerr and Katherina Walz
Centro de Estudios Cientı́ficos, Valdivia, Chile

Daniel Carpio
Hystopathology Division, School of Medicine, Universidad Austral de Chile, Valdivia, Chile

Raquel Rodriguez-Diez, Marta Ruiz-Ortega and Jesus Egido
Cellular Biology in Renal Diseases Laboratory, Universidad Autónoma Madrid, Madrid, Spain

Index

www.ingramcontent.com/pod-product-compliance
Lightning Source LLC
Chambersburg PA
CBHW080526200326
41458CB00012B/4347